Nucleon-Nucleon Interactions–1977
(Vancouver)

AIP Conference Proceedings

Series Editor: Hugh C. Wolfe

No. 41

Nucleon-Nucleon Interactions–1977

(Vancouver)

Editors

D.F. Measday

University of British Columbia

H.W. Fearing and A. Strathdee

TRIUMF

American Institute of Physics

New York   1978

L.C. Catalog Card No. 78-54249
ISBN 0-88318- 140-1
DOE CONF-770674

SPONSORS AND STAFF

Sponsors

This conference has been supported by the National Research Council of Canada, Atomic Energy of Canada Limited, the University of British Columbia, and TRIUMF.

Organizing Committee

Chairman   D.F. Measday
Secretary   C. Thorne

D.A. Axen   G. Jones
D.S. Beder   L.P. Robertson
D.V. Bugg   J.G. Rogers
M.K. Craddock

Program Committee

Chairman   A.W. Thomas

D.A. Axen   H.W. Fearing
D.S. Beder   G. Jones
M.K. Craddock   J.G. Rogers

Organizing Committee for Nucleon-Nucleon Bremsstrahlung Workshop

Chairman   H.W. Fearing

B.M.K. Nefkens
J.R. Richardson
J.G. Rogers

CONFERENCE PARTICIPANTS

J-C. ALDER, Univ. de Lausanne, Lausanne, Switzerland
Y. ALEXANDER, Univ. of Maryland, College Park, MD, USA
C. AMSLER, Queen Mary College, London, England
A.N. ANDERSON, Univ. of Alberta, Edmonton, Alberta, Canada
L.E. ANTONUK, Univ. of Alberta, Edmonton, Alberta, Canada
R.A. ARNDT, Virginia Polytech. Inst. & State Univ., Blacksburg, VA, USA
A. ASHMORE, Daresbury Lab, Warrington, England
I.P. AUER, Argonne National Lab., Argonne, IL, USA
E.G. AULD, Univ. of British Columbia, Vancouver, B.C., Canada
M. AUSTERN, Univ. of Pittsburgh, Pittsburgh, PA, USA
D.A. AXEN, Univ. of British Columbia, Vancouver, B.C., Canada
B.L. BAKKER, Vrije Univ., Amsterdam, The Netherlands
D.S. BEDER, Univ. of British Columbia, Vancouver, B.C., Canada
J.E. BENN, Physik-Institut der Univ. Zürich, Zürich, Switzerland
A.F. BERETVAS, Argonne National Lab., Argonne, IL, USA
D. BESSIS, CEN Saclay, Gif-sur-Yvette, France
P.R. BEVINGTON, Case Western Reserve Univ., Cleveland, OH, USA
G. BIZARD, Lawrence Berkeley Lab., Berkeley, CA, USA
J.S. BLAIR, Univ. of Washington, Seattle, WA, USA
G. BOHANNON, Massachusetts Inst. of Technology, Cambridge, MA, USA
M. BOLSTERLI, Los Alamos Scientific Lab., Los Alamos, NM, USA
B.E. BONNER, Los Alamos Scientific Lab., Los Alamos, NM, USA
G.E. BROWN, State Univ. of New York, Stony Brook, NY, USA
R. BRYAN, Texas A & M Univ., College Station, TX, USA
R. CARLINI, Univ. of New Mexico, Albuquerque, NM, USA
J.B. CARROLL, Lawrence Berkeley Lab., Berkeley, CA, USA
C. CERNIGOI, Univ. of Trieste, Trieste, Italy
A. CHATTERJEE, Saha Inst. of Nuclear Physics, Calcutta, India
C.Y. CHEUNG, Univ. of Washington, Seattle, WA, USA
A. CHISHOLM, Univ. of Auckland, Auckland, New Zealand
R.B. CLARK, Texas A & M Univ. College Station, TX, USA
E. COLTON, Argonne National Lab., Argonne, IL, USA
A.A. CONE, Langara College, Vancouver, B.C., Canada
H.E. CONZETT, Lawrence Berkeley Lab., Berkeley, CA, USA
E. COOPER, Univ. of Alberta, Edmonton, Alberta, Canada
B. CORK, Lawrence Berkeley Lab., Berkeley, CA, USA
M.K. CRADDOCK, Univ. of British Columbia, Vancouver, B.C., Canada
M.S. DE JONG, Univ. of Manitoba, Winnipeg, Manitoba
J. DE KAM, Univ. of Alberta, Edmonton, Alberta, Canada
J. DE SWART, Univ. of Nijmegen, Nijmegen, The Netherlands
B.D. DIETERLE, Univ. of New Mexico, Albuquerque, NM, USA
C.A. DOMINGUEZ, CIEA-IPN, Mexico, D.F.
J. DONAHUE, Los Alamos Scientific Lab., Los Alamos, NM, USA
R. DUBOIS, Univ. of British Columbia, Vancouver, B.C., Canada
O. DUMBRAJS, Univ. of Helsinki, Helsinki, Finland
G.G. DUTTO, TRIUMF, Vancouver, B.C., Canada
J.A. EDGINGTON, Queen Mary College, London, England
S. ELKATEB, Fraser Valley College, Abbotsford, B.C., Canada
G.N. EPSTEIN, Univ. of Pittsburgh, Pittsburgh, PA, USA

K.L. ERDMAN, TRIUMF and Univ. of British Columbia, Vancouver, B.C., Canada
W. FABIAN, State Univ. of New York, Stony Brook, NY, USA
K.K. FANG, Univ. of Saskatchewan, Saskatoon, Saskatchewan, Canada
H.W. FEARING, TRIUMF, Vancouver, B.C., Canada
D.H. FITZGERALD, Univ. of California, Davis, CA, USA
P. FRANZINI, Columbia Univ., New York, NY, USA
S. FURUICHI, Rikkyo Univ., Tokyo, Japan
R. GARRETT, Univ. of California, Davis, CA, USA
W.R. GIBBS, Los Alamos Scientific Lab., Los Alamos, NM, USA
D.R. GILL, TRIUMF, Vancouver, B.C., Canada
G. GLASS, Texas A & M Univ., College Station, TX, USA
L. GOLDZAHL, CEN Saclay, Gif-sur-Yvette, France
P. GRANGE, Centre de Recherches Nucléaires, Strasbourg, France
A.E.S. GREEN, Univ. of Florida, Gainesville, FL, USA
L.G. GREENIAUS, Univ. of Alberta, Edmonton, Alberta, Canada
F. GROSS, College of William and Mary, Williamsburg, VA, USA
W. HAEBERLI, Univ. of Wisconsin, Madison, WI, USA
M. HAFTEL, Naval Research Lab., Washington, D.C., USA
M.D. HASINOFF, Univ. of British Columbia, Vancouver, B.C., Canada
W. HAXTON, Los Alamos Scientific Lab., Los Alamos, NM, USA
R. HESS, Univ. de Genève, Genève, Switzerland
K. HOLINDE, Univ. Bonn, Bonn, W. Germany
C. HOLLAS, Univ. of Texas, Austin, TX, USA
H.R. HOOPER, Univ. of Alberta, Edmonton, Alberta, Canada
J.C. HOPKINS, Los Alamos Scientific Lab., Los Alamos, NM, USA
J. HUDOMALJ-GABITZSCH, Rice Univ., Houston, TX, USA
A.H. HUSSEIN, Univ. of Petroleum & Minerals, Dhahran, Saudi Arabia
D.A. HUTCHEON, TRIUMF, Vancouver, B.C., Canada
M. JAIN, Los Alamos Scientific Lab., Los Alamos, NM, USA
B. JENNINGS, State Univ. of New York, Stony Brook, NY, USA
K. JOHNSON, Los Alamos Scientific Lab., Los Alamos, NM, USA
R.R. JOHNSON, Univ. of British Columbia, Vancouver, B.C., Canada
G. JONES, Univ. of British Columbia, Vancouver, B.C., Canada
J. JOVANOVICH, Univ. of Manitoba, Winnipeg, Manitoba, Canada
J. KÄLLNE, Univ. of Alberta, Edmonton, Alberta, Canada
A.N. KAMAL, Univ. of Alberta, Edmonton, Alberta, Canada
R. KEELER, Univ. of British Columbia, Vancouver, B.C., Canada
F.C. KHANNA, Chalk River Nuclear Lab., Chalk River, Ontario, Canada
R.M. KLOEPPER, IBM, New York, NY, USA
W.M. KLOET, Los Alamos Scientific Lab., Los Alamos, NM, USA
L. KNUTSON, Univ. of Wisconsin, Madison, WI, USA
B.K.S. KOENE, Univ. of Manitoba, Winnipeg, Manitoba, Canada
M. KRELL, Univ. de Sherbrooke, Sherbrooke, Québec, Canada
P. KROLL, Univ. of Wuppertal, Wuppertal, Germany
R. KUNSELMAN, Univ. of Wyoming, Laramie, WY, USA
R.H. LANDAU, Oregon State Univ., Corvallis, OR, USA
J-L. LAVILLE, Univ. de Caen, Caen, France
J. LEE-FRANZINI, State Univ. of New York, Stony Brook, NY, USA
M.K. LIOU, Brooklyn College, Brooklyn, NY, USA
B. LOISEAU, Univ. Pierre et Marie Curie, Paris, France
E.L. LOMON, Massachusetts Inst. of Technology, Cambridge, MA, USA
G.A. LUDGATE, Univ. of British Columbia, Vancouver, B.C., Canada

R. MACHLEIDT, Univ. Bonn, Bonn, W. Germany
P. MACQ, Univ. de Louvain, Louvain- a-Neuve, Belgium
K. MALTMAN, Univ. of British Columbia, Vancouver, B.C., Canada
G.R. MASON, Univ. of Victoria, Victoria, B.C., Canada
T. MASTERSON, Univ. of British Columbia, Vancouver, B.C., Canada
L. MATHELITSCH, Univ. Graz, Graz, Austria
E.L. MATHIE, Univ. of British Columbia, Vancouver, B.C., Canada
W.J. McDONALD, Univ. of Alberta, Edmonton, Alberta, Canada
K. McFARLANE, Temple Univ., Philadelphia, PA, USA
B.H.J. McKELLAR, Univ. of Melbourne, Parkville, Victoria, Australia
D.F. MEASDAY, Univ. of British Columbia, Vancouver, B.C., Canada
A. MEKJIAN, Lawrence Berkeley Lab., Berkeley, CA, USA
C.A. MILLER, Univ. of Alberta, Edmonton, Alberta, Canada
G.A. MILLER, Univ. of Washington, Seattle, WA, USA
R.E. MISCHKE, Los Alamos Scientific Lab., Los Alamos, NM, USA
H.H. MISKA, Massachusetts Inst. of Technology, Cambridge, MA, USA
I.O. MOEN, Trent Univ., Peterborough, Ontario, Canada
M.J. MORAVCSIK, Univ. of Oregon, Eugene, OR, USA
G.S. MUTCHLER, Rice Univ., Houston, TX, USA
F. MYHRER, CERN, Geneva, Switzerland
B.T. MURDOCH, Univ. of Manitoba, Winnipeg, Manitoba, Canada
D. NAGLE, Los Alamos Scientific Lab., Los Alamos, NM, USA
B.M.K. NEFKENS, Univ. of California, Los Angeles, CA, USA
J.W. NEGELE, Massachusetts Inst. of Technology, Cambridge, MA, USA
C. NEWSOM, Univ. of Texas, Austin, TX, USA
J. NG, Univ. of Alberta, Edmonton, Alberta, Canada
L.C. NORTHCLIFFE, Texas A & M Univ., College Station, TX, USA
H.P. NOYES, SLAC, Stanford, CA, USA
W. OCHS, Max-Planck-Inst., München, Germany
C.J. ORAM, Univ. of British Columbia, Vancouver, B.C., Canada
D. OTTEWELL, TRIUMF, Vancouver, B.C., Canada
F. PARTOVI, Massachusetts Inst. of Technology, Cambridge, MA, USA
P. PAVLOPOULOS, CERN, Geneva, Switzerland
R.E. POLLOCK, Indiana Univ., Bloomington, IN, USA
J-M. POUTISSOU, Univ. de Montréal, Montréal, P.Q., Canada
F.N. RAD, Massachusetts Inst. of Technology, Cambridge, MA, USA
B. RAM, New Mexico State Univ. Las Cruces, NM, USA
L. RATNER, Argonne National Lab., Argonne, IL, USA
A. REITAN, Fysisk Institutt, Trondheim, Norway
J.R. RICHARDSON, Univ. of California, Los Angeles, CA, USA
T.A. RIJKEN, Univ. of Nijmegen, Nijmegen, The Netherlands
P.J. RILEY, Univ. of Texas, Austin, TX, USA
L.P. ROBERTSON, Univ. of Victoria, Victoria, B.C., Canada
R. ROCKMORE, Rutgers Univ., New Brunswick, NJ, USA
J.G. ROGERS, TRIUMF, Vancouver, B.C., Canada
A. ROSENTHAL, Univ. of Colorado, Boulder, CO, USA
E. RÖSSLE, Univ. Freiburg, Freiburg, Germany
V. ROSTOKIN, Moscow Physics Engineering Inst., Moscow, USSR
R. ROY, Univ. Laval, Québec, P.Q., Canada
R. RÜCKL, Univ. of California, Los Angeles, CA, USA
D.G. RYAN, McGill Univ., Montréal, P.Q., Canada
M. SALOMON, Univ. of British Columbia, Vancouver, B.C., Canada

J.T. SAMPLE, TRIUMF, Vancouver, B.C., Canada
P.U. SAUER, Technical Univ. Hannover, Hannover, Germany
D. SAYLOR, Worcester Polytechnic Inst., Worcester, MA, USA
E.W. SCHMID, Univ. Tübingen, Tübingen, Germany
P. SCHWANDT, Indiana Univ., Bloomington, IN, USA
H.S. SHERIF, Univ. of Alberta, Edmonton, Alberta, Canada
P. SIGNELL, Michigan State Univ., East Lansing, MI, USA
R.R. SILBAR, Los Alamos Scientific Lab., Los Alamos, NM, USA
J.E. SIMMONS, Los Alamos Scientific Lab., Los Alamos, NM, USA
R.S. SLOBODA, Univ. of Alberta, Edmonton, Alberta, Canada
H. SPINKA, Argonne National Lab., Argonne, IL, USA
D.W.L. SPRUNG, McMaster Univ., Hamilton, Ontario, Canada
D.G. STAIRS, McGill Univ., Montréal, P.Q., Canada
A.W. STETZ, Oregon State Univ., Corvallis, OR, USA
A. SZYJEWICZ, Univ. of Alberta, Edmonton, Alberta, Canada
C.E. THOMANN, Univ. Zürich, Zürich, Switzerland
A.W. THOMAS, TRIUMF, Vancouver, B.C., Canada
G.H. THOMAS, Argonne National Lab., Argonne, IL, USA
E.L. TOMUSIAK, Univ. of Saskatchewan, Saskatoon, Sask., Canada
T. UEDA, Osaka Univ., Osaka, Japan
E-V. VANAGAS, Institut of Physics, Vilnius, USSR
W.T.H. van OERS, Univ. of Manitoba, Winnipeg, Manitoba, Canada
R. van WAGENINGEN, Vrije Univ., Amsterdam, The Netherlands
V. VANZANI, Univ. di Padova, Padova, Italy
H. VERHEUL, Vrije Univ., Amsterdam, The Netherlands
B.J. VerWEST, Texas A & M Univ., College Station, TX, USA
R. VINH MAU, Univ. Pierre et Marie Curie, Paris, France
E.W. VOGT, Univ. of British Columbia, Vancouver, B.C., Canada
M. WADA, Nihon Univ., Funabashi, Japan
S. WAKAIZUMI, Hiroshima Univ., Hiroshima, Japan
P.L. WALDEN, TRIUMF, Vancouver, B.C., Canada
S.J. WALLACE, Univ. of Maryland, College Park, MD, USA
Y. WATANABE, Argonne National Lab., Argonne, IL, USA
W. WATARI, Osaka City Univ., Osaka, Japan
G. WATERS, TRIUMF, Vancouver, B.C., Canada
D. WEDDIGEN, Univ. Karlsruhe, Karlsruhe, W. Germany
M.S. WEISS, Lawrence Livermore Lab., Livermore, CA, USA
D.W. WERREN, Univ. de Genève, Genève, Switzerland
B.L. WHITE, Univ. of British Columbia, Vancouver, B.C., Canada
L. WILETS, Univ. of Washington, Seattle, WA, USA
H.B. WILLARD, Case Western Reserve Univ., Cleveland, OH, USA
R. WILSON, Harvard Univ., Cambridge, MA, USA
D.M. WOLFE, Univ. of New Mexico, Albuquerque, NM, USA
R. WOLOSHYN, Univ. of Pennsylvania, Philadelphia, PA, USA
C-S. WU, Univ. of Victoria, Victoria, B.C., Canada
A. YOKOSAWA, Argonne National Lab., Argonne, IL, USA
L. ZAMICK, Rutgers Univ., New Brunswick, NJ, USA
H.F.K. ZINGL, Univ. Graz, Graz, Austria

PREFACE

This conference came ten years after the First International Conference on the Nucleon-Nucleon Interaction which was held in Gainesville, Florida, March 1967, and the proceedings are available in Reviews of Modern Physics, vol 39, #3. We were privileged to have A.E.S. Green and R. Wilson, organizers of the first conference, attend this one as active participants; unfortunately M.H. MacGregor could not attend.

The most noticeable result of the passage of ten years was the remarkable increase in the quality and quantity of the data throughout the whole spectrum of energies. At the lower energies, experiments of very high precision are being performed and more can be anticipated. In the phase-shift analyses of neutron-proton scattering around 50 MeV there used to be problems with the ε_1 parameter, but that seems to have been cleared up by painstaking work of the Davis group. New observables have been measured (e.g. Ayy) and care and attention to detail have shown that the old differential cross-section data are most probably in error, but a third experiment by another group seems a wise precaution before we become too complacent.

At a few hundred MeV the few existing data have been deluged by new data from LAMPF, SATURNE, SIN and TRIUMF, and within a year or so it is reasonable to assume that we shall have a good analysis for both the T=1 and T=0 and phases up to about 800 MeV. At energies of a few GeV a remarkable program is under way in a mad rush to beat the imminent closure of the ZGS at Argonne National Laboratory. With SATURNE II coming into operation fairly soon, it is reasonable to anticipate that the gap will eventually be filled in and that for the next conference complete data will exist up to about 10 GeV! At this conference we had a glimpse of this situation with contributions from the ZGS group on their experimental and theoretical program.

Another experimental topic which has changed unrecognizably is the study of the $\pi\pi$ interaction. Ten years ago there were ambiguities and uncertainties, but now with the advent of high statistics experiments and improved analyses the phase shifts in the $\pi\pi$ interaction are, with a few exceptions, uniquely defined and fairly well known up to about 2 GeV.

On Friday, July 1, a special workshop was held on nucleon-nucleon bremsstrahlung, and about 60 of the conference attendees extended their visit to Vancouver. For many years the subject of bremsstrahlung has been marooned in the doldrums, with everyone predicting a dull future for the topic. However, a breath of fresh air was felt and our hopes raised a little by new theoretical calculations and a new experiment from TRIUMF, although admittedly there was not enough to set everyone off in full sail. The contributions to this workshop have been included in this proceedings.

The conference itself seemed to run without a serious hitch, and we are grateful to the many people at the University of British Columbia who contributed to the smooth functioning of all the arrangements. Most important was the co-operation of the weather bureau which helped to make the cruise of the Malibu Princess on Monday evening a most momorable event as we sailed in the beautiful fjords of the B.C. coast. On Tuesday we were fortunate to have a showing at the Vancouver Planetarium on the topic of the missing planet Krypton. Professor M.W. Ovenden gave a short explanation of his ideas on the evolution of our solar system and so gave a fine addition to the normal show. At the banquet on Wednesday the Minister of Education for British Columbia, Dr. P.L. McGeer, gave an interesting talk on the history of this part of North America. Unfortunately, Prof. E.P. Wigner was unable to attend because of the illness of his wife, and we missed his sense of perspective of the topic to which he has contributed so much.

No parallel sessions were held, but two poster sessions were organized in the residence. With the ploy of serving tea next to the room, the sessions went well and were a pleasant and informal way of passing on information. We heeded the warning from the organizers of the Santa Fe conference (1975) and had no competing events, so that is probably why our results are more positive than is often the case.

In any conference a lot of the problems fall onto the shoulders of one person. For us it was Claudia Thorne who made an enormous effort to ensure that the organization ran smoothly and that everyone had a comfortable and pleasant stay in Vancouver. We are all grateful to her for the way she tackled any and every problem with poise and confidence.

D.F. Measday
H.W. Fearing
A. Strathdee

Table of Contents

I. NUCLEON-NUCLEON ELASTIC SCATTERING AND PHASE-SHIFT ANALYSIS

II. NUCLEON-NUCLEON POTENTIALS AND THEORETICAL DEVELOPMENTS

VI.   PAPERS PRESENTED IN POSTER SESSIONS

NUCLEON-NUCLEON INTERACTION: LOW ENERGY EXPERIMENTS

W. Haeberli[†]
University of Wisconsin, Madison, Wisconsin 53706

ABSTRACT

The results of recent low-energy nucleon-nucleon scattering
are summarized. New p-p cross section experiments below 20 MeV
have resolved some of the inconsistencies in earlier cross section
data and lead to a low energy behavior of the ^3P_C phase shift combination consistent with the analysis of higher energy data. Accurate polarization measurements in p-p scattering permit a model
independent determination of all p-wave phase shifts at 10 MeV.
New measurements of the n-p cross section at 50 MeV resolve an inconsistency in the $^1P_1$ phase shift at that energy. New n-p polarization measurements near 25 MeV and measurements of the spin correlation parameter at 50 MeV resolve the problem of the low energy
behavior of the S-D coupling parameter ε_1. The last section mentions other types of experiments related to the nucleon-nucleon
interaction, i.e. study of the three-body problem and of the deuteron D-state.

INTRODUCTION

New experiments on the nucleon-nucleon interaction at low energies (E < 100 MeV) continue to be of considerable interest. The
justification for these experiments lies in the fact that the direct determination of the interaction from experiments is still inaccurate and incomplete, so that the knowledge of the low-energy
phase shifts depends on the correctness of the extrapolation methods used in energy-dependent phase shift analyses in which certain
theoretical assumptions are used to anticipate the behavior at low
energies from the analysis of data at higher energies. The current
experiments on low-energy nucleon-nucleon scattering thus provide
a test of these theoretical assumptions. In this talk I should
like to discuss recent experiments on nucleon-nucleon scattering
and occasionally will point out what improvements in experimental
techniques have made these experiments possible. The last section
mentions other experiments which are related to the low energy
nucleon-nucleon interaction.

PROTON-PROTON SCATTERING

Cross Sections and low-energy behavior of the ^3P_C *phase shift
combination:* Measurements of the absolute cross section in p-p
scattering at low energies have reached astonishing accuracies. In
a paper submitted to this conference, Thomann *et al.*[1] report cross

[†] Work supported in part by the United States Energy and Research
Development Administration.

section measurements between 0.35 MeV and 1.0 MeV to an absolute accuracy of 0.2%, while the energy of the interference minimum at 382.48 keV was determined to an accuracy of 100 eV. Measurements of even higher accuracy below 2 MeV had been reported earlier by Muhry *et al.*[2,3]. These data could be fitted only by reducing the strength of the vacuum polarisation interaction by an arbitrary factor.[3] Recently, the same group[4] made new experiments to reinvestigate sources of possible systematic errors in the measurements, such as the accuracy in angular position of the detectors, effects of thermal motion of the target molecules, dissociation of the target molecules by the beam, multiple scattering, target contaminants and beam integration. In the analysis, they also considered, besides vacuum polarization which has a 0.4% effect at $\theta_{cm} = 46°$, relativistic corrections and corrections due to screening by atomic electrons. The conclusion of this effort is that the *relative* cross sections are perfectly consistent with the effective-range S-wave phase shift of Noyes and Lipinsky[5] [assuming any reasonable value for the p-wave phases from other sources]. However the absolute normalization of the measured cross sections differs by 0.3-0.4% from the calculated cross sections, while the experimental normalization uncertainty is thought to be ~0.1%. The new investigation of possible normalization errors produced an upward normalization of only 0.06±0.03% so that an unexplained discrepancy still exists.

During the last two years, the p-p phase shifts below about 20 MeV have been subject to considerable revisions, at least as far as the p-wave phase shifts are concerned. We are dealing with three p-wave phase shifts $^3P_0$, $^3P_1$ and $^3P_2$. As long as they are small, it is convenient to replace them by three linear combinations, namely the central p-wave parameter:

$$^3P_C = \frac{1}{9} \left(^3P_0 + 3 \, ^3P_1 + 5 \, ^3P_2 \right)$$

the $\ell \cdot s$ p-wave parameter:

$$^3P_{LS} = \frac{1}{12} \left(-2 \, ^3P_0 - 3 \, ^3P_1 + 5 \, ^3P_2 \right)$$

and the tensor p-wave parameter:

$$^3P_T = \frac{5}{12} \left(-2 \, ^3P_0 + 3 \, ^3P_1 - \, ^3P_2 \right).$$

At low energies, the dominant contribution to the cross section comes from the s-wave phase shift $^1S_0$ and from the central p-wave combination ^3P_C through the interference with Coulomb and s-wave scattering. At low energies the cross section is relatively insensitive to the other p-wave combinations and to higher partial waves.

There has been considerable question about the energy dependence of ^3P_C below 10 MeV. The results of the multienergy analysis of Sherr, Signell and Heller[6] which are shown by the band labeled Sh70 in Fig. 1, differ considerably from the energy dependent analysis of McGregor, Arndt and Wright[7] who analyzed data up to 450 MeV. Sherr *et al.* pointed out inconsistencies in the experimental data at low energies and mentioned that all data between 1 and 10 MeV are questionable. The problem is also evident in the more recent energy dependent analysis of Arndt, Hackman and Roper[8], where the analysis of low energy data (1-27.6 MeV) indicates a minimum in ^3P_C near 6 MeV while the analysis of the entire 1-500 MeV data set leads to more positive values of ^3P_C (see Fig. 1). However, the latter result was obtained by ignoring the absolute normalization of the low-energy cross section data. In particular the precision measurements from Los Alamos[9] near 10 MeV were arbitrarily renormalized by about 2%. These problems recently led Imai *et al.*[10] to remeasure the p-p cross section at 5, 7 and 8 MeV with an absolute overall accuracy of ±0.4%. Fig. 2 compares the resulting values of ^3P_C (solid dots) with other work. The dashed curve is calculated from a one-pion exchange model, while the dot-dashed curve shows the prediction of a one-boson exchange calculation. Clearly there is considerable discrepancy between this new experiment and earlier ones. Fortunately, another experiment which is relevant to this problem has just been completed by Hegland *et al.*[11] at the University of Minnesota who have obtained new precision measurements of the *relative* cross section at seven energies between 5.957 MeV and

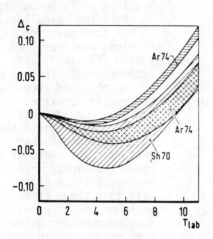

Fig. 1. Low-energy behavior of the ^3P_C phase shift combination according to Refs.[6,8].

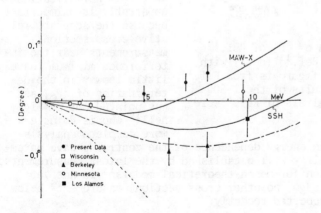

Fig. 2. Comparison of ^3P_C of Ref.[10] with earlier work. The figure is from Ref.[10].

19.700 MeV for c.m. scattering angles between about 14° and 90°. These measurements were made with a solid hydrocarbon target, with high-resolution detectors and good detector geometry, the idea being that angle-dependent errors are bound to be different for a solid target from those encountered in absolute cross section measurements which necessarily depend on gas targets. The values of ^3P_C obtained from their results (solid dots, Fig. 3) completely support the conclusions of Imai *et al.*[10] [open circles]. For comparison I have added the phase shifts of the 1-500 MeV analysis of Arndt *et al.*[8] [rectangles] in which the normalization of the low-energy cross sections was freely adjusted (floated). The agreement with the values of ^3P_C obtained from the two recent experiments provides *a posteriori* justification to adjust the normalization of the earlier cross sections in Arndt's analysis, since it now appears almost certain that these cross sections are beset with some unknown normalization error. While the cross sections of Hegland *et al.*[11] were normalized to other cross sections for c.m. angles > 40°, this normalization apparently is unimportant because the accurate relative cross section measurements near the interference minimum leave little leeway in the determination of ^3P_C, as was pointed out by Signell[12] who is doing a more complete analysis of the results. The energy dependence of the central p-wave parameter is probably fairly well established by the latest measurements, but their implication for meson-theoretical models[10] is not yet clear. To my knowledge, no other cross section measurements below 100 MeV have been reported recently.

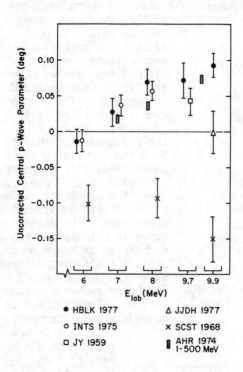

Fig. 3. Comparison of ^3P_C of Ref.[10] (open circles) and Ref.[11] (dots) with earlier work. The figure is from Ref.[11] except that the results of the 1-500 MeV phase shift analysis of Ref.[8] were added (rectangles).

Polarization in Proton-Proton Scattering: Low energy cross section measurements alone do not permit a determination of the

individual p-wave phase shifts. Even though the central p-wave combination ^3P_C is only of the order of 0.1° at 10 MeV, the individual p-wave phase shifts may in fact be quite large. Additional information can be gained from polarization experiments, which are
sensitive to the other two p-wave combinations, $^3P_{LS}$ and ^3P_T. The
Yale group[13] pointed out many years ago that on purely phenomenological grounds, cross section measurements at low energy do not
exclude phase shift solutions with large $^3P_{LS}$ or ^3P_T and therefore
large polarization, in fact polarization as large as 0.04 to 0.05
at 3 MeV proton energy. This led Alexeff and me[14] almost 20
years ago to measure the polarization in p-p scattering in a very
difficult double-scattering experiment, in which the accuracy was
pushed to something like $\Delta P = \pm 0.003$. Essentially no effect was
found. Only at one out of three angles did the measured polarization differ by two standard deviations from zero, but recent much
more precise measurements at 10 MeV suggest that at 3 MeV the p-p
polarization is very small indeed.

Even now p-p polarization measurements are much less plentiful
than cross section measurements. Good measurements of the angular
dependence $P(\theta)$ at 66 MeV[15] and near 100 MeV have been available
for some time. Results at 97.7 MeV[16] are illustrated in Fig. 4.

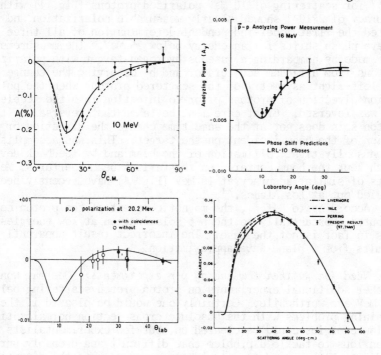

Fig. 4. Polarization in p-p scattering (see text).

At these relatively high energies the polarization is large enough
and the rate of energy loss in the target is small enough that adequate statistical accuracy can be obtained in double-scattering
experiments. Other than for measurements at isolate angles, until
recently the only additional data below 100 MeV was an angular distribution at 20.2 MeV (Fig. 4). These results[17] were obtained by
measurement of the left-right asymmetry in the scattering of a polarized proton beam from the Saclay cyclotron by a CH_2 target. Even
though the errors are only a fraction of a percent, essentially no
deviation from zero polarization was detected, in agreement with
phase shift predictions.[7] Of course, the absence of a measurable
polarization does not make the results useless since they eliminate
possible phase shift solutions giving larger polarization. In addition the experiment was important because of an earlier report[18]
of polarization as large as 2% at 20 MeV. The expectation that at
lower energies the polarization would be even smaller discouraged
further low energy experiments. In fact, Noyes and Lipinsky[3] argued
that polarization measurements near 10 MeV are not likely to be useful since the required accuracy of $\pm 4 \times 10^{-4}$ seemed impractical to
achieve. However, what seemed impractical ten years ago has more
recently been shown to be quite feasible. Measurements at Wisconsin[20] for scattering of 10 MeV polarized protons (Fig. 4) with an
accuracy of 2×10^{-4} showed clearly measurable polarization and permitted the first model-independent determination of all three
p-wave phase shifts for an energy below 25 MeV. The measurements
were made by bombarding a gaseous hydrogen target with polarized
protons from a tandem accelerator and by observing the change in
the left-right asymmetry of the scattered protons when the polarization direction of protons prior to injection into the accelerator was reversed. Great care must be taken that reversal of the
proton spin does not at the same time change the direction or position of the beam incident on the target. This was accomplished
by unusually tight collimation of the beam and by feedback devices
which kept the beam centered on the various beam apertures. Measurements of similar accuracy at 16 MeV (Fig. 4) have recently been
carried out at Los Alamos.[21]

According to Ref.[10], the magnetic-moment of the proton contributes substantially to the p-p polarization at low energies.
This in fact is not the case. The incorrect result apparently
results from a plane-wave approximation.

Need for further low-energy p-p experiments: One may wonder
whether additional experiments on proton-proton scattering below
100 MeV are worthwhile. Certainly one would be pleased if the
remaining problems with the absolute cross section normalization
at low energies could be cleared up, but few experimentalists will
be anxious to tackle a problem that difficult and probably unre-
warding. With regard to polarization measurements the situation
is different. We now have measurements at 10 MeV and 16 MeV which
are an order of magnitude more accurate than the available data
at higher energies. While the two low-energy experiments were

carried out with beams from tandem accelerators, polarized beams of high intensity and readily reversed polarization are also available from variable-energy cyclotrons. At the moment, the main question facing the experimenter is not whether greatly improved polarization measurements for instance at 25 MeV and 50 MeV are possible, but whether the experiments are useful. A recent analysis of 20-30 MeV p-p and n-p data by Bohannon, Burt and Signell[22] shows that the p-p phase shifts at 25 MeV are already quite well determined by the present data set, which includes not only cross section and polarization measurements, but also measurements of two triple scattering parameters $R(\theta)$ and $A(\theta)$ and of spin-correlation parameters at $90°$. In order to assess the possible merit of additional polarization measurements, Signell[23] provided me with calculations of the reduction in the phase shift uncertainties which would be obtained if polarization measurements were made at the same angles and with the same uncertainties as were obtained by the Wisconsin group at 10 MeV. The results for 25 MeV and 50 MeV energy, which are shown in Table I, show that the accuracy of the tensor and $\ell \cdot s$ p-wave combinations would be very substantially improved by such measurements. But this does not answer the essential question whether there is any real interest in reducing these uncertainties. The fact that various model calculations[22] of the $\ell \cdot s$ p-wave combination differ from each other by $0.3°$ or less, while the present experimental value has an uncertainty of $\pm 0.15°$ suggests that a reduction of the experimental uncertainty by a factor of three may be of interest.

TABLE I

pp phase shift uncertainties in degrees

	$E_p = 25$ MeV		$E_p = 50$ MeV	
	present	with improved $P(\theta)$	present	with improved $P(\theta)$
$^1S_0$	0.24	0.12	0.3	0.1
$^1D_2$	0.029	0.027	0.076	0.06
^3P_C	0.034	0.034	0.034	0.033
^3P_T	0.107	0.024	0.07	0.017
$^3P_{LS}$	0.15	0.042	0.098	0.035

There also is the possibility that the uncertainties in the present phase shifts are larger than assumed. One notes that the recent p-wave phase shifts of Arndt, Hackman and Roper (Table 6, Ref.[24]) differ from those of Bohannon[22] considerably more than the assigned uncertainties permit. Should it turn out that improved accuracy of the data set is desirable, the possibility of remeasuring triple-scattering parameters should also be considered, since the polarized-beam technology has improved immensely since the last measurements[25] in 1965 and angular distributions of only two parameters (R and A) have been measured.

NEUTRON-PROTON SCATTERING

Cross Section and the $^1P_1$ *phase shift at 50 MeV:* Experiments on neutron-proton scattering are inherently less accurate than for proton-proton scattering, but considerable progress has been made recently in improving the data set at 50 MeV and below. The most accurate information on the n-p system is the total cross section, which, according to the evaluation by Hopkins and Breit[26] is known below 30 MeV with an estimated standard deviation of ±1%. The values given in this evaluation agree with the measurements of Davis and Barschall[27] between 1.5 MeV and 27.5 MeV. The cross sections given by Hopkins and Breit also are consistent with the recent evaluation of the effective range parameters by Lomon and Wilson.[28]

There has been a problem for some time with the n-p $^1P_1$ phase shift at 50 MeV, which was well determined by the data, but at variance with values at neighboring energies and with values predicted by meson-theoretical models as pointed out by Arndt *et al.*[29] They found that the value of $^1P_1$ was primarily dependent on the differential cross section measurements near 50 MeV. In a later paper Blinstock and Bryan[30] investigated what particular observable would be most effective to tie down the $^1P_1$ phase shift. Their calculations suggested that a remeasurement of the differential cross section would be more effective in this respect than the measurement of more exotic observables. Measurements of $\sigma(\theta)$ at 50 MeV by the Davis group[31] resolved the anomaly in the $^1P_1$ phase shift. Very recently, the same group[32] made additional improvements in the 50 MeV forward-angle data and also measured an angular distribution at 25.8 MeV. Neutrons with an energy spread of about 2 MeV were produced by the $^7Li(p,n)$ reaction and their mean energy was measured to 0.1 MeV by the time of flight. Cross sections at backward angles were obtained by detecting the recoil protons from a CH_2 target with counter telescopes, while for forward angle measurements neutrons scattered by a hydrogenous scintillator were detected in coincidence with the recoil protons. The latter measurements depend critically on the knowledge of the neutron detection efficiency as a function of energy. After much work the relative efficiency of the detectors used is now known to ±3%. The overall accuracy of the relative n-p cross section angular distributions at these two energies is now thought to be about ±2% for the backward hemisphere and about ±3% at forward angles. The results, some of which are shown in Fig. 5, represent the current state of the art. The measurements at 25.8 MeV are consistent with the results of Barschall's group[33] at nearby energies and confirm the for-aft asymmetry observed in their 24 MeV data.

Polarization measurements and the S-D coupling parameter: Even though the art of n-p experiments has improved, the current thinking is to forget about a determination of the n-p phase shifts indepen-

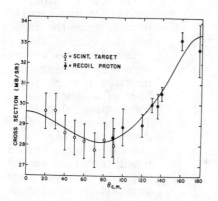

Fig. 5. n-p Cross Section at 25.8 MeV (Ref.[32]).

dent of the p-p phase shifts. Rather the idea is to obtain the isospin-one n-p phases for $\ell \geq 1$ from p-p experiments and to apply corrections for Coulomb effects. In two recent papers the n-p and p-p data sets between 20 and 30 MeV[22] and between 1 and 450 MeV[24] are analyzed simultaneously, but in fact the isospin-one phases are influenced little by the n-p data since in general the p-p data are of higher accuracy. Thus the n-p data in effect are used to find primarily the isospin-zero phase shifts which do not enter in the p-p system, namely $^3S_1$, $^1P_1$, and $^3D_{1,2,3}$ and the parameter ε_1 which couples the J = 1 states $^3S_1$ and $^3D_1$.

The behavior of the S-D coupling parameter ε_1 at low energies has long been a problem because the straightforward phase shift analysis by McGregor, Arndt and Wright[7] yielded values of the S-D wave coupling parameter that were negative at low energies. From the sign of the quadrupole moment, this coupling parameter is thought to be positive. Consequently, the same authors constrained the coupling parameter to positive values at low energies and obtained the so-called constrained phase shift set. Substantial progress on this problem has recently been made. The availability of new high-precision n-p polarization measurements as a function of angle near 21 MeV by Jones and Brooks[34] and by Morris et al.[35] led to the already mentioned recent reevaluation of the entire n-p and p-p data set between 20 and 30 MeV by Bohannon et al.[22]. The inclusion of the new data noticeably changed the central values of the isospin-zero phase shifts and reduced their uncertainties. In particular the analysis illustrates the usefulness of accurate polarization data in the determination of the D-wave $\ell \cdot s$ phase shift combination. In addition the coupling parameter ε_1 is no longer negative, the value being $\varepsilon_1 = +1.03° \pm 0.57°$. The measured polarization near 21 MeV has a maximum value of about 0.05 while the uncertainty in the two recent determinations is about a tenth of that value. Thus, the relative accuracy of the n-p polarization measurements now approaches that of the most accurate p-p polarization measurements. Unfortunately there is still an unexplained discrepancy between the two recent n-p experiments, as shown in Fig. 6. The experiment of Morris et al.[35] used a conventional technique: polarized neutrons (P $\simeq$ 0.35) produced by the $^3H(d,n)$ reaction were scattered from a hydrogenous liquid scintillator. The scattered neutrons were detected by two symmetrically placed plastic scin-

10

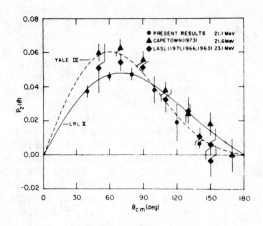

Fig. 6. n-p Polarization near 21 MeV (Ref.[35]).

tillators in coincidence with the recoil protons from the target and the spin of the incident neutrons was reversed by passing the neutrons through the bore of a superconducting solenoid. The measurements of Jones and Brooks[34], on the other hand, used a novel technique employing only a single monocrystalline anthracene scintillator in which recoil protons are detected. The direction of the recoil protons is deduced by exploiting the fact that the light-output depends on the orientation of the recoil with respect to the crystal axis. The pulse height spectra are then divided into bins, each bin corresponding to a certain range of recoil angles. The advantage of the method lies in a simple detector arrangement, high count rates and lack of multiple scattering problems because of the relatively small required size of the sample. The disadvantage is that scattering angle and azimuthal angle are determined indirectly from pulse height information.

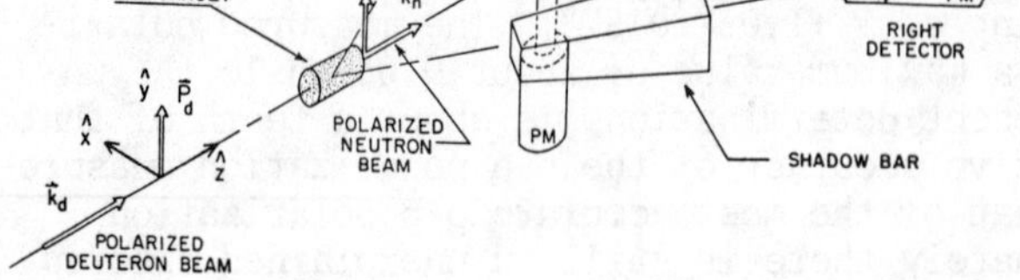

Fig. 7. Arrangement used by Walter et al.[37] to measure the polarization in n-p scattering.

The most accurate n-p polarization measurements are new results at 14 MeV submitted to this conference by Brock et al.[36] from Auckland and results between 13.5 and 16.9 MeV which I obtained from Walter's group[37] at TUNL. The combined statistical and estimated systematic uncertainty of the TUNL measurements is 0.002 or less for most of the points. In their experimental arrangement (Fig. 7) polarized D(d,n) neutrons at 0° are produced by bombardment of a deuterium gas target with vector-polarized deuterons. In this way high neutron polarization can be obtained at 0° where the cross section has a maximum. The neutron polarization is reversed by reversing the polarization of the deuterons prior to injection into the accelerator. Time-of-flight measurements were used to discriminate against breakup neutrons. Considerable effort was spent to study multiple-scattering in the sample by Monte-Carlo calculations, and experimentally by adding a graphite mantel around the scatterer. The concern here is that the polarization in n-^{12}C scattering is large compared to the n-p polarization, so that multiple scattering of neutrons in the hydrocarbon target may cause problems. The conclusion of their work is that an earlier experiment[38] which showed large changes in the apparent n-p polarization with scatterer dimensions must be in error. The angular distribution obtained at 16.9 MeV (Fig. 8, open circles from Ref.[35]) does not agree well with phase shift predictions. The deviations are qualitatively the same for the precision 14.1 MeV data of the Auckland group.[36] Those measurements also made use of the large neutron polarization from a polarization transfer reaction, in this case the ^{3}H(d,n) reaction at 150 keV. A summary of P-wave and D-wave spin-orbit phase shift combinations, $^{3}P_{LS}$ and $^{3}D_{LS}$, between 10 and 35 MeV which was prepared by A. Chisholm[39] is shown in Fig. 9.

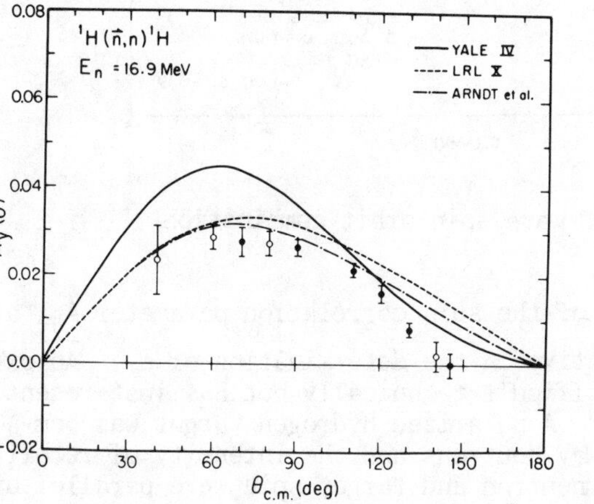

Fig. 8. Polarization in n-p scattering at 16.9 MeV (Ref.[37]). The open circles are from Ref.[35].

The already mentioned analysis[22] of the entire 20-30 MeV data set shows a splitting $\Delta^{3}P_{LS}$ between n-p and p-p scattering consistent with zero, but if the splitting is as large as expected from various models [~ 0.08°] it should be possible to detect it if the best currently available techniques are applied to p-p and n-p polarization measurements at 25 MeV.

As mentioned above, the new polarization data have permitted a better determination of the coupling parameter ε_1 at 25 MeV. At 50 MeV, ε_1 was very poorly determined by the n-p data set and again showed a preference for negative values. Blinstock and Bryan[30]

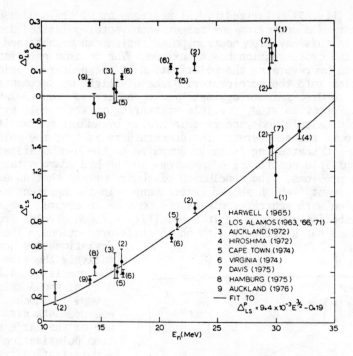

Fig. 9. P- and D-wave spin-orbit combination
(Ref.[39]).

showed that a measurement of the spin correlation parameter A_{yy} at
50 MeV would be very effective in the determination of ε_1. An experiment of this type is difficult technically but has just recently
been completed at Davis.[40] A polarized hydrogen target was bom-
barded with polarized 50 MeV neutrons and the intensity of recoil
protons was compared when neutron and target spin were parallel or
antiparallel to each other. One of the main problems of polarized-
target experiments is that the target polarization can be measured
only with moderate accuracy, in this case to 25% of its value.
Since a report[41] on improvements of this experiment will be presented at this conference I will not discuss it further, other than to
say that the coupling parameter ε_1 is now also found to be positive
at 50 MeV ($\varepsilon_1 = 2.6°\pm1°$). The problem of negative ε_1 at low energies now seems a thing of the past.

DISPERSION RELATIONS AS A TEST OF LOW-ENERGY PHASE SHIFTS

The analysis of the NN data set at a given energy or in a
narrow range of energies is essentially model-independent, since at
most some of the higher partial waves are deduced from theory. The
energy-dependent phase shift analyses lead to smaller uncertainties

of the phase parameters but the results then depend critically on the parametrization of the energy dependence. Viollier, Plattner and Alder[42] point out that a sensitive consistency test of the low-energy behavior in energy-dependent phase shift analyses can be obtained by the application of a forward-angle dispersion relation to the various channel-spin amplitudes. In their analysis, all amplitudes are modified to eliminate Coulomb singularities, and the spin-flip amplitudes are divided by $\sin^{|\Delta m|} \theta_{cm}$ to obtain a non-vanishing amplitude in the forward direction. As a result of their analysis they calculate a residue function R(E) for each amplitude. At the pion-exchange energy $E_{p,lab}$ = -9.71 MeV all six residue functions should assume the same value $G^2_{pp\pi^0}$, independent of additional cut contributions (these additional contributions will simply lead to an energy dependence of the residue function). The results for two phase shift analyses are shown in Fig. 10. The

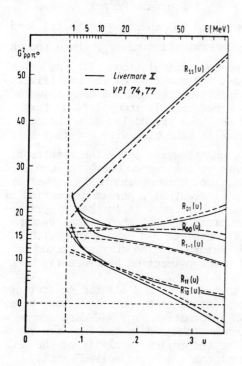

curves in general converge to a reasonable value of the coupling constant, but the Livermore triplet phase shifts show an incorrect trend below 20 MeV, caused by the neglect of Coulomb penetration effects in their parametrization of the phase shifts with ℓ > 0. In the more recent phase analysis by Arndt *et al.* this deficiency was eliminated, as evidenced by the convergence of the triplet residue functions toward the expected value of $G^2_{pp\pi^0}$ = 15. However, in this case the singlet residue function [i.e. essentially the $^1S_0$ partial wave] extrapolates to a value of G^2 which is much too small. Indeed, the singlet effective range parameter used in this analysis [r = 2.687±0.0146] is appreciably smaller than the best value recommended by Noyes[43] [2.794±0.015]. The scattering length is slightly different as well, but this has little effect on the residue function.

Fig. 10. Application of forward dispersion relations to p-p scattering (Ref.[42]).

RELATED EXPERIMENTS

Instead of studying NN elastic scattering in more and more precise detail, it may be more interesting to look at the NN problem
in alternate ways, even if the interpretation of such experiments
in many cases is still speculative, requiring the further development of appropriate theoretical tools. One such avenue is the study
of the few-body problem, the other the study of nuclear processes
which are sensitive to the D-state wave function of the deuteron.

Study of the N-d System: The proceedings of two recent international symposia on the few-nucleon problem[44,45] give an excellent
account of recent progress in this field. One of the nucleon-
nucleon parameters which is now believed to be well established
from the analysis of few-body reactions is the neutron-neutron scattering length a_{nn} for which various kinematically complete experiments give quite consistent values, leading to an average value and
estimated uncertainty[46] of a_{nn} = -16.2±0.6 fm. Recent reviews of
the determination of a_{nn} are contained in papers by Zeitnitz[46] and
by Slaus[47] who also discuss the determination of r_{nn}. Even though
recent analyses use "exact" three-body calculations, i.e. numerical
solutions of the Fadeev equations, the question of the sensitivity
to the details of the assumed NN interaction has not been answered
entirely. Also, the results of kinematically incomplete deuteron
break-up experiments[47] have not yet been reconciled with the accepted value of a_{nn}. Clearly this is connected with the fact that
one does not yet understand in a quantitative sense the results of
kinematically complete experiments in the region where the cross
section is not strongly enhanced, i.e. between the peaks. Whether
further progress in the theory will eventually permit extraction of
off-shell NN information remains to be seen. If further experimental information is needed to constrain or test theoretical assumptions, the study of kinematically complete N-d experiments with
polarized particles may prove of interest. The current status of
polarization effects in the N-d breakup reaction has recently been
reviewed by Conzett.[48]
It is obvious, of course, that data on N-d elastic scattering
contains information about the NN interaction, but for a long time
the possibility of extracting NN information from N-d data seemed
extremely remote. This situation has changed in the last few years
by the dramatic improvement in the theoretical analysis of the N-d
problem for which the Fadeev equations are being solved exactly
with increasingly more realistic NN interactions. A nice account
of this progress is contained in the papers by Conzett[48,49] at the
two recent few-body conferences. While no useful NN information
has been extracted yet from N-d scattering, it is already clear that
the N-d data, and in particular the multitude of N-d polarization
parameters, are sensitive for instance to the NN tensor interaction.
Fig. 11 shows recent calculations by Stolk and Tjon[50] for elastic

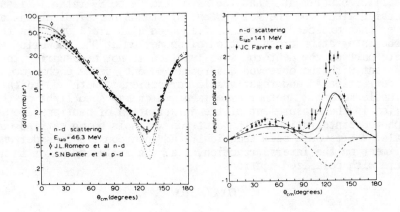

Fig. 11.   Three-body calculations for n-d scattering
(Ref.[50]).

n-d scattering with the Reid soft-core potential. The Fadeev
equations are solved exactly for the s-wave component of the nuclear force and by a perturbation treatment for the higher partial
waves. The short-dashed curve is for pure s-wave interaction,
while the dot-dashed curve also includes p-waves. The deep minimum
in the cross section is raised substantially by including the
d-wave component of the deuteron (dashed curve: $^1S_0$, $^3S_1$-$^3D_1$ and
deuteron d-state, but no p-waves). Finally, the solid curve results from use of the full Reid potential. While the calculations
are not yet entirely satisfactory, they clearly illustrate the sensitivity to the various components of the NN interaction.

The Deuteron D-state: Experiments which address themselves
specifically to the D-state part of the deuteron wave function are
discussed in reviews by Gibbs[51] and by Thomas at this conference.
Large deuteron D-state effects show up in unexpected places, like
the 700 MeV $^{12}C(p,d)^{11}C$ cross section which, according to Rost and
Shepard[52] arises primarily from the deuteron D-state because of
the poor angular-momentum matching for the S-state part of the deuteron. Similarly, the study of tensor analyzing powers in d,p
stripping reactions with $E_d \sim 10$ MeV deuterons for intermediate-
mass nuclei[53] revealed that the analyzing powers are strongly affected by the deuteron D-state. In both cases one is bothered by
the uncertainty in the nuclear interaction potential. In the case
of d,p stripping this problem can be overcome by choosing a sub-
coulomb reaction as discussed by Knutson and Haeberli.[54] In this
case the tensor analyzing powers arise entirely from the deuteron
D-state. The experiment is now being improved by lowering the beam

energy and increasing the accuracy of the beam polarization calibration. It remains to be seen whether the tentative conclusion of the earlier experiment, that the asymtotic D- to S- ratio is less than calculated from the Reid potential stands up. Indications of a lower than expected D-state contribution also are found[55] in the photodesintegration cross section of deuterium at 0°. Lowering the D-state part of the deuteron wave function is not necessarily inconsistent with the observed quadrupole moment, since exchange current corrections[56] and relativistic corrections[57] both add to the quadrupole moment. Another experiment which may shed light on this question is the measurement of the polarization of neutrons from the photodesintegration of the deuteron which is being carried out at Yale.[58] It would seem technically advantageous to consider an experiment on the inverse reaction, i.e. the analyzing power in n-p capture with polarized neutrons.

CONCLUSION

A number of recent experiments on nucleon-nucleon scattering at low energies have cleared up several disturbing problems. Among these are the energy dependence of the central p-wave combination, the low-energy behavior of the S-D coupling parameter ε_1 and the $^1P_1$ n-p phase shift at 50 MeV. It is concluded that the application of the best currently available techniques to measurements of the p-p and n-p polarization would lead to a further substantial decrease in the phase shift uncertainties to the point where an expected charge splitting in the $^3P_{LS}$ phase shift combination between n-p and p-p of $\sim 0.1°$ could be detected. Also, improved measurements of triple-scattering parameters are feasible, but it is unclear whether there is sufficient motivation to pursue such experiments. The tendency at present is to look at other aspects of the NN problem which have in the past not received as much attention as NN scattering, e.g. studies which are related to the properties of the deuteron, and the investigation of the three-body problem.

ACKNOWLEDGMENTS

I should like to thank Drs. E. Baumgartner, R.E. Brown, A. Chisholm, P. Signell and R.L. Walter for permission to quote their results prior to publication. I am also thankful to Drs. H.E. Conzett, G.R. Plattner and D. Saylor for discussions during the preparation of the manuscript.

REFERENCES

1. Ch. Thomann, J.E. Benn and S. Münch, paper submitted to this conference.
2. H. Mühry, H. Wassmer and E. Baumgartner, Helv. Phys. Acta 46, 604 (1973).
3. H. Wassmer and H. Mühry, Helv. Phys. Acta 46, 626 (1973).

4. J. Birchall, E. Baumgartner, H.M. Friess, H. Mühry, F. Rösel and D. Trautmann, to be published.
5. H.P. Noyes and H.M. Lipinski, Phys. Rev. C4, 995 (1971).
6. M.S. Sher, P. Signell and L. Heller, Ann. Phys. (N.Y.) 58, 1 (1970).
7. M.H. McGregor, R.A. Arndt and R.M. Wright, Phys. Rev. 182, 1714 (1969).
8. R.A. Arndt, R.H. Hackman and L.D. Roper, Phys. Rev. C9, 555 (1974).
9. N. Jarmie, J.L. Jett, J.L. Detch and R.L. Hutson, Phys. Rev. Lett. 25, 34 (1970).
10. K Imai, K. Nisimura and N. Tamura, Nucl. Phys. A246, 76 (1975).
11. P.M. Hegland, R.E. Brown, J.S. Lilley and J.A. Koepke, Phys. Rev. Lett. 39, 9 (1977),and private communication.
12. P. Signell, private communication, June 1977.
13. M.H. Hull and J. Shapiro, Phys. Rev. 109, 846 (1958).
14. I. Alexeff and W. Haeberli, Nucl. Phys. 15, 609 (1960).
15. J.N. Palmieri, A.M. Cormack, N.F. Ramsey and R. Wilson, Ann. Phys. (N.Y.) 5, 299 (1958).
16. M.R. Wigan, R.A. Bell, P.J. Martin, O.N. Jarvis and P.N. Scanlon, Nucl. Phys. A114, 377 (1968).
17. P. Catillon, J. Sura and A. Tarrats, Phys. Rev. Lett. 20, 602 (1968).
18. R.J. Slobodrian, J.S.C. McKee, H. Bichsel and W.F. Tivol, Phys. Rev. Lett. 19, 704 (1967).
19. P. Noyes and H.M. Lipinsky, Phys. Rev. 162, 884 (1967).
20. J.D. Hutton, W. Haeberli, L.D. Knutson and P. Signell, Phys. Rev. Lett. 35, 429 (1975).
21. P.A. Lovoi, G.G. Ohlsen, N. Jarmie, C.E. Moss and D.M. Stupin, Proc. Fourth Int. Symp. on Polarization Phenomena, W. Grüebler and V. König, eds (Birkhäuser 1976) p. 450.
22. G.E. Bohannon, T. Burt and P. Signell, Phys. Rev. C13, 1816 (1976).
23. P. Signell, private communication, 1976.
24. R.A. Arndt, R.H. Hackman and L.D. Roper, Phys. Rev. C15, 1002 (1977).
25. A. Ashmore, B.W. Davies, M. Devine, J.J. Joey, J. Litt and M. Shephard, Nucl. Phys. 73, 256 (1965).
26. J.C. Hopkins and G. Breit, Nuclear Data Tables A9, 137 (1971).
27. J.C. Davis and H.H. Barschall, Phys. Rev. C3, 1798 (1971).
28. E. Lomon and R. Wilson, Phys. Rev. C9, 1329 (1974).
29. R.A. Arndt, J. Blinstock and R. Bryan, Phys. Rev. D8, 1397 (1973).
30. J. Blinstock and R. Bryan, Phys. Rev. D9, 2528 (1974).
31. T.C. Montgomery, F.P. Brady, B.E. Bonner, W.B. Broste and M.W. McNaughton, Phys. Rev. Lett. 31, 640 (1973).
32. T.C. Montgomery, B.E. Bonner, F.P. Brady, W.B. Broste and M.W. McNaughton, Phys. Rev. C, to be published.
33. L.N. Rothenberg, Phys. Rev. C1, 690 (1972); T.G. Masterson, Phys. Rev. C6, 690 (1972); T.W. Burrows, Phys. Rev. C7, 1306 (1973).
34. D.T.L. Jones and F.D. Brooks, Nucl. Phys. A222, 79 (1974).

35. C.L. Morris, T.K. O'Malley, J.W. May, Jr. and S.T. Thornton, Phys. Rev. C9, 924 (1974).

36. J.E. Brock, A. Chisholm, J.C. Duder and R. Garret, paper submitted to this conference.

37. W. Tornow, P.W. Lisowski, R.C. Byrd and R.L. Walter, to be published.

38. B. Th. Leeman, R. Casparis, M. Preiswerk, H. Rudin, R. Wagner and P.E. Zuprański, Proc. Fourth Int. Symp. on Polarization Phenomena, W. Grüebler and V. König, eds (Birkhäuser 1976), p. 437.

39. A. Chisholm, private communication.

40. S.W. Johnsen, F.B. Brady, N.S.P. King, M.W. McNaughton and P. Signell, Phys. Rev. Lett. 38, 1123 (1977).

41. D.H. Fitzgerald, contribution to this conference.

42. R.D. Viollier, G.R. Plattner and K. Alder, Phys. Lett. 48B, 99 (1974), and G.R. Plattner, private communication.

43. P. Noyes, Ann. Rev. Nucl. Sci. 22, 465 (1972).

44. Few Body Problems in Nuclear and Particle Physics, R.J. Slobodrian, B. Cujec and K. Ramavataram, eds (University Laval Press, Quebec 1975).

45. Few Body Dynamics, A.N. Mitra, I. Slaus, V.S. Bhasin and V.K. Gupta, eds (North Holland 1976).

46. B. Zeitnitz, in Interaction Studies in Nuclei, H. Jochim and B. Ziegler, eds (North Holland 1975), p. 499.

47. I. Slaus, Proc. Int. Conf. on the Interactions of Neutrons with Nuclei, E. Sheldon, ed (Technical Information, U.S. ERDA report CONF-760715-P2, 1976), p. 273.

48. H.E. Conzett, in Few Body Dynamics, A.N. Mitra, I. Slaus, V.S. Bhasin and V.K. Gupta, eds (North Holland, 1976), p. 612.

49. H.E. Conzett, in Few Body Problems in Nuclear and Particle Physics, R.J. Slobodrian, B. Cujec and K. Ramavataram, eds (University Laval Press, Quebec 1975), p. 566.

50. C. Stolk and J.A. Tjon, Phys. Rev. Lett. 35, 985 (1975).

51. W.R. Gibbs, Proc. Fourth Int. Symp. on Polarization Phenomena, W. Grüebler and V. König, eds (Birkhäuser 1976), p. 61.

52. E. Rost and J.R. Shepard, Phys. Lett. 59B, 413 (1975).

53. L.D. Knutson, E.J. Stephenson, N. Rohrig and W. Haeberli, Phys. Rev. Lett. 31, 392 (1973).

54. L.D. Knutson and W. Haeberli, Phys. Rev. Lett. 35, 558 (1975).

55. R.J. Hughes, A. Ziegler, H. Wäffler and B. Ziegler, Nucl. Phys. A267, 329 (1976).

56. M. Gari and H. Hyuga, Nucl. Phys. A264, 409 (1976); A. Jackson, A. Landé and D.O. Riska, Phys. Lett. 55B, 23 (1975).

57. F. Gross, in Few Body Problems in Nuclear and Particle Physics, R.J. Slobodrian, B. Cujec and K. Ramavataram, eds (University Laval Press, Quebec 1975), p. 782.

58. F.W.K. Firk, Proc. Int. Conf. on the Interactions of Neutrons with Nuclei, E. Sheldon, ed (Technical Information Center, U.S. ERDA report CONF-760715-P2, 1976), p. 389.

TRIPLE SCATTERING PARAMETERS

J. A. Edgington
Queen Mary College, (University of London)
Mile End Road, London E1 4NS

ABSTRACT

This paper describes measurements, made at TRIUMF, of triple scattering parameters in proton-proton and neutron-proton elastic scattering.

First, my apologies to those of you who bought tickets to this conference hoping to hear David Bugg give this talk. I shall try to do justice to his contributions to the experiments I am going to describe. These relate to measurements of Wolfenstein parameters in p-p and n-p elastic scattering, which have been performed at TRIUMF by an Anglo-Canadian group known by the acronym BASQUE (Table I). We have been at TRIUMF for three years now, and I should like to take this opportunity to express the gratitude of the British members for the welcome we have received. Notwithstanding our sobriquet, we feel very much an integral part of the laboratory.

Table I   Past and present members of BASQUE group

C.Amsler	R.Dubois	C.J.Oram
D.A.Axen	J.A.Edgington	J.R.Richardson
J.Beveridge	L.Felawka	L.P.Robertson
I.M.Blair	D.Gibson	K.I.Shakarchi
R.C.Brown	S.Jaccard	N.M.Stewart
D.V.Bugg	R.Keeler	J.Va'vra
A.S.Clough	G.A.Ludgate	G.L.Waters

TRIUMF's polarised beam of variable energy makes it an ideal means of studying the spin-dependence in nucleon-nucleon scattering. Our initial proposals took the isovector phases as known and concentrated on elucidating the isoscalar phases through n-p scattering. We had no polarised target and besides, the usual normalisation problems of target polarisation made correlation experiments less inviting than triple scattering measurements. Peter Signell kindly ran the MSU phase shift program and convinced us that measurements near 90° of D, R, and A, plus the corresponding transfer parameters, would alone

ISSN: 0094-243X/78/019/$1.50 Copyright 1978 American Institute of Physics

be adequate to determine isoscalar phase shifts uniquely
and precisely. Table II shows our measurements so far.
We will eventually have measured all six parameters at
each of four energies where there is enough other data
for a full phase shift analysis: 210, 325, 425 and 495
MeV. Only the D/D_t data, taken last fall, have been
fully analysed, and this analysis, as I shall report,
has already resolved the major ambiguity which existed
here. Our aim was a precision of ± 0.03 on each parameter,
at each of about half a dozen angles.

Table II Experimental program of BASQUE group

| | n-p Wolfenstein parameters | | |
	D and D_t	A and A_t	R and R_t
210 MeV	data taking completed		
325 MeV	analysis completed	data taking completed	
425 MeV	data taking completed		
495 MeV	data taking completed		

angular range: $30^{\circ} - 100^{\circ}$ for D, R, A
$80^{\circ} - 150^{\circ}$ for D_t, R_t, A_t

precision : ± 0.03 in six angular bins

We soon realised that if we were succesful then
the errors on the existing isovector phase shifts would
begin to dominate our analysis, so we decided to refine
the p-p data by similar measurements at similar energies,
and also 380 MeV. David Bugg modified his πN partial
wave analysis program to deal with the NN system and
with its aid we determined that D and R at four angles
in the range $10^{\circ} \leqslant \theta \leqslant 50^{\circ}$, and R' at one angle near
30°, would be the most useful measurements, and we
completed all of these last year. You will recall that
it was the triple-scattering measurements at Rochester
and Chicago in the late 1960's that pinned down unique
partial wave solutions at 210 and 425 MeV. I am glad
to report that we believe we now have a well-defined and
unique solution at 515 MeV also, as well as substantial
error reductions at lower energies. There are surprises
in some partial waves, as I shall discuss.
First, though, I must describe the experiments and
convince you we've done them properly. The major piece

of equipment is a carbon polarimeter comprising twelve
MWPCs, six of (0.5 m)^2 in front of, and six (1 m)^2
behind, a 6 cm carbon scatterer. The chambers' wires
have a 2 mm spacing but they are read out as adjacent
pairs, yielding 4 mm resolution. The polarimeter was
calibrated using polarised protons, namely those
scattered elastically from hydrogen at 24° (lab).
Figure 1 shows the arrangement. This was our first
measurement, using the unpolarised extracted beam which
was all that was available for the first year. The
scattered protons' spin direction was precessed using
a 6 Tm superconducting solenoid. This was constructed
at Rutherford Laboratory, was quenched twice there during
'training', and in three years at TRIUMF has been
virtually trouble-free. It consumes roughly 1½ litres
of liquid helium per hour, with no recovery.

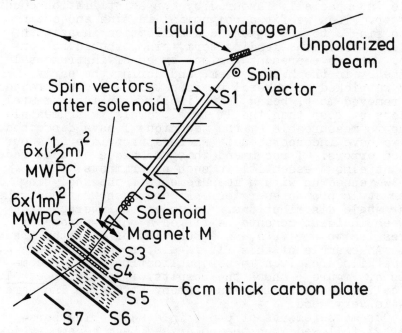

Figure 1. Layout for polarimeter calibration.
Scintillation counter trigger is S$_{1234567}$

We separated elastic from inelastic protons, and
from pions and assorted junk by a combination of time-of-
flight and momentum analysis. Note the small steering
magnet M; it deflects elastic protons by only some 7 mrad,
so there is no spatial separation of the various momenta.
Instead we reconstruct the track, off line, from

information from the front MWPCs, and project it back
to the source. The angular dispersion results in pions
and protons of the same velocity giving separate projected
images. All this is necessary because we are passing the
protons through ports in the collimator, intended eventually for neutrons.

Ideally the analysing power would be the observed
scattering asymmetry divided by the p-p polarisation
parameter at 24°. Spin reversal with the solenoid
eliminates instrumental asymmetries to first order,
but to be sure, we analyse the data off-line as follows.
First we rotate the trajectory of the scattered particle
about that of the incident proton, and check that all
along its circumference the resulting cone intercepts
the back scintillation counters, and misses the veto
counter S7. This 'cone test' ensures that no azimuth
angle is especially favoured by our scintillation counter
trigger. Then we fit a polynomial in sine and cosine
up to $\sin^{16}\phi$ to the observed distribution in azimuth
angle ϕ. Only the leading term, $\sin\phi$, should be non-
zero. Others represent various types of instrumental
asymmetry in the MWPC system, and during the early
days we picked up several biasses this way. We traced
and removed most, being left finally with instrumental
effects at less than the 1% level. This instrumental
noise is measureable by the technique I have described,
and we have incorporated it where appropriate into our
stated errors. I recommend this technique of data reduction as almost essential in such experiments.

We ended up with a measurement of the analysing
power at 14 proton energies. Figure 2 shows two of the
polar angle distributions. In order to interpolate
between incident momenta we fitted a three-parameter
expression to our data, as a function of momentum transfer. An example of this fit is also shown in Figure 2,
and details are given in our publication[1]. For comparison, Figure 3 shows the results of Aebischer et al.[2]
who calibrated a similar instrument at CERN. The agreement is very good.

Our measurements of p-p triple scattering parameters[3] followed when the polarised beam became available. The solenoid was moved so as to precess the
primary beam's polarisation, and the latter was monitored
by two double arm counter telescopes set to record p-p
elastic scatters at 26° (lab) from a polythene foil
(Figure 4). The polarimeter array was moved successively
to laboratory scattering angles of 6°, 9°, 15° and 24°.
With the solenoid off, and spin reversal accomplished
at the ion source, the array measured a vertical
polarisation of $(P \pm DP_{beam})/(1 \pm PP_{beam})$; whereas

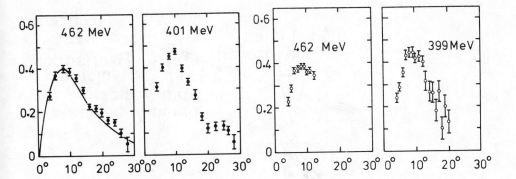

Figure 2. Analysing power of our polarimeter as a function of proton scattering angle in the 6 cm carbon plate. The curve illustrates our three-parameter representation of these data

Figure 3. Analysing power of Aebischer et al.'s carbon polarimeter, with no constraint on the proton's energy loss. The carbon plate was 5 cm thick

with the solenoid set to precess the beam's polarisation by ±90°, the measured horizontal polarisation component was ±RP_{beam}, the vertical component being +P. To measure R' this latter spin orientation was used, but the small magnet M was replaced by a considerably larger one (actually the old UCLA cyclotron magnet) to precess the longitudinal spin component of the scattered proton through 90°. This measurement was made at only one angle.

The beam polarisation was about 78% and its intensity was limited to 2 nA to ensure that the accidental count rates were negligible. The method of data analysis was similar to that just described and results are shown in Figures 5-8. We show also the existing triple scattering data of comparable precision[4,5], that is about ± 0.02 on most points, including the allowance for instrumental noise. Most of the other reported measurements have errors of order ±0.1. Some points marked 'counter data' appear on the graphs of P. These come from an independent measure-ment of the counting rate asymmetry, on reversal of the incident polarisation, in three counter telescopes viewing the hydrogen target at 9°, 15° and 24°; this measurement, together with the simultaneous asymmetry measurement at 26° in the polarisation monitor, was used to determine accurately the relative values of P at these four angles. It serves, additionally, as a check on the method of extracting values of the spin parameters from the polarimeter data.

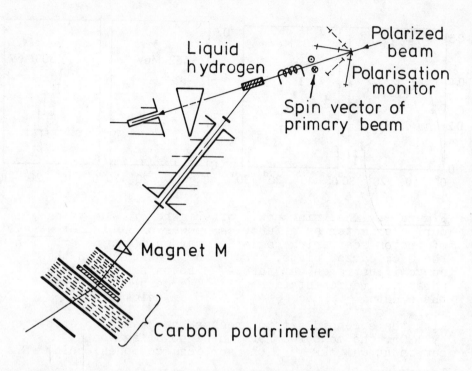

Figure 4. Layout for p-p Wolfenstein parameter measurements

Figure 5. D para-
meter in p-p scatter-
ing as a function of
c.m. scattering
angle. Curves are
from our phase shift
analysis.

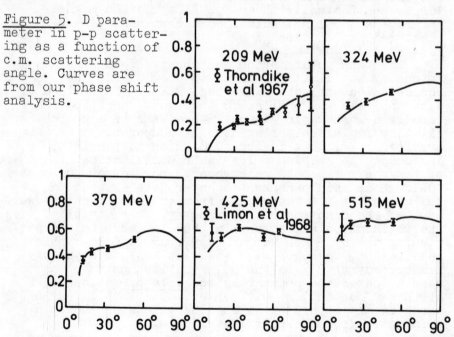

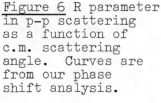

Figure 6 R parameter in p-p scattering as a function of c.m. scattering angle. Curves are from our phase shift analysis.

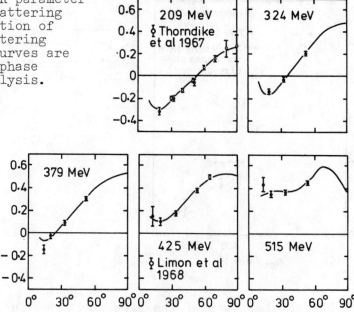

Figure 7 P parameter in p-p scattering as a function of c.m. scattering angle. Curves are from our phase shift analysis.

● – MWPC data

O – counter data

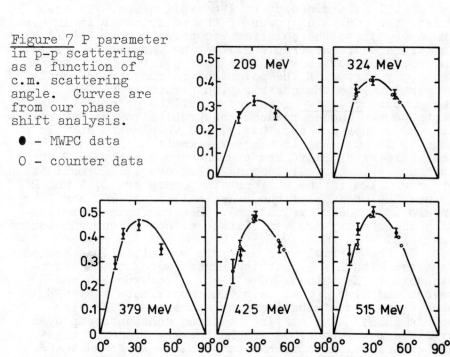

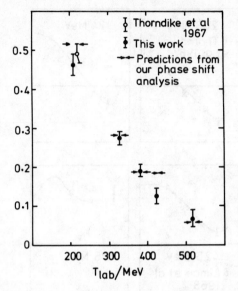

Figure 8. R' at $\theta_{lab} = 15^{\circ}$ in p-p scattering as a function of energy.

Notice that each of the experiments so far described has an ill determined overall normalisation. In the first, the analysing power of the polarimeter is inversely proportional to the polarisation parameter at 24° (lab); in the second, the triple scattering parameters are inversely proportional to the beam polarisation, and hence, by virtue of the polarisation monitor, directly proportional to the polarisation parameter at 26°(lab). Our precise relative measurement of the polarisation at these two angles reduces the problem to the determination of either one absolutely, and we plan to do this soon by a double scattering experiment at 24°(lab). Meanwhile we estimate the present uncertainty in P at this angle to be ±6%, leading to an overall normalisation error of ±12% in the Wolfenstein parameters D, R and R' (since the product of the polarisations at 24° and 26° enters into their evaluation). However, this normalisation uncertainty has very little effect on our subsequent analysis.

In the partial wave analysis we initially searched on waves with L ≤ 5, setting higher waves to one-pion-exchange (OPE) values. One might also expect G and H waves, and also $\bar{\mathcal{E}}_4$, to be given approximately by OPE plus a small contribution from exchanges of heavier bosons (HBE). The varying coupling constants employed by different authors[6-11] make the magnitude of these HBE contributions somewhat uncertain. We found that $\bar{\mathcal{E}}_4$, which has a strong effect on the angular dependence of D, is very well determined and is at all energies close to its OPE value, the HBE contribution being small.

We therefore fixed $\bar{\varepsilon}_4$ at its (OPE + HBE) value and
re-searched. Had the calculated, but uncertain, contributions of HBE to G and H waves been substantially less
than the experimental errors on their phase shifts, we
might have felt justified in fixing these phases too at
their (CPE + HBE) values. Such was not the case, and
some of the phases also differed markedly from these
(OPE + HBE) values. Table III shows the results for one
energy, 425 MeV. Only for $^1G_4$ is there agreement with
(OPE + HBE). The H waves are clearly different, $^3H_4$
remarkably so. This is the case at other energies also.
If the H waves are fixed at (OPE + HBE) values, the
quality of the fits worsens at all energies, χ^2 per
datum increasing by an average of 33%. This seems
unacceptable and indicates to us that there are inadequacies
in the simple (OPE + HBE) prescription we have used.

Table III Contributions of (OPE + HBE) to higher partial
wave phase shifts at 425 MeV, and experimental values

Contribution from HBE according to:	$\bar{\varepsilon}_4$	$^1G_4$	$^3H_4$	$^3H_5$	$^3H_6$
Bryan and Scott[6]	.041	.842	-.011	.142	.288
Scotti and Wong[7]	.066	1.462	.043	.299	.443
Ino et al.[8]	.063	.726	.072	.203	.314
Arndt et al.[9]	.096	1.162	.243	.377	.437
Köpp and Söding[10]	-.055	1.157	.228	.297	.406
Bugg[11]	.047	.676	-.080	.073	.209
Average value	.043 $\pm.052$	1.004 $\pm.306$	.082 $\pm.129$	.232 $\pm.113$	.349 $\pm.094$
OPE Contribution	-1.930	1.000	.727	-1.645	.326
(OPE + HBE)	-1.887	2.004	.809	-1.413	.675
Experiment ($\bar{\varepsilon}_4$ fixed)	-1.887	1.970 $\pm.220$	-.499 $\pm.203$	-.599 $\pm.338$	.151 $\pm.167$

All values in degrees;
Pion coupling constant taken as $g_o^2 = 14.43$

We therefore left G and H waves free and arrived at the results shown in Table IV. Our analyses achieve a value of χ^2 close to one per datum point at all energies. Our new data have reduced the errors on the phase shifts by factors varying from roughly two (at 209 MeV) to as much as seven (in some phases at 515 MeV). Our solution at 380 MeV clearly does not interpolate well between neighbouring energies. Some data at this energy are suspect, giving a very large contribution to χ^2 . We present this solution as our best guess, with the proviso that the data is incomplete at this energy. The solution at 515 MeV is now very stable but could be improved further, in particular by triple scattering measurements at forward angles, where the predicted errors on D and R are still of the order of ± 0.05 to ± 0.1. Fortunately the Geneva group are working on this and we shall be hearing some of their results at this Conference[12].

Note that at this energy we find that $\bar{\mathcal{E}}_2$ and $^3F_2$ have both become positive. This somewhat unexpected behaviour in the F waves can be studied further by decomposing the F wave phase shifts into their central, tensor and spin-orbit components (Figure 9). At 515 MeV $F_{central}$ becomes strongly attractive, and F_{tensor} weakly so. What has theory to say about this? We show predictions from the dispersion-theoretic calculations of the Paris group[13], and also from Bohannon and Signell[14]. Clearly the latters' central potential is much too attractive, though the tensor and spin-orbit potentials look about right. The Paris group come quite close to our results. Tantalisingly they have not yet extended their analysis to 515 MeV.

To indicate that our solution at this energy is not grossly wrong, I show in Figure 10 a similar breakdown of the P waves. Nothing untoward occurs at 515 MeV. Our treatment of inelasticity may be too naive (absorption in $^1D_2$ only), or perhaps there is some real physical effect trying to emerge. The ZGS measurements of total cross sections in longitudinal spin states, which will be discussed later at this Conference[15], also show something unexpected starting to happen in the triplet states at this energy.

We have attempted to use our analysis to derive a value for the πNN coupling constant, g_0^2. We varied g_0^2, thus changing the OPE contribution to the 'fixed' phases such as $\bar{\mathcal{E}}_4$. At the three lowest energies $^3H_5$ also agrees with (OPE + HBE), and it has a large OPE contribution; we therefore fixed it as well as $\bar{\mathcal{E}}_4$ at (OPE + HBE) values. The values of g_0^2 which result from minimising χ^2 are given in Table V. Adding

Table IV Experimental phase shifts for I = 1 system, in degrees

Phase	210 MeV	325 MeV	380 MeV	425 MeV	515 MeV
$^3P_0$	-2.18±0.51	-13.10±1.63	-10.93±3.11	-20.62±1.47	-17.68±3.38
$^1S_0$	4.85±0.52	-9.05±0.62	-16.08±0.87	-18.12±0.74	-22.67±3.25
$^3P_1$	-22.46±0.22	-31.35±1.04	-33.63±0.79	-35.99±0.57	-34.67±1.97
$^3P_2$	15.96±0.17	16.70±0.37	17.76±0.68	17.51±0.35	21.33±0.46
$\bar{\varepsilon}_2$	-2.81±0.12	-2.80±0.25	0.04±0.37	-1.99±0.38	2.67±0.73
$^3F_2$	-1.25±0.26	-0.81±0.40	-2.36±0.67	-0.84±0.46	-0.97±0.81
$^1D_2$	7.33±0.28	9.10±0.48	12.36±0.31	12.48±0.45	15.58±0.70
$^3F_3$	-2.53±0.17	-2.90±0.59	-2.64±0.60	-2.75±0.36	-2.38±0.54
$^3F_4$	1.96±0.16	2.85±0.14	2.64±0.56	3.03±0.20	5.00±0.34
$\bar{\varepsilon}_4$	(-1.21)	(-1.58)	(-1.83)	(-1.94)	(-2.31)
$^3H_4$	-0.15±0.20	0.55±0.17	-0.25±0.47	-0.38±0.22	-3.13±0.54
$^1G_4$	1.14±0.10	1.12±0.20	2.36±0.19	1.98±0.25	4.78±0.56
$^3H_5$	-0.84±0.17	1.67±0.34	1.63±0.45	-0.88±0.38	-1.83±0.46
$^3H_6$	-0.02±0.13	0.55±0.12	0.60±0.33	0.22±0.17	-0.94±0.38
$\theta(^3P_1)$					
$\theta(^1D_2)$		3.52±0.17	6.82±0.63	10.71±0.41	22.60±0.60
g_0^2	14.05±0.91	13.25±0.94	13.91±0.92	(13.85)	(13.85)

Values in parentheses are (OPE + HBE) values. Coulomb amplitudes and phase shifts have been used with a dipole form factor for each proton. Inelasticities are given by $\eta = \cos 2\theta$.

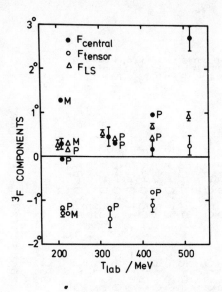 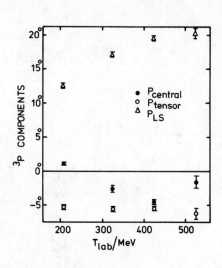

Fig.9 Central,tensor and spin-orbit
components of F waves,compared with
calculations of the Paris Group (P)
and Bohannon and Signell (M).

Fig.10 Central,tensor
and spin-orbit compo-
nents of P waves.

Table V. Values of g_o^2 with $\bar{\mathcal{E}}_4$ only, and $\bar{\mathcal{E}}_4$ plus

$^3H_5$, fixed at their (OPE + HBE) values

	209 MeV	325 MeV	380 MeV
$\bar{\mathcal{E}}_4$ fixed	14.05±.91	13.25± .94	13.91± .92
$\bar{\mathcal{E}}_4 + {}^3H_5$ fixed	14.05±.80	13.86±1.02	13.87±1.16

Weighted mean $\boxed{g_o^2 = 13.84 \pm 0.65}$: error includes
allowance for possible systematic effects.

uncertainties in the HBE contribution,and also due to a
possibly incorrect treatment of inelasticity,to the statistical error,we derive a value $g_o^2 = 13.84+0.65$. This value
turns out to be very insensitive to the absolute normalisation of the polarisation.

Now I shall turn to our n-p experiments. We use the
arrangement of Figure 11, producing the neutron beam by
charge exchange scattering of polarised protons on deuterium. The polarisation parameter R_t is large for this reaction, particularly near the forward direction. Fig.12
shows measurements we have made of R_t at an incident proton energy of 343 MeV, at various neutron production angles.

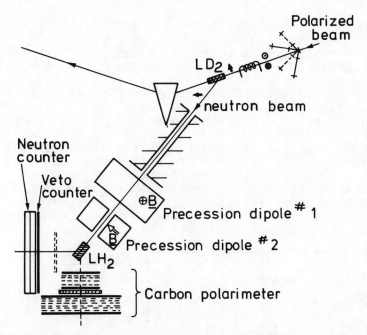

Fig.11.Layout for n-p Wolfenstein parameter measurements

David Measday had suggested[16] that this reaction would be
a useful means of obtaining polarised neutron beams, having the advantage of a narrow energy spread of order 10
MeV at these small angles. We were gratified to confirm
his predictions, and have carried out all our subsequent
neutron work at 9°(lab.). Figure 13 shows the variation
of R_t with energy at this angle; measurements of the
neutron beam's polarisation were carried out with two
different spin analysers and are in reasonable agreement.

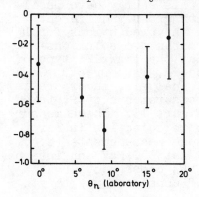

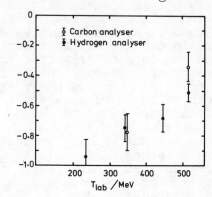

Fig.12.R_t in D(p,n)2p at
        343 MeV

Fig. 13. R_t in D(p,n)2p
         at 9°(lab).

As you see we begin to lose polarisation rather rapidly
at the highest energies.

   The neutron flux in the quasi-elastic peak at our
target position is about 10^4 MeV^{-1} cm^{-2} μA^{-1} s^{-1}, which
translates into a useful intensity of about 3.10^5 s^{-1}
through our target under typical running conditions
(50 nA polarised beam).

   Note that by juggling the scintillation counter
trigger we can require that neutrons are detected
by charge exchange interaction in the carbon of the
polarimeter, while protons are detected in the 'neutron
counter' array. Thus we can measure the D, R and A
parameters simultaneously with the transfer parameters,
though at a lower rate as the neutron's conversion
efficiency in the carbon is low. We have these data
but haven't analysed them yet.

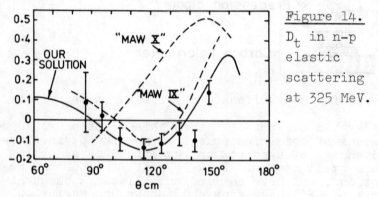

Figure 14.
D_t in n-p
elastic
scattering
at 325 MeV.

   The analysis is relatively straightforward,
following the lines already outlined. The values of D_t
at 325 MeV are shown in Figure 14. You will recall that
previous isoscalar analyses at this energy have oscil-
lated between two solutions, with grossly different
values of $^3S_1$ and $\bar{\mathcal{E}}_1$. We show the predictions for D_t
obtained using the Livermore group's two solutions[17]
and our data clearly confirm that the 'MAW IX' type of
solution, as favoured by most models, is correct. We have
carried out our own phase shift analysis, incorporating
both these D_t results and our measurements of P, taken
at the same time and shown in Figure 15. These measurements enable us to be slightly more selective in our
data set and we have excluded some measurements (e.g.,
the P values of Siegel et al.[18] at 350 MeV) on internal
grounds, and others (e.g., the PPA cross section data[19]
at backward angles) on grounds of consistency with other
measurements.

   In our analysis we set $\bar{\mathcal{E}}_4$, and also the H waves,
to (OPE + HBE) values. This is one of the energies at

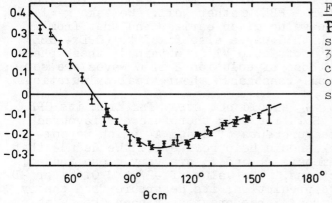

Figure 15.
P in n-p elastic
scattering at
325 MeV. The
curve is from
our phase shift
solution.

Table VI: Experimental phase shifts for I=0 system at
325 MeV, in degrees. Values in parentheses
are (OPE + HBE) values.

Partial wave	Solution 1	Solution 2
$^3S_1$	-2.4 ± 1.3	4.8 ± 1.4
$^1P_1$	-27.4 ± 0.8	-39.3 ± 1.8
$\bar{\varepsilon}_1$	8.4 ± 0.6	6.6 ± 0.6
$^3D_1$	-27.4 ± 0.4	-29.5 ± 0.8
$^3D_2$	23.9 ± 0.9	19.1 ± 1.4
$^3D_3$	1.9 ± 0.5	2.7 ± 0.8
$^1F_3$	-7.0 ± 0.5	-4.1 ± 0.4
$\bar{\varepsilon}_3$	7.2 ± 0.4	8.2 ± 0.3
$^3G_3$	(-4.7)	-2.5 ± 1.0
$^3G_4$	(7.9)	2.7 ± 1.0
$^3G_5$	(-0.5)	-0.8 ± 0.4
χ^2	255.1	241.4
# data points	282	282
# degrees of freedom	259	256

which the HBE contributions are small, and we feel we
can trust the (OPE + HBE) rather well. We take g^2=14.25,
an average of the value given earlier and that from
πN scattering. With G waves also fixed to (OPE + HBE)
values, solution 1 of Table VI is obtained; these phase
shifts drift to those of solution 2 if G waves are set
free. However, all reasonable theoretical calculations

require[20] that $^3G_4$ should not stray far from its OPE
value of 7.9^0. Solution 1 is therefore our favoured
solution, and measurements of A and A_t which we completed last week should help refine it. We deduce that
the mixing parameters $\bar{\varepsilon}$, and $\bar{\varepsilon}_3$ are very close indeed
to model predictions. The value of $-27.4^0 \pm 0.4^0$ for $^3D_1$
is also close to prediction; its departure from theory
in earlier analyses[17] has previously been used[20] as an
argument against solutions of type 1. Our data seem to
have constrained an acceptable solution.

REFERENCES

1. C.Amsler et al., Nucl. Instr. and Meth., in press
2. D.Aebischer et al.,Nucl.Instr. and Meth. 124,49(1975)
3. D.A.Axen et al., Lettere al Nuovo Cimento,to be
     published
4. E.H.Thorndike, Rev.Mod.Phys. 39,513(1967)
5. P.Limon et al., Phys.Rev.169,1026(1968)
6. R.A.Bryan and B.L.Scott, Phys.Rev. 135B,434(1964)
7. A.Scotti and D.Y.Wong, Phys.Rev. 138B,145(1965)
8. T.Ino et al.,Prog.Theor.Phys.(Kyoto) 33,489(1965)
9. R.A.Arndt et al., Phys.Lett.21,314(1966)
10. G.Köpp and P.Söding, Phys.Lett.23,494(1966)
11. D.V.Bugg, Nucl.Phys.B5,29(1968)
12. D.Besset et al., contribution to this Conference
13. M.Lacombe et al.,Phys.Rev.D12,1495(1975), and
     private communication
14. G.Bohannon and P.Signell,Phys.Rev.D10,815(1974)
15. I.P.Auer et al.,Phys.Lett.67B,113(1977),and ANL-HEP-PR-
     77-29 (1977),and contribution to this Conference
16. F.Folkmann and D.F.Measday,CERN MSC report C-17/675(1968)
17. M.H.MacGregor,R.A.Arndt and R.M.Wright,Phys.Rev.173,
     1272(1968)('MAW IX');ibid.,Phys.Rev.182,1714(1969)
     ('MAW X')
18. R. T. Siegel et al.,Phys.Rev.101,838(1956)
19. P.F.Shepard et al., PPAR-10(1969)
20. R.A.Bryan, Phys.Rev.Letters 35,967(1975), and private
     communication.

ELASTIC np SCATTERING

Byron D. Dieterle
University of New Mexico, Albuquerque, NM 87131

ABSTRACT

Elastic differential cross-sections for neutron-proton scattering in the range 200-1000 MeV are reviewed. Recent high statistics,
backward-angle experiments at Saclay and LAMPF are in agreement, but
disagree with previous measurements from PPA and Dubna. With few exceptions, data come from older, lower statistics experiments done
with poor quality neutron beams. The study of neutron-proton interactions requires more high-rate measurements of complete angular
distributions, which can be performed using recently developed
accelerator facilities. Preliminary measurements at LAMPF of 800 MeV
np scattering at back angles and forward angles are presented. The
latter is the first such high statistics experiment at medium
energies.

INTRODUCTION

First, I would like to make some historical comments. If history
is a list of events with time, then many events per unit time often
corresponds to periods of historical interest. In physics, a large
number of measurements per unit time would correspond to periods of
physics interest.

In medium-energy nucleon physics one can see this effect by looking at Fig. 1,[1] which shows the total neutron-proton cross-section

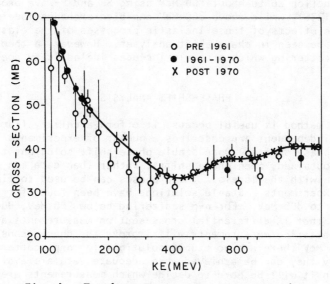

Fig. 1. Total neutron-proton cross-section.

data points for different decades. There are no points from the
40's, a large number from the 50's, almost none from the 60's, and a
revival period in the 70's. One could say that a wave of interest
swept through this energy range from low to high energies (although
the energy range was always said to be 'high' at the time). This
initial survey of total cross-sections missed the double plateau
caused by I=1 single-pion production and I=0 two-pion production.

Physicists are now returning to medium energy with better
machines and detector systems to make more accurate measurements. It
is hoped that a quantitative understanding of nucleon-nucleon interactions will result.

THEORY

A description of nucleon-nucleon forces in terms of pion exchange alone has proved to be inadequate. The forces between nucleons are now thought to be due to the virtual exchange of additional
bosons such as the ρ, ω, η, ϕ and possible scalar mesons. One of
the benefits of the higher-energy physics has been to determine the
properties of these particles so that meaningful masses and widths
can be used for the exchange amplitudes of the heavier bosons.

A reasonably good description of N-N scattering below 400 MeV
results if one uses a sum of one-boson-exchange terms to describe
the interaction.[2] The data on cross-section, polarization and other
parameters have been fit using several different combinations of exchanged bosons. Usually scalar terms are found necessary.

In the energy range above 400 MeV, where single-pion production
becomes important, there are several models which are useful for
calculating inelastic reactions. The Mandelstam model[3] describes
pion production up to about 800 MeV using S- and P-wave production
of the $\Delta(1236)$, while above 800 MeV one-pion-exchange models[4] are
used. The effects of these inelastic processes of the elastic amplitudes can be seen in phase-shift analyses. However, a theory of
elastic scattering which properly includes inelasticity does not
exist.

PHASE-SHIFT ANALYSIS

This method is useful because it offers a unitary framework that
is model independent and describes a number of experiments in terms
of a few phase shifts. Once stable phase-shift solutions to existing
data are obtained, then phase shifts, rather than data, can be used
to compare with the models. These phases can be used to find the
crucial experiments. Stable solutions have been found for p-p scattering up to 800 MeV. For n-p scattering below 500 MeV, Bugg[5] has
predicted that 1% differential cross-section measurements and 3%
double scattering measurements will improve the phases considerably.
Above 500 MeV there are no stable solutions for n-p scattering, but
presumably they can be achieved when adequate data are available.
Until then it will be hard to decide which measurements are the most
important.

TECHNICAL PROGRESS

Errors can be reduced in the newer experiments because of improved apparatus which is now available:

1) External primary beams with high intensity and small phase space are available at Saclay, TRIUMF, LAMPF and SIN. Saclay and TRIUMF have variable energy. LAMPF and TRIUMF have polarized beams.

2) Detectors and electronics improvements such as high-density fast NIM electronics, MWPC, polarized targets, CAMAC, and small computer systems have improved data-collection rates and resolution.

3) Neutron secondary beams at the new facilities are greatly improved over the older cyclotron beams. Saclay has a monoenergetic beam from deuteron stripping by protons, while TRIUMF and LAMPF have an H⁻ chopped beam on deuterium, which gives a very narrow neutron energy spectrum plus some low-energy neutron background from pion production.

ELASTIC n-p SCATTERING

Table I shows the relevant experiments[6] and the dates published. Again we see the period of interest in the 50's and a revival in the 70's. The errors for the later experiments are usually for a much smaller angular region and do not properly reflect the vast increase in total counts.

I will review these data and make some comparisons where data points overlap. Some emphasis will be put on two recent experiments at LAMPF.

Table I. Summary - $\frac{d\sigma}{d\Omega}$(np) experiments

T_{lab} (MeV)	θ^* (deg)	Statistical error (%)		
200	6-180	2-20	Dubna	(1963)
260	40-180	5-20	Berkeley	(1950)
300	10-180	10-15	Berkeley	(1954)
350	114-174	1.5	Liverpool	(1962)
400	13-180	4-10	Carnegie	(1954)
580	5-180	5-10	Dubna	(1956)
630	12-180	4-20	Dubna	(1959)
710	159-180	7-10	Berkeley	(1960)
991 (p,n)	15-150	5-30	Birmingham	(1967)
1250-6000	20-100	10-15	Berkeley	(1970)
58-391	11-54	5-20 (10% syst.)	PPA	(1976)
180-1270	45-180	5 (3-20% syst.)	PPA	(1974)
430-1270	140-180	4 (5% syst.)	Saclay	(1975)
647	55-180	3 (7% syst.)	LAMPF	(1976)
a800	60-180	3 (7-10% syst.)	LAMPF	
a800	9-60	4 (3-10% syst.)	LAMPF	

aPreliminary data. Systematics will be reduced by corrections.

1) forward $\frac{d\sigma}{d\Omega}(\theta^* > 60°)$ - Univ. of New Mexico, Temple Univ., and LASL (MP-10, P-11)

2) backward $\frac{d\sigma}{d\Omega}(\theta^* < 60°)$ - Univ. of New Mexico, Texas A&M, Univ. of Texas, and LASL (PDOR)

Quantities immediately available from the data are:

differential cross-section, $\frac{d\sigma}{d\Omega}$

total elastic cross-section, $\sigma_{el} = \int \frac{d\sigma}{d\Omega} \, d\Omega$

total inelastic cross-section, $\sigma_{inel} = \sigma_{tot} - \sigma_{el}$

Extrapolation of the forward differential cross-section gives the real part of the forward scattering amplitude f_n,[7]

$$\alpha = \frac{Re \, f_n(0)}{Im \, f_n(0)} = \left[\left(\frac{d\sigma}{d\Omega}(0) \cdot \frac{4\pi}{k\sigma_{tot}} \right)^2 - 1 \right]^{1/2}, \tag{1}$$

assuming no spin effects.

The slope parameter b is obtained from a fit of the forward data to the invariant form

$$\frac{d\sigma}{dt} = Ae^{bt}. \tag{2}$$

The backward data can be fit to the form

$$\frac{d\sigma}{dt} = \alpha_1 e^{\beta_1 t} + \alpha_2 e^{\beta_2 t}. \tag{3}$$

This is useful in separating pion exchange effects (term with steep slope) from other boson exchanges. Extrapolation of the cross-section[8] inside the region for pion exchange ($-u < 0.02$ (GeV/c)2) to the pole at $u = +m_\pi^2$ gives the pion coupling constant f^2.

General features of the data at these energies are seen in Fig. 2. These old data show:

1) poor measurements at forward angles,

2) peaks at 0° and 180°, and

3) a decrease in backward cross-sections as the energy increases.

This is unlike p-p scattering, which has a flat angular distribution up to 400 MeV and is always symmetric about 90°.

RECENT BACK-ANGLE EXPERIMENTS

Table II compares the three recent back-angle experiments. The beams are different and the normalization technique PPA uses (calibrated neutron counter) is different. However, they are all similar in that they detect the proton in a magnetic spectrometer and have beams with calibrated flux and energy. In addition they overlap the old data at 630 MeV from Dubna. The Saclay and PPA data have been published.

The LAMPF spectrometer was used in Experiment 125[9] to measure n-p backward elastic scattering at 650 MeV and 800 MeV. Since then

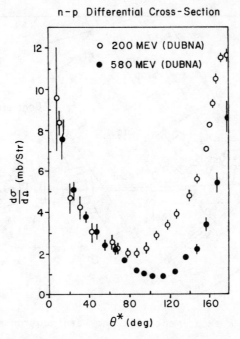

Fig. 2. n-p differential cross-sections.

Table II.   Recent back-angle experiments

	PPA	SACLAY	LAMPF
Beam (primary)	3 GeV/c	2-3.8 GeV/c	1.2-1.45 GeV/c
Production target	Pt	Be	LD_2
Reaction	nuclear	stripping	CEX
Neutron spectrum	white 0.5-2 GeV/c	$\Delta p/p = 4\%$ 1-1.9 GeV/c	$\Delta p/p = 0.5\%$ 1.2-1.45 GeV/c
Flux	4×10^3	5×10^7	10^6
Normalization	neutron counter	$\begin{cases} np \rightarrow d\pi^0 \\ \frac{d\sigma}{d\Omega}(np) = \frac{1}{2}\frac{N(prot.)}{N(neut.)}\frac{d\sigma}{d\Omega}(pp \rightarrow d\pi^+) \end{cases}$	$np \rightarrow d\pi^0$
Detector	Mag. Spect. Mag. Sp. Ch.	Mag. Spect. Scint. Hodo.	Mag. Spect. MWPC

40

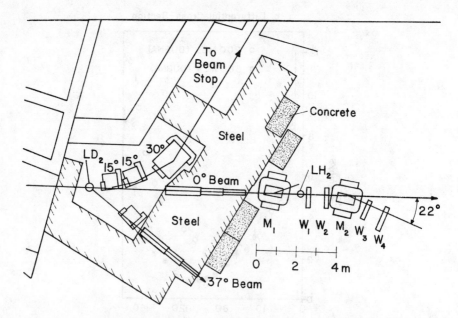

Fig. 3. LAMPF experiment 125, backward neutron scattering.

it has been used at 460 MeV.[10] Only the 650 MeV data in the extreme
back angles have been published.[11] Figure 3 shows the apparatus on
the floor of Area B in the Nucleon Physics Lab. The spectrometer
points out to about 60° scattering angle. Beyond this angle there
is interference with the sweep magnet M1, but this occurs only a few
degrees away from the angles where the protons are ranged out by the
target.

Differential cross-section data are shown in Fig. 4. LAMPF and
Saclay data agree well except for the last few angles inside 176°.
The Dubna and PPA data are each in disagreement with all other
measurements. A simple normalization factor would not correct this.
Therefore, we believe the back-angle Dubna and PPA results are incorrect. The forward-angle Dubna data were matched to the backward data
at about 50° and may have the correct shape. The total elastic
cross-section quoted by Dubna and the inelastic total cross-section
derived from that number will also be wrong.

Extreme back-angle LAMPF data[11] are shown in Fig. 5, along with
a double exponential fit which shows the sharp peak due to pion exchange. Extrapolation to the pole in the pion-exchange amplitude by
Chew's method[8] gives a coupling constant of

$$f^2 = 0.073 \pm 0.003,$$

which is compatible with values found in other ways.[12]

Using the Saclay data, the LAMPF authors found that a good value
of f^2 could only be found if the Saclay points at $\theta^* < 176°$ were
rejected.

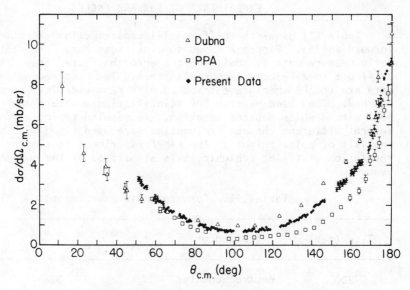

Fig. 4. Differential cross-sections near 650 MeV.

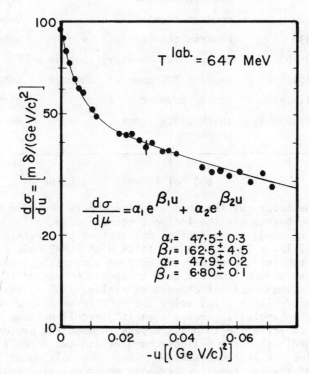

Fig. 5. Fit to invariant cross-section
measurements from experiment 125.

EXPERIMENTS AT FORWARD ANGLES

Table III gives the differential cross-section experiments at forward angles. Figure 2 shows some of these data. The totality of early measurements is small and the error bars are large. It is difficult to make comparisons with them. The most recent published data are the PPA measurements at 20 different energies between 60 and 400 MeV. They used neutron TOF scintillation counters and a chopped beam with a white neutron spectrum. To resolve timing ambiguities several different chopper frequencies were used along with off-line analysis of pulse heights. The LAMPF experiment is similar, but resolution of timing ambiguities is simplified by the monoenergetic beam.

Table III. Forward angle experiments

T_{lab} (MeV)	Detector	Beam Structure		
200	neutron counter	-	Dubna	(1963)
300	cloud chamber	-	Berkeley	(1954)
400	neutron counter	-	CERN	(1954)
580	neutron counter	-	Dubna	(1956)
630	neutron counter	-	Dubna	(1959)
60-400	neutron TOF counter	chopped	PPA	(1976)
800	neutron TOF counter	chopped	LAMPF	
991(p,n)	proton counter	-	Birmingham	(1967)
1250-6000	thick plate spark chamber	-	Berkeley	(1970)

800 MeV FORWARD SCATTERING

The apparatus for this experiment[13] is shown in Fig. 6. The main components are the incident neutron beam, a liquid hydrogen target 30 cm long, and a neutron detector which consists of a carbon block 12 in. × 24 in. × 4 in. thick with 4 MWPC planes and a scintillation counter to measure the neutron conversion time relative to the rf signal coming from the beam chopper. A counter in front of the carbon vetoes incident charged particles. The conversion location is determined within 2 in. using the MWPC information. When the neutron scattering angle is greater than 15°, recoil protons have enough energy to leave the target and pass through the proton arm. Three X-Y coordinates from helical chambers and two scintillation counters determine the proton track and time. The collimator immediately in front of the LH_2 target intercepts neutrons which scatter from the inside wall of the other collimator and head directly toward the

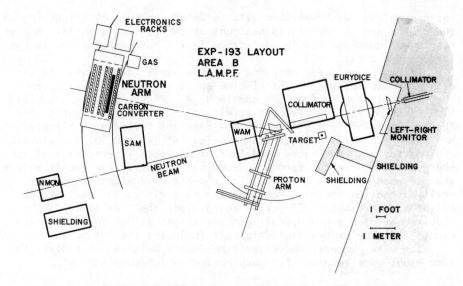

Fig. 6. LAMPF experiment 193, forward neutron scattering.

neutron counter. This is very important at small angles, where this
rate can exceed the rate of scattered neutrons from the target. The
face of the collimator was cylindrical, canted with respect to the
beam, and located close to the beam edge and the target. It inter-
cepted off-beam line neutrons for detector locations as small as
3° lab. A sweep magnet and numerous monitors are also shown:

NMON - downstream neutron monitor (CH_2 + range telescope)
SAM - small-angle target monitor (proton telescope)
WAM - wide-angle target monitor (proton telescope)
L-R - upstream neutron monitor (CH_2 + range telescope)
primary beam monitor (toroid) - not shown

With the exception of the toroid these monitors tracked to
better than 1% over a wide range of intensities. This was necessary
for our method of normalization.

Absolute normalization was achieved by counting with the detector in the beam and out of the beam. The ratio of these count rates
N_{in} and N_{out} is

$$R = N_{out}/N_{in} = \frac{e_{out} \; N_B N_T (d\sigma/d\Omega) \Delta\Omega}{e_{in} \; N_B}, \tag{4}$$

where   e = efficiency (in or out of beam)
N_B = beam counts/monitor
N_T = target areal density
$\Delta\Omega$ = effective solid angle.

The absolute beam flux cancels in this expression, and the factor
e_{out}/e_{in} becomes 1 after appropriately mapping the efficiency across
the face of the counter. This was done in the beam with 1% statistics. At large angles the neutron energy is smaller than the beam
energy, and a correction factor of roughly 15% must be applied at the

44

larger angles. We have taken data to determine this by triggering on the proton arm and detecting neutrons at the conjugate angle. Analysis is not complete and no correction factor has been applied to the large-angle data.

The value of $d\sigma/d\Omega$ obtained from the above formula will be the elastic cross-section if the rf time spectrum is cut so as to remove the inelastic events from the sample N_{out}. Figure 7 shows a typical time spectrum at 17.5° lab angle, with time going from right to left. The peak labeled 'γ' is from 800 MeV neutrons following the reaction $n + p \rightarrow n + p + \pi^0$, and the peak labeled 'N' is from 800 MeV elastic n-p scattering. The background is from the same interactions of low-energy neutrons in the target. From the few early time events before the 'γ' peak one sees that the timing ambiguity results in a very small correction factor. Neutrons arriving with a 40 nsec longer flight time are about 150 MeV in energy, and the flux is low. The requirement of a proton recoil removes almost all of the gamma-time events and much of the inelastic tail following the 'N' peak.

The wire numbers, converter time, pulse heights and scalars for each event were recorded for each neutron which consisted of

$$\overline{(\text{VETO})} \cdot (\text{MWPC})_{3/4} \cdot (\text{TOF SCINTILLATOR})$$

from the neutron arm. The recoil proton information was recorded but not included in the trigger. CAMAC was used to interface PDP-11 computer and data were written on tape. Typical trigger rates were several per second. Many on-line distributions were collected and displayed as diagnostics during the course of data-taking. Short turn-around times at MOTHER, an off-line computer, allowed some data to be analyzed during the experimental runs. Those early cross-

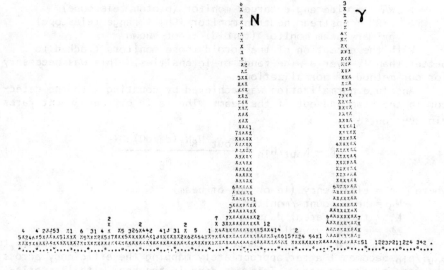

← time

Fig. 7

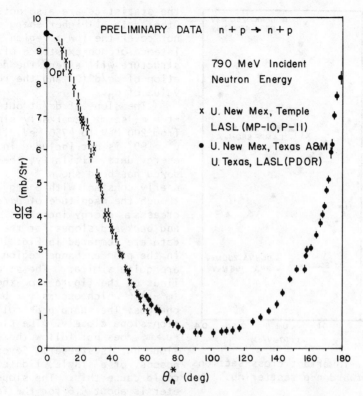

Fig. 8. Preliminary differential cross-sections for n-p scattering.

section measurements agree quite well with the data which were more
carefully analyzed later and are presented here.

Figure 8 shows the preliminary data for forward and backward
angles at 800 MeV. The mismatch at 60° will get smaller due to known
corrections that must be applied to both sets of data. These corrections are of the order of 10% and in the right direction to increase
agreement. Integration of the curve gives the inelastic cross-
sections

$$\sigma_{el} \approx 21.5 \text{ mb}$$

$$\sigma_{inel} \approx 37.8 - 21.5 = 16.3 \text{ mb}.$$

RESULTS AND DISCUSSION

The peak in the backward direction due to π^+ exchange leads to
the question, is there structure in the forward direction due to π^0
exchange? The exchange amplitude is down a factor of only two from
the Clebsch-Gordon coefficients. A peak or dip should be obvious for
$-t < m_\pi^2$. Looking at Fig. 9, we see that the 800 MeV data points extend a little into this region, but do not show a peak. The 200 MeV
results from Dubna and PPA extend into the pion exchange region and
tend to show peaking. PPA data at other energies show this also, but

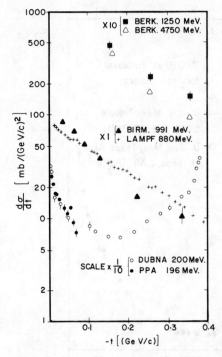

Fig. 9. Invariant cross-sections for forward n-p scattering.

the statistics are also not good. Other data at higher energies do not cover the low t region. The existence or non-existence of forward structure will affect the determination of $d\sigma/d\Omega(0)$ and the resulting value of α_n.

The slope of $d\sigma/dt$ outside $-t = m_\pi^2$ seems remarkably similar from 200 MeV to 4750 MeV, if $\theta^* > 90°$ is not included in the low-energy data. Similarly, the slope of $d\sigma/du$ has been shown[14,15] to be nearly constant with energy, although the magnitude of $d\sigma/du$ decreases as energy increases. Forward and backward slopes for the 800 MeV data are compared in Fig. 10. Except in the pion exchange region, they are quite similar. The straight lines on the figure show the term $(m^2-t)^{-2}$ which occurs for boson exchanges. The sharp peak follows the pion-slope closely. The slope for $-u > m_\pi^2$ does not follow the rho-slope but is steeper. Several bosons, or a single, lighter boson could cause this. The slope parameter is about 5.6 for the forward data and 8.0 for the backward at 800 MeV—quite similar to the forward slope at higher energies, which is about 7.0.[1]

To summarize, the invariant cross-sections near the forward and backward regions have shapes that vary little with energy. The forward and backward slopes are similar outside the pion exchange region. There are not enough data to say there is structure for $-t < m_\pi^2$ and, if so, whether it is energy dependent. An extrapolation of the 800 MeV data to 0° is somewhat suspect without data in the pion exchange region.

Nevertheless, we have done this and obtain an intercept of $d\sigma/d\Omega(0) = 9.5$ mb/sr. We have used $\sigma_{tot} = 37.8$ mb to get the optical point of 8.6 mb/sr. Figure 11 shows our result at 800 MeV to be compatible with previous results.

The error on the extrapolated point was arbitrarily set about twice the experimental error. Using this value we obtain (β_n is due to spin effects)

$$\sqrt{\alpha_n^2 + \beta_n^2} = 0.3 \pm 0.1.$$

Several comments are in order:
1) β_n is thought to be 'near zero', but is not well determined.[7]
2) The extrapolation to t=0 from outside is questionable.
   Several points on Fig. 11 are affected by this statement.

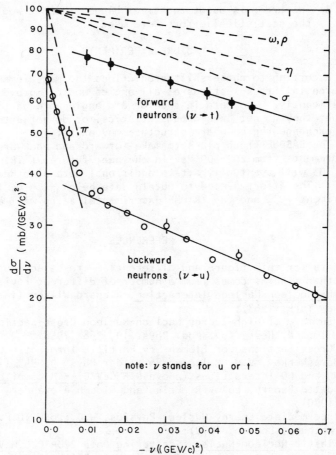

Fig. 10. Invariant backward and forward cross-sections at 800 MeV (preliminary).

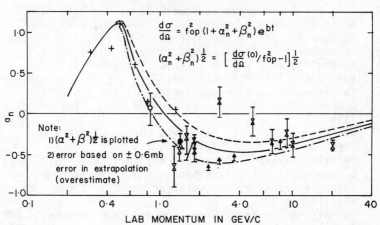

Fig. 11. Dependence of α_n on lab momentum.

3) The error is about two times bigger than it would be if only the statistical errors were used.

FUTURE EXPERIMENTS

According to phase-shift predictions the measurement of n-p differential cross-sections at all angles would improve the phases significantly. The data in the forward angles are particularly poor in that they do not have good statistics, or do not extend into the pion exchange region where structure may exist.

The BASQUE group plans to make forward and backward angle measurements from 200-500 MeV in the near future at TRIUMF. Experiment 193 will eventually attain additional data at other energies at LAMPF. The effort needed to obtain data below $-t = m_\pi^2$ is small for experiment 193, and the TRIUMF experimental setup will also reach this region.

REFERENCES

1. Data for this figure and differential cross-sections plotted in other figures comes from a number of different review sources:
    The Nucleon-Nucleon Interaction, Richard Wilson (Interscience, New York, 1963);
    "Summary of High-Energy Nucleon-Nucleon Cross-Section Data", Wilmot N. Hess, Rev. Mod. Phys. 10, 368 (1958);
    "Cross Sections for Elementary Particle Interactions", V.S. Barashenkov, V.M. Maltsev, Fort. der Phys. 9, 549 (1961);
    NN and ND Interactions (above 0.5 GeV/C) - A compilation, Odette Benary, LeRoy R. Price and Gideon Alexander, UCRL-20000 NN (1970);
    Intermediate Energy Nuclear Physics, W.O. Lock, D.F. Measday (Methuen, London, 1970);
    Elastic Nucleon-Nucleon Scattering Data 270-3000 MeV, Jiri Bystriky, Franz Lehar, Zdenek Janout, CEA-N-1547(E) (1972);
    D.F. Measday and J.N. Palmieri, Nucl. Phys. 85, 142 (1966)
    T.J. Devlin et al., Phys. Rev. D 8, 136 (1973).
2. R.A. Bryan and Bruce L. Scott, Phys. Rev 164, 1215 (1967);
    A. Scotti and D.Y. Wong, Phys. Rev. 138, B145 (1965);
    T. Ueda and A.E.S. Green, Phys. Rev. 174, 1304 (1968).
3. S. Mandelstam, Proc. Roy. Soc. A244, 491 (1958).
4. U. Amaldi, Jr., Rev. Mod. Phys. 39, 649 (1967).
5. D.V. Bugg, TRIUMF report, TRI-75-5 (1975), unpublished.
6. Key to experiments measuring $d\sigma/d\Omega$ (see Table I):
    Dubna (1963) - Yu.M. Kazarinov and Yu.N. Simonov, JETP 16, 24 (1963);
    Berkeley (1950) - E. Kelly, C. Leith, E. Segrè and C. Wiegand, Phys. Rev. 79, 96 (1950);
    Berkeley (1954) - J. De Pangher, Phys. Rev. 95, 578 (1954);
    Liverpool (1962) - A. Ashmore et al., Nucl. Phys. 36, 258 (1962);
    Carnegie (1954) - A.J. Hartzler and R.T. Siegel, Phys. Rev. 95, 185 (1954);
    Dubna (1956) - Yu.M. Kazarinov and Yu.N. Simonov, JETP 31, 169 (1956); and N.S. Amaglobeli and Yu.M. Kazarinov, JETP 34, 53 (1958);

Dubna (1959) - N.S. Amaglobeli and Yu.M. Kazarinov, JETP 37, 1587 (1959);
Berkeley (1960) - K.R. Larsen, Nuovo Cimento 18, 1039 (1960);
Birmingham (1967) - T.A. Murrey et al., Nuovo Cimento 49, 261 (1967);
Berkeley (1970) - M.L. Perl et al., Phys. Rev. D 1, 1857 (1970);
PPA (1976) - A.J. Bersbach et al., Phys. Rev. D 13, 535 (1976);
PPA (1974) - P.F. Shepard et al., Phys. Rev. D 10, 2735 (1974);
Saclay (1975) - G. Bizard et al., Nucl. Phys. B85, 14 (1975);
LAMPF (1976) - M.L. Evans et al., Phys. Rev. Lett. 36, 497 (1976); and M.L. Evans, private communication (for angles <145°).
LAMPF 800 MeV backward data were taken by the same group as above, and are preliminary.
LAMPF 800 MeV forward data were taken by the group involved with Experiment 193 [K. MacFarlane, V. Highland, L. Auerbach, K. Johnson (Temple Univ.), D. Wolfe, B. Dieterle, J. Donahue, T. Rupp, R. Carlini, C. Leavitt (Univ. of New Mexico), R. Bentley, P-11 (LASL), and J. Pratt, MP-10 (LASL).

7. L.M.C. Dutton et al., Phys. Lett. 25B, 245 (1967).
8. G.F. Chew, Phys. Rev. 112, 1380 (1958).
9. LAMPF Proposal No. 125, B. Dieterle (spokesman).
10. LAMPF Proposal No. 255, L. Northcliffe (spokesman).
11. M.L. Evans et al., Phys. Rev. Lett. 36, 497 (1976)
12. M.H. McGregor et al., Phys. Rev. 173, 1272 (1968).
13. LAMPF Proposal No. 193, B. Dieterle and K. MacFarlane (spokesmen).
14. G. Bizard et al., Nucl. Phys. B85, 14 (1975).
15. R.E. Mischke, "Particle Physics Research at LAMPF", LAMPF Users Group Newsletter 9, #1, 45 (1977), unpublished.

MEASUREMENT OF P,D,R, AND A PARAMETERS AT SMALL ANGLES FOR P-P ELASTIC SCATTERING AT 310,390 AND 490 MEV*

D. Besset, Q.H. Do, B. Favier, G. Greeniaus[+],
R. Hess, C. Lechanoine, D. Rapin, D.W. Werren
DPNC, University of Geneva,
Geneva, Switzerland

Ch. Weddigen
Kernforschungszentrum und Universität Karlsruhe
Institut für Experimentelle Kernphysik
Federal Republic of Germany

ABSTRACT

An experiment for the determination of the Wolfenstein parameters P, D, R and A in p-p elastic scattering in an angular domain covering the Coulomb interference region was running at SIN. Preliminary results are presented at 310, 390 and 490 MeV incident kinetic energy for $2^0 \leq \theta_{lab} \leq 8^0$. These are in fair agreement with the phase shift predictions.

INTRODUCTION

P-P elastic scattering between $300 \leq E_{inc} \leq 600$ MeV incident energy was studied at small scattering angles including the Coulomb interference region. The program was originated at the SC-CERN by DF Measday and E Heer. In the first step the differential cross-section has been measured at 8 different energies and the ratio

$$\alpha_p = \text{Re}(a_N+b_N)/\text{Im}(a_N+b_N) \big|_{\theta=0^0} \qquad (1)$$

for the spin independent nuclear amplitude (a_N+b_N) was derived[1]. The polarization parameter $P(\theta)$ was also measured[2]. Directly from these measurements a rough estimation of the amplitudes $\text{Im}(a_N+b_N)\big|_{\theta=0^0}$ and ε with

$$\text{Im}(e_N) = \varepsilon \sin\theta \qquad (2)$$

could be obtained independently of a phase shift analysis.

*Work supported in part by the Swiss National Funds for Scientific Research and by the Swiss Institute for Nuclear Research - SIN
+Present address : University of Alberta, Canada.

This program is being continued and extended at SIN, where measurements of P,D,R and A parameters have been carried on at 310, 390, 490 and 590 MeV. Preliminary and partial results are presented.

PRINCIPLE OF THE EXPERIMENT

A polarized proton beam is scattered on a liquid H_2-target. The transverse polarization of the scattered particles is then observed using a second scattering on a carbon target. The experimental layout is shown schematically in Fig. 1

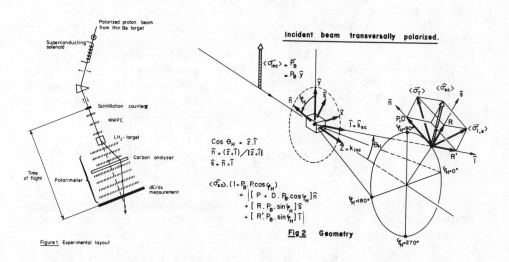

Figure 1. Experimental layout

Fig 2   Geometry

A transversally polarized beam is produced either by accelerating up to 595 MeV the protons issued from the SIN polarized source ($P \approx 65\%$) or by scattering at 8^0 the main unpolarized beam on a Be-target (P=0.412±0.010). The beam energy can be lowered to 300 MeV by using a Cu-degrader. A spin rotating solenoid and a magnetic deflection allow to obtain also a longitudinally polarized beam.

The scatterings on H_2 and on carbon are observed with three telescopes consisting of multiwire proportional chambers (MWPC). A total of 10 x-y MWPC, 2 mm wire spacing are located directly in the beam of $\approx 3 \cdot 10^5$p/sec intensity. A fast decision logic attached to the MWPC selects all events with the scattering angles $\theta_H \gtrsim 2^0$ and $\theta_c \gtrsim 5^0$. The full azimuthal distribution of the hydrogen scattering is observed up to $\theta_H \approx 15^0$. For each event (θ_H,ϕ_H), the full azimuthal distribution on carbon is also determined in the

range $5^0 < \theta_C < \theta_{MAX}^{Carbon}$. Scintillation counters are associated with the trigger logic and with the TOF and dE/dx meas-
urements. Information on TOF and dE/dx as well as VETO
counters placed around the H_2-target, allow a good separation between the elastically and inelastically scattered
events : $p-p \to \pi^+ d$, $p-p \to pn\pi$. A more elaborate description of the beams and the experimental layout can be found
in ref. 3.

THEORETICAL CONSIDERATIONS

The parameters D and R are usually determined with a transversally polarized beam the polarization of which is either
normal to the scattering plane (measurement of D) or in
the scattering plane (measurement of R). In the SIN experiment the full azimuthal distribution is observed and the
scattering plane can have any orientation with respect to
the transverse polarization of the incident protons. With
the definitions given in Fig. 2 for the coordinate systems
attached to the beam $(\hat{x}, \hat{y}, \hat{z})$ and to the scattered proton
$(\hat{n}, \hat{s}, \hat{\ell})$, the scattered proton polarization $\langle \vec{\sigma}_{sc} \rangle$ is
given by the relation[4]

$$I(\theta_H, \phi_H)\langle \vec{\sigma}_{sc} \rangle = I_0(\theta_H)\left\{ \left[P+D\ c_1\right] \hat{n} + \left[Rc_2+Ac_3\right]\hat{s} + \left[R'c_2+A'c_3\right]\hat{\ell} \right\} \quad (3)$$

$$I(\theta_H, \phi_H) = I_0\left[1+P\ c_1\right] \quad (4)$$

c_1, c_2 and c_3 are the polarization components of the inci-
dent particle along the axes $\hat{n}$, $\hat{y}' = \hat{n}x\hat{z}$ and $\hat{z}$. $\hat{n}$ is the
normal to the scattering plane.

$$c_1 = \langle\vec{\sigma}_{inc}\rangle \cdot \hat{n} \; ; \; c_2 = \langle\vec{\sigma}_{inc}\rangle \cdot (\hat{n}x\hat{z}) \; ; \; c_3 = \langle\sigma_{inc}\rangle \cdot \hat{z} \quad (5)$$

Note that the coordinate system in which $\langle\vec{\sigma}_{sc}\rangle$ is defined,
is changing with angles θ_H, ϕ_H. The scalar products c_1
and c_2 are also varying with the azimuthal angle ϕ_H.
Thus it may be interesting to place an observer in the
laboratory system in order to visualize simultaneously the
full distribution of the scattering and the polarization
pattern. His observation shown in Fig. 3 to 8 are computed
a) with the equ. (3) and some trivial geometry transformations, b) using the PSA of Saclay[5] for the predictions
of the parameters P,D,R,R' A and A' and c) with a fully
polarized beam $|\langle\vec{\sigma}_{inc}\rangle| = 1.0$. The arrows indicate the
magnitude and the orientation of $\langle\vec{\sigma}_{sc}\rangle$. The scattering at
small angle is dominated by the Coulomb interaction which
has almost no effect on the polarization. The arrow at
$\theta=0^0$ indicates also $\langle\vec{\sigma}_{inc}\rangle$. In Fig. 3 one has considered
a longitudinally polarized beam with the observer located
on the side. He is looking at the polarization component

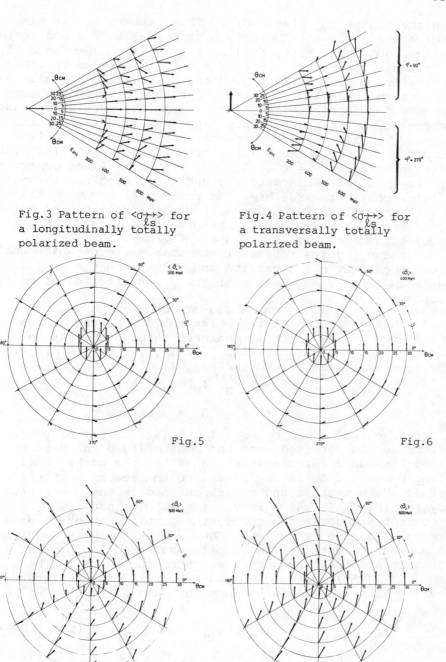

Fig.3 Pattern of $\langle\sigma_{\ell s}^{\rightarrow\rightarrow}\rangle$ for a longitudinally totally polarized beam.

Fig.4 Pattern of $\langle\sigma_{\ell s}^{\rightarrow\rightarrow}\rangle$ for a transversally totally polarized beam.

Fig.5

Fig.6

Fig.7

Fig.8

Fig. 5 to 8 Pattern of $\langle\sigma_T\rangle$ for a transversally totally polarized beam at 300, 400, 500 and 600 MeV incident kinetic energy.

in the scattering plane $<\vec{\sigma}_{\ell s}>$. This pattern is the same for every angle ϕ_H. The component normal to the scattering plane, not shown here, follows the angular dependence of the parameter $P=P(\theta)$. In figure 4 to 8 the beam is transversally polarized. In Fig. 4 the observer is again located on the side and is looking for the polarization component in the scattering plane $<\vec{\sigma}_{\ell s}>$. The particular case where $<\vec{\sigma}_{inc}>$ is also in this plane ($\phi_H = 90^0$, 270^0) is drawn. The azimuthal dependence of this pattern varies roughly as $\cos\phi_H$, the polarization in the scattering plane disappearing at $\phi_H = 0^0$, 180^0. In Fig. 5 to 8 the observer is in front of the scattering and is looking for the particles coming towards him. The polarization components $<\vec{\sigma}_T>$ perpendicular to the scattered particle trajectories (components along $\hat{n}$ and $\hat{s}$) are presented in an azimuthal projection for 300, 400, 500 and 600 MeV.

These figures represent the spin patterns which could be detected by our apparatus and illustrate the very large spin and energy dependence of the p-p scattering at medium energy.

ANALYSIS AND NORMALISATION

In the experiment the polarization pattern of the scattered proton $<\vec{\sigma}_{sc}>$ is measured through the carbon scattering. The distribution $I_C(\theta_H, \phi_H, \theta_C, \phi_C)$ is described by a relation similar to the equ. 4:

$$I_C(\theta_H, \phi_H, \theta_C, \phi_C) = I(\theta_H, \phi_H) \cdot I_C(\theta_C) \cdot \left[1 + c_4 \cdot P_C(\theta_C)\right] \quad (6)$$

$$c_4 = <\vec{\sigma}_{sc}> \cdot \hat{n}_C = <\sigma_T> \cos \phi_C \quad (7)$$

$I(\theta_H, \phi_H)$ is given by equ. 4., $I_C(\theta_C)$ is the proton distribution obtained from the scattering of an unpolarized beam on a carbon-target, $P_C(\theta)$ is the carbon analyzing power and $\hat{n}_C$ is the normal to the second scattering plane. Note that ϕ_C and θ_C are defined in the coordinate system ($\hat{n}$, $\hat{s}$, ℓ) which is varying with θ_H and ϕ_H. Equation 6 describes properly the double scattering data recorded by the experiment. However the analysis requires a convenient method. The results presented in the next section are obtained with a weighted sums method. The sums $S_{\alpha\beta\gamma\delta}(\theta_H)$ that are computed over all the registered events have the following form for each bin $\Delta\theta_H$:

$$S_{\alpha\beta\gamma\delta}(\theta_H) = \int_{\Delta\theta_H} d\theta_H \int_0^{2\pi} d\phi_H (\sin\phi_H)^\alpha (\cos\phi_H)^\beta \int_0^{2\pi} d\phi_C (\sin\phi_C)^\gamma (\cos\phi_C$$

$$\times \left[\int_{5^0}^{\theta_{max}^C(\theta_H)} d\theta_C \, I_C(\theta_H, \phi_H, \theta_C, \phi_C)\right] \quad (8)$$

With α, β, γ, δ = 0,1. The sums may also be calculated analytically by introducing equ. 3 and 6 in equ. 8. Table 1
gives the analytic form thus obtained for the sums which
are statistically independent. In the case of a transversally polarized beam, one sees from Table 1 that S_{0100}
gives the beam polarization P_B, S_{0001} the carbon mean analyzing power $\bar{P}_C$, S_{0101} the D parameter and S_{1010} the R parameter if the parameter P is known from another experiment.

  Carbon data. To avoid this problem one has also
measured the single scattering of the beam on carbon (thus
with $\theta_H = 0^0$) in between the double scatterings runs and
at regular time intervals. These data allow to determine
the product $P_B \cdot \bar{P}_C$ in the same conditions for the beam and
the carbon analyzer as in the main experiment. Putting
this quantity ($P_B \bar{P}_C$) in S_{0101} and S_{1010} one obtains the D
and R parameter free from normalization errors other than
statistical. Then from S_{0100}, S_{0001} and ($P_B \cdot \bar{P}_C$) one has
three relations to find three unknowns, P, P_B and $\bar{P}_C$ also
without need of external data.

  It may be surprising that one does not need a separate
experiment to calibrate the beam polarization. In fact a
deeper analysis shows that the sum S_{0100} is a measurement
of the analyzing power of hydrogen P, and S_{0001} is the
measurement of the polarization parameter P', P and P' are
of course equal by time reversal and parity conservation.
Thus in the experiment one performs automatically a usual
beam polarization calibration based on double scattering
and on the identity of P and P'.

$S'_{\alpha\beta\gamma\delta}$	S_{0000} = $\Sigma 1$	S_{0100} = $\Sigma\cos\phi_H$	S_{1000} = $\Sigma\sin\phi_H$	S_{0001} = $\Sigma\cos\phi_C$	S_{0010} = $\Sigma\sin\phi_C$	S_{0101} = $\Sigma\cos\phi_H\cos\phi_C$	S_{1001} = $\Sigma\sin\phi_H\cos\phi_C$	S_{0110} = $\Sigma\cos\phi_H\sin\phi_C$	S_{1010} = $\Sigma\sin\phi_H\sin\phi_C$
Case a) transversally polarized beam with $\langle\vec{\sigma}_{inc}\rangle = P_B \cdot \hat{y}$									
Q		$P \cdot P_B$	0	$P \cdot \bar{P}_C$	0	$D \cdot (P_B \bar{P}_C)$	0	0	$-R(P_B\bar{P}_C)$
Case b) longitudinally polarized beam $\langle\vec{\sigma}_{inc}\rangle = P_B \cdot \hat{z}$									
Q		Q	Q	$P \cdot \bar{P}_C$	$-A\, P_B \cdot P_C$	0	0	0	0
Factor F	-	$(\Sigma 1)/2$		$(\Sigma 1)/2$		$(\Sigma 1)/4$			
Statistical error $(\Delta Q)^2$		$2/(\Sigma 1)$		$2/(\Sigma 1)$		$4/(\Sigma 1)$			

Table 1. Results of the weighted sums computation (equ.8) for a transversally and a
longitudinally polarized beam. The quantities Q are defined as $Q \cdot F = S_{\alpha\beta\gamma\delta}$

RESULTS AND DISCUSSION

All results presented here have been obtained with the scattered polarized beam. Table II gives the results of the beam polarization. These data show that the polarization is not affected when the beam goes through a Cu-degrader.

Table II. Beam polarization calibration

Beam energy {MeV}	310	390	490
Beam polarization	0,420±0,015	0,415±0,012	0,405±0,012

Errors : All quoted errors are statistical only. The data have not been corrected for the binning and for the angular resolution of the system. These effects are important in the Coulomb region for the experimental points near $\theta_{CM} \approx 5^0$.

The inefficiencies in the MWPC and the different cuts in the selection of the events may introduce same systematic errors. The study of these effects is in progress. However the good compatibility observed between the data with different beam polarization orientations indicates that if systematic errors are present, they are less than the quoted statistical errors and likely negligible.

The results for P,D, R and A are compared with the phase shift analysis predictions of SACLAY [5]. Measurements at 310 and 390 MeV are compared with PSA 300 and 490 MeV with PSA 500. The recent data of TRIUMF are also indicated in the figures.

Results at 310 MeV are presented in Fig. 9 to 12. In Fig. 9, our carbon data are compared with the best fit of the BASQUE measurement as indicated by the broken line. An excellent agreement is obtained for the angular dependence but the normalization factor is a little too low : $P_B = 0.397^{\pm}0,004$ instead of $P_B = 0.420±0,015$. The P,D and R parameters are shown in Fig, 10 to 12. Fig. 13 to 16 show our results at 390 MeV for P, A, D and R. Fig. 17 to 19 show our data for P, D and R at 490 MeV.

From these figures one sees that our data are in fair agreement with the PSA and with the TRIUMF measurements. The parameter P at 390 and 490 MeV agrees also very well with our previous experiment[2].

The final analysis will include data at 590 MeV and the parameter A at 310 and 490 MeV. Also the angular range will be larger than now $3^0 \lneq \theta_{CM} \lneq 30^0$. For this purpose one should analyze carefully events with incomplete azimuthal distribution on hydrogen.

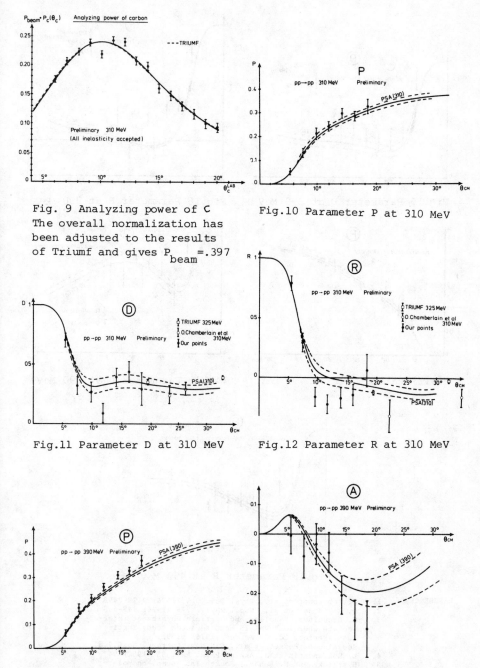

Fig. 9 Analyzing power of C
The overall normalization has
been adjusted to the results
of Triumf and gives P_{beam} =.397

Fig.10 Parameter P at 310 MeV

Fig.11 Parameter D at 310 MeV

Fig.12 Parameter R at 310 MeV

Fig.13 Parameter P at 390 MeV

Fig.14 Parameter A at 390 MeV

58

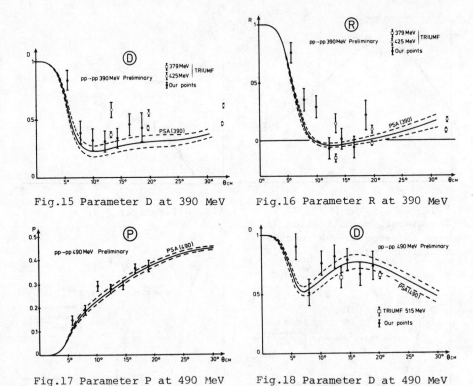

Fig.15 Parameter D at 390 MeV   Fig.16 Parameter R at 390 MeV

Fig.17 Parameter P at 490 MeV   Fig.18 Parameter D at 490 MeV

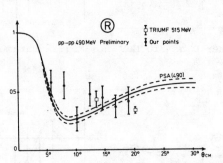

Fig.19 Parameter R at 490 MeV

REFERENCES   1. D. Aebischer et al., Phys. Rev. D13 (1976) 2478-2498
             2. D. Aebischer et al., Nucl. Phys. A276 (1976) 445-465
             3. G. Greeniaus, "Nuclear and Particle Physics at inter-
                 mediate energies" Vol.15 NATO Advanced Study Insti-
                 tutes Series. Ed. JB.Warren (1976) p.591.
                 See also D. Besset et al., SIN News Letters, No.6,7,8.
             4. L. Wolfenstein et J. Askin, Phys. Rev. 85 (1952) 947
             5. Bystricky and F. Lehar., Fourth Int. Symp. on polar-
                 ization phenomena in nuclear reaction, August 1975,
                 Zurich. (Eds. W. Gruebler and V. König, Birkhäuser-
                 verlag, Basel-Stuttgart, 1976).
                 See also Nucl. Phys. B, to be published.
             6. D. Axen et al., TRIUMF, private communication.

ASYMMETRY MEASUREMENTS IN NUCLEON-NUCLEON SCATTERING
WITH POLARIZED BEAMS AND TARGETS
AT ZGS TO FERMILAB ENERGIES*

A. Yokosawa
Argonne National Laboratory, Argonne, Il. 60439

ABSTRACT

Results of various asymmetry measurements in nucleon-nucleon scattering with polarized beams and targets at ZGS energies are presented. A possible direct-channel resonance in the pp system is discussed. Most of the discussion above ZGS energies are aimed at future measurements.

INTRODUCTION

In this talk I mainly cover asymmetry measurements in proton-proton elastic scattering and measurements of the total cross-section difference, although many other measurements have been carried out using polarized-proton beams at the Argonne ZGS during the last few years.

I discuss i) pp scattering-amplitudes measurements, ii) a striking energy dependence appearing in pp system, iii) pp elastic scattering at large p_T, iv) proton-neutron elastic scattering, and v) pp scattering above ZGS energies.

Elastic scattering is an old topic in hadronic physics. In spite of numerous elastic-scattering data available at various energies, our understanding of the nucleon-nucleon system has generally been poor and many theoretical assumptions were yet to be confirmed.

Ever since Wolfenstein's article,[1] followed by Schumacher and Bethe,[2] and Moravcsik,[3] on various observables with respect to nucleon-nucleon scattering amplitudes, many authors have discussed on the determination of amplitudes. At least nine measurements are needed to determine five amplitudes. Polarized beams and targets with three spin directions are essentially needed for those measurements. Using these facilities, for the

*Work supported by the U. S. Energy Research and Development Administration.

first time, we have accomplished amplitude measurements at Argonne. By knowing scattering amplitudes, one learns about diffraction processes, Pomeron exchanges, and π and A_1 exchanges, and might find some unexpected phenomena.

Large p_T phenomena are currently a topical subject, and asymmetry measurements in exclusive and inclusive channels should yield a new insight in the pp system, particularly in terms of quark-quark scattering.

Our goal is to establish the energy dependence of pp amplitude at ZGS energies to Fermilab energies. We hope to clarify the dip structure observed in the differential cross section at high energies, and to provide an under-standing of the rise in the total cross section.

Although the ZGS is capable of accelerating a polarized beam, this may not be done at other higher energy accelerators. At Fermilab energies, the produc-tion cross section of Λ and $\overline{\Lambda}$ is large enough so that one can produce useful polarized proton and antiproton beams.[4] Protons decaying from Λ particles are longitudinally polarized ($\approx 60\%$), and necessary spin rotations in the vertical or horizontal direction can be accomplished with standard magnets.

pp SCATTERING AMPLITUDES

We describe scattering amplitudes in three different ways:

i) Transversity amplitudes

Cross Section: $|T_1|^2 + |T_2|^2 + 2|T_3|^2 + 2|T_4|^2 + 2|T_5|^2$

ii) S-channel helicity amplitudes

$\left. \begin{array}{l} <++|++> = \phi_1 \\ <--|++> = \phi_2 \\ <+-|+-> = \phi_3 \end{array} \right\}$ net helicity nonflip amplitude

$<+-|-+> = \phi_4$ double flip

$<++|+-> = \phi_5$ single flip

iii)   t-channel exchange amplitudes[5]

Natural-parity exchange: N_o, N_1, N_2

$$N_o = \tfrac{1}{2}(\phi_1 + \phi_3), N_1 = \phi_5, \ N_2 = \tfrac{1}{2}(\phi_4 - \phi_2)$$

Unnatural-parity exchange: U_o, U_2

$U_o = \tfrac{1}{2}(\phi_1 - \phi_3)$, A_1-like exchange

$U_2 = \tfrac{1}{2}(\phi_2 - \phi_4)$, π exchange

Cross section: $|N_o|^2 + 2|N_1|^2 + |N_2|^2 + |U_o|^2 + |U_2|^2$

(The subscript gives the amount of t-channel helicity flip.)

EXPERIMENTAL OBSERVABLES

The spin directions N, L, and S of the polarized beam, the polarized target, and the recoil particles are defined as shown in Fig. 1. In practice, the measurements are made on the horizontal scattering plane.

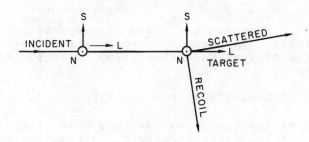

N: NORMAL TO THE SCATTERING PLANE
L: LONGITUDINAL DIRECTION
S = N x L IN THE SCATTERING PLANE

Fig. 1   Unit vectors N, L, and S.

Possible experiments obtaining spin direction of particles by means of polarized beam, polarized target, or spin determination of final state are listed below. We adopt the notation (Beam, Target; Scattered, Recoil) to express observables; * indicates that spin direction is known; 0 means that spin direction is not known. Assume that the spin direction of scattered particles cannot be obtained.

Observables	Description	Symbol
$(0,0;0,0)$	Cross section	σ
$(*,0;0,0)$ or $(0,*;0,0)$	Polarization	P
$(*,*;0,0)$	Correlation tensor	C_{jk}
$(*,0;0,*)$	Polarization transfer tensor	K_{jk}
$(0,*;0,*)$	Depolarization tensor	D_{jk}
$(*,*;0,*)$	High-rank spin tensor	H_{ijk}

These observables are listed in terms of $|t|$-channel exchange amplitudes in Table I. Old notations[1-3] are not used here because we find, at high energies, notations used by Halzen and Thomas[5] are more appropriate. We note that D_{NN}, D_{SS}, and D_{LS} are commonly called the "D parameter," "R parameter," and "A parameter," respectively. In three-spin measurements, we can determine three or four parameters simultaneously from one measurement.

DETERMINATION OF SCATTERING AMPLITUDES

The strategy of this work has been studied by various people, most recently by Johnson and Thomas.[6] In Table II, we illustrate the determination of scattering amplitudes vs. observables at small $|t|$ assuming $|N_0|$ is much larger than $|N_1|$, $|N_2|$, $|U_0|$, and $|U_2|$. Double-scattering measurements are required to determine the real part of the N_2, U_0, and U_2 amplitudes and the imaginary part of the N_1 amplitude; single-scattering measurements can determine the remaining parts.

TABLE I

Laboratory Observables (θ_R is the Laboratory Recoil Angle)

Observables (B,T;S,R)	Exchange Amplitudes

(Single Scattering)

$\sigma^{Tot} = (0,0;0,0)$ $\mathrm{Im} N_o$

$\Delta\sigma_T^{Tot} = (N,N;0,0)$ $\mathrm{Im} U_2$

$\Delta\sigma_L^{Tot} = (L,L;0,0)$ $\mathrm{Im} U_o$

$\sigma = (0,0;0,0)$ $|N_o|^2 + 2|N_1|^2 + |N_2|^2 + |U_o|^2 + |U_2|^2$

$P = (0,N;0,0)$ $-2\mathrm{Im}(N_o-N_2)N_1^*/\sigma$ (Also $(N,0;0,0)$)

$C_{NN} = (N,N;0,0)$ $2\mathrm{Re}(U_oU_2^*-N_oN_2^*+|N_1|^2)/\sigma$

$C_{SS} = (S,S;0,0)$ $2\mathrm{Re}(N_oU_2^*-N_2U_o^*)/\sigma$

$C_{SL} = (S,L;0,0)$ $2\mathrm{Re}(U_o+U_2)N_1^*/\sigma$

$C_{LL} = (L,L;0,0)$ $-2\mathrm{Re}(N_oU_o^*-N_2U_2^*)/\sigma$

(Double Scattering)

1) K_{jk} Measurement

$K_{NN} = (N,0;0,N)$ $-2\mathrm{Re}(U_oU_2^*+N_oN_2^*-|N_1|^2)/\sigma$

$K_{SS} = (S,0;0,S)$ $\left[-2\mathrm{Re}(U_2-U_o)N_1^* \sin\theta_R -2\mathrm{Re}(N_oU_2^* + N_2U_o^*)\cos\theta_R\right]/\sigma$

$K_{LS} = (L,0;0,S)$ $\left[-2\mathrm{Re}(N_oU_o^*+N_2U_2^*)\sin\theta_R - 2\mathrm{Re}(N_1^*(U_2-U_o))\cos\theta_R\right]/\sigma$

2) D_{jk} Measurement

$D_{NN} = (0,N;0,N)$ $\left[|N_o|^2+2|N_1|^2+|N_2|^2-|U_o|^2-|U_2|^2\right]/\sigma$

$D_{SS} = (0,S;0,S)$ $\left[-2\mathrm{Re}(N_o+N_2)N_1^*\sin\theta_R - (|N_o|^2-|N_2|^2+|U_2|^2-|U_o|^2)\cos\theta_R\right]/\sigma$

$D_{LS} = (0,L;0,S)$ $\left[(|N_o|^2-|N_2|^2-|U_2|^2+|U_o|^2)\sin\theta_R-2\mathrm{Re}(N_o+N_2)N_1^*\cos\theta_R\right]/\sigma$

3) Three Spin Measurement

Simultaneous Observables

$H_{SNS} = (S,N;0,S)$ $\left[2\mathrm{Im}(N_oU_2^*+N_2U_o^*)\sin\theta_R+2\mathrm{Im}(U_2-U_o)N_1^*\cos\theta_R\right]/\sigma$ $(0,N;0,N),(S,0;0,S)$

$H_{HSS} = (N,S;0,S)$ $\left[-2\mathrm{Im}(U_oU_2^*-N_oN_2^*)\sin\theta_R + 2\mathrm{Im}(N_o+N_2)N_1^*\cos\theta_R\right]/\sigma$ $(0,S;0,S),(N,0;0,N)$

$H_{SSN} = (S,S;0,N)$ $-2\mathrm{Im}(U_2+U_o)N_1^*/\sigma$ $(S,S;0,0),(0,S;0,S),(S,0;0,0)$

$H_{LSN} = (L,S;0,N)$ $2\mathrm{Im}(U_oN_o^* - U_2N_2^*)/\sigma$ $(0,S;0,0),(L,0;0,S)$

$H_{NLS} = (N,L;0,S)$ $\left[-2\mathrm{Im}(N_o+N_2)N_1^*\sin\theta_R + 2\mathrm{Im}(U_oU_2^*+N_oN_2^*)\cos\theta_R\right]/\sigma$ $(0,L;0,S),(N,0;0,N)$

$H_{SLN} = (S,L;0,N)$ $-2\mathrm{Im}(N_oU_2^*-N_2U_o^*)/\sigma$ $(S,L;0,0),(0,L;0,S),(S,0;0,S)$

TABLE II

Determination of Amplitudes vs. Lab Observables

Amplitude	Lab Observable
ReN_1	$P = (0,N;0,0)$, $H_{NLS} = (N,L;0,S)$
ImN_1	$D_{SS} = (0,S;0,S)$
ReN_2	$H_{NSS} = (N,S;0,S)$
ImN_2	$C_{NN}=(N,N;0,0)$, $K_{NN} = (N,0;0,N)$
ReU_0	$H_{LSN} = (L,S;0,N)$
ImU_0	$C_{LL} = (L,L;0,0)$, $K_{LS} = (L,0;0,S)$
ReU_2	$H_{SNS} = (S,N;0,S)$, $H_{SLN} = (S,L;0,N)$
ImU_2	$C_{SS} = (S,S;0,0)$

FACILITIES PROVIDING $\vec{N}$, $\vec{S}$, AND $\vec{L}$ SPIN DIRECTIONS

The spin of polarized protons emerging from the ZGS is in the N direction. A superconducting solenoid with a field of 12.0 T·m at 6 GeV/c, for example, is used to rotate the spin of the incident beam from the N to the S direction. The longitudinally-polarized beam is produced by a bending magnet with a vertical field to precess the proton spins in the S direction until their polarization is parallel to the beam momentum. This scheme of operation avoiding the use of a horizontal field does not require the vertical adjustment of polarized targets. The sign of beam polarization is flipped on alternate pulses, and this is essential to reduce systematic errors.

The polarized-proton target is 2 x 2 x 8-cm ethylene glycol doped with $K_2Cr_2O_7$ and maintained at ~0.4 K. For free protons in the target, the polarization is 0.8-0.9. Two kinds of superconducting magnets are used to provide three directions--N, S, and L--of target spins. A polarized target providing S and L spin directions is shown on the following page.

The spin analyses of recoil protons are done by a carbon polarimeter together with proportional wire-chamber detectors.

MEASURED OBSERVABLES TOWARD THE AMPLITUDE DETERMINATION

The following parameters in the region $|t| < 2.0$ $(GeV/c)^2$ have been completed (measurements at $|t| = 0$ are listed later):

Observables	Momenta (GeV/c)	Status
P	2-6	Ref. 7
C_{NN} (N,N;O,O)	2-6	"
C_{SS} (S,S;O,O)	6	Ref. 8
C_{LL} (L,L;O,O)	6	Ref. 9
C_{LL} (L,L;O,O)	1.2 to 2.5	Preliminary result
C_{SL} (S,L;O,O)	6	Data available
K_{NN} (N,O;O,N)	6	Being analyzed

Observables	Momenta (GeV/c)	Status
K_{SS} (S,O;O,S)	6	Data available
D_{NN} (O,N;O,N)	6	Ref. 10
D_{SS} (O,S;O,S)	6	Being analyzed, also Ref. 11
D_{LS} (O,L;O,S)	6	"
H_{SNS} (S,N;O,S)	6	Data available
H_{NSS} (N,S;O,S)	6	Being analyzed
H_{LSN} (L,S;O,N)	6	"

In Fig. 2, C_{NN} data are shown.[7] The energy dependence of C_{NN} is remarkable, and the Regge prediction shown in Fig. 2 is quite inadequate. Both C_{NN} and C_{SS} data[8] at 6 GeV/c are shown in Fig. 3. One can see from Table II that the C_{SS} data imply that the imaginary part of the U_2 term (corresponding to π exchange) is negative. For comparison we show two attempts by Field and Stevens[12] to describe the pp elastic-scattering process. The dashed curve, calculated by using the super-Regge Model, involves a large number of poles (P, f, ω, ρ, A_2, π, B) and corrections due to absorption (Regge cuts). The solid curve, calculated by using the Kane model involves the same Regge poles and absorption corrections calculated according to the Sopkovich prescriptions,[13] but does not require exchange degeneracy.[14] In addition, inelastic intermediate states play an important role in the Kane model. The difference between the two is primarily in the treatment of absorption correction $\pi_c(B_c)$. The data clearly favor the Kane model.

Results of the parameter C_{LL} = (L,L;O,O) are nonzero as shown in Fig. 4; this is rather unexpected, because we observe an A_1-like exchange amplitude. This behavior is shown to be consistent with the observed value of $\Delta\sigma_L \sim -1.0$ mb (See next section.), as described by E. Berger[15] in which the expectation from the behavior of an A_1-like trajectory is calculated. An A_1-type exchange term has also been identified in the $\pi p \uparrow \rightarrow \rho n$ reaction at CERN.[16] Recently the mass and width of the A_1 meson are determined to be ~1450 MeV

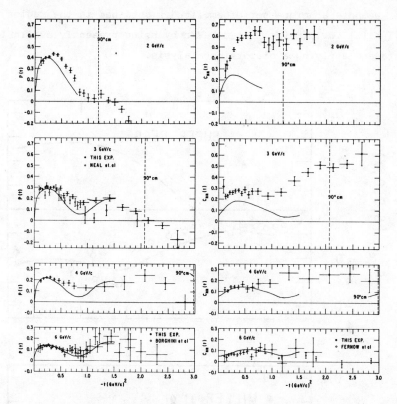

Fig. 2 P and C_{NN} plotted as functions of t at 2, 3, 4, and 6 GeV/c.
The smooth curves are predictions based on a Regge-pole model by
R. Field and P. Stevens (Ref. 12).

and ~350 MeV respectively with a resonant state of I = 1 and
$J^P = 1^+$.[17] Kane et al. point out that the A_1 exchange in pp does
occur with the strength expected from meson dominance of the axial
vector weak current.[40]

Our preliminary data of the parameter C_{SL} = (S,L;0,0) are con-
sistent with zero for the region of |t| <1.0. In the region of
0.2 < |t| <0.6, we obtain H_{SNS} ≈ (20 ± 10)% and D_{LS} > 60% as shown
in Fig. 5, and K_{SS} = (3 ± 10)% . Thus, the real part of the U_2
term (corresponding to π exchange) is positive. We expect the
results of H_{NSS} and H_{LSN} with much better accuracy.

While we are waiting for the data being currently analyzed, we
attempt to construct pp scattering amplitudes at 6 GeV/c using the
presently available data. At small |t|, say around 0.3, amplitudes
are given in Fig. 6.

We note a unique set of amplitudes obtained in the region $0.1 < |t| < 0.7$ $(GeV/c)^2$ at 6 GeV/c by using presently available data and applying a dispersion analysis.[18]

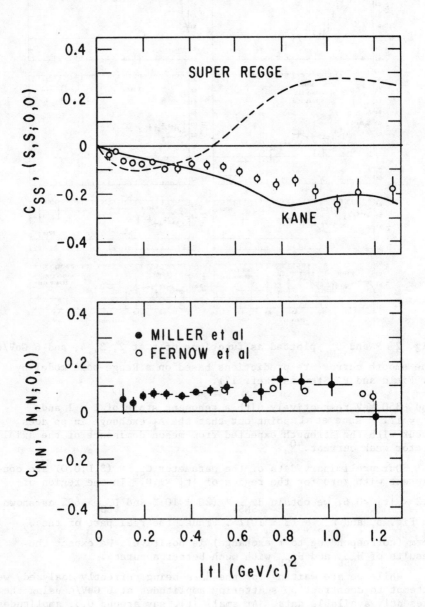

Fig. 3 C_{SS} and C_{NN} at 6 GeV/c.

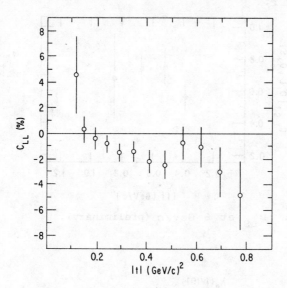

Fig. 4 C_{LL} at 6 GeV/c.

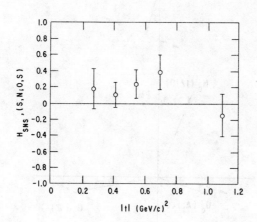

Fig. 5 H_{SNS} at 6 GeV/c.

70

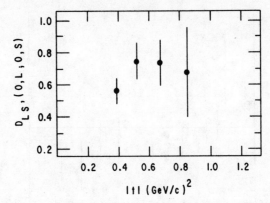

Fig. 5 cont. D_{LS} at 6 GeV/c (preliminary).

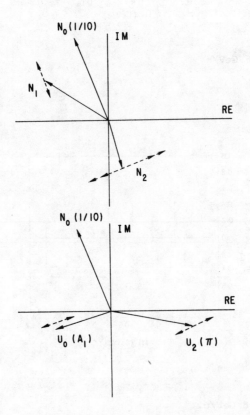

Fig. 6 Preliminary pp scattering amplitudes at 6 GeV/c.

TOTAL CROSS-SECTION DIFFERENCE FOR pp SCATTERING

As shown in Table I, measurements of σ^{Tot}, $\Delta\sigma_T^{Tot}$, and $\Delta\sigma_L^{Tot}$ are required to determine $Im\phi_1(0)$, $Im\phi_2(0)$, and $Im\phi_3(0)$. Measurements of $\Delta\sigma_T^{Tot}$ and $\Delta\sigma_L^{Tot}$ from 1.2 to 6.0 GeV/c were carried out.[19,20] A typical experimental setup for $\Delta\sigma_L$ and also C_{LL} is shown in Fig. 7a and 7b respectively. These measurements were done in a standard transmission experiment. First, I would like to remind you what the pp total cross sections[21] look like. (See Fig. 8.) A striking energy dependence in $\Delta\sigma_L$ is observed with a maximum difference of -17 mb at P_{lab} = 1.47 GeV/c, as shown in Fig. 9a. We observe that the structure is seen in $\sigma^{Tot}(\vec{\rightrightarrows})$, but not in $\sigma^{Tot}(\vec{\rightleftarrows})$. Figure 9b shows the elastic total cross section, σ_{el},[22] together with $\frac{1}{2}\sigma^{Tot}(\vec{\rightrightarrows})$; there is a similar structure in both (See p.16)

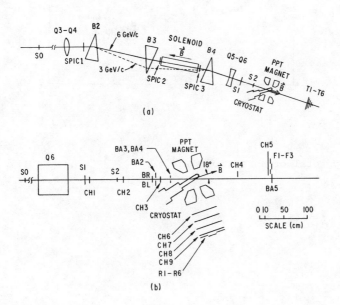

(a)

(b)

Fig. 7a Beam line experimental setup for the $\Delta\sigma_L$ measurement.

72

Fig. 7a cont.

SPIC 1-3 are segmented proportional ion chambers, SO-S2 are beam-telescope scintillation counters, and T1-T6 are transmission counters. Other beam-defining veto counters and position monitor counters are not shown. To obtain L-type polarized beam at 3 GeV/c, B3 and the solenoid were moved to the new beam line shown as a dashed line.

Fig. 7b Experimental setup for the C_{LL} measurement. CH1 to CH9 are multi-wire proportional chambers, SO, S1, S2, BR, BL, BA2, BA3, BA5, F1-F3, and R1-R6 are scintillation counters. The PPT magnetic field, hence the target spin direction, was at an angle of 18° to the beam line in order to increase the aperature for the recoil protons.

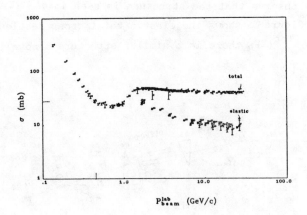

Fig. 8 pp total cross sections.

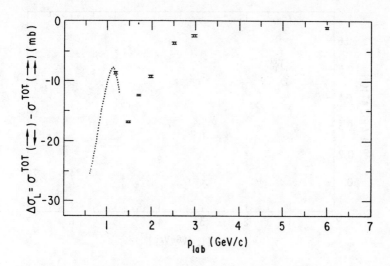

Fig. 9a Total cross-section difference $\Delta\sigma_L = \sigma^{Tot}(\overset{\leftarrow}{\rightarrow}) - \sigma^{Tot}(\overset{\rightarrow}{\rightarrow})$. The errors shown are statistical only. The dotted curve is calculated by using a solution from one of existing phase-shift analyses.

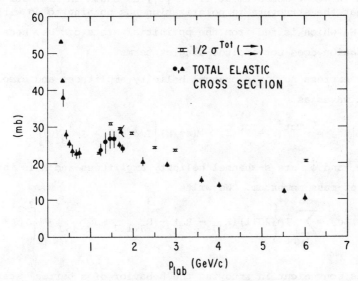

Fig. 9b Elastic total cross section and $\frac{1}{2}\sigma^{Tot}(\overset{\rightarrow}{\rightarrow})$. Values shown by closed circles are calculated by integrating the elastic differential cross sections of Ref. 22.

74

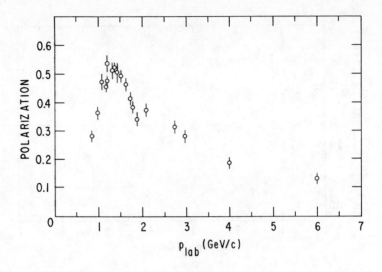

Fig. 9c Polarization at $0.1 < |t| < 0.2$.

curves. There is also a structure in the plot of polarization[23] against incident momentum at fixed $|t|$, as shown in Fig. 9c. We note that the structure in polarization has nothing to do with an S-wave $N\Delta$ which is fed from the pp initial state of $^1D_2$, because the polarization does not include singlet terms.

We express $\Delta\sigma_L$ in s-channel helicity amplitudes and also in partial waves as

$$\Delta\sigma_L = \sigma^{\text{Tot}}(\rightleftarrows) - \sigma^{\text{Tot}}(\rightrightarrows) = (4\pi/k)\left[\text{Im}\phi_1(0) - \text{Im}\phi_3(0)\right],$$

where ϕ_1 and ϕ_3 are s-channel helicity amplitudes and k is the center-of-mass momentum. We write

$$-(k^2/4\pi)\Delta\sigma_L = \sum_{J=0}^{\infty} \text{Im}\left\{(2J+1)(R_{JJ} - R_J) - R_{J+1,J} + R_{J-1,J} - 4\left[J(J+1)\right]^{\frac{1}{2}}R^J\right\},$$

which is convenient in studying the behavior of a partial scattering amplitude. A plot of $(-k^2/4\pi)\Delta\sigma_L$ is shown in Fig. 10. The partial amplitude that is characteristic to ϕ_3 and not to ϕ_1 is R_{JJ}

$(J = L = \text{odd for } I = 1 \text{ amplitude}).$[24]

Our interest then is to investigate if an R_{JJ} partial wave follows a Breit-Wigner formula.

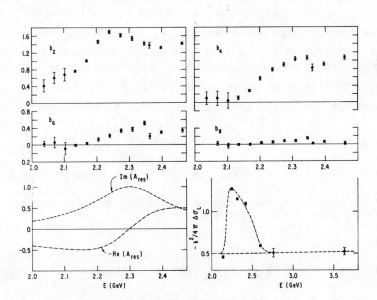

Fig. 10   Results of Legendre coefficient analysis.

We present the results of a Legendre coefficient analysis[25] using differential-cross-section[22] and polarization[23] data at 1.0-2.0 GeV/c. The analysis was carried out by looking at the energy dependence of the coefficients a_n and b_n in the expansions

$$d\sigma/d\Omega = \lambdabar^2 \sum_n a_n P_n (\cos\theta), \tag{1}$$

and

$$P \cdot d\sigma/d\Omega = \lambdabar^2 \sum_n b_n P_n^{(1)} (\cos\theta). \tag{2}$$

These coefficients are related to various partial waves, and we show only relevant relations here.

The coefficients a_n, obtained by fitting differential-cross-section data to Eq. 1, mainly tell us that the highest significant value of J is four; a_8 and higher coefficients are nearly zero, and we ignore those terms with J > 4 and L > 4. Figure 10 shows the coefficients b_n obtained by fitting the product of differential-cross-section and polarization data, plotted against incident momentum and energy. All coefficients up to and including b_6 have a remarkable energy dependence around $p_{Lab} \approx 1.5$ GeV/c. We need to know if such a rapid change is due to one or two resonant partial waves while the other amplitudes vary slowly with energy; our particular attention is on the R_{JJ} partial waves, $^3P_1$ and $^3F_3$. We determine how the behavior compares with the Breit-Wigner formula,

$$A_{res} = (\varepsilon + i)(\Gamma_{el}/\Gamma)/(\varepsilon^2+1), \text{ where } \varepsilon = 2(E_o - E)/\Gamma .$$

The energy dependence of this formula is illustrated in Fig. 10.

In general, the coefficient with higher order is easier to interpret because fewer terms are involved.

Coefficient b_6:

The coefficient b_6 is related to the partial-wave amplitude by

$$b_6 = 1.8(Im^3F_3 Re^3F_4 - Re^3F_3 Im^3F_4) + \text{nonresonant terms.} \quad (3)$$

A rise in b_6 is consistent with $^3F_3$ following the Breit-Wigner formula when the nonresonant $^3F_4$ is varying slowly with energy. The first term is about equal before and after resonance, say at 2100 and 2400 MeV, respectively, and the difference in b_6 at these momenta is due to the second term, which has ReA_{res} changing plus to minus. We note that $^3P_1$, another possible resonance candidate, in R_{JJ} is absent in b_6.

These results suggest the existence of a dibaryon resonance with the following properties: mass ≈ 2250 MeV, charge = 2, $J^P = 3^-$ for the resonance. We are not aware of any predictions of such a dibaryon resonance.[26] We plan to make additional measurements in order to clarify the nature of the structure.

The value of $\Delta\sigma_L$ at 6 GeV/c is ~-1.0 mb, implying the existence of a significant A_1-like quantum-number exchange term. Traditionally this term has been assumed to be zero since the existence of the A_1 resonance has never been established.

I would like to point out other independent work on the triplet-dibaryon resonance:

i) A dispersion analysis of proton-proton amplitudes using the data for total cross sections in pure-spin states by Grein and Kroll;[27] A resonant-like behavior in a triplet state with a mass of 2250 MeV is shown. A structure with a mass of 2160 MeV (often called $^1D_2$ pp state) is strictly due to a threshold effect.[28] Their conclusion is also clear by comparing $\Delta\sigma_T$ and $\Delta\sigma_L$ data[19,20] as shown in Fig. 11.

ii) Possible p-p resonance in the $^3F_3$ state by N. Hoshizaki;[29] their earlier phase-shift results[30] are updated by using new experimental data.

iii) Our preliminary data of C_{LL} near $\theta_{c.m.} = 90^\circ$ in the momenta region of 1.2 to 2.5 GeV/c seem consistent with the interpretation of the triplet-dibaryon resonance.

So far, we have discussed structures in the pp system. There is no doubt that one should carry out a similar investigation in the pn system using either a polarized-deuteron beam on a polarized-proton target or a polarized-proton on a polarized-deuteron target.

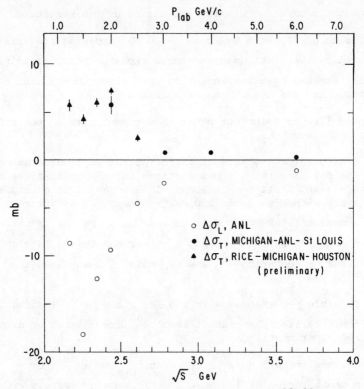

Fig. 11 Energy dependence of $\Delta\sigma_T$ and $\Delta\sigma_L$ data.[19,20]

P AND D_{NN} MEASUREMENTS AT LARGE p_T

These measurements have been typically carried out with a proton beam of $\sim 10^{10}$ particles/pulse by A. Krisch et al.[31-35] They detected elastic events using a double-armed spectrometer. The spectrometer measured the angle and momentum of both the scattered proton and the recoil proton using four analyzing magnets and several scintillation counters. By varying the magnet currents, they covered the angular range without moving the detectors. Inelastic and non-hydrogen events were less than 3%. They managed to measure a large p_T region with great accuracy. Figure 12 shows differential cross sections at large p_T.[36] Changes in slopes are seen at several places. In polarization and C_{NN} at 6 and 12 GeV/c, a remarkable dip and bump structure is seen as shown in Figs. 13 and 14. We note the sharp zero at $|t| \approx p_T^2 = 0.9$, the broad maximum centered near $|t| = 1.8 (p_T^2 = 1.7)$, and the zero again at $|t| = 3.6$

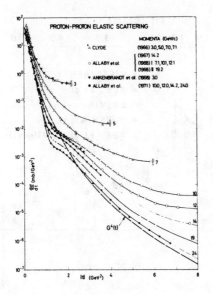

Fig. 12 pp differential cross section at large p_T.

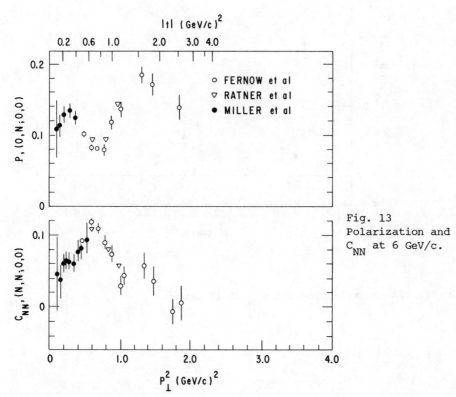

Fig. 13 Polarization and C_{NN} at 6 GeV/c.

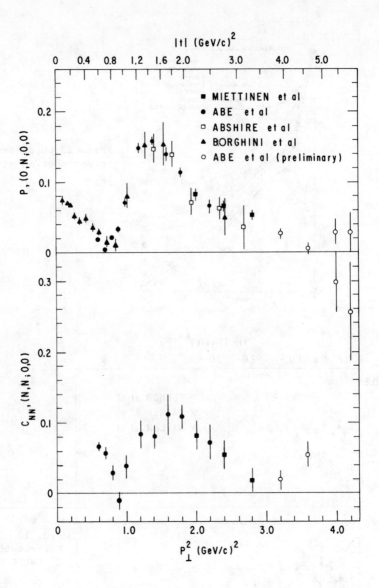

Fig. 14 Polarization and C$_{NN}$ at 12 GeV/c.

$(p_T^2 = 3.0)$ $(GeV/c)^2$. The same group is currently pursuing the measurement of a higher $|t|$ region at 12 GeV/c.

pn ELASTIC SCATTERING

To study the isospin effect, we need to investigate the pn system, which consists of I = 0 and I = 1 states. We have a long way to go before we can determine pn amplitudes.

Typically, the differential cross section in pp and pn below 10 GeV/c behaves in accordance with

$$(\frac{d\sigma}{dt})_{pp} \approx (\frac{d\sigma}{dt})_{pn} \text{ at } |t| < 0.3$$

and

$$(\frac{d\sigma}{dt})_{pp} < (\frac{d\sigma}{dt})_{pn} \text{ at } 0.3 < |t| < 2.0 .$$

The ANL EMS Group measured polarization parameters from 2 to 12 GeV/c.[37] The results are shown in Fig. 15. As shown in Table II, the polarization gives the real part of N_1 under the condition assumed there. The I = 0 and I = 1 contributions to ReN_0 were separated by using both the pp and pn data.[37] The energy dependence of ReN_1 (I = 0) was shown to be much larger than that of ReN_1 (I = 1). This may lead us to speculate on a new low-lying trajectory.

Our present plan on investigation of the pn system would start with $\Delta\sigma_T^{Tot}$ and $\Delta\sigma_L^{Tot}$ measurements; in particular, we look for possible structures as observed in the pn system. Then, with a polarized-neutron beam or a polarized-neutron target, various asymmetry measurements will be carried out.

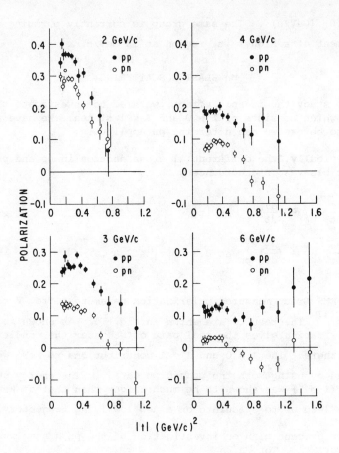

Fig. 15 Polarization in pn elastic scattering.

pp SCATTERING ABOVE ZGS ENERGIES

The rising total cross section at high energies is an extremely interesting phenomenon. Recently C. N. Yang[38] suggested that an understanding of such a phenomenon might come from the measurement involving a spin effect; in particular, he described interaction between the incident particle and hadronic current in the proton. Experiments at ISR and Fermilab in p-p elastic scattering revealed a remarkable dip structure in the differential-cross-section data at $|t| \approx 1.3$ $(GeV/c)^2$; as shown in Fig. 16, the slope changes from 12 to 1.8 at $|t| \approx 1.4$ $(GeV/c)^2$.

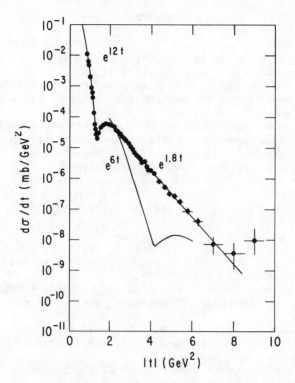

Fig. 16 pp differential cross section at ISR energies.

Our preliminary polarization data in the small $|t|$ region at [39] 300 GeV/c are consistent with zero, implying that the Pomeron term is dominant and $\text{Re}N_1 \approx$ zero. The results are indeed expected by theorists. However, anything is possible where and after the slope change takes place.

At Fermilab energies, polarized beams (also polarized targets) should be very useful for investigating the interactions between hadronic constituents. We are aware that it is practically impossible to accelerate polarized protons at Fermilab. However, the high production rate of hyperons is an attractive source for enriched-antiproton, polarized-p, and polarized -$\bar{\text{p}}$ beams.[4]

REFERENCES

1. L. Wolfenstein and J. Ashkin, Phys. Rev. 85, 947 (1952);
   L. Wolfenstein, Phys. Rev. 96, 1654 (1954).
2. C. R. Schumacher and H. A. Bethe, Phys. Rev. 121, 1534 (1961).
3. M. J. Moravcsik, The Two-Nucleon Interaction, Clarendon Press,
   Oxford (1963).
4. Proceedings of the symposium on experiments using enriched antiproton, polarized-proton, and polarized-antiproton beams at
   Fermilab energies, June 10, 1977, ANL-HEP-CP-77-45.
5. Halzen and Thomas, Phys. Rev. D10, 344 (1974).
6. P. W. Johnson, R. C. Miller, and G. H. Thomas, Phys. Rev. 15D,
   1895 (1977).
7. D. Miller, C. Wilson, R. Giese, D. Hill, K. Nield, P. Rynes,
   B. Sandler and A. Yokosawa, Phys. Rev. Lett. 36, 763 (1976);
   D. Miller et al., to be published in Phys. Rev.
8. I. P. Auer, D. Hill, R. C. Miller, K. Nield, B. Sandler,
   Y. Watanabe, A. Yokosawa, A. Beretvas, D. Miller and C. Wilson,
   Phys. Rev. Lett. 37, 1727 (1976).
9. I. P. Auer, A. Beretvas,   E. Colton, D. Hill, K. Nield,
   H. Spinka, D. Underwood, Y. Watanabe, and A. Yokosawa, to be
   published.
10. G. W. Abshire et al., Phys. Rev. D12, 3393 (1975).
11. J. Deregel et al., Nucl. Phys. B103, 269 (1976); J. Deregel
    et al., Phys. Lett. 43B, 338 (1973).
12. R. Field and P. Stevens, ANL/HEP CP 75-73, p. 28 (1975).
13. N. J. Sopkovich, Nuovo Cimento 26, 186 (1962).
14. G. L. Kane and A. Seidl, Rev. Mod. Phys. 48, 309 (1976). (The
    whole amplitude rather than the pole term is approximately
    degenerate.)
15. E. L. Berger, Argonne Report ANL/HEP 77-08, to be published;
    E. L. Berger and J.L. Basdevant, Proceedings of the Vth
    International Conference on Experimental Meson Spectroscopy,
    Boston, April 1977.
16. H. Becker et al., High Energy Physics with Polarized Beams
    and Targets, AIP Conference Proceedings No. 35, Argonne 1976,
    p. 243.
17. R. S. Longacre and R. Aaron, Phys. Rev. Lett. 38, 1509 (1977).
18. P. Kroll, E. Leader, and W. von Schlippe, C30 submitted to the
    Conference.
19. W. deBoer et al., Phys. Rev. Lett. 34, 558 (1975); Rice-
    Michigan-Houston Collaboration, J. B. Roberts (private
    communication).
20. I. P. Auer et al., Phys. Lett. 67B, 113 (1977); also see Ref. 9.
21. A compilation of NN and ND interactions, UCRL-20000 NN (1970).
22. To obtain the total elastic cross section from 1.2 to 1.7 GeV/c,

we have integrated the differential cross section data by
B. A. Ryan et al., Phys. Rev. $\underline{3}$, 1 (1971). These data were
used because of internal consistency.

23. M. G. Albrow et al., Nucl. Phys. $\underline{B23}$, 445 (1970); See Ref. 7
for data at 3, 4, and 6 GeV/c.

24. For the notation of R_{JJ}, see e.g., M. H. MacGregor et al.,
Annual Review of Nuclear Science $\underline{10}$, 291 (1960).

25. For more details, see K. Hidaka et al., to be published.

26. For instance, see R. L. Jaffe, Phys. Rev. Lett. $\underline{38}$, 195 (1977).

27. W. Grein and P. Kroll, WU B 77-6, to be published.

28. Their conclusion is consistent with C14 paper submitted to this
conference by G. N. Epstein and D. O. Riska.

29. N. Hoshizaki, preprint.

30. N. Hoshizaki and T. Kadota, Prog. Theor. Phys. $\underline{57}$, 335 (1977);
N. Hoshizaki, Prog. Theor. Phys. $\underline{57}$, 1099 (1977).

31. E. F. Parker et al., Phys. Rev. Lett. $\underline{31}$, 783 (1973); $\underline{32}$, 77
(1974); $\underline{34}$, 558 (1975).

32. R. C. Fernow et al., Phys. Lett. $\underline{52B}$, 243 (1974).

33. K. Abe et al., Phys. Lett. $\underline{63B}$, 239 (1976).

34. L. G. Ratner et al., Phys. Rev. $\underline{D15}$, (1977).

35. H. E. Miettinen et al., to be published in Phys. Rev. (1977).

36. A. Diddens, Proceedings of the XVII International Conference,
London, July 1974, p. I-41.

37. R. Diebold et al., Phys. Rev. Lett. $\underline{35}$, 632 (1975).

38. T. T. Chou and C. N. Yang. Nucl. Phys. $\underline{B107}$, 1 (1976).

39. Fermilab E-61 experiment, ANL-Berkely-FNAL-Harvard-Yale
Collaboration.

40. H. E. Haber and G. L. Kane, UMHE 77-9 (1977).

PROTON - PROTON SCATTERING FROM 0.35 TO 1.0 MEV

Ch. Thomann, J.E. Benn, S. Münch
Physics Department, University of Zürich, Zürich, Switzerland

We present here new experimental cross section data of
the elastic proton-proton scattering in the very low energy range from 0.35 to 1 MeV. The experiment was performed with an absolute accuracy of ± 0.2 %. There are
several reasons that gave us the idea for this measurement. Namely, the data at low energies contain not only
essential information on the effective range parameters,
but give also detailed information on higher electro-
dynamic effects such as vacuum polarization, there are
very few experimental data below 5 MeV and in the existing data between 1 and 10 MeV inconsistencies were
found by Signell et al. [1]. Measurements in the low energy range contain also a number of delicate experimental
problems that may have been overlooked in previous works.

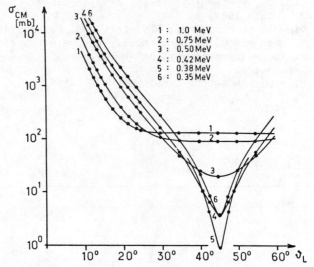

Fig. 1. The p-p cross section below 1 MeV. The
measured data set is indicated by dots.

An important feature of the measured energy range is the
so called interference minimum, where the Coulomb and the
nuclear amplitudes nearly cancel each other. That results
in an extremely small cross section at 90° center-of-mass
which is shown in figure 1. The very accurate value of
the scattering length a_{pp}^E in the literature is mainly deduced from the energy value of the interference minimum,

and, in the past, this energy value was determined only by the experiment of Brolley, Seagrave and Beery [2]. One of the six relative cross section values measured by Brolley et al. lay outside the expected range and was eliminated from the analysis without a strong argument. Therefore we decided to measure the cross section near the interference minimum with an improved accuracy. We found that at least the energy value E_{IM}=382.43±0.20 keV by Brolley et al. is in a good agreement with the value deduced from our new data: E_{IM} = 382.48 ± 0.10 keV. Signell and Rijken deduced from our data a preliminary value a_{pp}^E = - 7.8192 ± 0.0015 fm.

A second feature of this energy range is the strong influence of vacuum polarization. Near the interference minimum, the vacuum polarization amplitude f_{vp} reaches 6 % of the remaining sum of the Coulomb and nuclear amplitudes corresponding to a 10 % contribution in the cross section. Relative cross section values from Brolley's experiment gave little information about the vacuum polarization. Therefore we measured absolute cross section values around the interference minimum and also angular distributions at 0.35 and 0.42 MeV.

We also measured angular distributions at 0.5, 0.75 and 1.0 MeV for the determination of the effective range parameters. An accurate analysis of our data will be made by Signell and Rijken in connection with the other existing data.

From the experimental point of view, an important feature of this energy range was the failure of the usual method with which one measured the number of protons in the incident beam, namely the charge integration in which one uses a Faraday cup located behind a thin foil. This thin foil is necessary to separate the hydrogen gas target from the high vacuum which is needed in the Faraday cup. The effects caused by this foil such as Rutherford scattering, charge exchange, secondary electrons and ionization grow for low energies. We found experimentally that these effects are not sufficiently reproducible for the required accuracy. Therefore a new precision calorimeter was developed which measures the heat created by the stopped protons in the Faraday cup with an accuracy of ± 0.07 % [3]. The calorimeter works also in the presence of the gas. No foil was therefore used.

For that reason we suspect that the cross section data by

Muery et al. from Basel [4], which are the only other
existing data in this energy range, are questionable.
They used the charge integration with a foil correction
factor of a rather doubtful accuracy. The data show also
systematic deviations from our data, especially at 0.5 MeV.
As was mentioned by Haeberli at the beginning of this
conference, the Basel data are also in disagreement with
vacuum polarization. We believe the reason is a systematic error in the Basel data due to the beam current
measurement.

Table I   Comparison of the Basel [4] and the
           Zürich measurement. The angles and
           the cross section values are given
           in the center-of-mass system.

E_0(MeV)	Θ_{CM}	σ_{Basel}(mb)	$\sigma_{Zürich}$(mb)	$\sigma_{Ba} - \sigma_{Zü}$
0.49925	24°	9579.9 ±0.27 %	9655 ±0.25 %	−0.78 %
	50°	318.66 ±0.30 %	320.3 ±0.27 %	−0.51 %
	90°	20.213±0.80 %	20.64±0.31 %	−2.11 %
0.99190	24°	2110.8 ±0.12 %	2117.1 ±0.23 %	−0.30 %
	50°	151.25 ±0.11 %	151.78±0.29 %	−0.35 %
	90°	133.90 ±0.15 %	134.36±0.24 %	−0.34 %

Besides these data, there exists below 5 MeV only the
Wisconsin data from 1.4 to 3.0 MeV [5], which also may
contain a small systematic error due to the same reason.
It has also to be mentioned that they had troubles with
the energy calibration. To check these data we have just
started further measurements up to 5 MeV. At 5 MeV and
higher energies there are data by Imai et al., Kyoto [6].

I shall now present some of the experimental details.

The feature of the calorimeter is a special compensation
of the heat produced by the protons such that the cup is
always at the same temperature as the environment. The
error due to heat exchange with the environment therefore becomes very small and the required accuracy could

be reached. In vacuum without foil the beam current could be measured simultaneously by standard charge integration. There was an excellent agreement of the two totally independent methods.

Fig. 2. View of the opened scattering chamber.

Special care was taken to establish the energy calibration near the interference minimum. This calibration was performed with a 180° analysing magnet, with an accuracy of ± 25 eV. In addition the energy loss in the gas target was measured separately.

Another problem was the extremely small cross section near the interference minimum resulting in a very small counting rate. Therefore errors due to background events became important, namely background by cosmic rays. On the other hand the p-p counting rate cannot be increased, because the gas pressure is limited by the error due to multiple scattering and energy loss in the gas. Also the G factor cannot be increased because the extremely strong variation of the cross section in angle requires a small angular range for the apertures. The accuracy that could be reached near the interference minimum was therefore limited by the proton current that could be obtained from the accelerator. We tried to optimize the accuracy by the choice of the G factor and the gas pressure.

The background problems were avoided by Brolley et al. by measuring the coincidence rate of opposite ($\pm$ 45°) detectors. But the coincidence rate is more influenced by multiple scattering in the gas target than the single rate. We also measured the coincidence rate and found a pressure dependence of - 4 %/torr due to multiple scattering. In addition, multiple scattering is strongly energy dependent. A measurement in the arrangement of Brolley would therefore not provide the required accuracy for absolute cross section values.

That was a brief description of our experiment. The data and details of the experiment will be published in Nuclear Physics.

References

[1] M.S. Sher, P. Signell, L. Heller, Ann. of Phys.,
           58 (1970) 1
      J. Holdeman, P. Signell, M. Sher, Phys. Rev. Lett. 24
           (1970) 243

[2] J.E. Brolley, J.P. Seagrave, J.G. Beery, Phys. Rev.
           135 (1964) B1119

      M. Gursky, L. Heller, Phys. Rev., 136 (1964) B1963

[3] Ch. Thomann, J.E. Benn, Nucl. Instr. and Meth., 138
           (1976) 293

[4] H. Mühry, H. Wassmer, E. Baumgartner, Helv. Phys.
           Acta, 46 (1973) 604
      H. Wassmer, H. Mühry, Helv. Phys. Acta, 46 (1973) 626

[5] D.J. Knecht, S. Messelt, E.D. Berners and L.C. North-
           cliffe, Phys. Rev. 114 (1959) 550
      D.J. Knecht, P.F. Dahl, S. Messelt, Phys. Rev., 148
           (1966) 1031

[6] K. Imai, K. Nisimura and N. Tamura, Nucl. Phys. A246
           (1975) 76

NEUTRON-PROTON MEASUREMENTS AND PHASE SHIFT ANALYSES NEAR 50 MeV[*]

D.H. Fitzgerald, S.W. Johnsen, F.P. Brady, R. Garrett, J.L. Romero,
T.S. Subramanian, J.L. Ullmann and J.W. Watson
University of California, Davis, CA 95616

Peter Signell
Michigan State University, East Lansing, MI 48823

ABSTRACT

We report on our measurements of n-p scattering observables at
50 MeV, most notably the recent measurement of the spin-correlation
parameter, A_{yy}. This measurement is sensitive to the phase shift
parameters $\bar{\varepsilon}_1$ and $\delta(^1P_1)$, for which early phase shift analyses gave
poorly determined or "unphysical" values. The results of our
measurements are incorporated in a phase shift analysis to determine
current best-fit values of 50 MeV n-p phase shifts.

INTRODUCTION

In the past, phase shift analyses[1,2] of n-p scattering data
have revealed that at 50 MeV the $^3S_1-^3D_1$ mixing parameter, $\bar{\varepsilon}_1$, was
undetermined in the range $-10°$ to $+3°$ and the $^1P_1$ phase shift
obtained disagreed with values expected from models. A more recent
energy-dependent analysis[3] ranging from 0 to 500 MeV gives $\bar{\varepsilon}_1 \approx -1°$.
However, negative values of $\bar{\varepsilon}_1$ are inconsistent with the predictions
of most models of the nucleon-nucleon force.

It has been shown[2] that the observable A_{yy} is sensitive to $\bar{\varepsilon}_1$
and, to a lesser extent, $\delta(^1P_1)$. We have made two essentially
independent sets of measurements of A_{yy} at 50 MeV. The first
measurements contained large uncertainties due to uncertainty in the
proton target polarization and dispersion in the run-to-run A_{yy}
values. The latest measurements have been made to reduce these
uncertainties. These A_{yy} measurements, together with our other n-p
data near 50 MeV, have had a considerable impact on the values
obtained for $\bar{\varepsilon}_1$ and $\delta(^1P_1)$.

EXPERIMENT

Because the experimental methods for the A_{yy} measurements have
been discussed elsewhere[4], only the cogent details are presented
here. As shown in Fig. 1, the polarized neutron beam is produced
by the $T(d,\vec{n})^4He$ reaction with 38.0 MeV deuterons incident on a gas
tritium target. The 50 MeV neutrons emitted at $29.7°$ are collimated
to produce a 24 mm diameter beam with an intensity of 10^5 n/sec and
a polarization of 0.47 ± 0.02 at the target. The polarized proton
target consists of a lanthanum magnesium nitrate (LMN) crystal of
2 mm thickness and 25 mm diameter. Free hydrogen in the LMN is
polarized dynamically. The target polarization was continuously
monitored and averaged 0.28 ± 0.02 during the experiment. Recoil

ISSN: 0094=243X/78/091/$1.50 Copyright 1978 American Institute of Physics protons from the LMN crystal were detected in four plastic scintillator ΔE-E telescopes. A neutron beam monitor was used for beam flux normalization and the prompt neutron peak was selected by time-of-flight.

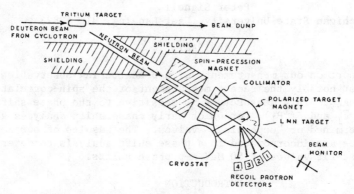

FIG. 1. Experimental configuration for the A_{yy} measurement.

If the target and beam polarizations, p_t and p_b respectively, are measured along the positive y-axis (perpendicular to the scattering plane) the intensity for scattering is given by[5]:

$$I(\theta, p_t, p_b) = I_0(\theta)[1 + (p_t + p_b) A_y(\theta) + p_t p_b A_{yy}(\theta)] .$$

$I_0(\theta)$ is the (unpolarized) n-p differential cross section and $A_y(\theta)$ is the analyzing power. Values of A_{yy} were determined from a combination of measurements with various p_t and p_b orientations:

$$A_{yy} = (|p_t||p_b|)^{-1} \frac{(\uparrow\uparrow) - (\uparrow\downarrow) + (\downarrow\downarrow) - (\downarrow\uparrow)}{(\uparrow\uparrow) + (\uparrow\downarrow) + (\downarrow\downarrow) + (\downarrow\uparrow)}$$

This combination eliminates the requirement of knowing $I_0(\theta)$ and $A_y(\theta)$ and reduces the effect of systematic errors. 60-70 hours of data were taken for each of the four spin orientations. The values obtained for A_{yy} are plotted in Fig. 2. The uncertainties shown are statistical and, on the average, these were reduced in the most recent measurement (open circles) to 70% of those seen in Ref. 4 (solid dots). More significantly, the 25% normalization uncertainty for the first measurement was reduced to 10% for the present results, owing to a better determination of the target polarization.

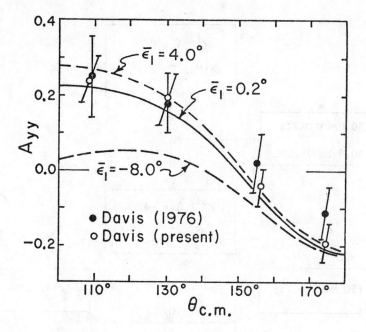

FIG. 2

PHASE SHIFT ANALYSES

Phase shift analyses have been carried out during the past few years at several stages as more n-p data became available near 50 MeV, as illustrated in Fig. 3. Analyses[1,2] which used mainly the Harwell n-p data[6] showed $\delta(^1P_1)$ near zero while boson exchange models predicted values between $-5°$ and $-10°$. A broad minimum in χ^2 vs $\bar{\varepsilon}_1$ suggested solutions for $\bar{\varepsilon}_1$ between $-10°$ and $+3°$. A later analysis showed that the addition of the Davis[8] (50 MeV) and Oak Ridge[6] (61 MeV) differential cross section data moved $\delta(^1P_1)$ to $-4.1°$, while the value of $\bar{\varepsilon}_1$ remained ambiguous. At stage 3, the Harwell data were excluded because of normalization problems. An analysis[4] including the first set of A_{yy} measurements gave a solution for $\bar{\varepsilon}_1$ of $0.2° \pm 1.7°$. It was noticed that one datum from the Oak Ridge (OR) set, $\sigma(90°)$, had a large effect on $\bar{\varepsilon}_1$ and its removal increased $\bar{\varepsilon}_1$ to $2.1°$. Thus, we carried out a measurement of $\sigma(\theta)$ at 63.1 MeV over the angular range from $35°$ to $165°$ (c.m.)[9]. As shown in Fig. 4, the shape of this data does not agree with the OR data. Its inclusion, together with the Davis 50 MeV $A_y(\theta)$ measurements[9], gives $\bar{\varepsilon}_1 = 2.6° \pm 1.0°$ and $\delta(^1P_1) = -4.5° \pm 1.2°$. Excluding the OR data allows $\bar{\varepsilon}_1$ to go to $4.0° \pm 1.0°$ and $\delta(^1P_1) = -6.4° \pm 1.2°$.

The second set of A_{yy} measurements have been added to the data set and a phase shift analysis carried out.[10] If the OR data are retained, then the values obtained are $\bar{\varepsilon}_1 = 2.9° \pm 1.0°$ and $\delta(^1P_1) = -6.5° \pm 1.1°$. Excluding the OR data, the values obtained are

94

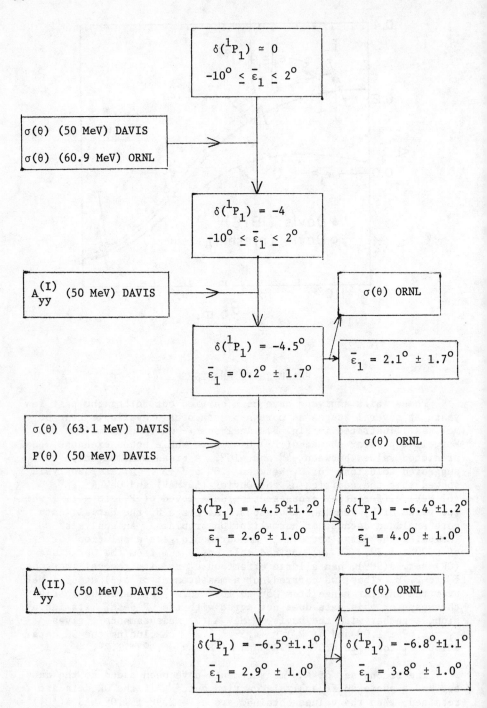

FIG. 3

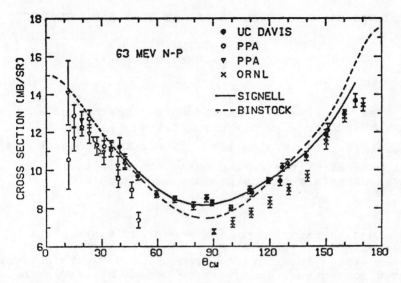

FIG. 4

$\bar{\epsilon}_1 = 3.8^o \pm 1.0^o$ and $\delta(^1P_1) = -6.8^o \pm 1.1^o$. Thus, as the 50 MeV n-p data base has broadened and improved, the results of phase shift analyses have converged near values predicted by models of the nucleon-nucleon interaction.

REFERENCES

*   Supported by the National Science Foundation Grant PHY 71-03499 and PHY 77-05301.

1. R.A. Arndt, J. Binstock and R. Bryan, Phys. Rev. D8, 1397 (1973).
2. J. Binstock and R. Bryan, Phys. Rev. D9, 2528 (1974).
3. R.A. Arndt, R.H. Hackman and L.D. Roper, Phys. Rev. C15, 1002 (1977). NOTE: a new analysis which inclues the Davis 63 MeV $\sigma(\theta)$ data and the first A_{yy} measurements yields $\bar{\epsilon}_1 = 2.97^o \pm 0.85^o$ as reported by R.A. Arndt at this conference.
4. S.W. Johnsen, F.P. Brady, N.S.P. King and M.W. McNaughton, Phys. Rev. Lett. 38, 1123 (1977) and references mentioned therein.
5. J. Raynal, Nucl. Phys. 28, 220 (1961).
6. See Ref. 1 for a table of the data.
7. R. Bryan and J. Binstock, Phys. Rev. D10, 72 (1974).
8. T.C. Montgomery, F.P. Brady, B.E. Bonner, W.B. Broste, and M.W. McNaughton, Phys. Rev. Letters 31, 640 (1973) and Phys. Rev. C, to be published (1977).
9. CNL-UCD 187 (U.C. Davis).
10. P. Signell (to be published). The computer output from the phase shift analysis, containing all necessary details for checking its accuracy, have been submitted to NTIS. Assuming acceptance, copies will be available from that source at the U.S. Dept. of Commerce, 5985 Port Royal Rd., Springfield, Va. 22161.

NUCLEON-NUCLEON INTERACTIONS AT FERMILAB ENERGIES *

Paolo Franzini
Columbia University, New York, N.Y. 10027

ABSTRACT

The experimental results on p-p scattering at Fermilab for σ_{total}, $\sigma_{elastic}$, $\sigma_{inelastic}$ and $\sigma_{diffractive}$ are synthesized. Their implications, as well as that of the future data inputs on the fundamental structure of the proton, are discussed.

I. INTRODUCTION

In this paper, I will discuss nucleon-nucleon interactions at Fermilab energies. In particular, proton-proton interactions from 8-500 GeV. This wide, nearly continuous, energy range became available to experimenters only through using the accelerating proton beam inside the accelerator, which involved large technical developments and will be discussed briefly. Most emphasis will be placed on what has been learned in the last five years and what can be learned in the future.

At Fermilab, we are in the high energy regime. Even at 8 GeV the available energy is ~ twice the proton's rest energy. And as the total energy in the c.m. of the two protons squared (s) is increased from $s = 16$ GeV^2 to about 1000 GeV^2, i.e., by about two orders of magnitude, which aspects of the collisions might one naively expect to change? The total cross section should remain constant while elastic scattering, the number of particles produced per collision (i.e., multiplicity), the transverse momentum of the secondaries (i.e., $<P_\perp>$ final), might all be expected to increase, maybe some even linearly with energy. In fact the total cross section and the multiplicity do increase but only logarithmically with s. The transverse momentum distribution is essentially s independent, with the average secondary transverse momentum being $\approx$ 300 MeV/c. Above 40 GeV the elastic cross section is essentially constant while its forward peak sharpens logarithmically with s.

The relative constancy of σ_{tot}, σ_{el} throughout this range, as well as the fact that the real part of the elastic scattering amplitude has been found to be small (< 0.05), suggests the validity of using optical

* Research supported by the National Science Foundation.

ISSN: 0094-243X/78/096/$1.50 Copyright 1978 American Institute of Physics

diffractive models[1] in the interpretation of pp elastic scattering. Note that in such models σ_{el} is the shadow of the almost black disk whose area is given by σ_{tot}; diffraction minima were expected in the elastic differential cross section and they were indeed seen. Hence a mapping of the measured $d\sigma(s)/dt$ distributions into impact parameter space yields the proton matter density profile function.

While the multiplicity increases slowly with s, it is a large number at the Fermilab energy range, the average number of final particles per collision is ≈ 13. The cross sections for most exclusive channels decrease rapidly with increasing s, while phase space available increases, but more channels open up. It is neither feasible nor meaningful to study each of these channels as a function of energy. Customarily most final states are summed over and one considers a quasi two-body, inclusive reaction, $a + b \rightarrow c + X$ where X is anything (at 500 GeV, it might be one nucleon plus fifty pions). For this reaction, we can define three Lorentz invariant parameters, which assuming cylindrical symmetry, fully define it. They are usually taken to be s (center of mass energy squared); $s = (p_a + p_b)^2$, t(four-momentum transfer squared), $t = (p_a - p_c)^2$, (< 0), and M^2 (invariant mass of system X), $M^2 = (p_a + p_b - p_c)^2$. In our case, particles a, b, and c are all protons. Schematically,

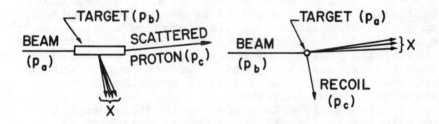

The analogy with optics becomes relevant, also for inelastic scattering,[2] at $s \gg m_p^2$ because the condition that, "incident and scattered waves are coherent across the thickness of a slab as long as the change in wavelength is smaller than the thickness of the slab", is now valid for the crossing of a single proton. For our reaction $p + p \rightarrow p + X$, using $r_p \simeq 1/m_\pi$, we obtain coherence up to $M_X^2 \leqslant (m_\pi/m_p)s$ which at Fermilab reaches the limit $M_X^2 < 140$ GeV^2. Thus coherent excitation of very heavy states is possible at Fermilab. The basic properties of such processes, usually called inelastic diffraction, are that (i) for fixed $M_x{}^2$ the cross section should be energy independent, (ii) the t distribution is determined by the transverse dimension of the scattering target (proton), (iii) it extends in mass up to a constant fraction of s.

Only recently, the theoretical description of inelastic diffraction has been put on a more rigorous basis, with interesting results. One consequence is that through unitarity it follows: (i) $\sigma_{el} + \sigma_{inel\ diff} < 1/2\ \sigma_{tot}$, which means in our energy range, $\sigma_{inel\ diff} < 38/2 - 7 = 12$ mb. (ii) Furthermore relation (i) is valid at any value of impact parameter, $\sigma_{el}(b) + \sigma_{inel\ diff}(b) \leq 1/2\ \sigma_{tot}(b)$.[3]

One anticipates that $\sigma_{inel\ diff}$ should be peripheral because the central region of the proton already saturates unitarity and absorption strongly attenuates coherent propagation. It should also be noted that for a completely black disk $\sigma_{el} = 1/2\ \sigma_{tot}$ and no diffractive inelastic processes are possible. In general, an increase in absorption reduces σ_{diff}. Since the increase in σ_{tot} could be thought of as being peripheral, $\sigma_{inel\ diff}$ was singled out as potentially being responsible for the rise. The s dependence of $\sigma_{inel\ diff}$ and whether it is peripheral or not are experimentally answerable questions.

Returning to my original question then, what can be learned about nucleon-nucleon interactions at Fermilab? The answer is, since we are in a region where m_p can almost be thought of as being ≈ 0 and proton-proton scattering can be thought of as hadron wave impinging upon an obstruction, a complete mapping in s, t and M^2 space yields the fundamental attributes about the structure of the proton. Of course, as long as the connection between scattering and structure is made through optical concepts, only average properties of such structure can be inferred.

It is still up to a new fundamental theory to compute the functions, which we are measuring at present, from prime principles. Notwithstanding the great recent theoretical success in obtaining newer insights in the "constituents" of elementary particles, no progress has been made so far in understanding the proton.

II. EXPERIMENTAL METHODS

The traditional method to study scattering since Lord Rutherford's days has been to direct a beam at a target and to observe the deflected beam particles. This method has followed the millionfold increase in energy of available beams, and is used extensively at Fermilab to study inclusive reactions. Typical setups include a beam transport system with particle identifying counters and spark chambers to improve knowledge of the incident angle. Following an interaction target, more spark chambers measure the deflection angle and finally magnets with more spark chambers give a measurement of the momentum of the deflected particle.

Because of the approximate constancy of transverse momenta in hadron scattering, the scaling with energy of deflection angles is inversely proportional to laboratory beam momentum. Thus at 300 GeV/c, the typical deflection angle is 300 (MeV/c)/300 (GeV/c) $\approx 10^{-3}$ rad, and to obtain good accuracy, angles must be measured to 10^{-5} rad. To first approximation, t is only a function of θ. M^2 is mostly determined by the change in momentum, $\delta(M^2) \approx 4m_p\Delta p$. Therefore, $\Delta p \approx \delta M^2/4$ and for $\delta(M^2)$ = 1 GeV^2, p = 300 GeV/c, $\Delta p/p \approx 10-3$.

At high energy, it is very convenient to look at the target recoil instead. The advantages are that:

1. The momentum transfer is directly the recoil momentum instead of a tiny fraction of the scattered particle's momentum, thus allowing accuracy equal to the measurement error.

2. The scattering angle's range is amplified from a few milliradians to 45°.

3. The mass M^2 is mostly a function of the recoil angle. Approximately $\delta M^2 = 2E\sqrt{((\delta T_r)^2 + |t|(\delta\theta_r)^2)}$, where T_r is the kinetic energy of the recoil. At the typical values of $|t| \approx 0.1$, from t = $-2m_pT$, T $\approx$ 0.05 GeV. For a 2% measurement of T (1 MeV) and a 2 x 10^{-3} rad angular resolution (2 mm counter at 0.8 m from the target), one gets $\delta M^2 \sim$ 0.5 (GeV)^2 at 300 GeV.

Our group has developed this technique to give the broadest coverage in mass squared and values of $|t|$ in the range 0.01 to 0.6 (GeV/c)^2.[4] In fact, the mass squared coverage 0 < M^2 < 0.25 s ($\sim$ 250 GeV^2 at 500 GeV P_{beam}) is the widest of any Fermilab experiment. The spectrometer technique has an edge with respect to the higher t values they can reach (up to 1 or 2 (GeV/c)^2), although since the cross section falls so rapidly with $|t|$, we typically already study 80% of σ, 10% being at $|t|$'s lower than 0.01.

The full advantage of the recoil method is realized in experiments using directly the internal beam. The latter is almost a necessity because in order to detect recoils of a few MeV kinetic energy, one needs very thin windowless targets and therefore a high intensity beam. There is also a great extra in using the internal beam, namely the continuous range of beam energy available.

The present internal beam intensity at Fermilab

reaches 10^{18} protons/sec, available for ~ 3 to 6 sec
during the acceleration and coasting time, every machine
cycle. The ideal target for such intensity is a gas jet.
We succeeded, for the first time, to develop a jet
working at room temperature producing a supersonic dis-
charge in the main ring vacuum with only a few degrees
divergence. Owing to the small divergence of the jet,
most of the gas is captures into a cone approximately
5 cm away from the nozzle and pumped at relatively high
pressure (0.1 torr) by a mechanical pump (300 ℓ/sec
Roots Blower). The beam intersects the H_2 jet halfway,
giving an effective source size of $\leqslant$ 1.5 mm. See Fig. 1.

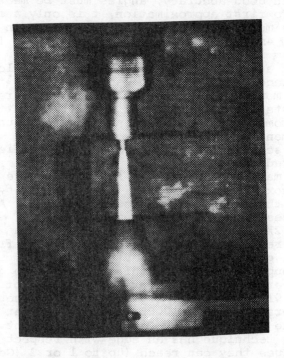

Fig. 1. Video photo of supersonic H_2 jet, showing
nozzle, catching cone and the jet itself. The bright
vertical plume is H_2 gas excited by rf field of
proton beam current.

The density of our jet is approximately 10^{-7} gm/cm^3 and
its thickness is ~ 2 x 10^{-8} gm/cm^2, which if anything,
is slightly too high.
    Prior to the installation of our jet, another jet
developed in Dubna, USSR, had been used in the Internal
Target Area at Fermilab. That jet had a half angle
divergence of 45°, giving a source size of ~ 20 mm and
was pumped cryogenically on liquid He cooled surfaces.

This pumping method had many disadvantages. There are now at Fermilab three operating jets, all pumped with standard pumping techniques, and all capable of operating with any gas from H_2 and He to Xe.

Elastic scattering experiments have been performed at ITA for $|t| = 0.0001$ to $2(GeV/c)^2$ and inelastic scattering experiments for $|t|$ from 0.01 to $0.6(GeV/c)^2$ and M^2 up to $200(GeV)^2$. A typical arrangement of an elastic and inelastic scattering experiment is our setup. Telescopes consisting of two solid state counters and two scintillation counters detect recoils, identify their nature (π's, p's, d's, etc.) and their kinetic energy. The main problem of recoil experiments is to be sure that only particles originating from the target are accepted (the ratio of target mass to surrounding mass is $< 10^{-10}$), just as in more conventional experiments. Two very sensitive diagnostic tools are available. They are (i) nuclear fragment productions, particularly deuterons and tritons, and (ii) detection of production of unphysical masses, i.e. $M^2 < 0$. We succeeded in reducing both backgrounds to unmeasurable levels by inserting midway between telescopes and target an anticounter with holes which allow each telescope to see only ~ 2 cm around the jet-beam intersection and a small amount of solid matter upstream from the jet and very far from the beam path.

The main disadvantage in working with the internal beam is the extremely limited access to the experimental setup, typically eight hours per week. This requires construction of very sophisticated remote control systems, as well as developing highly reliable electronics. A further constraint was posed, in our case, by our decision to avoid cables between solid state detector and preamp, as well as the use of 20 photomultipliers in the vacuum scattering chamber. We thus designed and constructed our own charge sensitive preamps for silicon detectors. We also use voltage divider on the photomultiplier with only 100 μA current and use current amplifiers.

The whole system is capable of 100 MHz rate for trigger logic and ~ 0.5 MHz rate for precise measurements of energy signals in solid state detectors. Typical event rates were $\sim 10^4$ analyzed events/sec. Analyzed refers to the fact that the data coming in as energy loss signals in 4 detectors from each of the eight telescopes was processed in real time with a hardwired computer designed for this purpose. The result of the processing was adding 1 in the appropriate cell of a three-dimensional map of $d^2\sigma(s)/dtdcos\theta$ stored in a very large memory. No pulse height information was otherwise recorded. This computer is ~ 1000

times faster than a general purpose computer and is especially valuable in giving immediate feedback on physically interpretable quantities ($d\sigma/dM^2$ for $M^2 < 0$; deuteron and triton production cross section etc.) and enabled us to rapidly achieve the stage of taking data virtually free of background.

Another recoil experiment at ITA (E-198)[5] extended the $|t|$ range for elastic scattering beyond $1 (GeV/c)^2$ by using a magnetic spectrometer to analyze recoil protons. The spectrometer was followed by a polarimeter to measure the polarization of recoil proton (E-313).[6] Both experiments are at the preliminary data analysis stage.

III. p-p ELASTIC CROSS SECTION AND TOTAL CROSS SECTION

While the first evidence for the appearance of strong structure in elastic scattering came from CERN ISR,[7] a wealth of experiments was performed at Fermilab up to t values of $-5 (GeV/c)^2$. Typical of the meson lab experiments are E-7, E-69, and E-96. The results obtained by these groups again confirm the complex structure of pp elastic scattering vs. t and s. If we write $d\sigma/dt \sim e^{\,b(t)t}$, then b at small t increases with s and at all s values decreases with t. At very high t and s, a dip appears in the cross section which has many characteristics of a diffraction minimum. The onset of this dip is very visible at $s \approx 200\ GeV^2$ and is fully visible at $s \approx 400\ GeV^2$.[8] See Fig. 2. The positions of such dips vs. s, and the values of the cross sections both before and after, can be accounted for at present by either of two ad hoc assumptions in the impact parameter calculations.

The Chou-Yang model fits quite well the data if the proton density function (in their model the convolution of the proton electric form factor with itself) is simply scaled at each value of s in such a way as to give agreement with the observed rise of the total cross section. In other words, the proton is getting blacker. In the so-called geometric scaling hypothesis,[9] it is the radius of the proton opacity function which is adjusted to follow σ_{tot}, i.e., the proton is getting bigger. In the second case, one usually works backwards, namely from data the opacity function is determined.

The small t value logarithmic slopes are also obtained by many experiments at ITA. The most extensive results were obtained by our group. Shown in Fig. 3a are slopes vs. s for the low t. Figure 3b shows that while the slope increases with s, the change in slope from t $\sim$-0.05 to -0.15 remains constant with s. This change in slope is clearly visible in Fig. 3c from E-69.[10] The study of pp elastic scattering at very small t values has been in the past the specialty of the Dubna group.[11] Such measurements initiated at Serpukhov up to 70 GeV were continued at Fermilab by the same group in collaboration with groups from Fermilab, Rochester and Rockefeller Universities.

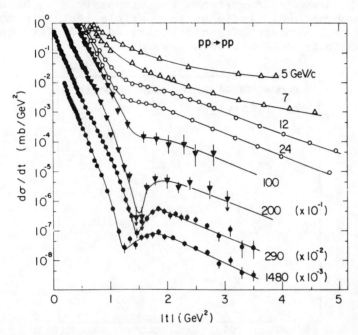

Fig. 2. Compilation of p-p elastic cross section, curves at 100 and 200 GeV are from Ref. 8.

At t values of $\approx$ -0.0025, the nuclear pp scattering amplitude becomes equal to the coulomb scattering amplitude. Observation of the amount of interference with the coulomb amplitude allows, under plausible assumptions for the spin dependence of the pp amplitude, to measure the ratio of the real to imaginary part of the elastic scattering amplitude at low t. Their results are shown in Fig. 4, together with some ISR results. E-69 at the Meson Lab has also measured the same region in t at specific s values with similar results. Note the extreme smallness of ReA which means that its contribution to σ_{el} is always < 1%.

There are two important consequences in general of good measurements of ρ = ReA/ImA. The first is that the diffractive explanation of σ_{el} requires the amplitude to be pure imaginary. The second is that ReA can be related by dispersion relation to the behavior of σ_{tot} up to infinite energy. Therefore accurate measurements of

ReA(s) up to s = 1000 or 2000 could, in principle, distinguish between different models' predictions for σ_{tot}
at s = 10,000 or 100,000 GeV^2. In practice, due to the
almost explosive instability of the continuation of an
analytic function, poorly known in some domain, by the
Cauchy Theorem, so far it has been impossible to accom-
plish such an end. It is quite possible that much more
careful measurements of ρ might give us good clues about
the asymptotic behavior of σ_{tot}.

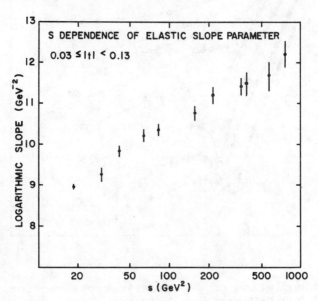

Fig. 3a. Logarithmic slope of elastic scattering as a
function of s for $0.03 \leq |t| \leq 0.13$. Ref. 4.

Total Cross Section. This type of measurement has
always been of first priority at any new accelerator.
Such was the case at ISR and in fact the first exciting
result in many years of strong interactions study was
the discovery of increasing σ_{tot} at s ≥ 1000 to s =
3600 GeV^2. [12] Very accurate measurements of σ_{tot} by
transmission method and by optical theorem method have
been performed at Fermilab.[13] Figure 5 gives a collection of such results. The data fit a phenomenological
parametrization of the form $4.91 \, \ln((p_{lab} + 541)/0.3)$
$+ 11.1/p_{lab}^{0.38}$ [14] whose significance escapes me.
As I said, this result was exciting and surprising
to everybody, but perhaps it was only that we were
growing complacent. It should be mentioned that Chen
and Wu had worked out a model in quantum electrodynamics
where σ_{tot} increases with s.[15]

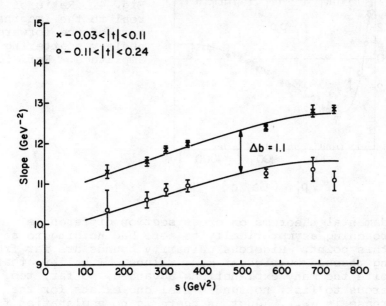

Fig. 3b.   The "break" at t ≈ 0.15 GeV^2 of the elastic slope, a function of s.   Ref. 4.

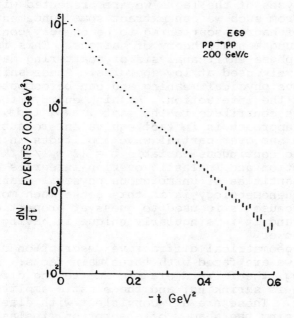

Fig. 3c.   DN/Dt vs. t for elastic scattering at 200 GeV. Ref. 10.

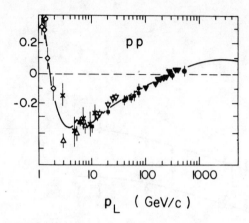

Fig. 4. Ratio of the real to the imaginary part of the forward nuclear scattering amplitude. Ref. 11.

p_L (GeV/c)

Fundamental theorems on cross section differences approaching asymptotically to zero had nothing to say on this point. Rigorous unitarity bounds are far from being reached by present σ_{tot} values. I believe that despite the many explanations advanced, so far, too numerous to list, no fundamental one exists for the increase in σ_{tot}. Just as there is no explanation for the value of σ_{tot}, nor for the mass of the proton.

It is necessary, however, to continue phenomeno-logical analyses of the facts we are presented with, as long as from such we can extract some fundamental properties of hadron scattering to be one day confronted with a new fundamental theory of hadrons. This is equi-valent to a phase shift analysis of scattering data, once extensively used at low energies. Phase shifts have an unambiguous physical meaning and can be computed from knowledge of the interaction. At high energy, too many partial waves contribute to the scatt. amplitude. The diffractive approach is in fact equivalent to substitu-ting for the sum over partial wave amplitudes an inte-gral over the continuous variable $b = (\ell + 1/2)/k$. The opacity function and inelastic overlap integrals are therefore quantities of unambiguous physical meaning. Regge Pole phenomenology is in this sense much more ambiguous because is is used to guess at properties of scattering, and it is singularly unique in having no predictive power.

In the geometrical diffractive description of pp scattering, we are faced with three main facts: σ_{tot} is increasing, the forward elastic scattering diffrac-tion pattern is shrinking, and the elastic amplitude is imaginary. These are all consistent with elastic scattering being the shadow of absorption (inelastic scattering). The s behavior can be described as either a blackening of the proton (increased absorption) or an increase in radius. In either case, the absorption must increase toward the periphery of the proton. The real situation is of course more complicated. For instance, we know that ReA = 0 only at t = 0 and since at t = 1 GeV, A(t=1)/A(t=0) ≈ 10⁻³, the amplitude has fallen off by ~ 10⁻³ and a very small ReA might become important. Combining all information available, one can however use the data on σ_{tot}, σ_{el} to obtain the opacity function for pp collisions. Using dispersion relations to obtain ReA(t), and the measured data for total and elastic scattering, the profile function of the proton at ISR energies has been obtained as a function of impact parameter.[16] Similar profiles were obtained at Fermilab for energies between 50 and 200 GeV and confirm that the increase in σ_{tot} is due to an increase in absorption at large impact parameter.[13]

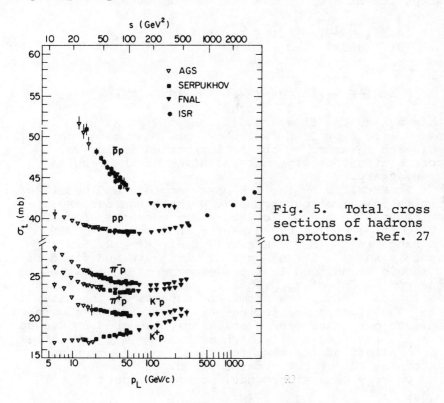

Fig. 5.   Total cross sections of hadrons on protons.   Ref. 27

IV. INELASTIC SCATTERING

We shall only discuss here the inclusive reaction p + p → p + X.   Exclusive reactions, as stated in the introduction, become one by one, smaller and smaller, while the possible channels increase. Thus, except for the investigation of special channels to study very limited phenomena, we are sort of forced to limit ourselves to the more global view. We note that if all phase space for pp → pX were to be covered, we would have studied one half of σ_{inel}, the remaining half being p + p → n + X. Of course, it is very rare that one or even a series of experiments cover all of phase space. Also, we are naturally prejudiced towards special configurations known to be more rich in features.

I wish to remark that the approximate description of inclusive p + p → p + X is given by

$$\frac{d\sigma(s)}{dtdM^2} = f(M,s)e^{bt} \quad , \text{ where } b = 6;$$

except for $M^2 = m_p^2$ and $M^2 \cong 2$ GeV^2. Thus,

$$\int_{-0.01}^{-0.2} dt \left(\frac{d^2\sigma}{dtdM^2}\right) / \int_0^{-\infty} dt \left(\frac{d^2\sigma}{dtdM^2}\right) = 64\%, \text{ and,}$$

$$\int_{-0.3}^{-1} dt \left(\frac{d^2\sigma}{dtdM^2}\right) / \int_0^{-\infty} dt \left(\frac{d^2\sigma}{dtdM^2}\right) = 16\%. \quad \text{Therefore,}$$

while again, the first results on pp → pX at high energy were from CERN ISR, they were most often for $|t| > 0.3$ and in order to obtain integrated inclusive cross section, massive extrapolation of the results is necessary.

At Fermilab by far the most complete and detailed results came from our group.[4] We have run our experiment from 8 to 500 GeV/c laboratory momentum, covering $0.01 \leq |t| \leq 0.5$, i.e., 85% of σ in t and up to 25% of the range in M^2 ($M_p^2 < M^2 \leq s$), i.e., $M^2 \leq 0.25$ s. Figure 6 shows a set of data at various s and fixed t. It should be noticed that our experiment in addition to measuring $\approx 25\%$ of σ_{inel} has so far the best M^2 and t resolution. Some very prominent aspects of the data are: the elastic scattering peak (left in the figure to indicate our sharp mass resolution), the sudden jump up of σ_{inel} reaching a very high peak at $M^2 \cong 2$ GeV^2, the rapid falloff of the mass spectrum, and the eventual flattening out of the cross section.

On very general grounds, we might expect

$$\frac{d\sigma(s)}{dtdM^2}\bigg|_{\text{fixed t}} = f(M^2) + \frac{g(M^2)}{s}, \text{ where we would call the}$$

first term diffractive and the second term scaling.

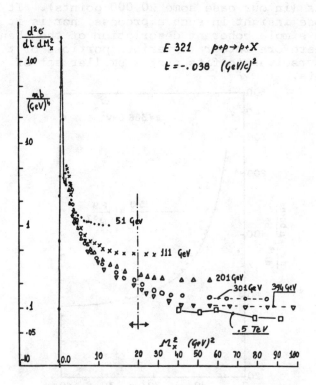

Fig. 6. $d^2\sigma/dtdM^2$ at $|t| = 0.038(GeV/c)^2$ for p_{lab}
from 51 GeV to 0.5 TeV.

For $M^2 \gtrsim 20$ GeV^2, it appears that $g(M^2) \approx$ constant.
This of course is far from surprising if one remembers
that in the "scaling term", we are summing over all
inelastic channels ($<n> = 13$). Furthermore, we cannot
anymore distinguish whether the observed "recoil" is
indeed the initial target proton at rest or rather one
of the fragments of X.
    There is one experiment at Fermilab[17] in the
proton area which covers in M^2 from 0.3 s onwards and
might come up with broad details of $g(M^2)$. The fact,
however, that $g(M^2)$ is approximately constant does
simplify the understanding of inelastic diffraction.
This point I would like to discuss more fully.
    Ideally, the experimental goal would be to deter-
mine a series of functions or parameters, such as
defined for example by

$$\frac{d^2\sigma(s)}{dtdM^2} = f(M^2,t)e^{b(t,M^2)t} + \frac{g(M^2,t)}{s}e^{b(t,M^2,s)t}$$

by parametrizing the above function and fitting to all

data points (in our case some 20,000 points). It is easy to lose insight in such a process, nor is it likely to yield a simple coherent description of the phenomena. We will therefore integrate various portions of t and M^2 to limit ourselves to fits over a smaller set of measurements.

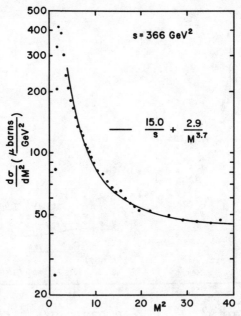

Fig. 7. Inelastic $d\sigma/dM^2$ at s = 366 GeV^2 and fitted curve.

In addition, some very relevant questions refer specifically to the values of the cross section for specific portion of the inelastic cross section. In particular, we would like to determine the total value of $\sigma_{inel\ diff}$. Not only are bounds now known for it, but plausible suggestions have been made that the rise of σ_{tot} is due to the rise of $\sigma_{inel\ diff}$. In addition, the mass spectrum of $d\sigma/dM^2$ after integrating over t, is of interest both for $\sigma_{scaling}$ and $\sigma_{diffractive}$. Of course, direct fits to single spectra are a guiding handle.

Our spectra show an extremely fast drop with increasing M^2, and give direct evidence for an energy independent contribution extending down to $M^2 \sim 2$ GeV^2 and precipituously dropping at $M^2 < 2$ GeV to the value of $d\sigma/dM^2$ at large M^2. This happens at values of M^2 well above the kinematical threshold (1.17 GeV^2). We have performed fits to our spectra with powers of (1/M) +

constant. Figure 7 is an example. When we reach the
highest energies, mass resolution broadening smears
somewhat the peak at 2 GeV^2. At lower energies, single
resonances are visible and make fits very poor. In the
spirit of duality, it is still very important to ask
about the average behavior of the mass spectra. This can
be answered (while avoiding the above mentioned complications) by integrating the measured distributions over t
and various portions of the M^2 spectrum, for instance, to
fixed fractions of s.

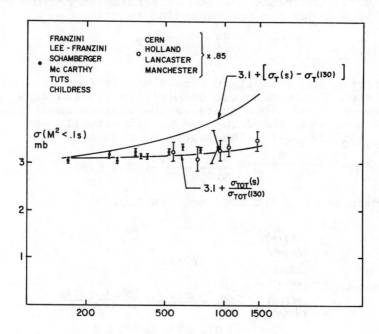

Fig. 8. σ_{inel}(M^2 < 0.1s) as a function of s.

Figure 8 shows such integrated cross sections vs. s.
The points with very small errors are our own and have
been obtained by integrating over t and M^2 our measure-
ments.[18]   Our data extend in |t| between 0.02 and 0.4,
hence to obtain total σ, we extrapolated the full t
range by multiplying our integrals by 1.15. The corresponding extrapolation needed for ISR[19] data (which covers
the range 0.25 < |t| < 0.45) is multiplying by a factor
of 7. Where we overlap with their results, there is
good agreement, aside from a small normalization difference of 15%. Hence to put both sets of data on one
graph, I have scaled their results by 0.85, they are the
open circle on the graph. Also shown are two lines, one
the expectation that σ_{diff} rises as σ_{tot}, the other that

σ_{diff} rises fractionally like σ_{tot}. The conclusion is obviously that σ_{diff} is not the origin of the rise in σ_{tot}. This is quite obvious from our data alone.

The next question is the separation of σ_{diff} (energy independent) and $\sigma_{scaling}$. From the data, it is apparent that the low M2 bump is energy independent. To complete the separation, one needs the shape of the low M2 bump. We found that it is quite possible to fit each spectrum with just two terms, $b(1/M^2)^\gamma + a$ with γ typically ≈ 2. A global fit to all the data is to simply fit the cross section integrated in M^2 to various fractions αs to the corresponding function

$$F(\alpha,s) = \int_{1.2}^{\alpha s} \frac{a}{s} + b \int_{2}^{\alpha s} (1/M^2)^\gamma dM^2 \ .$$

No information is lost in such a way as long as s and α span enough values to cut deeply into the low mass enhancement. And since there is the obvious small rise trend observed, the form used to fit actually was $(\sigma_{tot}/\sigma_{tot}^{min}) * F(\alpha,s)$. The result of such a fit is:

$$\frac{d\sigma}{dM^2} = (\frac{15.4 \pm 0.2}{s} + \frac{3.54 \pm 0.3}{4.06 \pm 0.2}) \times (\frac{\sigma_{tot}}{\sigma_{tot}^{min}}) \ .$$

The data and the fit are shown in Fig. 9.

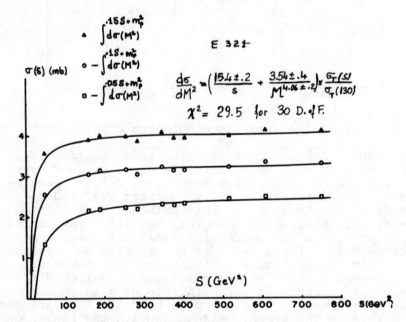

Fig. 9. $\sigma(\alpha s, s)$ vs. s and fit.

The simplicity of this result is remarkable. Consider the scaling term. If it were valid for the whole range of M^2, $1 < M^2 < s$ (we measured it for a quarter of that range), its integral would be 15.4 mb. Adding the contribution for $pp \to n + X$, we obtain for $\sigma_{inel} = 29.0$ mb from the scaling term.

The energy independent term is given by $(d\sigma/dM^2) \sim 3.54/M^4$, therefore it is dimensionally correct without introduction of a new length or mass. This is very satisfactory because we expect this term to be due to diffraction excitation which depends on the proton opacity, itself a dimensionless quantity. Its integral over phase space

$$\int_2^s \frac{3.54}{M^4} \, dM^2 \sim \frac{3.54}{2} = 1.77 \text{ mb}$$

gives an unambiguous value for the diffractive cross section. Finally, because of pp symmetry, an equal value must come from $pp \to Xp$ (as opposed to $pp \to pX$). We obtain therefore $\sigma_{diff} = 3.54$ mb. In the total σ_{inel}, this value has to be added only once, $pp \to pX$ being already counted. Therefore we obtain $\sigma_{inel} = 29.0 + 1.77 = 30.8$ mb, as compared to the measured value being 31.3 mb, not bad at all.

We must at this point mention that most other experiments claim the spectra at low mass to be of the form $1/M^2$ (the exception being bubble chamber experiments at 60, 100, 400 GeV).[20] We are convinced that some authors were plainly misled to this conclusion by the limited M^2 range of their measurement. It should be remarked that a $1/M^2$ spectrum gives values rising with s as $\log(s/(2 \text{ GeV}^2))$ for the integral,

$$\int_2^{\alpha s} (d\sigma/dM^2) \, dM^2,$$ for any value of α, in complete disagreement with the data.

While the results presented so far are beautiful in their simplicity, nature in detail is never simple. If we look carefully at the low mass enhancement, we quickly discover that it is ridden with bumps.[21] They are, in fact, mostly proton resonances and particularly those with the same isospin and parity (apart from a factor of $(-1)^\ell$) as the proton. That is, they are those resonances which can be diffractively produced. In addition, there is the special bump at 1400 MeV ($M^2 = 2$ GeV^2). This is well known not to be a resonance (no sharp phase change has ever been observed). It has also the peculiar property of being produced within an extremely narrow forward cone in t. In Fig. 10, we see an example of the resonance structures in the M spectra. We have also determined their production cross section and logarithmic slopes for $s = 20$ to 750 GeV^2.

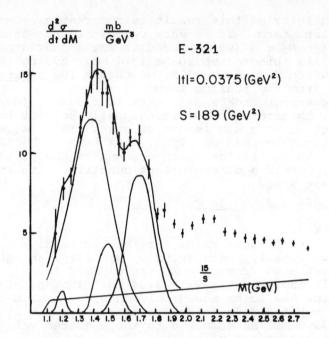

Fig. 10. $d^2\sigma/dtdM$ vs. M. Contributions of scaling term; first 3 resonances and "1400" are shown.

In particular, $\sigma(1400) = 0.80 \pm 0.02$, $b(1400) = 20.5 \pm 1$, $\sigma(1520) = 0.45 \pm 0.02$, $b(1520) = 4 \pm 1$, $\sigma(1688) = 0.80 \pm 0.04$, $b(1688) = 6.6 \pm 1$. In Fig. 11 we compare our measured cross sections with those measured previously.[22-26]

Going back to the diffractive model, the Pumplin Bound stated: $\sigma_{el} + \sigma_{diff} \leq (1/2)\sigma_{tot}$ and this is valid for the unmeasurable $\sigma(b)$ at each impact parameter, $\sigma_{el}(b) + \sigma_{inel\ diff}(b) \leq 1/2\ \sigma_{tot}(b)$. From this and from the analysis of the proton absorption function, one obtains a more stringent limit because unitarity is saturated by elastic scattering at $b \approx 0$. That is, the bound for $\sigma_{inel\ diff}$ is peaked toward impact parameters ≈ 1 fermi and its total integral is bounded to $\leq 8-9$ mb. Of course, our value of $\sigma_{diff} = 3.54$ mb is far from saturating the bound, contrary to the situation when the whole integral $\int^{0.1s}(d\sigma/dM^2)dM^2$, which is $\approx 7-8$ mb (and contains a scaling contribution), was identified with σ_{diff}.

It would be most interesting to determine $\sigma_{diff}(b)$ from data. For example, given that b for the (1400) is ≈ 20, b for $\sigma_{el} \approx 13$, for $\sigma_{resonance}$ is ≈ 6.5 the following composite picture emerges: the 1400 is peripherally produced, resonances from the center, and elastic scattering in between, etc. It is clearly just becoming possible to study such details of the opacity function of the proton. Our own data contain much more information than presented here and new data are forthcoming from many places. It is also extremely important to go to higher laboratory energy (1000 GeV at Fermilab) because colliding beams offer no hopes to study these phenomena ($\Delta M^2 = 50$ GeV^2 at 500 GeV + 500 GeV colliding beams).

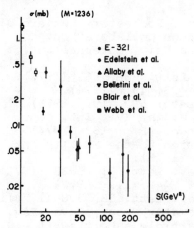

Fig. 11a.
σ(1236) vs s.

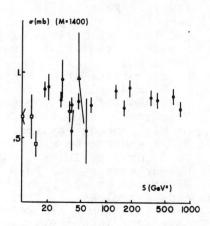

Fig. 11b.
σ(1400) vs s.

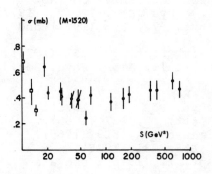

Fig. 11c.
σ(1520) vs s.

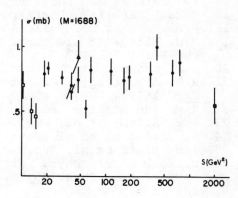

Fig. 11d.
σ(1688) vs s.

References

1. T.T. Chou, C.N. Yang, Phys. Rev. 170, 1591 (1968).
2. M.L. Good, W.D. Walker, Phys. Rev. 120, 1857 (1960).
3. J. Pumplin, Phys. Rev. D8, 2899 (1973).
4. S. Childress et al, Phys. Rev. Lett. 32, 389 (1974).
   R.D. Schamberger, Jr. et al, Phys. Rev. Lett. 34, 1121 (1975).
   S. Childress et al, Phys. Lett. 65B, 177 (1976).
   R.D. Schamberget, Jr. et al, submitted to Phys. Rev. (1977). Fermilab, E-321.
5. Fermilab E-198, University of Rochester, Rutgers U., and Imperial College of London Collaboration.
6. M.D. Corcoran et al, Indiana Univ. Internal Rpt. IUHEE #9.
7. G. Barbellini et al, Phys. Lett. 39B, 663 (1972).
   A. Bohm et al, Phys. Lett. 49B, 491 (1974).
8. C.W. Akerlof et al, Phys. Lett. 59B, 197 (1975).
9. V. Barger, Rapporteur Talk at the XVII Int. Conf. on High Energy Physics, London, P. I 193 (1974).
10. C. Arkendrant et al, submitted to the XVIIIth Int. Conf. on HEP, Tblisi, USSR (1976).
11. C. Bartenev et al, Phys. Rev. Lett. 31, 1088 (1973).
12. U. Amaldi et al, Phys. Lett. 44B, 112 (1973).
13. A.S. Carrol et al, Phys. Rev. Lett. 33, 928 (1974).
    Fermilab Single Arm Spectrometer Group, Phys. Rev. Lett. 35, 1406 (1975) and article submitted to P.R.
14. R.E. Hendrick et al, Phys. Rev. D11, 536 (1975).
15. H. Cheng, T.T. Wu, Phys. Rev. D3, 2195 (1971).
16. W. Grein et al, Nucl. Phys. B89, 93 (1975).
17. E-284, Fermilab, Northern Illinois Univ. Collab.
18. P.M. Tuts et al, Bull. Amer. Phys. Soc. 22, 21 (1977).
19. M.G. Albrow et al, Nucl. Phys. B72, 376 (1974) and PPF-Print-76-0238.
20. J.W. Chapman et al, Phys. Rev. Lett. 32, 257 (1974).
21. R.D. Schamberger, Jr. et al, Bull. Amer. Phys. Soc. 22, (1977).
22. R.M. Edelstein et al, Phys. Rev. D5, 1073 (1972).
23. J.V. Allaby et al, Nucl. Phys. B52, 316 (1975).
24. G. Belletini et al, Phys. Lett. 18, 167 (1965).
25. I.M. Blair et al, Nuovo Cimento 63A, 529 (1969).
26. R. Webb et al, Phys. Lett. B55, 331 (1975).
27. Rev. of Mod. Phys. 48, 2, pt. II (1976); see also Ref. 13a.

EFFECTS OF RECENT MEASUREMENTS ON PHASE
SHIFT ANALYSIS OF NUCLEON-NUCLEON SCATTERING

Richard Arndt
Virginia Polytechnic Institute and State University
Blacksburg, Virginia 24061

Four recent measurements in p-p and n-p scattering below 520 MeV are used to indicate the influence that new experiments can have upon phase parameters derived from the expanded data base. The cases are described separately and the collective effect upon energy dependent analyses is discussed. The representations used in this discussion are from the recent VPI analyses of 1135 pp data from 1 to 500 MeV, and 1386 np data from 0 to 425 MeV.[1] The analyses consisted of an Energy Dependent Analysis (EDA) involving 46 variable parameters and producing an overall χ^2 of 2815 for the 2670 data. The EDA was complemented by Single Energy Analyses (SEA) of data binned around 7 energies. The local energy dependence within a bin was that given by the EDA; some results of those analyses are given in Table I of Ref. 1 which summarizes the binning and fitting through SEA's and Table II which gives the resultant phase parameters at 7 energies. Some indication of how well the SEA minima could be described by a quadratic expansion in the parameter space (the usual error matrix approximation) can be seen in the last column of Table I, where the 2n derivative matrix was used to predict χ^2 for the binned data and the EDA fit; it compares favorably to the actual (data) values. Figure 4, Ref. 1, indicates the distribution of n-p scattering data which is used for the determination of I=0 phases. It is certainly the weaker component of the data base consisting almost entirely of $\sigma(\theta)$, σ_T and P with larger experimental uncertainties than for corresponding p-p quantities.

Although we use these particular analyses to measure the effect of newer data, the conclusions are certain to be generally valid when applied to other analyses. The results of these studies, which are based upon preliminary data, should not be taken as a final indication of the phase parameters which will be obtained when the data is properly incorporated into the data base.

I. p-p Cross Sections and Polarizations 285-497 MeV

This data by Aebischer et al.[2] consist of 162 values of $\sigma(\theta)$ and 50 values of polarization. Table I indicates the influence which this data has upon the EDA. The cross-sections are very small angle ($\theta_{cm} < 20°$), very precise, but renormalizable data. They produce a very large χ^2 mismatch with the published solution ($\chi^2 = 2146$) but are easily accommodated ($\chi^2 \rightarrow 272$) with a relatively small change in the variable parameters. The polarization was reasonably fitted by the old EDA ($\chi^2 = 84$); searching reduces this to 61.

II. Axen et al.[3] P, D, R, R' (pp) from 209 MeV to 518.4 MeV
71 data, 27 experiments

This data is not well fitted by the published EDA ($\chi^2 = 204$)

ISSN: 0094-243X/78/117/$1.50

with the fit becoming, as expected, progressively worse with increasing energies (χ^2 of 26 for the 4 pol @ 518 MeV). This was expected since the published analyses were done upon a data base which stopped at 500 MeV and which was very weak above 400 MeV. The situation improved somewhat upon reanalysis ($\chi^2 \rightarrow 135$); however, there remains a mismatch with the "high" energy polarization ($\chi^2 = 19$ for the 518 MeV data). The Axen data was incorporated with the published data in order to do the "binned" analyses shown in Table II. An analysis was performed at 500 MeV using 94 data between 460 MeV and 519 MeV; the results are shown in Table III and indicate a disparity between the EDA and SEA which needs to be resolved through further careful studies. There is no question, however, that this data will have a substantial effect upon the p-p phase parameters above 400 MeV.

III. Davis n-p $\sigma(\theta)$ @ 63.1 MeV[4]

These data fell somewhat outside of the previous single energy binning limits (40-60 MeV) so the bin limits were extended to just include this data in the hope of shedding some light upon the (ε_1, $^1P_1$) indeterminacy.[5] This brings in an additional 24 "old" data. Analysis results are shown in Table IV. The error on $^1P_1$ is reduced by a factor of 2 but the value is virtually unaffected. An ε_1 mapping done upon the expanded "old" data is shown in Fig. 1. When the new $\sigma(\theta)$ was included, the results, shown in Fig. 2, reveal a double solution but with the preferred one having a χ^2 about 20 below the second, clearly unphysical, solution. Since the binning is quite large and we are in a region where things are changing rapidly, it would appear necessary to re-do the EDA and then use the improved local energy dependence to do a better job with this data.

IV. n-p Pol, D_t at 325 MeV Amsler et al.[6]

This data, primarily the 8 D_t data, promises to have a dramatic influence on the phase shifts at 325 MeV. The data was incorporated with the older data and an I=0 analysis performed with the I=1 phases fixed from the EDA as published. The 9(I=0) parameter analyses are given in Table V. The dominant change is in ε_1, which drops to half its previous value and which is seemingly well defined around the new minimum (see Fig. 3). Some detailed mappings were then done to establish whether the new solution was unique or whether the multiple structure previously reported upon[5] still persists. Such mappings were done in a number of variables (ε_1, $^1P_1$, $^3G_4$, $^1S_{np}$); the most interesting results were obtained in the $^3G_4$, $^1S_{np}$ maps, shown in Fig. 4. They reveal a very strong correlation between these two parameters with $^1S_{np}$ changing by 8° from -25.6 to -17.4 as $^3G_4$ changes from 11.25° to 7°, resulting in χ^2 increase of only 2. It is apparent that the new D_t data does not completely clarify the situation at 325 MeV. Table VI presents three of (probably) numerous solutions (grad $\chi^2 = 0$) which can still be found in an analysis of the 325 MeV data. Hopefully the new R_t data being taken by the TRIUMF group will remedy this situation. Figure 5 illustrates a fit (through SEA) to the new D_t data. The dashed curve is the prediction of the recent VPI analysis which did not include this data and which obtained a much larger value for ε_1 at 325 MeV.

V.   Effects of New Data on Energy Dependent Analysis

Incorporation of the new np data into the data base and subsequent reanalysis gives results summarized in Table VII.   Some phase parameters are included in Table VII to indicate the general effects which the new data will have.   Figure 6 illustrates the changes found in some np phases.   Substantial changes are indicated in ε_1 and in $^1S_{np}$ (which gives a value very close to $^1S_{pp}$ at higher energies) while the $^1P_1$ phase was virtually unaffected by the new data.

Summary

It is very clear that recent data on pp and np scattering will have a substantial influence upon phase parameters derived from analysis of the expanded data base.   These preliminary studies have indicated the types of change which will be forthcoming and they are far from negligible.

References

1. "Nucleon-Nucleon Scattering Analyses.   II.   Neutron-Proton Scattering from 0 to 425 MeV and Proton-Proton Scattering from 1 to 500 MeV", R.A. Arndt, R.H. Hackman and L.D. Roper, Phys. Rev. C 15, 1002 (1977).
2. "Small Angle p-p Elastic Scattering at Energies between 285 and 572 MeV", D. Aebischer, B. Favier, L.G. Greeniaus, R. Hess, A. Junod, C. Lechanoine, J.C. Niklès, D. Rapin, C. Richard-Serre and D. Werren, Phys. Rev. D 13, 2478 (1976); "Polarization Measurements in p-p Elastic Scattering between 398 and 572 MeV", Nuclear Physics (1977).
3. "D, R, R' and P in p-p Elastic Scattering from 209 to 515 MeV", D. Axen, L. Felawka, S. Jaccard, J. Va'vra, G. Ludgate, N. Stewart, C. Amsler, R. Brown, D. Bugg, J. Edgington, C. Oram, K. Shakarchi and A. Clough, Nuovo Cimento Lett. 20, 151 (1977).
4. "Neutron-Proton Differential Cross-Section Measurements at 63 MeV", S.W. Johnsen, F.P. Brady, N.S.P. King and M.W. McNaughton, submitted to Phys. Rev. Lett.
5. "Nucleon-Nucleon Scattering Analyses.   III. Phase-Shift Analyses: Complex Structure and Multiple Solutions at 50 and 325 MeV", R.A. Arndt, R.H. Hackman and L.D. Roper, Phys. Rev. C 15, 1021 (1977).
6. "Neutron-Proton Scattering at 325 MeV", C. Amsler, D. Axen, J. Beveridge, D.V. Bugg, A.S. Clough, J.A. Edgington, S. Jaccard, G. Ludgate, C. Oram, J.R. Richardson, L. Robertson, N. Stewart, J. Va'vra, Phys. Lett. 69B, 419 (1977).

Table I. Effect of recent p-p data on Energy-Dependent Representation 1-500 MeV. Bracketed quantities are numbers of data; table entries are χ^2.

	Before Search	After Search
"Old" data (1284)	1135	1201
Axen (71)	204	135
Aebischer P (50)	84	61
Aebischer σ (162)	2146	272
Total	3569	1169

Table II. Single energy p-p analyses with recent Axen Data at 200, 325, 425 and 500 MeV. Table entries are χ^2. Numbers of data are given below slashes and the χ^2 from earlier analyses are given in brackets.

T_{LAB}	Nδ	Old	Axen	ε_4
200 Mev	0	59/64	13/10	-.95
(209-213)	9	53 (51)	7	-.95
	11	53	6	-1.03±.074
325 MeV	0	145/159	18/17	-1.59
(305-345)	9	138 (137)	11	-1.59
	11	132	6	-1.63±.14
425 MeV	0	156/161	42/17	-2.21
(394-460)	9	149 (141)	23	-2.21
	11	151	20	-2.06±.18
500 MeV	0	56/77	94/17	-2.72
	9	49 (*)	19	-2.72

Table III. Single energy pp phases at 500 MeV (460-520) with recent p-p data included. The columns labeled (Sch), (before Sch) are the phase parameters from EDA after and before incorporation of the new data.

$^1S_0$	-18.5 ± 3.6	-24.5 (Sch)	-24.5 (before Sch)
$^1D_2$	$14.3\pm.9$	14.6	13.9
$^3P_0$	-37 ± 4	-26.6	-26.7
$^3P_1$	-34.5 ± 2	-37.5	-38.3
$^3P_2$	$18.9\pm.8$	17.2	16.3
ε_2	$-.54\pm.5$	$-.7$	$-.25$
$^3F_2$	$.27\pm1$	$-.26$	$.52$
$^3F_3$	-3.1 ± 1.4	-2.8	-3.69
$^3F_4$	$4.8\pm.28$	4.7	4.17

Table IV. n-p analyses at 50 (40-63.1) MeV with new Davis $\sigma(\theta)$. Columns labeled SE (SE*) are from single energy analyses without (with) the new $\sigma(\theta)$ and extra data included from bin widening. Columns labeled ED (ED*) are from Energy Dependent Analyses without (with) new np data.

	SE	SE*	ED	ED*
χ^2 (old)	176	190	196	200
χ^2 (extra)		30	38	39
χ^2 (Davis)		42	85	63
$^1S_{n_p}$	43.7 ± 5.7	45.9 ± 4.3	40.5	40.34
$^1P_1$	-4.21 ± 1.58	$-4.16\pm.70$	-6.6	-6.67
$^3S_1$	61.37 ± 2.69	58.01 ± 1.56	63.02	63.1
ε_1	$-.92\pm2.2$	$2.97\pm.85$	$-.8$	$.23$
$^3D_1$	-5.36 ± 1.21	$-7.76\pm.64$	-6.25	-6.17
$^3D_2$	9.17 ± 1.19	$11.93\pm.78$	9.35	9.19
$^3D_3$	$1.26\pm.62$	$-.03\pm.35$	$.65$	$.73$

Table V. n-p analyses at 325 (300-350) MeV. Columns labeled SE (SE*) are single energy analyses without (with) new $P(\theta)$ and D_b. Columns labeled ED (ED*) are from Energy Dependent Analyses without (with) new np data.

	SE	SE*	ED	ED*
χ^2 (old)	83	86	117	101
χ^2_p	–	46	160	77
$\chi^2_{D_t}$	–	5	96	7
$^1S_{np}$	(-4)	-19.4 ± 2.2	-4	-11.9
$^1P_1$	-35.8 ± 1.1	-33.4 ± 1	-34.2	-35.1
$^3S_1$	-2.2 ± 2.4	2.8 ± 1.8	-1.3	-.15
ϵ_1	15.7 ± 1.5	$6.6\pm.96$	12.3	6.92
$^3D_1$	-21.9 ± 1.3	-25.6 ± 1.1	-19.8	-23.4
$^3D_2$	21 ± 1.2	21.9 ± 1.1	25.4	24.69
$^3D_3$	$2.9\pm.8$	$2.75\pm.58$	4.3	3.38
$^1F_3$	$-2.7\pm.54$	$-5.8\pm.47$	-2.9	-4.2
ϵ_3	$7.11\pm.52$	$8.1\pm.4$	6.5	7.39

Table VI. Three of many np solutions at 325 MeV with recent np $P(\theta)$ and $D_t(\theta)$ data.

	A	B	C
	137.99	138.17	140.12
$^1S_{np}$	-25.86 ± 4	-22.53 ± 3.8	-38.5 ± 4.5
$^1P_1$	-26.4 ± 3	-29.1 ± 3.7	-21.1 ± 2.1
$^3S_1$	$.14\pm2$	$.86\pm2.4$	1.5 ± 1.5
ϵ_1	7.9 ± 1.1	7.6 ± 1.1	8.1 ± 1
$^3D_1$	-24.5 ± 1.2	-24.9 ± 1.26	-25.2 ± 1.1
$^3D_2$	22.9 ± 1.1	23.02 ± 1.15	19.2 ± 1.6
$^3D_3$	$2.8\pm.63$	$2.71\pm.65$	$1.93\pm.66$
$^1F_3$	$-7.57\pm.91$	-6.8 ± 1.1	$-9.9\pm.68$
ϵ_3	6.05 ± 1.28	6.9 ± 1.2	4.29 ± 1.1
$^3G_4$	11.23 ± 1.59	9.8 ± 1.9	13.9 ± 1.1
$^3G_3$	-4.6 ± 1.02	-4.06 ± 1	-4.8 ± 1

Table VII. Effects upon energy dependent representation of the incorporation of recent np measurements. The "Duke" data are 7 preliminary polarizations which were included in the analysis but which are not discussed in the text. Columns marked old (new) are results of analyses without (with) new np data.

	Old	New
χ^2_{old} (1390)	1683	1723
χ^2_{Duke} (7)	21	21
χ^2_{Davis} (16)	85	63
$\chi^2_{Amsler\ P}$ (42)	160	77
$\chi^2_{Amsler\ D_t}$ (8)	96	7
ε_1 (50)	-8	.2
ε_1 (350)	13.3	7.4
$^3D_1$ (50)	-6.3	-6.2
$^3D_1$ (350)	-19.7	-23.6
$^3D_3$ (50)	.65	.73
$^3D_3$ (350)	4.3	3.3
ε_3 (50)	1.73	1.76
ε_3 (350)	6.63	7.6

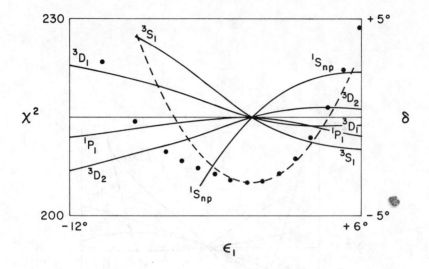

Fig. 1. ϵ_1 mapping at 50 (40-63.1) MeV based upon "old" data. The dots are actual χ^2's while the dashed curve is the quadratic approximation. Some of the correlated phase are shown.

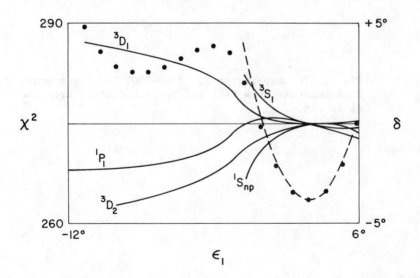

Fig. 2. ϵ_1 mapping at 50 (40-63.1) MeV based upon "old" data plus new $\sigma(\theta)$ at 63.1 MeV. Symbols are the same as in Fig. 1.

126

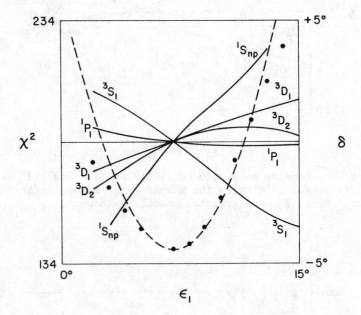

Fig. 3. ϵ_1 mapping at 325 (300-350) MeV based upon data including new $P(\theta)$ and $D_t(\theta)$. Symbols are same as in Fig. 1

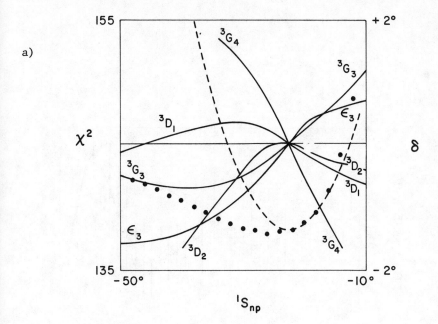

a)

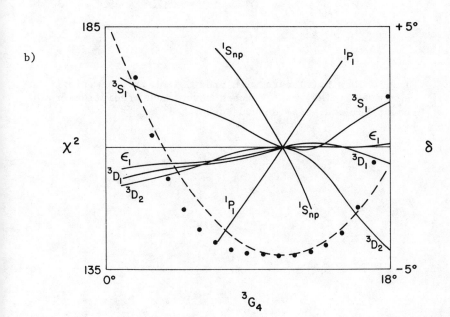

b)

Fig. 4. $^1S_{np}$ and $^3G_4$ mappings at 325 (300-350) MeV. Data include
new $P(\theta)$ and $D_t(\theta)$. Symbols are same as in Fig. 1.

128

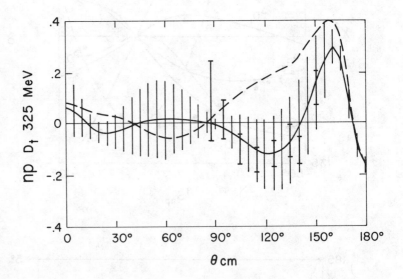

Fig. 5. New $D_t(\theta)$ data with predictions from new single energy analysis (cross hatched curve) and prediction from recent VPI analysis (dashed curve).

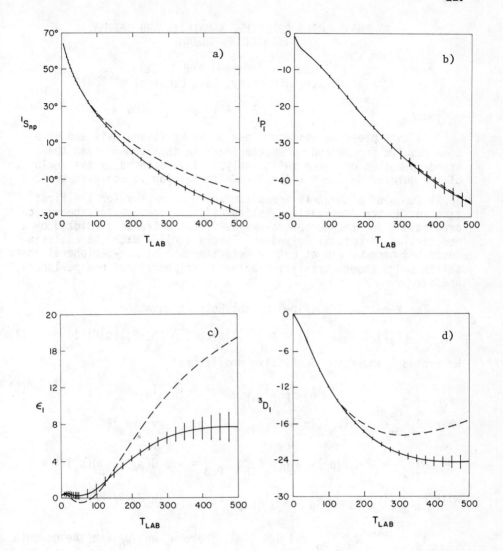

Fig. 6. Some energy dependent phases derived from analysis of expanded (including recent np data) data base. The cross hatched curves are new values with "errors", while the dashed curves are values from recent VPI analysis.

NN PHASE SHIFT ANALYSIS, K-MATRIX, AND OPTIMAL POLYNOMIAL THEORY

T. Rijken, P. Signell and T. Burt
Michigan State University, East Lansing, MI 48824

ABSTRACT

In this paper we discuss a new phase shift analysis and give some results for pp and np scattering. An improvement over the standard method of phase shift analysis is presented by the inclusion of the cut for $4m_\pi^2 \leq t \leq 45m_\pi^2$. Unitarization is achieved through application of a K-matrix formalism where we employ for the first time in NN, the "derivative amplitudes". We compare the theoretical predictions for the peripheral waves with the trends predicted by a new version of Optimal Polynomial Theory (OPT). With the analysis described here we aim at better determination of non-peripheral phase shifts and a smooth transition between peripheral and non-peripheral phases.

The K-matrix is defined by the Heitler equation

$$(f|T|i) = (f|K|i) - i \sum_c (f|T|c)(2\pi)^4 \delta^4(P_c - P_i)(c|K|i). \tag{1}$$

We expand T and K in "derivative amplitudes"[1]

$$T = \sum_{\alpha,\beta} I^*_{Q,\alpha} I_{P,\beta} T_{\alpha,\beta} \quad , \quad K = \sum_{\alpha,\beta} I^*_{Q,\alpha} I_{P,\beta} K_{\alpha,\beta} \tag{2}$$

where $I_{P,0} = -c_0 \bar{\tilde{u}}(p_b) u(p_a)$, $I_{P,1} = -ic_1 \bar{\tilde{u}}(p_b) \gamma_5 \gamma_\mu u(p_a) P^\mu$,

$$I_{P,2} = -ic_2 \bar{\tilde{u}}(p_b) \gamma_5 \gamma_\mu u(p_a) L_P^\mu, \quad I_{P,3} = -ic_3 \bar{\tilde{u}}(p_b) \gamma_5 \gamma_\mu u(p_a) T_P^\mu$$

with $P_\mu = \frac{1}{2}(p_a - p_b)_\mu$, $W_\mu = (p_a + p_b)_\mu$, $L_P^\mu = \epsilon^{\mu\nu\rho\sigma} W_\nu P_\rho (\partial/\partial P^\sigma)$,

$T_P^\mu = \epsilon^{\mu\nu\rho\sigma} W_\nu P_\rho L_{P\sigma}$ and $\tilde{u} \equiv C^T u^T$. Here p_a and p_b are the momenta of the initial state and C is the charge conjugation matrix. In (2) Q refers to the final state and the $I_{Q,\alpha}$ are defined analogously to the $I_{1,\alpha}$. The factors c_α are chosen such that the following ortho-normality relations are satisfied:

$$\sum_{\text{spins}} I^*_\alpha I_\beta = n_\alpha \delta_{\alpha\beta} (\alpha, \beta = 0,1,2,3) \tag{3}$$

with $n_0 = n_1 = 1$, $n_2 = n_3 = \ell^2 = -\frac{d}{dx}(1-x^2)\frac{d}{dx}$, where $x = \cos\theta_{CM}$.

Due to parity and total spin conservation we need only compute the

ISSN: 0094-243X/78/130/$1.50

amplitudes $T_{\alpha,\beta} = \{T_{0,0}, T_{2,2}, T_{1,1}, T_{3,3}, T_{1,3}\}$ and $K_{\alpha,\beta} = \{K_{0,0}, K_{2,2}, K_{1,1}, K_{3,3}, K_{1,3}\}$.

The amplitudes $T_{\alpha,\beta}$ and $K_{\alpha,\beta}$ are free of kinematical singularities in x^1 and the partial wave expansions read

$$T_{\alpha,\beta}(E,x) = \sum_J (2J+1) T^J_{\alpha,\beta}(E) P_J(x) \tag{4}$$

and similarly for $k_{\alpha,\beta}$. The connection with the parity-conserving helicity partial waves[2] is

$$f_0 = T^J_{0,0}, \quad f^J_1 = J(J+1) T^J_{2,2}$$

$$f^J_{11} = T^J_{1,1}, \quad f^J_{22} = J(J+1) T^J_{3,3}, \quad f^J_{12} = \sqrt{J(J+1)} \, T^J_{1,2}. \tag{5}$$

Due to (3) and (4) equation (1) can be reduced to

$$T^J_{\alpha,\beta} = K^J_{\alpha,\beta} - i \frac{p}{\sqrt{s}} \sum_\gamma T^J_{\alpha,\gamma} n_\gamma(J) K^J_{\gamma,\beta} \tag{6}$$

with $n_0(J) = n_1(J) = 1$, $n_2(J) = n_3(J) = J(J+1)$. In (6) p = CM momentum and $s = 4 E^2 = W^2$. Introducing

$$g^J_0 = K^J_{0,0}, \quad g^J_1 = J(J+1) K^J_{2,2}$$

$$g^J_{11} = K^J_{1,1}, \quad g^J_{22} = J(J+1) K^J_{3,3}, \quad g^J_{12} = \sqrt{J(J+1)} \, K_{1,3} \tag{7}$$

one has

$$f^J_0 = g^J_0/(1 - i \frac{p}{\sqrt{s}} g^J_0), \quad f^J_1 = g^J_1/(1 - i \frac{p}{\sqrt{s}} g^J_1) \tag{8}$$

and similar relations for the triplet coupled waves.

The functions $K_{0,0}(x)$, $\ell^2 K_{2,2}(x)$, $K_{1,1}(x)$, $\ell^2 K_{3,3}(x)$, and $K_{1,3}(x)$ are analytic in $x = \cos\theta_{CM}$, having one-pion-exchange poles and cuts starting at the 2π-thresholds. Removing the one-pion-exchange poles and introducing for the remainders the signatured functions $\tilde{K}^\pm(x)$, which have only a right hand cut in x, we map the cut x-plane onto an ellipse and make the Optimal Polynomial Expansion:

$$\tilde{K}^\pm_{0,0}(x) = \sum_n a^\pm_n z(x)^n \quad \text{etc.} \tag{9}$$

Using these expansions we compare the trends for peripheral waves (referred to as OPT in the figures) with the corresponding theoretical ones. The input for the new OPT consists of the MAW-X phases.[3]

In the phase shift analysis for each energy we describe the amplitudes by

$$K(x) = K_{1\pi}(x) + K_{2\pi}(x) + \tilde{K}(x) \tag{10}$$

where $K_{1\pi}$ = One-Pion-Exchange, $K_{2\pi}$ = Two-Pion-Exchange, and $\tilde{K}$ denotes the rest. $K_{2\pi}$ describes the contributions of the 2π-cut for $4m_\pi^2 \leq t \leq 45m_\pi^2$ and in this paper we use the cuts from the dispersion calculation by Bohannon and Signell[4](BS). The input in the latter work consists of $\pi\pi \to \pi\pi$, $\pi N \to \pi N$, and γNN-form factor data.

We parameterize $\tilde{K}(x)$ by expansions similar to (9) with coefficients determined by the searched non-peripheral waves. However in this paper we neglect the 3π-, and 4π- cuts below $t = 45m_\pi^2$, which makes the contribution of $\tilde{K}(x)$ in (10) to the peripheral waves very small.

In Figures 1-3 we investigate for what energies one can fix the peripheral phases by $K_{1\pi} + K_{2\pi}$ (denoted by BS). It appears that we can fix $^1G_4$, $^3F_4$, ε_4, ... and $^1F_3$, ε_3, $^3G_3$, ... for $T_{Lab} \lesssim 150$ MeV.

Taking into account the 3π- and 4π- thresholds via $\tilde{K}(x)$ in (10) one can improve these results, extending them to energies beyond 210 MeV. In Figure 4 we show the $^3P_0$ phase. The figures also compare our results with those of Arndt et al.[5]

Hopefully we can apply the present method at higher energies and for example eliminate multiple solutions. Also we hope to extend this analysis above the one pion production threshold where the necessity of a reduction of the number of free parameters in a phase shift analysis is much more compelling.

REFERENCES

1. V. DeAlfaro et al, Ann. of Phys. 44, 165 (1967).
   C. Rebbi, Ann. of Phys. 49, 106 (1968).
2. G. Goldberger et al, Phys. Rev. 120, 2250 (1960).
3. M.M. McGregor et al, Phys. Rev. 182, 1714 (1969).
4. G. Bohannon and P. Signell, Phys. Rev. D10, 815 (1974).
5. R.A. Arndt et al, Phys. Rev. C15, 1002 (1976). We also checked our phase shift analysis program against that of Arndt et al. We used input from a current (6/10/77) computer run of Arndt et al and then compared our program's output to Arndt's. The predictions checked to high accuracy. In a similar manner we simultaneously checked the input $A(\theta)$ output presented graphically in Binstock and Bryan, Phys. Rev. D8, 1397 (1973): the results checked to the accuracy with which we could read the graphs.

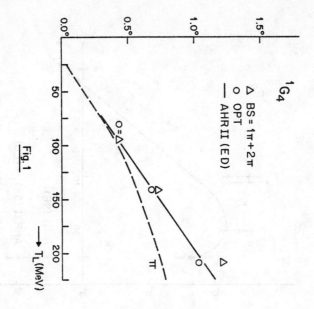

Fig. 1

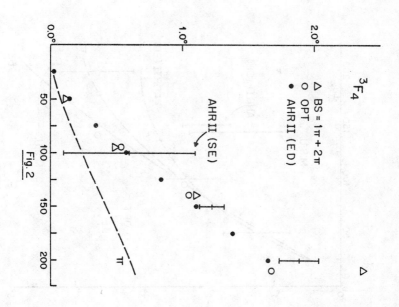

Fig. 2

134

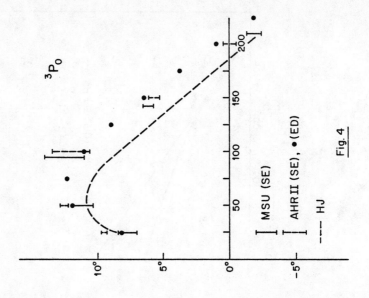

Fig. 4

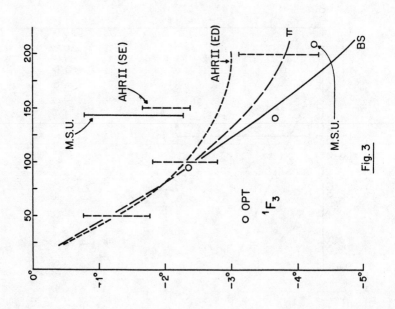

Fig. 3

ON THE SPIN DEPENDENCE OF THE NUCLEON-NUCLEON INTERACTION

P. Kroll
Phys. Dept., University of Wuppertal

E. Leader and W. von Schlippe
Westfield College, London NW3

ABSTRACT

By utilizing a dispersion theoretic calculation of the isospin zero helicity non-flip amplitude to supplement the experimental data, and by making physically reasonable assumptions about isospin one dominance of certain helicity flip amplitudes we have been able to obtain a unique, stable set of amplitudes in the region $-0.7 \leq t \leq -0.1$ GeV^2 at p_L = 6 GeV. We further demonstrate the importance of the phase of the non-flip amplitudes in understanding the large t behaviour of P, C_{NN} and C_{SS}.

INTRODUCTION

With the polarized proton beam at the Argonne National Laboratory it has become possible - at least in principle - to measure all observables of proton-proton scattering at medium energies. A great number of observables have already been measured and a lot of new results have been presented at this conference[1].

However, despite the great surge of activity there are still not enough measurements to allow an unambiguous model-independent amplitude analysis. However, believing in certain reasonable theoretical ideas, one can find a unique set of amplitudes at p_L = 6 GeV. From these amplitudes predictions for observables not yet measured can be made. Future measurements of them will provide crucial tests of the theoretical input.

We have also discussed the shape and the energy dependence of polarization and correlation parameters at large $|t|$, say for $|t|$ between 0.6 and 2.2 GeV^2. Relying on Regge-like behaviour of helicity flip amplitudes it can easily be demonstrated that the appearance of dips with rising energy is due to the rapid vanishing of terms proportional to the real part of the non-flip amplitude.

BASIC ASSUMPTION

For $p_L \geq 3$ GeV the pp differential cross-section is dominated by the non-flip amplitude N_o +). As can be seen by inspection of the
_____

+)We use the combinations N_i and U_i of s-channel helicity amplitudes. The index i denotes the net helicity flip whereas N(U) indicates that at high energies only exchanges with natural(unnatural) parity contribute.

ISSN: 0094-243X/78/135/$1.50 Copyright 1978 American Institue of Physics

spin dependent data the approximation $d\sigma/dt = |N_o|^2$ seems to hold on the percent level. This allows a determination of N_o independently of all other amplitudes by means of a phase-modulus dispersion relation. This has been done by Grein et al[2] and we make use of their results. The knowledge of N_o and its dominance is of great help in the amplitude analysis as will be seen in the following.

Our second assumption is the decoupling of the Pomeron from flip amplitudes and the neglect of ω and f contributions to double flip amplitudes. This assumption is in agreement with SU (3) symmetry arguments and with results of amplitude analysis in πN and KN scattering. The most important consequence of it is that only $I = 1$ exchanges are allowed in the double flip amplitudes, hence charge exchange data can be included in the analysis.

AMPLITUDE ANALYSIS AT 6 GeV

At 6 GeV taking into account the above assumptions we have 8 experimental quantities at our disposal to determine the t-channel isospin one amplitudes, namely the differential cross-sections and polarizations for both charge exchange reactions $pn \rightarrow np$ and $p\bar{p} \rightarrow n\bar{n}$ respectively, and three components of amplitudes - parallel ($\parallel$) or perpendicular ($\perp$) with respect to N_o from elastic pp and pn data

$$N^1_{1\perp} = -\frac{1}{4}\sqrt{\frac{d\sigma}{dt}} (P_{pp} - P_{pn}); \quad N_{2\parallel} = \frac{1}{2}C_{NN}\sqrt{d\sigma/dt}; \quad U_{2\parallel} = \frac{1}{2}C_{SS}\sqrt{d\sigma/dt} \quad (1)$$

and finally the C_{LL} data

$$C_{LL}\sqrt{d\sigma/dt} = -2\text{Re }(N^*_o U_o) - 2\text{Re }(N^*_2 U_2) \quad (2)$$

As an additional constraint we included the charge exchange differential cross-sections at 4o GeV. It turned out that this constraint strongly stabilizes the results.

The analysis has been performed in terms of amplitudes with definite t-channel quantum numbers. For these we assumed the usual Regge phases. So, such an amplitude has been parametrized as

$$N_i (R,t) = (\sqrt{-t})^i \beta_R(t) \xi_R p_L^{\alpha_R-1} \quad (3)$$

and similarly for the U_i. (ξ_R being a signature factor). At selected t values we searched for the parameters β_R. The energy dependence is active only via the 4o GeV constraint. The π contribution was parametrized according to the "poor man's absorption model"

$$N_2(\pi,t) = -\beta_\pi(t) \xi_\pi p_L^{\alpha_\pi-1}; \quad U_2(\pi,t) = (\frac{2t}{t-\mu^2} -1)\beta_\pi(t)\xi_\pi p_L^{\alpha_\pi-1} \quad (4)$$

(μ being the π mass).

Besides the π- which is the most important contribution to N_2 and U_2 - the ρ and $A_2 (N_o(\rho) = N_o(A_2) = 0)$ we also need

i) an A_1 which is required by the C_{LL} data.
(compare fig. 1)

ii) the B and a new piece – let us call it C – which has to be
dominantly real and crossing odd. C might be interpreted as a
mixture of ρ and B cuts. C and B are forced upon us by the fact
that C_{NN} is roughly equal to $-C_{SS}$ (compare fig. 2) and by the
large difference between the charge exchange differential
cross-sections.

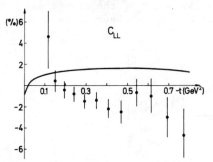

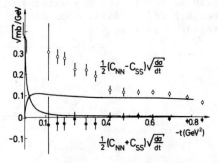

Fig. 1 The C_{LL} data[3] at 6 GeV
The π contribution (see eq. (4)
with $\beta_\pi \approx \exp(3.2t)$ is shown for
comparison.

Fig. 2 $1/2(C_{NN} \pm C_{SS})\sqrt{d\sigma/dt}$ at
6 GeV compared with the π contribution. Data taken from refs. 4,5.

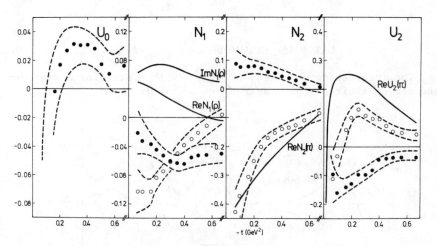

Fig. 3 The I=1 amplitudes (in $\sqrt{mb}$/GeV) for a version and with
$N_2(C) = U_2(C)$ and a free phase of C but without a B.
(Trajectories $\alpha_\rho = \alpha_A = 0.48 + 0.93t + 0.21t^2$, $\alpha_B = \alpha_\pi = \alpha_{A_1} = \alpha_C = 0$)
o real parts, • imaginary parts. The dashed lines indicates the
error bands.

138

C cannot be cleanly isolated from B, therefore we have performed several versions of the analysis. For instance coupling of C only to N_2 or to N_2 and U_2 and with or without a B. We found that the π, ρ, A_2 and A_1 amplitudes are practically independent of the particular coupling of B and C. We also checked the influence of the assumed Regge trajectories on the resulting amplitudes which turned out to be weak. In fig. 3 a set of amplitudes is shown. Note the strong ρ-A_2 exchange degeneracy breaking in particular in N_1 which is due to the approximate mirror symmetry of the charge exchange P $d\sigma/dt$ data. Some predictions for other observables are presented in fig. 4

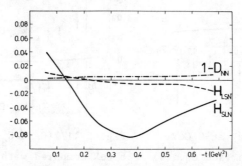

Fig. 4 Predictions for H_{SLN} (solid line) H_{LNS} (dashed line) and $1-D_{NN}$ (dashed-dotted line).

LARGE |t| BEHAVIOUR OF P, C_{NN} AND C_{SS}

The interesting phenomena of dips - with a striking energy dependence of their depths - near t =-1 GeV^2 in

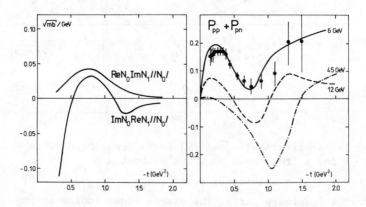

Fig.5 Contributions to N_1^O (at 6 GeV) and the sum of the elastic pp and pn polarization at several energies. Data are taken from ref.4 and 6.

$$N_{1\perp}^{o} = -1/4 \ (P_{pp} + P_{pn}) \ \sqrt{d\sigma/dt} \tag{5}$$

and in C_{NN}, whereas $N_{1\perp}^{1}$ and C_{SS} at least at 6 GeV behave smoothly, can be understood as an effect of the phase of N_{o}. We parametrized the t dependence of the flip amplitudes in agreement with the results for our I=1 amplitudes and fitted the large $|t|$ data at several values of p_{L}. As an example the results for N_{o}^{o} are shown in fig. 5. The polarization combination is expressed in terms of amplitudes by

$$P_{pp} + P_{pn} = -4(\text{Im } N_{o} \text{ Re } N_{1}^{o} - \text{Re } N_{o} \text{ Im } N_{1}^{o})/ |N_{o}|^{2} \tag{6}$$

As can be seen from the figure there is a compensation between both terms in eq.(6) at medium t values. Since Re N_{1}^{o} and Im N_{1}^{o} have the same energy dependence whereas Re N_{o}/Im N_{o} rapidly tends to zero with rising energy (roughly as $1/p_{L}$) the compensation is strongly energy dependent, producing so the dip.

REFERENCES

1. A. Yokosawa, these proceedings

2. W. Grein, R. Guigas and P. Kroll, Nucl. Phys. B89. 93 (1975)

3. I.P. Auer et al, to be published in Phys. Lett.

4. R.C. Fernow et al, Phys. Lett. 52B, 243 (1974)
   D. Miller et al, Phys. Rev. Lett. 36, 763 (1976)

5. I.P. Auer et al, Phys. Rev. Lett. 37, 1727 (1976)

6. R. Diebold et al, Phys. Rev. Lett. 35, 632 (1972)

AN OVERVIEW OF THE NUCLEON-NUCLEON INTERACTION

R. Vinh Mau
LPTPE, Université P. et M. Curie - Paris 75230 and
Division de Physique Théorique*, I.P.N. - ORSAY 91406

1.   INTRODUCTION

The question of the nature of the force between two nucleons is central to physics and a great deal of effort has been devoted to this problem for more than four decades. Yet, from a purely theoretical point of view, the progress in the subject has been rather slow, the most significant advance being achieved more than forty years ago when Yukawa (1935) proposed his meson theory of nuclear forces. The main result of this theory, namely the one-pion-exchange (OPE) contribution to the nucleon-nucleon interaction, still remains unquestioned. However, when one performs a detailed analysis of the wealth of nucleon-nucleon data, one realises quickly that the OPE contribution gives a good description of the interaction only at very large distances (r>2 fm) and it has to be supplemented by further contributions.

If the field theory viewpoint with specific fundamental Lagrangian models is to be adopted, the OPE contribution can be regarded, in a perturbation theory, as a 2nd order process of a fundamental pion nucleon coupling (Fig. 1). In this scheme, it is natural to consider also the next-order processes, in particular the 4th order terms (Fig. 2), and indeed, many authors attempted such calculations (see for example, Partovi and Lomon (1973), for a complete list of references). Even though the early calculations are uncertain and affected by ambiguous approximations, it must be stressed that diagrams like those represented in Fig. 2, i.e. the exchange of uncorrelated pions, exist independently of perturbation theory and should be accounted for properly in any realistic theoretical model.

With the discovery of the ρ and ω resonances, another line of research was developed which consists in considering the exchange of non strange mesonic resonances as responsible for the medium and short range forces, in the same way as the OPE is responsible for the long range forces. This generalization of the OPE leads to models (the one

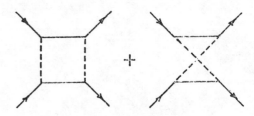

Fig. 1: The one-pion-exchange contribution to the nucleon-nucleon interaction

Fig. 2: The "4th order contribution" to the nucleon-nucleon interaction

*Laboratoire associé au C.N.R.S.

boson exchange models) which are very appealing because of their simplicity. However, to fit the data in these models, one always needs the exchange of a fictitious scalar particle, with a low mass at 400-500 MeV, to provide some intermediate range attractive forces that are not accounted for by the exchange of observed resonances. Unfortunately, the existence of such a particle is not supported by any experimental evidence.

These two competing extensions of the OPE possess their own merits and their own imperfections: even if one discards the question of the validity of using perturbation theory in strong interaction processes, the perturbative calculations have to face the problem of treating the meson-meson interaction. On the other hand, besides the uncertainties related to the existence of such and such bosons, the one boson exchange (OBE) models suffer from the defect of treating the massive composite systems, which decay very rapidly, as stable particles (zero width approximation).

The past two decades have also witnessed many purely phenomenological studies of nuclear forces. Given the rapid progress in experimental work, the viewpoint here is to take advantage of it and to abandon, maybe temporarily, the difficult task of searching for a satisfactory theoretical model. Instead, more or less arbitrary empirical parametrisations of phase shifts (Livermore group 1969, Yale group 1962) or potentials (Yale 1962, Hamada and Johnston 1962, Reid 1968, Sprung and de Tourreil 1973) are proposed to fit the two nucleon data. Once the free parameters are determined by this fit, these potentials are, in turn, used in nuclear structure calculations, sometimes with success. One such phenomenological potential, currently fashionable among nuclear theorists, is the Reid potential. This potential is designed to fit the low angular momentum (J < 3) nucleon-nucleon phase shifts, the J ≥ 3 phases being assumed to be all given by the OPE potential only. Even so, the number of free parameters is large compared to the number of fitted data and therefore no information about the interaction process itself is provided.

In this talk, we would like to report on another type of approach to the NN interaction which can be regarded as a synthesis of the previous ones. We try to develop a meson theory of nuclear forces that combines the best features of the field theoretical and the OBE calculations. Let us anticipate in saying that the method developed enables one not only to treat without any ambiguity the type of diagrams shown in Fig. 2, but also to take into account in a realistic way the mesonic resonances with their observed masses and their decay

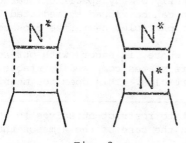

Fig. 3

widths even when these widths are very large, as in the case of the scalar and isoscalar "meson". Moreover, diagrams like the ones shown in Fig. 3 where N* refers to any nucleonic resonance, that occur neither in the field theory calculations (unless the N* is treated as an elementary particle) nor in the OBE models, are accounted for in the calculations presented below. Finally, diagrams like a), b), c),

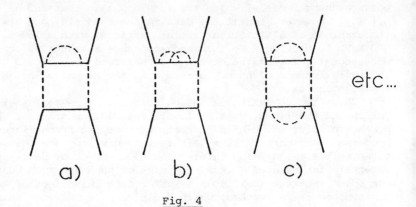

Fig. 4

etc... of Fig. 4 (which are of 6th, 8th and higher order in perturbation theory) are also included without their explicit calculation being necessary.

The prerequisite demands we wish to fulfil are:

i) Only nucleons and pions can be treated as particles* since they are stable as far as usual strong interactions are concerned. The rapidly decaying nucleonic and mesonic resonances must be regarded as composite systems. This requires that nucleon-nucleon forces should be studied in conjunction with the known properties of mesons, their interactions with themselves and with nucleons.

ii) Perturbation theory must be avoided. Without any available alternative procedures to deal with Lagrangian models, one should rather try to correlate well established information from various hadronic reactions instead of insisting on the search for a definite fundamental Lagrangian model.

2. NUCLEAR FORCES AND PARTICLE EXCHANGE

As already mentioned, the OPE contribution gives a good account of the peripheral nucleon-nucleon phase shifts and, therefore, of the long range part of the interaction. Furthermore, everyone now agrees on its actual form and it appears in all theoretical models of nuclear forces. This success is even more important in its conceptual implications since it strengthens the general concept that particle exchanges are responsible for nuclear forces.

As a matter of fact, independently of any specific interaction models, the two nucleon interaction can be viewed as mediated by the exchange of particles or systems of particles as in the crossed (nucleon-antinucleon) channels.

This is represented in Fig. 5 where, in accord with our requirement i), the intermediate states are states composed by stable (with respect to strong interactions) particles. This does not mean that

---

*For our purposes, it is sufficient to restrict ourselves to non strange particle. We also discard the case of the η meson that seems to couple weakly with nucleons.

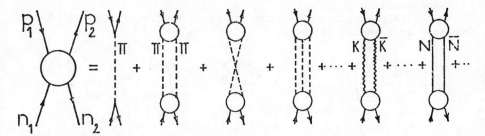

Fig. 5: Particle exchange contribution to the nucleon-nucleon interaction. The initial and final nucleons are denoted by their
four-momenta n_1, p_1 and n_2, p_2.

resonant intermediate states are neglected, but rather that these
states are taken into account through the interaction between the
intermediate particles. For example, in the two pion exchange (2ΠE)
contribution the ε "resonance" and the ρ resonance exchanges are here
described in terms of strong interactions between the two intermediate
pions in the J = I = 0 and J = I = 1 states. Also, the "blobs" of
each term in Fig. 5 can contain anything allowed by conservation laws.
For example, in the TPE exchange term, each blob can include one
nucleon, one nucleon + any number of pions, a nucleonic resonance,
etc... Actually, in the nucleon-antinucleon channel, the sum that
appears in Fig. 5 represents nothing else but the unitarity condition:

$$\mathrm{Im} < \overline{p}_1 p_2 |T| n_1 \overline{n}_2 > = \sum_\alpha <\overline{p}_1 p_2 |T^+|\alpha><\alpha|T|n_1 \overline{n}_2> \tag{2.1}$$

where the $|\alpha>$ form a complete set of states having the same quantum
numbers as $|n_1 \overline{n}_2>$ and $|\overline{p}_1 p_2>(|\Pi>,\ |2\Pi>,\ |3\Pi>,\ ...,\ |K\overline{K}>,\ |N\overline{N}>$ etc..).
Knowing the imaginary part of the nucleon-antinucleon amplitude, it
is natural to use a dispersion relation to get the amplitude itself
and then crossing relations to pass from the N$\overline{\text{N}}$ amplitude to the NN
amplitude. The procedure sketched above dispenses with the use of
perturbation theory and allows us to compute the nucleon-nucleon scattering amplitude in terms of the N$\overline{\text{N}}$ → 2Π, 3Π amplitudes, etc... which
in turn can be obtained from the ΠN → ΠN, and ΠN → ΠΠN amplitudes,
etc... by crossing properties and analytic continuation.
    More precisely, the scattering amplitude can be represented by
the following dispersion relation:

$$M(w,t) = \frac{g^2}{t-\mu^2} + \frac{1}{\Pi} \int_{4\mu^2}^{\infty} \frac{\rho_{2\Pi}(w,t')}{t'-t} + \frac{1}{\Pi} \int_{9\mu^2}^{\infty} \frac{\rho_{3\Pi}(w,t')}{t'-t} \quad \text{etc...} \tag{2.2}$$

where $w = (n_1 + p_1)^2$
       $t = (p_1 - p_2)^2$
       μ = the pion mass
       g = pion nucleon coupling constant $(\frac{g^2}{4\Pi} = 14.5)$
and where the functions $\rho_{2\Pi}(w,t')$, $\rho_{3\Pi}(w,t')$, etc... correspond to the

imaginary part of the amplitude with two pions, three pions, etc... in the intermediate state.

Equivalently, in configuration space, the interaction can be written as a series of terms:

$$V \simeq \frac{g^2}{r} \frac{e^{-\mu r}}{r} + \int_{4\mu^2}^{\infty} \rho_{2\Pi}(w,t') \frac{e^{-\sqrt{t'}r}}{r} dt' + \int_{9\mu^2}^{\infty} \rho_{3\Pi}(w,t') \frac{e^{-\sqrt{t'}r}}{r} dt' + \text{etc...} \qquad (2.3)$$

The successive terms of eqs. (2.2) and (2.3) correspond to the one pion exchange (OPE) contribution, the two pion exchange (2ΠE) contribution, etc... The whole dynamics of the problem is contained in the spectral function $\rho_{2\Pi}$, $\rho_{3\Pi}$, etc... and their actual expressions depend, of course, on the contents of the "blobs" in Fig. 5. For example, if one keeps only the one nucleon state in each "blob" of the 2ΠE contribution one gets the previously mentioned 4th order term in perturbation theory.

Although at present no definite a priori statement can be made about the relative sizes of the various spectral functions, one can say, using phase space factor arguments, that the slope of the functions ρ_i, at various thresholds, decreases with the mass of the exchanged systems. Consequently, if no unexpected strong enhancement occurs in the many particle exchanges, the size of the spectral functions ρ_i may also decrease with the mass of the exchanged systems (Fig. 6). Thus, the series represented in Fig. 5 is a series of contributions that have shorter and shorter ranges.

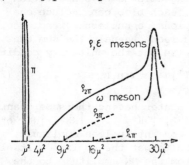

Fig. 6: The spectral functions ρ_i

3. THE LONG AND MEDIUM RANGE FORCES

After the long range forces given by the OPE, the next longest range forces are due to the two pion exchange (2ΠE). In the dynamical calculations (Amati et al. 1963, Cottingham and Vinh Mau 1963, Brown and Durso 1971, Chemtob, Durso and Riska 1971, Vinh Mau et al. 1972, Bohannon and Signell 1974, Epstein and McKellar 1974) that I am going to describe, the underlying belief is that most of the medium range forces are correctly given by the 2ΠE when the latter is properly determined. Also the hope is that the long and medium range forces if accurately known provide strong enough constraints to leave little freedom for an eventual phenomenological determination of the short range part of the interaction.

The main task is therefore to perform an accurate calculation of the 2ΠE contribution i.e. the spectral function $\rho_{2\Pi}$ that appears in eq. (2.2).

3.1 - The Two Pion Exchange Contribution (2ΠEC)

As the actual formalism is rather intricate and in order to provide a simple insight into the methods and the physical inputs, I will, in the following, leave out heavy technical algebraic details and deliberately choose a mode of presentation which takes some liberties with mathematical rigor. The interested reader is referred to ref. 18 for the complete treatment. The argument may be broken down into the following steps:

i) In the t channel ($N\bar{N} \to N\bar{N}$), perform a partial wave expansion of $\rho_{2\Pi}$:

$$\rho_{2\Pi} = \sum (2\ell+1) \; \mathrm{Im} F_\ell^{N\bar{N} \to 2\pi \to N\bar{N}} (t') \; P_\ell (\cos \theta_w) \qquad (3.1)$$

where $F_\ell^{N\bar{N} \to 2\Pi \to N\bar{N}} (t')$ are partial wave amplitudes for $N\bar{N}$ scattering with two pions as intermediate state. Here $w = -2p^2 (1 + \cos \theta_w)$ in the CMS of the $N\bar{N}$ channel.

ii) Express $\mathrm{Im} F_\ell^{N\bar{N} \to 2\Pi \to N\bar{N}} (t')$ in terms of $N\bar{N} \to 2\Pi$ helicity amplitudes $f_\ell^{N\bar{N} \to 2\Pi}$ through the unitarity condition

$$\mathrm{Im} F_\ell^{N\bar{N} \to 2\Pi \to N\bar{N}} (t') = \left(\begin{array}{c} \text{kinematical} \\ \text{factor} \end{array} \right) \; \left| f_\ell^{N\bar{N} \to 2\Pi} (t') \right|^2 \qquad (3.2)$$

which can be assumed to hold for $4\mu^2 < t' < 50\mu^2$ since inelasticity is very weak up to the $K\bar{K}$ threshold (the 2Π system is very weakly coupled to the 4Π, 6Π systems, etc...).

iii) Use the analyticity properties of $f_\ell^{N\bar{N} \to 2\Pi}$ to write the dispersion relation

$$\ell n \; f_\ell^{N\bar{N} \to 2\Pi} (t') = \frac{1}{\Pi} \int_{4\mu^2}^{50\mu^2} \frac{\delta_\ell^{\Pi\Pi \to \Pi\Pi} (t'')}{t'' - t'} dt'' + \frac{1}{\Pi} \int_{-\infty}^{N\bar{N} \to 2\Pi} \frac{\Phi_\ell^{N\bar{N} \to 2\Pi} (t'')}{t'' - t'} dt'' \qquad (3.3)$$

where $\Phi_\ell^{N\bar{N} \to 2\Pi}$ is defined by $f_\ell^{N\bar{N} \to 2\Pi} = \left| f_\ell^{N\bar{N} \to 2\Pi} \right| e^{i\phi_\ell^{N\bar{N} \to 2\Pi}}$ and

where the unitarity condition has been used again to identify $\Phi_\ell^{N\bar{N} \to 2\Pi}$ with the $\Pi\Pi$ phase shifts $\delta_\ell^{\Pi\Pi \to \Pi\Pi}$, for $4\mu^2 < t'' < 50\mu^2$.

In the first integral of eq. (3.3), $\delta_\ell^{\Pi\Pi}$ can be taken from theoretical models or from experimental studies ($\Pi N \to \Pi\Pi N$, $K_{\ell 4}$ decay, $e^+ e^- \to \Pi\Pi$, nucleon form factors, etc...) for the low angular momentum (S and P waves) and are known to be negligible for the high angular momentum (D or higher waves).

In the second integral of eq. (3.3), $f_\ell^{N\bar{N} \to 2\Pi}$ can be obtained from the Froissart-Gribov formula since $t'' < 0$, namely

$$f_\ell^{N\bar{N} \to 2\Pi} (t'') = \frac{1}{\Pi} \int_{m^2}^{\infty} Q_\ell \left(\frac{2s + t'' - 2m^2 - 2\mu^2}{2 (t'' - 4m^2)^{1/2} (t'' - 4\mu^2)^{1/2}} \right) \mathrm{Im} M^{\Pi N \to \Pi N} (s, t'') \; ds \qquad (3.4)$$

where $M^{\Pi N \to \Pi N}(s,t'')$ is the physical ΠN scattering amplitude since $s > m^2$ and $t'' < 0$.

Schematically, one can summarize i) to iii) by:

$$\rho_{2\Pi}(w,t') \sim \left| f_0^{N\bar{N} \to 2\Pi}(t') \right|^2 + \left| f_1^{N\bar{N} \to 2\Pi}(t') \right|^2 \cos\theta_w +$$

$\Pi\,\Pi$ S wave phase shift (J=I=0) + ΠN phase shifts

$\Pi\,\Pi$ P wave phase shift (J=I=1) + ΠN phase shifts

$$+ \sum_{\ell > 2} (2\ell+1) \left| f_\ell^{N\bar{N} \to 2\Pi}(t') \right|^2 P_\ell(\cos\theta_w)$$

ΠN phase shifts

$\qquad(3.5)$

Note, that in the OBE models, the last sum in eq. (3.5) is set to zero and $\left| f_0^{N\bar{N} \to 2\Pi} \right|^2$ and $\left| f_1^{N\bar{N} \to 2\Pi} \right|^2$ are approximated respectively by $g_\sigma^2\,\delta(t'-m_\sigma^2)$ and $g_\rho^2\,\delta(t'-m_\rho^2)$.

The same procedure could, in principle, be repeated for $\rho_{3\Pi}$ and $\rho_{4\Pi}$ with, however, formidable complications in general. In the case of $\rho_{3\Pi}$, the ω meson exchange can be easily taken into account, and other specific processes can also be included.

The set of equations (2.2) and (3.3) to (3.5) forms the bulk of the relationships which, via analyticity properties, unitarity and crossing relations, connect the $2\Pi E$ contribution with the pion-nucleon and pion-pion interactions. The pion-nucleon scattering is very accurately known by various phase shift analyses. These are similar qualitatively, but have some quantitative differences. We have taken some account of these differences by using the CERN solution (Donnachie, Kirsopp, Lovelace 1968) and the Glasgow A solution (Moorehouse 1970). Of course, if needed, one can use other solutions. The S and P wave pion-pion interaction has also been extensively studied in the last few years (cf. for example Basdevant et al. 1973).

By using these analyses as inputs, one automatically includes all the nucleonic resonances (they are contained in ΠN phase shifts) as well as the realistic S and P wave pion-pion interaction, and besides these, the smooth background of both pion-nucleon and pion-pion scattering is also taken into account.

3.2 - The ω Meson Exchange Contribution

The ω meson (J=1, I=0) is a narrow 3Π resonance, has a width of only 10 MeV and its exchange can be considered, in a good approximation, as a single particle exchange. The strength of the coupling of this particle to the nucleon is not yet well determined. Current evidence suggests that the ω meson is responsible for much of the isoscalar nuclear electromagnetic form factors. We take

$$G_\omega^T/G_\omega^V = -0.12 \text{ and } (G_\omega^V)^2/4\Pi = 4.65 \qquad (3.6)$$

which follows from the ρ coupling constant, SU(3) prediction and iso-scalar nucleon magnetic moment. G_ω^T and G_ω^V are the tensor and vector coupling constants. However, SU(3) symmetry can be badly broken here and therefore the value of G^V chosen in eq. (3.6) should be considered only as a working hypothesis. It should be noted that the dominant forces provided by the ω meson exchange contribution are repulsive.

3.3 - The Nucleon-Nucleon Peripheral Phase Shifts from the ($\Pi+2\Pi+\omega$) Contribution

The ability of the OPE, of the above calculated 2ΠE and of ω exchange contributions (as part of the three pion exchange) to describe the long and intermediate range forces can be checked by calculating the low energy (up to 330 MeV) peripheral (J>2) NN phase shifts (Vinh Mau et al. 1972) and by comparing them with the empirical ones.

The inputs of these calculations come from several different sources and we list them below in order of decreasing reliability

i)   The "fourth-order contribution" (see Fig. 2) that depend only on the pion-nucleon coupling constant. We have taken $g^2/4\Pi = 14.5$.

ii) The "double spectral contributions" (see Fig. 7). These contributions along with i) contain the exchange of two pions in D waves and above. They depend upon the ΠN phase-shift analysis. We have taken here the two different analyses: the CERN and the GLASGOW solutions.

iii) The exchange of two pions in a P state (the $\overline{NN}{\rightarrow}2\Pi$ P wave amplitude) which is dominated by the ρ meson. The model used here is the one given by Hohler et al. (1968).

iv) The exchange of two pions in a S state (the $\overline{NN}{\rightarrow}2\Pi$ S wave amplitude) which is related to the Π Π S wave. We have used here the results of Nielsen and Oades (1974).

v)   The ω exchange with the coupling constants described in the previous Section.

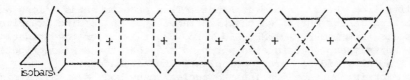

Fig. 7: The "double spectral contributions" in the case where the ΠN scattering is approximated by the isobaric model. Heavy lines refer to isobars.

A typical example of the results for the unitarized partial wave phase parameters is displayed in Fig. 8. We also plot the contributions from the one-pion-exchange (OPE) alone and with the fourth order (OPE + 4th). The results are compared with the phenomenological energy independent and energy dependent Livermore phase-shift analysis (MacGregor et al. 1969, referred to as MAW).

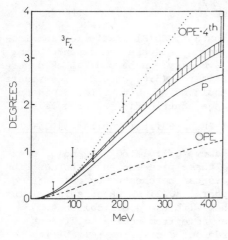

Fig. 8

In general and in the energy range considered, one can draw the following conclusions:

i)   In most cases, the results obtained from the CERN and GLASGOW ΠN phase-shift analyses differ very little, especially for energies less than 200 MeV. Only the low partial waves show some differences and then only at energies greater than 250 MeV.

ii)   When compared to phenomenology, the results, in most cases, give a definite improvement over OPE + 4th. This is especially true at the higher energy.

iii) The calculation depends on the ω coupling constant G_ω^V which is taken here from SU(3) prediction, from isoscalar nucleon electromagnetic form factors and Π Π P wave parameters (see eq. (3.6)). In any case, an increase of this constant would somewhat improve the fit since the dominant part of the ω exchange is repulsive.

iv)   Of the nineteen calculated phase parameters thirteen high partial waves are in good agreement with the Livermore phase-shift analysis over the whole energy range. The two exceptions ($^1H_5$, $^3G_5$) are in the isospin T=0 which is only accessible in neutron-proton scattering. Experimentally they are not too well determined and our calculation actually fits much better a previous Livermore phase-shift analysis (MacGregor et al., (1968).

For further progress two possibilities present themselves:

i)   For the nucleon-nucleon scattering itself, one can perform a phase-shift analysis similar to those done by the Livermore group, now with extra constraints on lower angular momentum phase shifts from the 2ΠE and ω exchange. This program needs a long and exacting labour but is worth while in view of the coming experimental results from LAMPF, TRIUMF, SIN, etc... It is under investigation by the Paris group. It is also worth noting that the 2ΠE contribution is imaginary and can provide inelasticity parameters.

ii)   For further applications to nuclear many body systems, it is useful to define an equivalent potential which, once supplemented by a short range core, can be used in various nuclear calculations.

3.4 - The Two Pion Exchange Potential (2ΠE)

While the previous Sections are devoted to the derivation of the two pion exchange contribution to the relativistic two nucleon S matrix, we would like to derive here an equivalent potential. Of course, such a potential is not necessary when one is only concerned with the two nucleon system and could even be ill defined, especially at high energies. However, a potential which has a reasonably sound basis outside the core region would be useful in various nuclear calculations. This is especially true since the various phenomenological potentials which fit the low-energy two nucleon data differ even at rather large distances. Also, a potential, which is reliable for distances smaller than those where the one pion exchange potential dominates, would help in reducing the degree of arbitrariness of the core region.

We wish to define a potential which, when inserted into a Schrödinger equation, gives, in an energy range below the meson production threshold, the same T matrix as the scattering amplitude derived in the previous Sections. Again in this Section, we disregard spin and isospin complications for the sake of simplicity of presentation. A complete treatment can be found in ref. 18.

The T matrix is related to the potential V by the Lippmann-Schwinger equation

$$\langle \vec{p}_2 | T | \vec{p}_1 \rangle = \langle \vec{p}_2 | V | \vec{p}_1 \rangle - \int d^3 q \; \langle \vec{p}_2 | V | \vec{q} \rangle \; \frac{m}{\vec{q}^2 - p^2 - i\varepsilon} \; \langle \vec{q} | T | \vec{p}_1 \rangle \qquad (3.7)$$

and is normalized by

$$\frac{d\sigma}{d\Omega} = (2\Pi)^6 \left(\frac{m}{4\Pi} \right)^2 \left| \langle \vec{p}_2 | T | \vec{p}_1 \rangle \right|^2 \qquad (3.8)$$

we therefore require

$$\langle \vec{p}_2 | T | \vec{p}_1 \rangle = - \frac{1}{(2\Pi)^3} \; \frac{m}{E} \; M \qquad (3.9)$$

where M is the relativistic scattering amplitude defined by eq. (2.2) of Section 2.

Because of its analytic structure M can be written as

$$M = M^{OPEC} + M^{2\Pi EC} + \dots, \qquad (3.10)$$

where M^{OPEC} and $M^{2\Pi EC}$, as functions of t, have respectively a pole at $t = \mu^2$ and a cut beginning at $t = 4\mu^2$ due to the two-pion exchange contribution; the other terms of this sum correspond to 3Π and further cuts.

On the other hand, it has been shown (Bowcock and Martin 1959, Blankenbecler et al. 1960) that for superpositions of Yukawa potentials, the T matrix defined by the Lippmann-Schwinger equation, eq. (3.7), is an analytic function of t which has a pole at $t = \mu^2$ and branch points at $t = 4\mu^2, 9\mu^2, \dots$; we therefore can write T as

$$T = T^{OPEP} + T^{2\Pi EC} + \dots, \qquad (3.11)$$

where again T^{OPEP} and $T^{2\Pi EC}$ are associated, respectively, with the pole at μ^2 and the cut with branch point at $4\mu^2$, etc... We make the same decomposition of the potential, namely,

$$V = V^{OPEP} + V^{2\Pi EP} + \dots \qquad (3.12)$$

Taking into account the proper normalization factors as given by eqs. (3.8) and (3.9), and identifying terms having the same singularities, yields

$$\langle \vec{p}_2 | T^{OPEP} | \vec{p}_1 \rangle \ = - \ \frac{1}{(2\Pi)^3} \ \frac{m}{E} \ M^{OPEC} \tag{3.13}$$

for the pion pole,
and

$$\langle \vec{p}_2 | T^{2\Pi EC} | \vec{p}_1 \rangle \ = - \ \frac{1}{(2\Pi)^3} \ \frac{m}{E} \ M^{2\Pi EC} \tag{3.14}$$

for the two-pion branch point.
The corresponding one-pion-exchange potential is

$$V^{OPEP}(\vec{\Delta}) \ = - \ \frac{1}{(2\Pi)^3} \ \frac{m}{E} \ M^{OPEC} \ = \ \frac{1}{(2\Pi)^3} \ \frac{1}{mE} \ \frac{g^2}{\mu^2 - t} \tag{3.15}$$

Since

$$\langle \vec{p}_2 | T^{2\Pi EC} | \vec{p}_1 \rangle \ = \ \langle \vec{p}_2 | V^{2\Pi EP} | \vec{p}_1 \rangle \ - \ \langle \vec{p}_2 | V^{OPEP,2} | \vec{p}_1 \rangle \tag{3.16}$$

where

$$\langle \vec{p}_2 | V^{OPEP,2} | \vec{p}_1 \rangle \ = \ \int d^3 q \ \frac{\langle \vec{p}_2 | V^{OPEP} | \vec{q} \rangle \ m \ \langle \vec{q} | V^{OPEP} | \vec{p}_1 \rangle}{q^2 - p^2 - i\varepsilon} \tag{3.17}$$

the two-pion-exchange potential is defined by

$$\langle \vec{p}_2 | V^{2\Pi EP} | \vec{p}_1 \rangle \ = \ - \ \frac{m}{E} \ \frac{1}{(2\Pi)^3} \ M^{2\Pi EC} + \langle \vec{p}_2 | V^{OPEP,2} | \vec{p}_1 \rangle \tag{3.18}$$

This equation gives the two-pion-exchange potential in terms of our $M^{2\Pi EC}$ with the iterated one-pion-exchange potential $\langle \vec{p}_2 | V^{OPEP,2} | \vec{p}_1 \rangle$ subtracted out. If the iterated OPEP is recast in the form

$$\langle \vec{p}_2 | V^{OPEP,2} | \vec{p}_1 \rangle \ = \ \frac{1}{(2\Pi)^3} \ \frac{m}{E} \ \frac{1}{\Pi} \ \int_{4\mu^2}^{\infty} \frac{\xi(w,t)}{t' - t} \ dt \tag{3.19}$$

the two-pion-exchange potential can be rewritten as:

$$\langle \vec{p}_2 | V^{2\Pi EP} | \vec{p}_1 \rangle \ = \ - \ \frac{1}{(2\Pi)^3} \ \frac{m}{E} \ \frac{1}{\Pi} \int_{4\mu^2}^{\infty} \frac{\rho_{2\Pi}(w,t') - \xi(w,t)}{t' - t} \ dt' \tag{3.20}$$

From eq. (3.20), one expects an energy dependence (w dependence) for the 2ΠE potential. However, as it can be shown (ref. 17), cancellations occur between terms that have the strongest energy dependence (the energy dependence of the box diagram (Fig.2a) part of $\rho_{2\Pi}$ and that of ξ exactly cancel) leaving some weak but significant energy dependence only in the central component and, to a lesser extent, the SO component of the 2ΠE potential. This energy dependence comes mostly from the "double spectral contributions".

The various components (central, spin-spin, spin-orbit, tensor and quadratic spin-orbit) of the 2Π exchange potential have been calculated in this way using the same ΠN and ΠΠ inputs as those described previously in Section 3.3. With the addition of the OPE and the exchange potentials, we show, for illustration, in Fig.9, the shapes of this (Π+2Π+ω) exchange potential at energy threshold ($w=4m^2$) and in configuration space.

For comparison, the Yale (1962) and Hamada-Johnston (1962) phenomenological potentials are also displayed in Figs. 9. Without going into a detailed discussion (see Cottingham et al. 1973), one can say that the agreement is satisfactory especially for $r > 0.8$ fm, if one keeps in mind that our central potentials are significantly energy dependent (which implies a certain non locality in coordinate space) whereas the Yale and Hamada-Johnston are local potentials.

One of the striking features that appear in our studies, both on the NN phase shifts and on the equivalent NN potential, is the importance of the uncorrelated 2ΠE contribution. We have shown that it cannot be approximated either by a 4th order contribution or by some low angular momentum ΠΠ resonances.

This uncorrelated 2Π exchange is calculated here in the framework of dispersion relations from the presently known properties of Π mesons, their interactions with themselves and with nucleons. The fact we have obtained, without adjusting any parameter, a good consistency between experimental data on the Π-nucleon, ΠΠ, and nucleon-nucleon systems does give us confidence in the dispersion relation approach to the fundamental problem of nucleon-nucleon forces. Conversely, within the same framework, better nucleon-nucleon data could help in removing some ambiguities of the Π-nucleon or ΠΠ interactions.

At this stage, one may ask if these theoretical long and medium range forces can be utilized as a basic ingredient in constructing an entire low energy nucleon-nucleon interaction which is realistic enough to fit both the two-body and the many-body data. We will try to answer this question in the next Section.

4. THE SHORT RANGE FORCES

In the general framework I just discussed, the description of the short range NN forces requires calculation of the three pion and higher mass exchanges. At present, these calculations are, however, beyond our abilities, apart from some specific 3Π exchange diagrams which are under investigation by the Paris group. In any case, up to now no satisfactory theoretical model for the short range forces has been found, despite persistent efforts by many people.

This situation led us to adopt a phenomenological viewpoint for the description of these short range forces. We add to the (Π+2Π+ω)

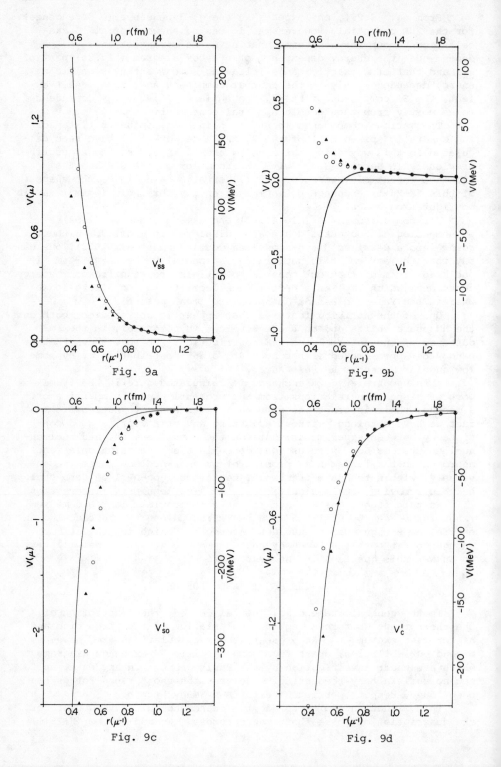

Fig. 9a

Fig. 9b

Fig. 9c

Fig. 9d

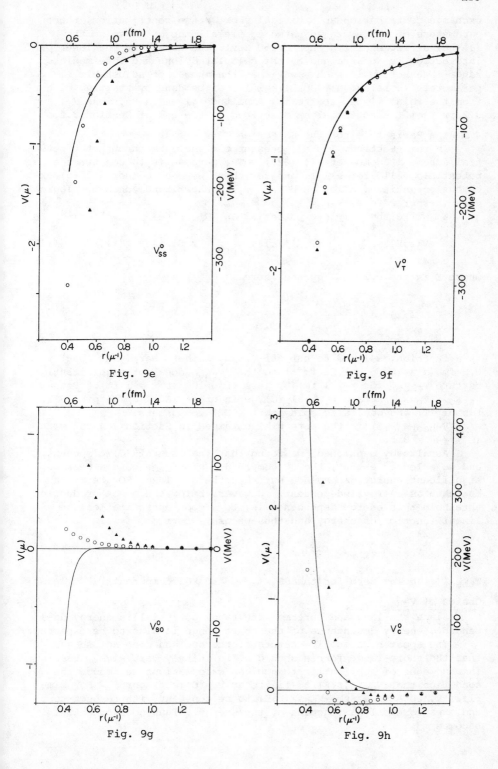

Fig. 9e

Fig. 9f

Fig. 9g

Fig. 9h

exchange contribution an empirical short range contribution, the parameters of which are adjusted to fit all the NN phase shifts (J < 6) at low energies (up to 350 MeV). We hope that the inner part interaction is constrained by the calculated long and intermediate range interaction in such a way that the number of adjustable parameters is relatively small. This can be done by starting either from the relativistic scattering amplitude given in Sections 3.1 and 3.2 or from the potential as described at the end of Section 3.4.

4.1 - The Paris Model of the SR Forces

In this Section, I shall present the analysis through the potential scheme (Lacombe et al. 1975). The purpose is to construct a potential, valid for all r, and for this, we add to the (Π+2Π+ω) exchange potential (called here $V_{theor.}$) a short range phenomenological core (referred to as $V_{phen.}$).

We write the complete potential as

$$V(r,E) = V_{theor.}(r,E) f(r) + V_{phen.}(r,E) \left[1 - f(r)\right] \qquad (4.1)$$

where E is the c.m. energy and where

$$f(r) = \frac{(pr)^{\alpha}}{1 + (pr)^{\alpha}}$$

is a function designed to cut off V_{theor} rather sharply at about 0.8 fm ($\alpha = 10$, $p = 1.25$ fm^{-1}) so that V_{phen} does not significantly perturb V_{theor} for r > 1 fm (by less than 10% at r = 1 fm). Both V_{theor} V_{phen} contain central (C), spin orbit (SO), spin-spin (SS), tensor (T) and quadratic spin orbit (SO2) components.

V_{theor} (r,E) is the potential obtained in Section 3.4 but with $(G_{\omega}^{V})^{2}/4\Pi = 9.5$.

As already mentioned, we found that the central (C) component, and to a lesser extent, the SO component, of V_{theor} have a weak but significant energy dependence but that the SS, T and SO2 terms can be taken as energy independent. However, below the meson production threshold this energy dependence is, to a good approximation, a linear function of energy and thus we use a form

$$V_{theor}(r,E) = U_{theor}(r) + EW_{theor}(r) \qquad (4.2)$$

$W_{theor}(r)$ being zero for the SS, T, SO and SO2 components.

Choice of V_{phen}

Since the long and intermediate range potential is energy dependent, energy dependence of the short-range is also to be expected.

The apparent composite nature of the nucleon does not suggest that the short-range forces should be infinitely repulsive. Bearing in mind ease of application to nuclear calculations we impose the condition that $V_{phen}(r,E)$ be finite at r = 0 (soft core). For simplicity, we also assume $V_{phen}(r,E)$ to be constant with respect to r and therefore only a function of the energy. Thus, our potential becomes

$$V(r,E) = \left[U_{theor}(r) + EW_{theor}(r) \right] f(r) + V_{phen}(E) \left[1-f(r) \right] \tag{4.3}$$

A first fit of the Livermore energy-independent phase-shift analysis (MAW) at 25, 50, 95, 142 and 330 MeV reveals that $V_{phen}(E)$ is a linear function of E for the C component and almost constant for the SS, SO, T and SO2 components.

We therefore take $V_{phen}(E)$ to be of the form $C + C' E$ for the central component and constant for the SS, SO, T and SO2 terms, so that the complete potential $V(r,E)$ can now be written as

$$V(r,E) = U(r) + EW(r) \tag{4.4}$$

with

$$U(r) = U_{theor}(r) \ f(r) + C \left[1 - f(r) \right] \ ,$$

$$W(r) = W_{theor}(r) \ f(r) + C' \left[1 - f(r) \right] \ , \tag{4.5}$$

C' being zero for the SS, T, SO, and SO2 components. $V(r,E)$ now contains six free parameters for each isospin state, I = 0 or I = 1, namely C_C, C_{SS}, C_T, C_{SO}, C_{SO2} and C'_C.

Fit of the data

Our best fit (which is obtained by minimizing a χ^2 defined as a sum of squares of the differences of empirical and calculated phase parameters divided by the diagonal elements of the error matrix) gives for the C_i and C'_i the values displayed in Table I. The quality of our fit is shown in Table II and Fig. 10. For proton-proton scattering, the total χ^2 is 160 for 64 data points, i.e., χ^2/data = 2.5. When one takes into account the small number of free parameters involved in our fit, this is very satisfactory if compared with the χ^2 of the well-known phenomenological fits. It is worth noting here that, in some cases, the small uncertainties quoted for the empirical J = 4 phases are responsible for half the value of our χ^2. For example, at 210 and 330 MeV, our $^1G_4$ is somewhat higher than the energy-independent MAW solution but closer to the energy-dependent MAW solution or the energy-independent solution of Signell and Holdeman (1971); similarly, our partial χ^2 for $^3F_4$ and ε_4 seem artificially large. For neutron-proton scattering, the total χ^2 is 171 for 46 data points, i.e., χ^2/data = 3.7. Also here, at 330 MeV, there is a discrepancy between MAW values and the solution 1 of Signell and Holdeman (1971), which has been used in this work. Again, the corresponding χ^2 represents a significant part of the total χ^2.

It is important to notice that this fit was obtained with the simplest parametrisation of our class of potentials (six free parameters for the five components of the potential) and it can be improved when some small energy dependence is allowed for the SS, T, SO and/or SO2 terms. In particular, if, instead of linearizing $V_{theor}(r,E)$ with respect to E, we keep its whole energy dependence and allow also some small energy dependence for the T potential inside the core ($C'_T = -0.17$ for I = 1 and $C'_T = 0.22$ for I = 0), then the χ^2/data drop respectively to 1.7 and 3 for pp and np scattering; the other values of the core parameters are quite close to those

156

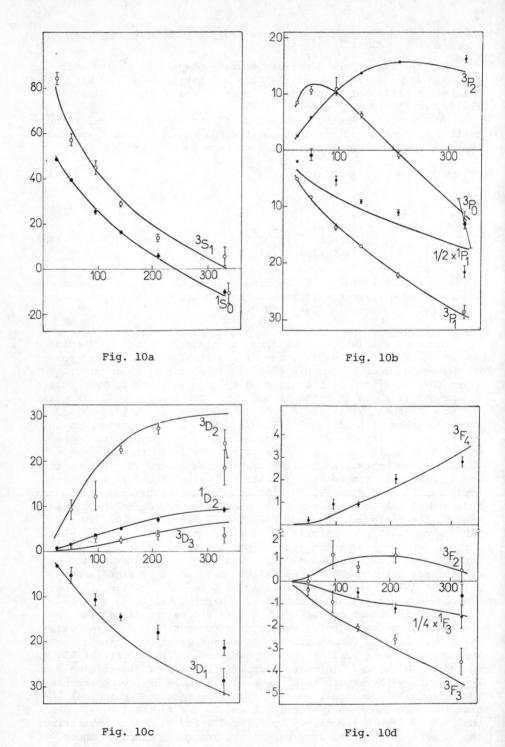

Fig. 10a

Fig. 10b

Fig. 10c

Fig. 10d

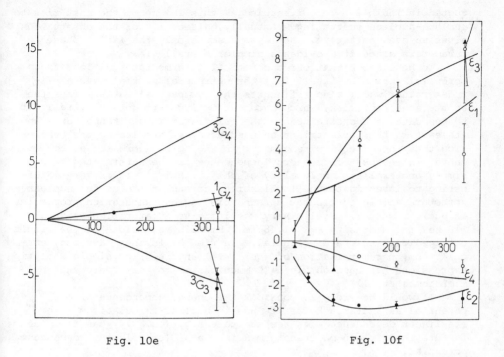

Fig. 10e   Fig. 10f

Table I : The values of the phenomenological parameters found
in the present fit.

	Isospin	
	1	0
C_C (MeV)	68.8	58.2
C'_C	0.952	0.998
C_{SS} (MeV)	8.6	-47.3
C_T (MeV)	29.9	-29.7
C_{SO} (MeV)	-286	50
C_{SO2} (MeV)	33	25

Table II : The deuteron and effective-range parameters.

$E_D = -2.2246$	$a_{np} = 5.4179$
$Q_D = 0.2900$	$r(-E_D, 0) = 1.753$
$P_D = 6.75$ %	$a_{pp} = -7.817$
$\mu_D = 0.8392$	$r_{pp} = 2.747$
$D/S = 0.0293$	

quoted in Table I. To our knowledge, this is one of the best fits to
nucleon-nucleon scattering. Actually, direct fit of the observables
gives an even better χ^2. Here, we discuss only the fit of phase
parameters since it provides a simpler overall view.

We have shown that our theoretical intermediate and long-range
potential (which by itself fits the peripheral partial-wave amplitudes rather well) also influences the low partial waves. Thus the
short-range repulsion, which has been included to get an overall fit
to the data, is constrained by the calculated longer range part, and,
at least in the parametrisation used here, we are led to the conclusion that the effective repulsion, in the core, increases as the
energy increases. This energy dependence can be interpreted as a
simple and explicit nonlocality of the potential. It is very interesting to investigate the implications of such features for nuclear
structure and in particular for the binding and saturation properties
of nuclear matter. Preliminary results give a saturation point at
∿10 MeV for a density k_f ∿ 1.48 fm, which is very close to the result
obtained with the Reid potential. In view of further many body calculations, parametrisation of eqs. (4.4) and (4.5) by simple analytic
expressions has been obtained (Communication to this Conference by
M. Lacombe et al.)

It would be wrong to consider the energy dependence of our
potential as a drawback for its use in nuclear calculations, since
its linear dependence with respect to energy (eq. 4.4) has the nice
feature that it can be transformed into a p^2 (or velocity) dependent
potential.

4.2 - The p^2 Dependence of the Paris Potential

The linear dependence with respect to energy of the complete
potential should not be regarded as an approximation that could break
down at large energies (∿500 MeV): it is true that for $V_{th}(r,E)$ this
linear dependence arises from an approximation (which holds even at
large energies) but for $V_{phen}(r,E)$ it is not due to an approximation
but given by the data fit.

Let us now show that our energy dependent potential can be
transformed into a p^2 dependent potential. Define the potential in
configuration space as

$$V(\vec{x},\vec{x}') = \int d^3k_f \, d^3k_i e^{i\vec{k}_f \cdot \vec{x}' - i\vec{k}_i \cdot \vec{x}} \tilde{V}(w,t) \qquad (4.6)$$

where

$$\vec{k}_i, \vec{k}_f = \text{initial and final momenta in the C.M.S.}$$
$$w = (\text{total energy})^2$$
$$t = (\vec{k}_f - \vec{k}_i)^2$$

Our central potential $\tilde{V}_C(w,t)$ is linearly dependent on w and can
be written as:

$$\tilde{V}_C(w,t) = \tilde{V}_0(t) + (w-4m^2)\tilde{V}_1(t) \qquad (4.7)$$

When the nucleons are on their mass shell, $w-4m^2 = 4k_i^2 = 4k_f^2$, and a simple off mass shell extrapolation is given by

$$w-4m^2 = 2k_i^2 + 2k_f^2 = -t + 4\vec{k}_i \cdot \vec{k}_f \tag{4.8}$$

Then $\tilde{V}_C(w,t)$ can be rewritten as

$$\tilde{V}_C(w,t) = \tilde{V}'_o(t) + 4\vec{k}_i \cdot \vec{k}_f \tilde{V}_1(t) \tag{4.9}$$

Inserting this equation into eq. (4.6) gives

$$V_C(\vec{x},\vec{x}') = V'_o\left(\frac{\vec{x}+\vec{x}'}{2}\right) \delta(\vec{x}'-\vec{x}) + \nabla_x V_1\left(\frac{\vec{x}+\vec{x}'}{2}\right) \delta(\vec{x}'-\vec{x}) \nabla_{x'} \tag{4.10}$$

This equation expresses nothing else but a hermitian p^2 (or velocity) dependent potential and this type of potential can be handled without any difficulty in many body calculations.

4.3 - The Stony Brook Model of the SR Forces

The TPE of the Stony Brook potential (Jackson et al., 1975) contains the 4th order contribution, the ρ exchange and a $f_0^{N\overline{N}\rightarrow 2\Pi}$ found by educated guess. The ω exchange is included with a coupling constant $g_\omega^2/4\Pi = 6.24$. This potential is regularized for high momentum transfer t (or small r) by a function obtained from certain multiple neutral vector meson exchange processes treated in the eikonal approximation. Namely, the regularized potential in momentum space has the following expression

$$V_R(\vec{p}',\vec{p}) = e^{2i[\chi(t) + \chi(u)]} V(\vec{p}',\vec{p})$$

with

$$\chi(t) = \frac{2\gamma(t-2m^2)}{\sqrt{t(t-4m^2)}} \log\left[(-\frac{t}{4m^2})^{1/2} + (1 - \frac{t}{4m^2})^{1/2}\right]$$

The calculated phase shifts are in good qualitative agreement with the experimental results. Some results are displayed for illustration in Figs. 11a and 11b.

5. COMPARISON OF THE PRESENT APPROACH
WITH THE TRANSITION POTENTIALS APPROACH

The main assumption that makes the genuine OBE models simple and attractive is that the dynamics of the nucleon-nucleon interaction are determined by the lowest order diagrams with no hadron closed loops. Accordingly, there is room neither for multiparticle exchanges nor for nucleonic excited states effects. The scattering amplitude is real and unitarity is to be satisfied through the Schrödinger equation (or an equivalent prescription such as the N/D equation) which generates iterated diagrams representing higher order

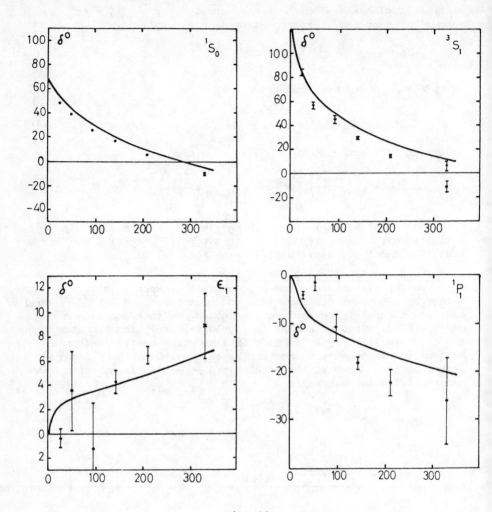

Fig. 11a

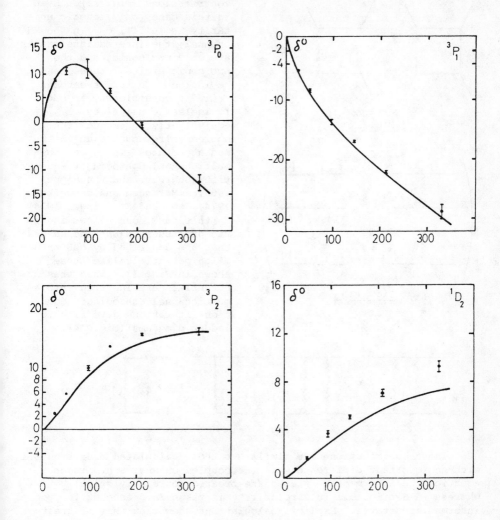

Fig. 11b

162

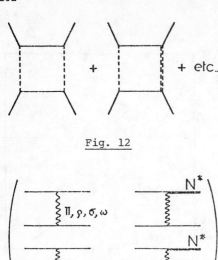

Fig. 12

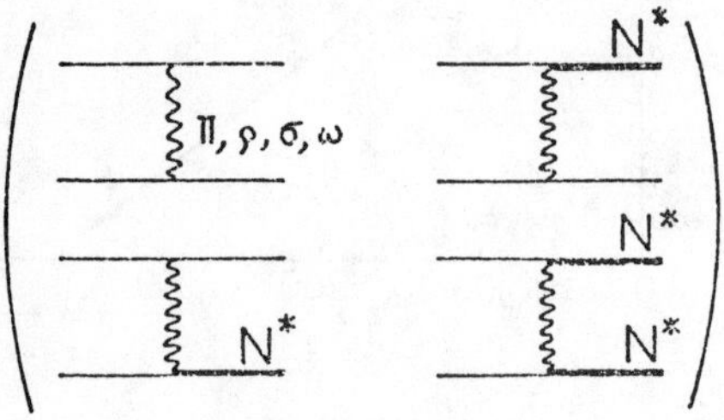

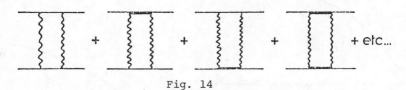

Fig. 13

contributions (Fig. 12). Even in this case, only elastic unitarity is fulfilled : in the NN channel, the intermediate states are only nucleonic, and the nucleonic excited states cannot be included in this treatment.

One possible way to satisfy inelastic unitarity or equivalently to include the nucleonic resonances as intermediate states, is to use the multichannel Schrödinger equation (Sugawara and von Hippel 1968, A.M. Green and co-workers 1974, Jena and Kisslinger 1974). In this case, one allows inelastic processes NN→NN* and NN→N*N* that are described by OBE transition potentials like those shown in Fig. 13. When these potentials are inserted into a multichannel Schrödinger equation, iterations give rise to ladder diagrams (Fig. 14).

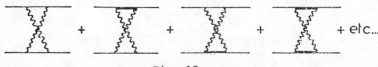

Fig. 14

These contributions are similar to those calculated in Section 3.1 with an important difference: in the coupled channel Schrödinger equation the intermediate states are treated nonrelativistically, whereas in Section 3.1 full relativity is taken into account for these intermediate states. Explicit calculations show that this mistreatment gives rise to important differences. Another important distinction between the two approaches is that the multichannel treatment generates only ladder diagrams and no crossed diagrams (Fig. 15); the latter come out naturally in the studies of Section 3.1 and are known to be just as important as the ladder ones.

Fig. 15

Last but not least, the multichannel approach is easily tractable only in the case of two channels, and up to now in this context only the Δ(1236) isobar effects have been considered in the literature. Again, it was known long ago (Amati et al. 1963, Cottingham and Vinh Mau 1963) that the inclusion of the Δ isobar alone together with the nucleon as intermediate states cannot provide a realistic nucleon-nucleon interaction if the ΠNΔ coupling constant is truly deduced from the Δ decay width. This gives rise to too much attraction especially in the isosinglet central potential. Later work showed that it is necessary to take full account of the ΠN interaction, meaning that higher resonances, $N_{11}(1400)$, etc..., play a significant role. Accordingly, multichannel calculations should be performed with transition potentials including higher isobars, but then one presumably runs into formidable calculational complications.

To summarize this Section, one can say that:

i)   not only the Δ resonance but also the other isobars give significant contributions to the NN forces

ii)   as they occur in intermediate states, they must be treated relativistically

iii)   not only ladder exchanges but also crossed exchanges should be included.

All these features arise naturally in the dispersion theoretic approach right from the start whereas in the multichannel treatment the original models (OBE models) have to be altered successively in an ad hoc way to account for some of these effects.

6. EXTENSION TO THE NUCLEON-ANTINUCLEON INTERACTION

During the last few years, several experiments have been reported (Cline et al. (1968), Kalbfleisch et al. (1969), Abrams et al. (1970), Gray et al. (1971), Benvenutti et al. (1971), Carroll et al. (1974), Kalogeropoulos et al. (1975), Chaloupka et al. (1976), Brükner et al. (1976)) indicating the existence of nucleon-antinucleon bound states and resonant states at energies near threshold. The widths of these states are found to be rather small (4-30 MeV), but their spin, parity, etc... are for the most part still undetermined.

Theoretically, the nucleon-nucleon and nucleon-antinucleon forces are closely related. Especially, when they are induced by particle exchange, one can get, in an unambiguous way, the $N\bar{N}$ amplitude from the NN one, simply by applying the G parity rule which changes the sign of the interaction if the exchanged particle has an odd G parity and leaves the sign unchanged in the even G parity case. This rule has been applied, in the past, to deduce $N\bar{N}$ potentials from the NN potentials provided by the one boson exchange (OBE) models (e.g. Bryan and Phillips (1968)).

In this Section are reported some results[38] on the $N\bar{N}$ bound states and resonances near threshold calculated with the $N\bar{N}$ potential that is deduced via the G parity rule from the above mentioned (Π+2Π+ω) exchange potential. For illustration, in Fig. 16 the central component of this potential (referred to as V^P) is displayed and for comparison, the Bryan-Phillips $N\bar{N}$ potential (referred to as V^{BP}) is also plotted. As it can be seen, the attraction is weaker in the V^P potential than in the Bryan-Phillips potential. The reason is that, in the NN case, this OBE model needs a strong ω exchange repulsion

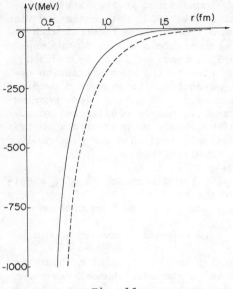

Fig. 16

$(g^2/4\Pi = 23.7)$ to partly compensate for the σ exchange attraction and that strong repulsion becomes, in the $N\bar{N}$ case, a strong attraction which adds to the σ attraction. In the V^P potential, one does not need such a strong ω coupling (our $g^2/4\Pi$ is only 9.5).

In Section 4.1, the long and intermediate range $(\Pi + 2\Pi + \omega)$ potential is cut off at internucleon distance $r \sim 0.8$ fm and the short range $(r < 0.8$ fm) is described by a phenomenological constant soft core. In the $N\bar{N}$ case, because of annihilation, the short range forces must contain also an absorptive component which can be described, for example, by a phenomenological complex potential. Here, one wished to examine the ability of the potential V^P to produce bound and resonant states near threshold with the underlying assumption that the short range part of the interaction will not cause drastic changes to the qualitative features of the spectrum.

The Schrödinger equation was solved with the following simple prescription for the short range potential

$$\alpha) \quad V_i(r) = V_i^P(r_c) \quad \text{for} \quad r < r_c$$
$$= V_i^P(r) \quad \text{for} \quad r > r_c$$

$$\beta) \quad V_i(r) = \frac{C_i}{1+(r/a)^{10}} + V_i^P(r)\,\frac{(r/a)^{10}}{1+(r/a)^{10}}$$

In both cases, the index i refers to the central, spin-spin, tensor, spin-orbit and quadratic spin orbit components of the potential. In model α, the parameter r_c is taken to be in the range from 0.6 fm to 0.8 fm and in model β, $a = 0.6$ fm and the parameters $C_i = 0$ except for the central component C_1.

The calculated spectrum is shown in Tables III and IV. From the results obtained and restricting ourselves to energies near threshold, we can draw the following conclusions:

i) The results are quantitatively sensitive to variations of the core parameters (r_c or C_1), but the bound states and resonances appear always in the same partial waves.

ii) Variations of the core parameters (r_c or C_1) do not alter the "spectroscopic order" of the levels.

Table III : The CMS energies (with respect to threshold) in MeV for the I = 1 NN̄ bound states and resonances calculated with our models α) and β).

	Model α			Model β			
	$r_c=0.6$ (fm)	$r_c=0.7$ (fm)	$r_c=0.8$ (fm)	$C_1=-1000$ (MeV)	$C_1=-500$ (MeV)	$C_1=-200$ (MeV)	$C_1=0$
$^1S_0$	-534	-262	-123	-589	-294	-168	-109
$^1P_1$	-293	- 55	30	-260	- 86	- 24	-0.3
$^3P_1$	-389	-116	- 17	-302	-130	- 64	- 34
$^3S_1$	(94%)	(94%)	(94%)	(98%)	(96%)	(95%)	(94%)
	-419	-202	- 92	-547	-256	-137	- 84
$^3D_1$	(6%)	(6%)	(6%)	(2%)	(4%)	(5%)	(6%)
$^1D_2$	- 74	165	>200	35	130	200	>200
$^3D_2$	-161	85	>200	-6.3	70	.115	150
$^3P_2$				(98%)			
	25			- 81			
$^3P_2$				(2%)			
$^3P_3$	200	>200	>200	>200	>200	>200	>200

Table IV : The CMS energies (with respect to threshold) in MeV for the I = 0 NN̄ bound states and resonances calculated with our models α) and β). (G) denotes a ground state, (E) an excited state.

	Model α			Model β			
	$r_c=0.6$ (fm)	$r_c=0.7$ (fm)	$r_c=0.8$ (fm)	$C_1=-1000$ (MeV)	$C_1=-500$ (MeV)	$C_1=-200$ (MeV)	$C_1=0$
$^1S_0$				-404	- 87	- 3	
$^3P_0$ (G)	<-800	<-800	-642	<-800	<-800	-801	-743
$^3P_0$ (E)	-545	-102	- 3	- 28	- 8	- 2	
$^1P_1$				125			
$^3S_1$			(49%)			(40%)	(39%)
(G)	<-800	<-800	-530	<-800	<-800	-710	-654
$^3D_1$			(51%)			(60%)	(61%)
$^3S_1$	(42%)	(57%)		(65%)	(88%)		
(E)	-386	- 39		-157	- 3		
$^3D_1$	(58%)	(43%)		(35%)	(12%)		
$^3P_2$		(45%)	(59%)	(49%)	(47%)	(46%)	(46%)
(G)	<-800	-788	-285	-753	-597	-517	-469
$^3P_2$		(55%)	(41%)	(51%)	(53%)	(54%)	(54%)
$^3P_2$	(58%)						
(E)	- 73						
$^3P_2$	(42%)						

iii) The tensor forces are very important not only because of their diagonal matrix elements but also because of their nondiagonal ones, especially in the isospin I = 0 states. For example, in the I = 0 states, the big differences between the values given by models α) and β) are due to the fact that in model β) the tensor force is suppressed for r < 0.6 fm. Here, the tensor forces are fairly taken into account by using a very accurate program code provided by Beiner and Gara (1972).

iv) The S states as well as the other low lying I = 0 states should not be taken too seriously since they should be significantly affected by the short range forces.

Comparison with experimental results can only be qualitative since besides the experimental uncertainties a more accurate description of the short range interactions is necessary. Nevertheless, for the $\bar{p}n$ system the $^3D_2$ and $^1D_2$ states can be proposed as good candidates for the experimental bumps at CMS energies with respect to threshold of 19 MeV (M = 1897 MeV) (Kalogeropoulos et al., 1975) and of 54 MeV (M = 1983 MeV) (Cline et al. (1968), Carroll et al. (1974), Chaloupka et al. (1976)) and the $^1P_1$ state as a good one for the bound state at -83 MeV (Gray et al., 1971). For the above mentioned reasons, we did not try to fit the data by adjusting r_C or C_1. However, we find with model α) and r_C = 0.675 fm, a 55 MeV (M = 1933 MeV) $^3D_2$ state, a 115 MeV (M = 1993 MeV) $^1D_2$ state and a $^1P_1$ -90 MeV state. We can then compare them respectively either to the experimental 1897 MeV, 1932 MeV and -83 MeV states or to the experimental 1932 MeV, 1968 MeV (Cline et al. (1968) and Benvenutti et al. (1971)) and -83 MeV; in the latter case, we would not be in agreement with the value J = 1 for the 1968 MeV state as proposed by Benvenutti et al. (1971).

As compared with previous works (Bogdanova et al. 1974, Peaslee, 1974, Dover, 1975) performed along the same lines with the VBP potential, this calculation gives fewer bound states and resonances (the prescription C_1 = 0 for the short range in model β) is almost the same as the one used by Bogdanova et al. (1974). On the other hand, it has been recently found (Myhrer and Thomas, 1976) that the imaginary part of the Bryan-Phillips potential can destroy the bound states and resonances produced by its real part. However, the absorptive potential they used has important effects even at large distances and therefore affects all partial waves. The actual situation could be somewhat in between, for example, the short range complex potential could be such that it absorbs mainly the S and P waves (Ball and Chew, 1958) and only slightly the D and higher waves leaving still some room for bound and resonant states. Here, the widths of the states have not been calculated since, unlike the positions, they must be dependent on the absorptive potential and cannot be calculated before a careful treatment of the short range forces, both in their scattering and annihilation components, has been performed.

7.   CONCLUDING REMARKS

I trust that I have clearly demonstrated that it is possible to derive quantitatively and unambiguously the nucleon-nucleon forces from the well established information on other independent hadronic processes. So far, this program has been carried out thoroughly for the long and medium range forces : they have been derived via a chain of dispersion, unitarity and crossing relations, from the ΠΠ and ΠN interactions. By including a very simple and economical parametrisation for the short range forces one can construct a potential which is realistic enough to fit the data on the two nucleon system in a very satisfactory manner. In turn, this potential is being used in various nuclear physics calculations (properties of finite nuclei, of nuclear matter and even of neutron stars). If the outcome of these calculations turns out to be reasonable we will have shown that it is possible to establish a coherent link between our knowledge of simple systems studied in particle physics and that of more complicated systems like nuclei or neutron stars.

Indeed, numerous attempts have been made over the past few years to introduce the mesonic degrees of freedom explicitly in calculating the properties of nuclei. In the work described here, the properties of pions and their interactions both amongst themselves and with nucleons are included in the potential right from the start. This approach can thus be considered as a very simple way of embodying the mesonic as well as "isobaric" effects in nuclear structure calculations.

REFERENCES

1) R.J. Abrams, R.L. Cool, G. Giacomelli, T.F. Kycia, B.A. Leontic, K.K. Li and D.N. Michael, Phys. Rev. D1, 1917 (1970)
2) D. Amati, E. Leader and B. Vitale, Phys. Rev. 130, 750 (1963)
3) J.S. Ball and G.F. Chen, Phys. Rev. 109, 1385 (1958)
4) J.L. Basdevant, B. Bonnier, C.D. Frogatt, J.L. Petersen and C. Schomblond in the Proceedings of the 2nd Aix-en-Provence Conference on Elementary Particles (1973)
5) M. Beiner and P. Gara, Computer Phys. Commun. 4, 1 (1972)
6) A. Benvenutti, D. Cline, R. Rutz, D.D. Reeder and V.R. Scherer, Phys. Rev. Lett. 27, 283 (1971)
7) R. Blankenbecler, M.L. Goldberger, N. Khuri and S.R. Treiman, Ann. Phys. 10, 62 (1960)
8) G. Bohannon and P. Signell, Phys. Rev. D10, 815 (1974)
9) K. Bongardt, H. Pilkuhn and A.G. Schaile, Phys. Lett. 52B, 271 (1974)
10) J. Bowcock and A. Martin, Nuovo Cimento 14, 516 (1959)
11) G.E. Brown and J.W. Durso, Phys. Lett. 35B, 120 (1971)
12) W. Brückner, B. Granz, D. Ingham, K. Kilian, V. Lynen, J. Niewisch, B. Pietrzyk, B. Povh, H.G. Ritter and H. Schröder, CERN preprint (1976)
13) A.S. Carroll, I.H. Chiang, T.F. Kycia, K.K. Li, P.O. Mazur, D.N. Michael, P. Mockett, D.C. Rahm and P. Rubinstein, Phys. Rev. Lett. 32, 247 (1974)
14) V. Chaloupka, H. Dreverman, F. Marzano, L. Montanet, P. Schmid, J. Fry, H. Rohringer, S. Simopoulon, J. Hanton, F. Grard, V.P. Henri, H. Johnstad, J.M. Lescaux, J.S. Skura, A. Bettini, M. Cresti, L. Peruzzo, P. Rossi, R. Bizzarri, M. Iori, E. Castelli, C. Omero and Poropat, Phys. Lett. 61B, 487 (1976)
15) M. Chemtob, J.W. Durso and D.O. Riska, Nucl. Phys. B38, 141 (1972)

16) D. Cline, J. English, D.D. Reeder, R. Terrell and J. Twitty, Phys. Rev. Lett. 21, 1268 (1968)

17) W.N. Cottingham, M. Lacombe, B. Loiseau, J.M. Richard and R. Vinh Mau, Phys. Rev. D8, 800 (1973)

18) W.N. Cottingham and R. Vinh Mau, Phys. Rev. 130, 735 (1963)

19) A. Donnachie, R.G. Kirsopp and C. Lovelace, Phys. Lett. 26B, 161 (1968)

20) C.B. Dover, Proc. IV International Symposium on Nucleon Anti-nucleon Interactions, Syracuse University, 1975, Vol. 2, 37

21) G.B. Epstein and B.H.J. McKellar, Phys. Rev. D10, 1005 (1974)

22) L. Gray, P. Hagerty and T. Kalogeropoulos, Phys. Rev. Lett. 26, 1491 (1971)

23) A.M. Green and P. Haapakoski, Nucl. Phys. A221, 429 (1974)

24) T. Hamada and I.D. Johnston, Nucl. Phys. 34, 32 (1962)

25) G. Höhler, R. Strauss and H. Wunder, Karlsruhe preprint (1968)

26) A.D. Jackson, D.O. Riska and B. Verwest, Nucl. Phys. A249, 397 (1975)

27) S. Jena and L.S. Kisslinger, Ann. Phys. 85, 251 (1974)

28) G. Kalbfleisch, R. Strand, V. Vanderburg, Phys. Lett. 29B, 259 (1969)

29) T.E. Kalogeropoulos and G.S. Tzanakos, Phys. Rev. Lett. 34, 1047 (1975)

30) M. Lacombe, B. Loiseau, J.M. Richard, R. Vinh Mau, P. Pires and R. de Tourreil, Phys. Rev. D12, 1495 (1975)

31) Livermore 1969: M.H. MacGregor, R.A. Arndt and R.M. Wright, Phys. Rev. 182, 1714 (1969)

32) M.H. MacGregor, R.A. Arndt and R.M. Wright, Phys. Rev. 173, 1714 (1968)

33) Moorehouse 1970: Particle Data Group, UCRL Report n°20030. Also, A.T. Davies, Nucl. Phys. B21, 359 (1970)

34) H. Nielsen and G.C. Oades, preprint (1974)

35) F. Partovi and E.L. Lomon, Phys. Rev. D5, 1192 (1972)

36) D.C. Peaslee, Phys. Rev. D9, 272 (1974)

37) R.V. Reid, Ann. Phys. 50, 411 (1968)

38) J.M. Richard, M. Lacombe and R. Vinh Mau, Phys. Lett. 64B, 121 (1976)

39) M.D. Scadron and H.F. Jones, Phys. Rev. D11, 175 (1975)

40) P. Signell and J. Holdeman, Phys. Rev. Lett. 27, 1393 (1971)

41) D.W.L. Sprung and R. de Tourreil, Nucl. Phys. A201, 193 (1973)

42) H. Sugawara and F. Von Hippel, Phys. Rev. 185, 2046 (1968)

43) R. Vinh Mau, J.M. Richard, B. Loiseau, M. Lacombe and W.N. Cottingham, Phys. Lett. 44B, 1 (1972)

44) Yale 1962: K.E. Lassila, M.H. Hull, M. Ruppel, F.A. McDonald and G. Breit, Phys. Rev. 126, 881 (1962)

45) H. Yukawa, Proc. Phys. Math. Soc., Japan, 17, 48 (1935)

CHIRAL SYMMETRY AND THE NUCLEON-NUCLEON INTERACTION[†]

G. E. Brown
Department of Physics
State University of New York, Stony Brook, New York 11794

I. INTRODUCTION

The long-range (r>1fm) part of the nucleon-nucleon inter-action has been successfully calculated in the past few years using dispersion relations. A report on this has been given to this conference by Prof. R. Vinh Mau. It seems clear to me that dispersion relations are the most convenient vehicle for these calculations, since they employ physical data from other processes, such as pion-nucleon and pion-pion scattering, and as we achieve a better understanding of these data, we can improve our calculations of the nucleon-nucleon potential.

Yet, I would like to go beyond the dispersion-theoretical calculations, and understand the NN-interaction in terms of a dynamical model, for two main reasons:

(1) The input data, such as $N\bar{N}\to\pi\pi$ amplitudes[*] in the dispersion-theoretical calculation are not direct physical observables, but must be obtained by analytic continuation. Despite claims by experts that this analytic continuation is accurate, amplitudes obtained by different experts differ significantly, and those of any one group change with time. Thus, it would be good to see what constraints can be imposed on these amplitudes by symmetries, such as the chiral symmetry I shall discuss. This is conveniently done within the framework of a dynamical model.

(2) Dispersion theoretical calculations apply only to on-shell interactions. For the many-body problem, it is important to know how to go off-shell, and dynamical models are needed for this.

On the other hand, I shall adopt the somewhat novel approach of assuming the $N\bar{N}\to\pi\pi$ helicity amplitudes, discussed in detail by Prof. Vinh Mau, to be the physical data (although they, in reality, are only obtained by analytic continuation from physical data), and will require my dynamical model to reproduce these data, more or less.

[†]Work supported by USERDA Contract No. E(11-1)-3001.

[*]For (energy)^2, $t>4m_n^2$, where m_n is the nucleon mass, these would be observable, but they are needed in the pseudophysical region $4m_\pi^2<t<4m_n^2$.

II. CHIRAL SYMMETRY; THE σ-MODEL

Chiral symmetry and charge invariance are included in symmetry under a direct product of two SU(2) transformations

$$SU(2) \times SU(2).$$

The generators of the SU(2)'s are

$$Q_i^{\pm} = (1 \pm \gamma_5) \frac{\tau_i}{2}, \tag{1}$$

the Q_i^+ referring to one SU(2) and the Q_i^-, to the other. The simplest basis vectors, in isospin space, for each SU(2), are two-component spinors

$$\chi = \begin{pmatrix} \chi_1 \\ \chi_2 \end{pmatrix}. \tag{2}$$

The direct product* of spinors from each of the SU(2)'s.

$$M_\alpha^{\bar{\beta}} = \chi_\alpha \chi^{\bar{\beta}} = [\sigma + i\vec{\pi}\cdot\vec{\tau}]_{\alpha\beta} = [M]_{\alpha\beta} \tag{3}$$

contains the scalar σ-field and the isovector π-field, which together transform like a four-vector under chiral transformations, generated by**

$$U(\vec{\beta}) = \exp \{-i \gamma_5 \vec{\beta}\cdot\vec{\tau}/2\} , \tag{4}$$

the $\vec{\beta} = (\beta_1,\beta_2,\beta_3)$ being three arbitrary real numbers characterizing the transformation.

The invariant is

$$I = \sum_{\alpha,\beta} M_\alpha^{\bar{\beta}} (M_\alpha^{\bar{\beta}})^* = \mathrm{Tr} \ [M][M^+] \tag{5}$$
$$= 3(\sigma^2 + \vec{\pi}^2),$$

showing clearly that σ and $\vec{\pi}$ form a four-vector.

The importance, for us, of an underlying chiral symmetry is that it gives a natural explanation for the vanishing of the isospin-averaged pion-nucleon scattering amplitude at threshold. This amplitude is[1]

$$a_o = a_1 + 2a_3 = -0.045 \pm 0.045 \tag{6}$$

where a_1 and a_3 are the scattering amplitudes in isospin states

---

* Here χ_α transforms under the one SU(2), $\chi^{\bar{\beta}}$ under the other.
** These U's are formed by $\exp\{-i(Q_i^+ - Q_i^-)\beta_i\}$.

I=1/2 and I=3/2, respectively, Were it not for chiral invariance, this smallness would appear to be coincidental. Historically, it was found already several decades ago that pseudoscalar coupling of pions to nucleons produced large scattering lengths. Through the virtual pair terms shown in figure 1,

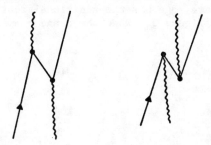

Figure 1: Contribution of virtual pair terms to π-nucleon scattering. Here the wavy lines represent the π-meson.

the coupling

$$\delta \mathcal{L} = g \, \bar{\psi} \, \gamma_5 \, \vec{\tau} \cdot \vec{\pi} \, \psi \tag{7}$$

gives a second order term

$$M_2 = \sum_n \frac{|<i|\delta\mathcal{L}|n>|^2}{E_i - E_n} \simeq \frac{g^2}{m_n} , \tag{7.1}$$

the energy denominator resulting because of the energy of the virtual pair $2m_n$; γ_5 has unit matrix element to the virtual pair state, and this gives a large M_2. Translated into a scattering length, (7.1) would give

$$a_1 + a_3 \simeq 2.8 \text{ fm} \tag{7.2}$$

which is orders of magnitude larger than (6). There is, then, a suppression of this amplitude by several orders of magnitude. The relation of these processes to the nucleon-nucleon interaction is shown in figure 2.

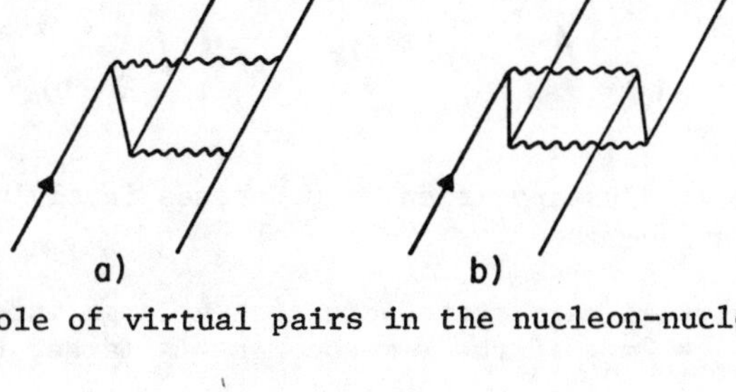

Figure 2: Role of virtual pairs in the nucleon-nucleon interaction.

It was realized historically that, since the process involving virtual pairs, figure 1, produces nonsense in the pion-nucleon scattering, the processes, figure 2, should be dropped in the nucleon-nucleon scattering. Thus, terms involving virtual pairs were dropped, with the invocation "pair suppression."

In the chiral game, one has the additional contribution of σ-exchange, shown by figure 3. When the σ and $\vec{\pi}$

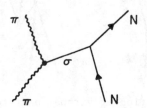

Figure 3: Role of the virtual σ-meson in π-nucleon scattering.

are built in through a four-vector, the underlying symmetry of the theory guarantees that the process, figure 3, will just cancel that of figure 1 at threshold; this cancellation will hold to all orders. Thus, the smallness of a_0 is no longer a coincidence, but guaranteed by the underlying symmetry. Within the framework of the σ-model, we shall next show how this comes about.

This role of the σ-meson in the π-nucleon scattering means that it will also play an important role in the nucleon-nucleon scattering. Each graph for pion interactions involving virtual pairs should be juxtaposed with one involving a σ-meson. For example, figure 2a should be put together with the graph involving a virtual σ-meson as shown in figure 4.

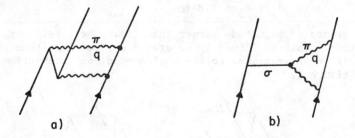

Figure 4: Juxtaposition of processes involving virtual pairs and σ-mesons.

Since the virtual pair state occurs at a large intermediate energy, $E_{int} \simeq 2m_n$, if the σ-meson mass is large, there will be an approximate cancellation between the processes, figure 4a and 4b for a wide range of intermediate momenta $\underset{\sim}{q}$, although the cancellation is exact only for $\underset{\sim}{q} \to 0$ (and, in fact $m_\pi \to 0$). This cancellation is a much more satisfactory description of processes involving virtual intermediate pairs than to just throw them away.

Chiral invariance gives a natural explanation of other important phenomena, among which are the Goldberger-Treiman relation, the Adler-Weisberger relation and the Weinberg formula for the individual S-wave π-nucleon scattering amplitudes. We shall not go into detail about these other relations.

III. THE SIGMA MODEL

The σ-model[2] is a very convenient model in which to illustrate chiral invariance. It is defined by the Lagrangian

$$\mathcal{L} = \mathcal{L}_o + \mathcal{L}_1 \tag{8}$$

where the chiral invariant $\mathcal{L}_o$ is

$$\mathcal{L}_o = \bar{\psi}[i\gamma_\mu \partial^\mu - g(\sigma + i\vec{\pi}\cdot\vec{\tau}\,\gamma_5)]\psi$$

$$+ \frac{1}{2}[(\partial_\mu \sigma)^2 + (\partial_\mu \vec{\pi})^2] - \frac{\mu^2}{2}[\sigma^2 + \vec{\pi}^2]$$

$$- \frac{\lambda^2}{4}[\sigma^2 + \pi^2]^2 \tag{8.1}$$

and the symmetry-breaking $\mathcal{L}_1$ is

$$\mathcal{L}_1 = c\sigma. \tag{8.2}$$

Since $\bar{\psi}\psi$ forms a scalar in the Lorentz group and $\bar{\psi}\vec{\tau}\gamma_5\psi$ transforms like a $\vec{\pi}$-meson, the term $\bar{\psi}(\sigma + i\vec{\pi}\cdot\vec{\tau}\,\gamma_5)\psi$ is the dot product of two four vectors; consequently, $\mathcal{L}_o$ can be seen to be made up out of dot products of four-vectors, and is obviously invariant under chiral transformations. The symmetry-breaking term $\mathcal{L}_1$, which contains only the σ-piece of the $\sigma, \vec{\pi}$ -four vector, obviously is not.

Note that the Lagrangian contains no nucleon mass term $\bar{\psi}m_n\psi$. Such a term is the 4th component of a four-vector, and is obviously not invariant under chiral rotations. How does the nucleon obtain its rest mass? We proceed to describe this.

It is convenient to rewrite $\mathcal{L}_o$ as

174

$$\mathcal{L}_0 = \bar{\psi}[i\gamma_\mu \partial^\mu - g(\sigma + i\vec{\pi}\cdot\vec{\tau} \gamma_5)]\psi$$

$$- \frac{\lambda^2}{4} \{[\sigma^2 + \vec{\pi}^2]^2 - f_\pi^2\}^2 + \frac{1}{2} [(\partial_\mu \sigma)^2 + (\partial_\mu \vec{\pi})^2] \tag{9}$$

with f_π a constant which can be expressed in terms of λ and μ. The $\mathcal{L}_0$ of (9) differs from that of (8.1) only by an inconsequential c-number.

For the moment, we take $\vec{\pi}=0$ and discuss the "potential energy"

$$V = \frac{\lambda^2}{4} [\sigma^2 - f_\pi^2]^2 \tag{10}$$

as function of σ. In this discussion we treat σ as a classical field (the so called semiclassical approximation), neglecting fluctuations, so that we can talk about V as a potential energy.

In figure 5 we plot V as a function of σ.

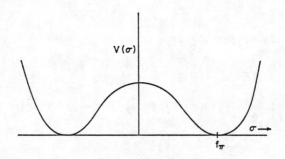

Figure 5: Potential energy V plotted as function of σ.

The minimum of V is at $\sigma=f_\pi$, so we can say that, to lowest order, $\sigma \cong f_\pi$. Identifying

$$gf_\pi = m_n, \tag{10.1}$$

we see that the σ-term in $\mathcal{L}_0$ is

$$-g \bar{\psi}\sigma\psi = - \bar{\psi} m_n \psi \tag{10.2}$$

and, in this way, we obtain the nucleon mass.

We now go on to discuss fluctuations in σ about the point f_π. The fluctuation field ϕ is defined by

$$\sigma = f_\pi + \phi. \tag{11}$$

In terms of ϕ,

$$V = \lambda^2 f_\pi^2 \phi^2 + \lambda^2 f_\pi \phi^3 + \frac{\lambda^4}{4} \phi^4. \tag{11.1}$$

We see that in addition to the quadratic term $\lambda^2 f_\pi^2 \phi^2$ which would occur in the ordinary Yukawa theory of scalar mesons, there are additional cubic and quartic terms which become important when ϕ is large.

In the Yukawa theory, the mass associated with the ϕ-meson enters into the Lagrangian through the term

$$\delta\mathcal{L} = -\frac{1}{2} m_\phi^2 \, \phi^2 \tag{12}$$

so we can identify

$$m_\sigma^2 = 2\lambda^2 f_\pi^2 = 2\lambda^2 m^2/g^2, \tag{12.1}$$

where this m_σ is the mass to be associated with small fluctuations, such as occur in the emission and absorption of quanta (scalar mesons) about the minimum of $V(\sigma)$.

We see that:

(i)   The curvature in $V(\sigma)$ at the minimum is proportional to m_σ.

(ii)   That λ is directly proportional to m_σ.

Thus, by making λ large, we can make the curvature in $V(\sigma)$ large.

Let us look now at the potential

$$V = \frac{\lambda^2}{4} \, [(\sigma^2 + \vec{\pi}^2) - f_\pi^2]^2, \tag{13}$$

inclusive of the $\vec{\pi}$ field. This will have the appearance of a roulette wheel, the bottom of the valley at radius f_π, (we take the $\vec{\pi}$-field to have only one component, so that we can plot V in three dimensions), as shown in figure 6.

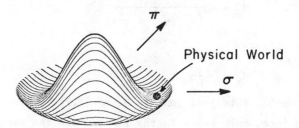

Figure 6:   The "potential" V as function of σ and $\vec{\pi}$-fields. The radius of the circle giving the minimum in V is f_π.

If m_σ is taken large enough, the curvature at the minimum will be sufficient to confine our physical system to that circle.

We now note that our symmetry-breaking term

$$\mathcal{L}_1 = c\sigma \overset{\sim}{=} c\sqrt{f_\pi{}^2 - \vec{\pi}^2} = cf_\pi(1 - \frac{\vec{\pi}^2}{2f_\pi{}^2} + ..)$$

$$= cf_\pi - \frac{1}{2}\frac{c}{f_\pi}\vec{\pi}^2 + ... \tag{14}$$

The term $(c/2f_\pi)\vec{\pi}^2$ can be equated with the mass term $(m_\pi{}^2/2)\vec{\pi}^2$ which would occur in the Lagrangian for a pion with mass so we can identify

$$c = f_\pi\, m_\pi{}^2. \tag{14.1}$$

Thus, the introduction of $\mathcal{L}_1$ has the effect of tilting our roulette wheel, so that the point $\vec{\pi}=0$, $\sigma=f_\pi$ is now the unique minimum, and this is where our physical world is assumed to sit, as shown in figure 6.

In order to calculate the contribution of σ-exchange, figure 3, we need the $\sigma\pi\pi$ interaction, or, rather, the $\phi\pi\pi$ interaction, since ϕ is the scalar fluctuation field. We can obtain this from eq. (13) by setting $\sigma=f_\pi+\phi$, the relevant term then being

$$\delta\mathcal{L}_{\sigma\pi\pi} = \lambda^2 f_\pi\, \phi\, \vec{\pi}^2 = \frac{m_\sigma{}^2}{2f_\pi}\, \phi\, \vec{\pi}^2. \tag{15}$$

We have the constant $m_\sigma{}^2/2f_\pi$ for the coupling at the $\sigma\pi\pi$ vertex, g for the NNσ vertex. Each emission or absorption of the σ-particle is accompanied by the boson normalization factor $1/\sqrt{2\omega_q}$, where $\omega_q = \sqrt{\underset{\sim}{q}^2 + m_\sigma{}^2}$; there is an energy denominator $-\omega_q$, one factor of 2 because of the two orders of emission and absorption and another factor of 2 from the π^2. Altogether, we have the matrix element

$$M_\sigma = -\frac{m_\sigma{}^2}{f_\pi}\frac{1}{(m_\sigma{}^2 + \underset{\sim}{q}^2)}. \tag{15.1}$$

Using (10.1) we find

$$M_\sigma = -\frac{g^2}{m_n} \tag{16}$$

in the limit $\underset{\sim}{q}\rightarrow 0$. This just cancels M_2 of (7.1). This cancellation is, in fact, only exact in the limit where the pion four-momentum goes to zero, because in the evaluation of M_2, eq. (7.1), we've neglected terms in the pion mass m_π compared with m_n, and these enter into M_2 in order $m_\pi{}^2/m_n{}^2$.

The above development does not mean that one should see, in experiments, effects from an elementary σ-meson of sharp mass. Whatever theoretical indications there are require a σ-mass of the order of m_n, so the exchange of an elementary σ-meson would

give an interaction of range $\sim \hbar/m_n c$, which would be hidden inside
of the repulsive core from ω-meson exchange. The $\sigma\pi\pi$ - coupling,
eq. (15), will immediately "pionize," the nucleon-nucleon interaction resulting from processes such as that shown in figure 7.

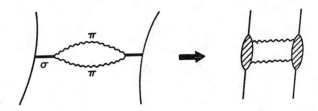

Figure 7: Origin of the long-range interaction mediated
by the σ-particle.

We can, in fact, go further and enclose the process of two-pion
emission or absorption by a "black box," as shown on the right of
figure 7, applying the chiral constraints to the left half and
right half of the figure, each viewed as a pion-nucleon
scattering, separately.

Assuming the σ-mass m_σ to be large, $m_\sigma \gg m_\pi$, an elegant way
to express all of the above[3] is to eliminate the σ-meson
completely through a canonical transformation; this converts
the pion-nucleon coupling from pseudoscalar to pseudovector.
Thus, we may use pseudovector coupling for the pion, forgetting
about the σ-particle.

IV. APPLICATION TO THE NUCLEON-NUCLEON INTERACTION

The chiral Lagrangian (8) is designed for the calculation
of on-mass-shell amplitudes involving pion emission and absorption by nucleons. We wish now to formulate the calculation of
the nucleon-nucleon interaction so that it can be applied.

The one-pion exchange potential is undoubtedly the most
"solid" part of our knowledge about the nucleon-nucleon interaction, and has been used for many years. Dispersion relations
have been applied[4] to the calculation of the two-pion-exchange
potential; exchange of two pions should give the interaction of
longest range, stretching out to $r \sim \hbar/2m_\pi c$, after the one-pion-
exchange potential.

The idea of the dispersion-relation calculation is to view
the two pions as connecting to the nucleons through "black boxes,"

178

shown in figure 8, which include all possible interactions
(including, for example, the process in which a σ-meson is
emitted by the nucleon, the σ splitting up into two pions).

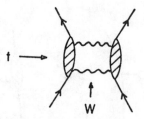

Figure 8: The two-pion-exchange interaction. Here the
wavy lines represent the pions. All possible interactions
are included in the cross-hatched boxes. The invariants
w (usually called s) and t are indicated.

The process figure 8 is then laid on its side and split into
possible on-shell intermediate states as shown in figure 9,

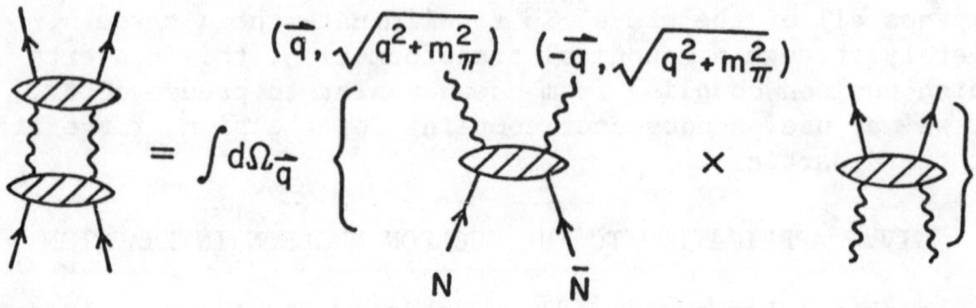

$$\left(\vec{q}, \sqrt{q^2 + m_\pi^2}\right) \quad \left(-\vec{q}, \sqrt{q^2 + m_\pi^2}\right)$$

$$= \int d\Omega_{\vec{q}} \left\{ \quad N \quad \bar{N} \quad \times \quad \right\}$$

Figure 9: The decomposition in the t-channel of the process
shown in figure 8. Here q is the magnitude of the pion
momentum in intermediate states

$$q = \sqrt{\tfrac{1}{4}t - m_\pi^2},$$

t playing the role of (energy)^2 in this channel. We work
in the c.m.s. of the N$\bar{\text{N}}$ system (or, equivalently, two-pion
system).

The next step is the expansion of the $N\bar{N} \to \pi\pi$ amplitudes into helicity amplitudes of definite angular momentum in the t-channel, shown in figure 10.

$$f_{\pm}^{J}(t) = \int d\Omega_{\vec{q}} \; P_J(\hat{q})$$

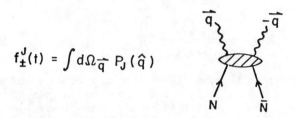

Figure 10: The expansion of the amplitude into helicity amplitudes of definite angular momentum in the t-channel. The lower suffices + - on the helicity amplitude f denote non-spin-flip and spin-flip amplitudes, respectively.

We shall concentrate particularly on the $f_{+}^{J=0}$ amplitude.

Knowing it, we can obtain the contribution of the J=0, t-channel to the nucleon-nucleon interaction

$$V^{J=0} = -C \int_{4m_{\pi}^2}^{\infty} \frac{q \, |f_{+}^{J=0}(t)|^2}{\sqrt{t} \; (m_n^2 - \tfrac{1}{4}t)^2} \; \frac{e^{-\sqrt{t}\, r}}{r} \; dt. \tag{17}$$

This is not quite the whole story, since part of this comes from the iteration of the one-pion-exchange term, and the latter piece should be subtracted out to obtain a genuine two-pion-exchange potential[5].

We see from eq. (17) that the exchange of J=0 systems of two-mesons leads to an interaction which is the same as would result from the exchange of a system of scalar mesons, of masses $m_s = 2\sqrt{m_{\pi}^2 + q^2}$, with amplitude $\sqrt{q} \; f^{J=0}(t)/\sqrt{t} \; (m_n^2 - \tfrac{1}{4}t)$, q and t being connected

$$q = \sqrt{\tfrac{1}{4}t - m_{\pi}^2} \; . \tag{17.1}$$

Thus, the low-q behavior of $f_{+}^{J=0}(t)$ determines the longest-range part of the two-pion-exchange interaction, and this brings us to soft pions, pions with small momenta!

Our prescription, then, is to calculate $f_{+}^{J=0}(t)$ from the Lagrangian in which the pion is coupled to the nucleons by pseudo-

180

vector coupling. This calculation is straightforward, and yields[6]

$$f_+^{J=0}(t) = \frac{g^2}{4\pi}\{m_n h \arctan h^{-1} - \frac{t}{4m_n}\}\tag{18}$$

where

$$h = \frac{q^2 + t/4}{2q\sqrt{m_n^2 - t/4}}\ .\tag{18.1}$$

It should be noted that the amplitude goes to zero as q and m_π both go to zero ($q_\mu \to 0$, all μ), although if only q goes to zero, m_π remaining fixed, one finds the value shown in figure 11 for $t = 4m_\pi^2$. The result (18) is the same as obtained by Brown and Durso[6] using other methods, but enforcing the soft-pion constraints.

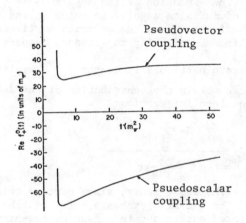

Figure 11: The soft-pion $f_+^{J=0}$ obtained from eq. (18) is shown by the upper solid line. The lower solid line shows the $f_+^{J=0}$ which would be obtained from pseudoscalar coupling of the pion to the nucleon. The general shift between the two curves is due to the contribution of virtual pairs in the curve from pseudoscalar coupling.

Had we used pseudoscalar coupling, we would have had an additional term

$$\delta f = (\frac{1}{4}t - m_n^2)/m_n\tag{18.2}$$

inside the curly brackets in eq. (18) and this would have produced the lower curve in figure 11. The large difference between these two curves comes from the virtual pair term which enters with pseudoscalar coupling and is represented by (18.2).

As can be seen from eq. (17) only $\left|f_+^{J=0}(t)\right|^2$ enters into
the nucleon-nucleon interaction, so that the different signs of
the $f_+^{J=0}(t)$'s calculated in pseudoscalar and in pseudovector
coupling play no role here and, in fact, for the low partial
waves in nucleon-nucleon scattering at ~300 MeV incident nucleon
energy, the two $f_+^{J=0}(t)$'s give roughly equivalent fits[7].

However, it should be noted that the pseudovector $f_+^{J=0}(t)$
is substantially smaller than the pseudoscalar one for small t;
furthermore, this is the regime in which the soft-pion
$f_+^{J=0}$ should be valid. This region can be tested by looking at
peripheral partial waves in the nucleon-nucleon interaction,
since low t implies large distances. The comparison here shows
that the two-pion exchange term, with the pseudovector $f_+^{J=0}$
helps in reproducing the peripheral partial waves.

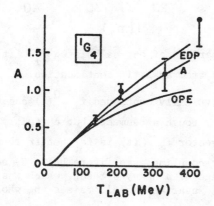

Figure 12: Results for the $^1G_4$ partial wave, as compared
with the Livermore energy-independent phase shifts[8]
and energy-dependent phase-shift analysis (curve labelled
EDP). The curve labelled OPE corresponds to the predictions
from one-pion exchange. Curve A is the result with the J=0
t-channel weighting function calculated from our pseudoscalar $f_+^{J=0}(t)$, eq. (18), and J=1 helicity amplitudes
corresponding to a two-pion continuum plus distributed-mass
ρ-exchange.[6] The main dependence is on $f_+^{J=0}(t)$.

182

The $f_+^{J=0}(t)$ can be obtained directly from analytical
continuation of the pion-nucleon scattering amplitudes, which are
determined empirically for negative t, to positive t. Assuming
this analytical continuation to be accurate - an assumption not
completely justified - we can view these analytically continued
f's to be empirical data, and ask how well our model reproduces
them. In figure 13 we show such a comparison

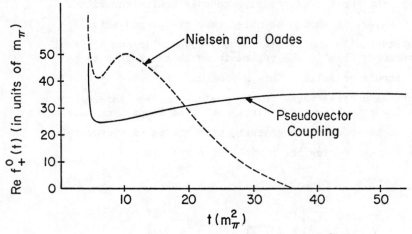

Figure 13: Comparison of our calculated $f_+^{J=0}(t)$ with
results obtained from analytic continuation.

For small t, the analytically continued $f_+^{J=0}(t)$ and that from our
model are in only very rough agreement, but certainly the agreement with our pseudovector $f_+^{J=0}(t)$ is far better than with the
one from pseudoscalar coupling (see figure 11). We see, therefore, that the exchange of a heavy σ-meson, which was built into
our underlying chiral invariant model, raises the whole curve of
$f_+^{J=0}(t)$ about the right amount.

The fact that the pseudoscalar curve is moved bodily over
the whole range of t up by about the same amount is consistent
with a heavy σ-meson. The exchange of a heavy object would give
a short-range interaction, whose contribution to the helicity
amplitude would be relatively independent of t.

Assuming that we know the $f_+^{J=0}(t)$ over the entire range of
t shown in figure 13, we may wish to add ingredients to our chiral
model so that we have a dynamical model which will reproduce this
amplitude, to a good approximation, up to larger t. The two
additional ingredients we wish to add are:

(i)   Intermediate $\Delta(1230)$ isobars.

(ii) $\pi\pi$ scattering.

The $\Delta(1230)$ isobar is most conveniently put in as a spin-3/2, isospin-3/2 particle, described by the theory of Rarita and Schwinger[9]. In this case, the $\pi N\Delta$ coupling is

$$\frac{f^*}{m_\pi} \bar{\psi}_\mu \, \vec{T} \cdot \frac{\partial \vec{\phi}_\pi}{\partial x_\mu} \, \psi(x) \tag{19}$$

where $\vec{T}$ is the transition spin, ψ_μ the Rarita-Schwinger isobar. The coupling constant f^* is given, in the constituent quark model, to be

$$\frac{f^{*2}}{f^2} = \frac{72}{25} \tag{19.1}$$

in terms of f, the pion-nucleon coupling. It may be more suitable to determine f_π^* from the decay width of the isobar, in which case one finds[10]

$$\frac{f^{*2}}{f^2} = 4, \tag{19.2}$$

the same value as would result in the Chew-Low theory of the isobar.

The fact that the isobar is coupled derivatively, eq. (19) means that none of the results obtained for $q_\mu \to 0$ will be changed. The effect on $f_+^{J=0}(t)$ of inclusion[11] of the isobar is shown in figure 14. It should be noted

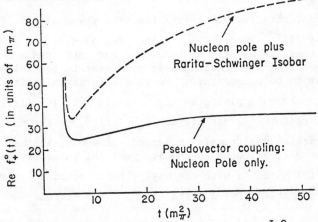

Figure 14: Effect of the $\Delta(1230)$ isobar on the $f_+^{J=0}(t)$ helicity amplitude.

that the contribution of the isobar goes to zero for small t. Indeed, the nucleon contribution also goes to zero as $q \to 0$; it was constructed that way. The nucleon pole in $f_+^{J=0}(t)$ lies so close to the soft-pion point that $f_+^{J=0}(t)$ is large already at $q^2 = -m_\pi^2$, $t = 4m_\pi^2$. The $\Delta(1230)$ isobar pole lies, however, further away, and the recovery from zero is smooth, as can be seen from figure 14.

We note in passing that the $\Delta(1230)$ contribution to the $f_\pm^{J=1}(t)$ is small , as is the contribution to helicity amplitudes of higher J, so that inclusion of the virtual $\Delta(1230)$ isobar affects chiefly the spin-zero, isoscalar t-channel. Said otherwise, inclusion of the isobar enhances the intermediate-range attraction in the nucleon-nucleon interaction, in just the same way as the exchange of scalar, isoscalar mesons (just the type foreseen in the old Yukawa theory) would do.

The result that intermediate $\Delta(1230)$ isobars contribute overwhelmingly to the S=0, I=0 t-channel is the result of the relativistic theory, and implies, in nonrelativistic terms, that the crossed meson exchange processes are just as important as the uncrossed ones. This is an important fact, which has not generally been taken into account in coupled-channel calculations employing the $\Delta(1230)$-isobar.

Thus for pion rescattering has been left out of the description, so that our $f_+^{J=0}(t)$ is completely real. In the region

$$4m_\pi^2 < t < 16m_\pi^2$$

elastic unitarity guarantees that the phase of $f_+^{J=0}(t)$ is given by $\exp(i\delta_{\pi\pi}^{J=0})$ where $\delta_{\pi\pi}^{J=0}$ is the pion-pion S-wave phase shift. This phase is usually assumed over the total range of t, not only up to $t = 16m_\pi^2$, since inelastic processes in the relevant channels appear to be unimportant. We thus enforce unitarity in an ad hoc fashion by multiplying our $f_+^{J=0}(t)$ by $\exp(i\delta_{\pi\pi}^{J=0}(t))$. Such a phase factor will not affect $|f_+^{J=0}(t)|^2$, so the nucleon-nucleon interaction will remain unchanged. However, inclusion of this factor will allow us to compare real and imaginary parts of our $f_+^{J=0}(t)$ directly with the analytically continued results; for this comparison, we use the same $\delta_{\pi\pi}^{J=0}$ as Nielsen and Oades[12] and the results are shown in figure 15[*].

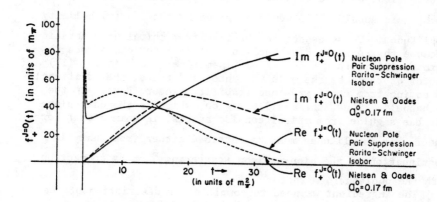

Figure 15: Comparison of helicity amplitudes Re $f_+^{J=0}$ and Im $f_+^{J=0}$ calculated from our model which includes the nucleon-pole term, eq. (18), calculated with pseudovector coupling and the Rarita-Schwinger isobar. Real and imaginary parts are obtained by assuming the phase of the amplitude to be exp i $\delta_{\pi\pi}^{J=0}$(t), where the pion-pion phase shifts of Nielsen & Oades[12] were used. For comparison, the amplitudes of Nielsen and Oades, obtained by analytic continuation of the πN→πN amplitudes are shown.

We have finally reached a fairly good agreement between our model curves and the analytically continued ones. Inclusion of the Σ-term (the chiral-symmetry breaking term) in our model would remove part of the discrepancy at small t, but the magnitude of this term is still very uncertain. We believe the differences between model curves and amplitudes from analytic continuation to be not much larger than uncertainties in the latter method, and that only experimental information of the

[*] In fact, we could obtain our $\delta_{\pi\pi}^{J=0}$(t), for small t at least, also from the σ-model[13] but it is not clear the the σ-model works well here, and, in any case, we use the $\delta_{\pi\pi}^{J=0}$ only for comparison with other amplitudes; they do not enter into our nucleon-nucleon interaction.

type shown in figure 12, will pin down $f_+^{J=0}(t)$ for small t, $t \lesssim 10$.

In order to build a dynamical model of the nucleon-nucleon interaction, we need to know also the interaction in the ρ-meson exchange channel. This would be given by the $f_\pm^{J=1}(t)$ helicity amplitudes. These cannot be obtained from chiral considerations, because the corresponding nucleon-nucleon interaction goes to zero rapidly with $q \to 0$ just from kinematics; i.e., in eq. (17) a factor q^3 replaces the q in the integrand there, the additional q^2 coming from the P-wave penetrability factor (P-wave in the t-channel). Thus, the small t region, which is the only place one has a guide from chiral considerations, is unimportant here. The $f_\pm^{J=1}(t)$ amplitudes must be obtained either from analytic continuation of the pion-nucleon scattering[12] or by connection with the nucleon form factor[14].

The ingredient we need to complete our description of the nucleon-nucleon interaction is the short-range repulsion. Traditionally, this has been thought to arise from ω-meson exchange. However, the largest coupling constant which would be allowed by SU(3) symmetry lies in the region

$$\frac{g_{\omega NN}^2}{4\pi} \simeq 5 \tag{20}$$

whereas in empirical nucleon-nucleon interactions, a number of ~ 10-20 seems to be preferred. Durso et al. showed that combined π- and ρ-exchange accompanied by virtual $\Delta(1230)$ isobar can explain the additional repulsion*.

In any case, it may be too ambitious to expect to explain the short-range nucleon-nucleon interaction in terms of particle exchange. One would expect the nucleon wave function to take a nosedive and go rapidly to zero as one moves in towards small distances, <1 fm, just as if there were a strong, short-range repulsion, from general arguments; namely, as many channels other than the two-nucleon one become virtually excited, the amplitude of wave function in the two-nucleon channel must become small, just because of the normalization condition (this is my version of a not-so-private argument of Herman Feshbach).

We see that chiral invariance gives important guidelines for constructing the nucleon-nucleon interaction, although one needs further arguments to obtain the short-range interaction. In fact,

*One can obtain some predictions about ρ-nucleon or $\rho N\Delta$ couplings from chiral symmetry, where the ρ is built in as a gauge particle (see, e.g., B.W. Lee and H.T. Nieh, Phys. Rev. 166 (1968) 1507.) These predictions relate, however, to only the vector coupling, whereas the tensor coupling of the ρ is mainly responsible for the effects studied here.

we are probably still far away from anything but a rough, semi-phenomenological description of the short range interaction and it may be that we shall have to make a theory in terms of interacting quarks, or something along these lines, until we have an adequate description.

One can view the above attempt to build a dynamical model of the nucleon-nucleon interaction as complementary to the work of the Paris group; we wish to know the <u>content</u> of their model, in terms of virtual particle excitation. We monitor our methods by requiring them to fit the same helicity amplitudes, which we treat as empirical data.

Were we to describe only nucleon-nucleon scattering in free space, such a dynamical model might appear to be a luxury, once we had a complete description of the interaction in terms of dispersion relations. However, if we wish to describe the nucleon-nucleon interaction in the presence of other nucleons, in the many-body system, then a dynamical model is necessary.

To see this, let us consider the attractive interaction between two nucleons arising from the scalar, isoscalar t-channel exchange. The dispersion theoretical model, when it gets to the level of the nucleon-nucleon interaction, can only describe this "black-box" interaction in the same manner as if it arose from exchange of systems of scalar,isoscalar mesons of variable masses as shown in figure 16. The dynamical model,

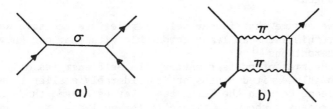

Figure 16: a) Description of the intermediate-range attraction in terms of σ-meson exchange; b) in terms of virtual isobar excitation.

as we have outlined, would describe most of the intermediate-range attraction in terms of virtual-isobar excitation, figure 16b.

In the many-body system, these two descriptions give quite different results. We shall discuss why, without getting lost in the complexities of current many-body formalismsm by very simple arguments. These arguments concern nucleon (and isobar) self energies in the many-body system. We show typical

contributions to the self energies in figure 17. For the purposes
of this

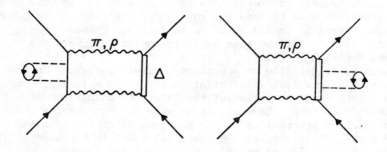

Figure 17: Typical energy interactions of the excited
nucleon or isobar in the many-body system. Whereas the
wavy lines represent the π or ρ mesons, the dotted lines
represent any of the mesons, including π and ρ, which can
be exchanged.

discussion, we assume that isobar self energies are the same as
nucleon ones; this seems at least compatible with what one knows
about the interactions [10].

In all theories of nuclear matter, the self energies of
nucleons in highly excited states are considerably smaller than
those of nucleons down in the Fermi sea. Let us assume that
through the processes shown in figure 16 the nucleons are in an
excited state a fraction κ_I of the time,

$$0 < \kappa_I < 1,$$

and in practice κ_I may be ∿0.1. Then, the change in nucleon
energy will be

$$\Delta\Sigma = (\bar{\Sigma}_{exc} - \bar{\Sigma}_{unexcited})\, \kappa_I.$$

In the Brueckner theory, where off-energy shell effects enter
into $\bar{\Sigma}_{exc}$, this correction is called the "dispersion correction."
In the hypernetted chain calculations[19] currently popular, a
similar effect arises from lack of commutation of potentials, and
of correlation operators and potentials. Very roughly speaking,

the effect is as we describe it: It is unfavorable for the nucleon to be excited, for during the time that it is, its interactions with the other nucleons are much less attractive than when it's in the unexcited state.

Complete calculations are very difficult to make, and have not been made, so we end our discussion with the above qualitative point.

REFERENCES

1. H. Pilkuhn et al., Nucl. Phys. B65, 460 (1973).
2. M. Gell-Mann and M. Lévy, Nuovo Cimento 16, 705 (1960).
3. S. Weinberg, Phys. Rev. Lett. 18, 188 (1967).
4. D. Amati, E. Leader and B. Vitale, Nuovo Cimento 17, 68 (1960); 18, 409 (1960).
5. This procedure is described in all gory detail in "The Nucleon-Nucleon Interaction" by G.E. Brown and A.D. Jackson, North-Holland Publ. Co., Amsterdam, Oxford 1976.
6. G.E. Brown and J.W. Durso, Phys. Lett. 35B, 120 (1971).
7. A.D. Jackson, D.-O. Riska and B. VerWest, Nucl. Phys. A249, 397 (1975).
8. M.H. MacGregor, R.A. Arndt, R.M. Wright, Phys. Rev. 182, 1714 (1969).
9. W. Rarita and J. Schwinger, Phys. Rev. 60, 61 (1941).
10. G.E. Brown and W. Weise, Phys. Repts. 22C, 281 (1975).
11. J.W. Durso, M. Saarela, G.E. Brown and A.D. Jackson, Nucl. Phys. A278, 445 (1977).
12. H. Nielsen and G. Oades, to be published.
13. S. Weinberg, Phys. Rev. Letts. 17, 616 (1966).
14. G. Höhler and E. Pietarinen, Nucl. Phys. B95, 210 (1975).

ROLE OF THE RANGE OF ISOBAR POTENTIALS
AND OF THE NΔρ-VERTEX IN NN-SCATTERING[+]

K. Holinde and R. Machleidt[++]
Institut fuer Theoretische Kernphysik, Bonn, Germany

ABSTRACT

The effect of modifying the range of isobar potentials according to a prescription given in a recent paper by Durso, Jackson, Brown and Saarela is demonstrated for some important NN-scattering phase shifts. Compared to the common choice, this new (shorter) range reduces the isobar contribution by about 50%. Furthermore, we study the effect of the NΔρ-vertex. This contribution not only reduces the cutoff-dependence of the transition potential, but also strongly improves the fit of certain phase shifts. The consideration of both effects allows the use of cutoff-parameters in the transition potentials which are in a much more reasonable range than before. Finally we investigate the effect of going to the static limit in the isobar potentials. It is shown that this effect is not negligible; however, the static limit seems to be a reasonable approximation provided (π+ρ)-exchange is taken into account.

INTRODUCTION

Nowadays, theoretical models for the two-nucleon interaction based on dispersion theory[1] are able to give a description of the NN-scattering and bound state data which is as good as that given by purely phenomenological models or semitheoretical models in the OBE-frame[2]. However, with regard to application in nuclear structure, all such models have a characteristic drawback: It is not possible to take into account modifications of the NN-interaction due to the presence of other nucleons, (e.g. Pauli- and dispersive corrections suppressing the intermediate-range attraction).

Part of these modifications can however be taken into account by introducing the Δ-isobar explicitly. Recently[3], we have presented a momentum-space nucleon-nucleon potential in the OBE-frame, in which part of the phenomenological σ-contribution describing the intermediate-range attraction is replaced by twice-iterated pion-range transition potentials, which couple the NN-channel with the NΔ and ΔΔ-channels. A reasonable description of NN-data was obtained which is sufficiently accurate to allow for a meaningful nuclear matter calculation. Due to Pauli- and dispersive effects in the transition potentials, a standard first-order Brueckner calculation yielded a strong reduction of

---

[+] Work supported in part by Deutsche Forschungsgemeinschaft

[++] Present address: Dept. of Physics, SUNY, Stony Brook, N.Y. 11794

the binding energy given by a pure OBEP. However, the results
suffer from two drawbacks: i) a strong cutoff is required in order
to sufficiently damp the contributions of the transition potentials; consequently, the results show a strong sensitivity to the
cutoff-parameter; ii) characteristic discrepancies show up in the
description of the NN-phase shifts.

In this paper we want to show how strongly the situation is
improved a) by including also ρ-exchange in the transition potentials, b) by using a modified range of the transition potentials
given in a recent paper by Durso, Jackson, Brown and Saarela[4].
Furthermore, since we stay throughout in momentum space, we are
able to study the quality of the approximation of going to the
static limit in the transition potentials. This approximation
is generally made by other authors in order to obtain an explicit
expression in r-space.

FORMALISM

Since our procedure is to neglect interactions between the
NΔ- and $\Delta\Delta$-channels, the R-matrix, R_{NN}, is given by a Lippmann-
Schwinger-type equation

$$R_{NN}(\underline{q}',\underline{q},q_o)=V_{eff}(\underline{q}',\underline{q},q_o)-MP\!\int d^3k \; \frac{V_{eff}(\underline{q}',\underline{k},q_o)R_{NN}(\underline{k},\underline{q},q_o)}{k^2 - q_o^2} \quad (1)$$

with

$$V_{eff}(\underline{q}',\underline{q},q_o)=V_{NN}(\underline{q}',\underline{q})-P\!\int d^3k \; \frac{V_{N\Delta}(\underline{q}',\underline{k})V_{\Delta N}(\underline{k},\underline{q})}{\dfrac{k^2}{2M}+\dfrac{k^2}{2M_\Delta}-\dfrac{q_o^2}{M}+M_\Delta-M} \quad (2)$$

V_{NN} is the interaction in the NN-channel, e.g. the usual OBEP;
$V_{N\Delta}$ describes the transition potential between the NN- and NΔ-
channel. Since, in this paper, we are only interested in studying
the qualitative effect of the range and of the ρ-exchange, we
neglect $V_{\Delta\Delta}$, i.e., the contribution arising from the double-Δ-
excitation.

The contribution of the πNΔ-vertex to $V_{N\Delta}$ is given in the
helicity state basis as (for isospin-one-states)

$$<\underline{q}'\Lambda_1'\Lambda_2'|V_{N\Delta}^\pi|\underline{q}\Lambda_1\Lambda_2^*> = \frac{g_{NN\pi}f_{N\Delta\pi}}{2\pi^2 m_\pi}\sqrt{\frac{8}{3}}\; F_\pi \Delta_\mu \; \frac{\bar{u}_{\Lambda_2'}(-\underline{q}')u_{\Lambda_2^*}^\mu(-\underline{q})\bar{u}_{\Lambda_1'}(\underline{q}')\gamma^5 u_{\Lambda_1}(\underline{q})}{\Delta^2 + m_\pi^2} \quad (3)$$

$g_{NN\pi}$, $f_{N\Delta\pi}$ are appropriate coupling constants; $F_\pi=(\Lambda_\pi^2-m_\pi^2)^2/(\Lambda_\pi^2+\Delta^2)^2$,
($\underline{\Delta}$=momentum transfer), is a form factor with Λ_π being a parameter,
the so-called cutoff mass. u denotes the positive-energy Dirac

spinor describing the nucleon, whereas u^μ is the conventional Rarita-Schwinger spinor describing the Δ-isobar. $V^\rho_{N\Delta}$ is found to be

$$<\underline{q}'\Lambda'_1\Lambda'_2|V^\rho_{N\Delta}|\underline{q}\Lambda_1\Lambda^*_2> = \frac{4\pi}{(2\pi)^3}\frac{f_{N\Delta\rho}}{m_\rho}\sqrt{\frac{8}{3}}\;F_\rho\bar{u}_{\Lambda'_2}(-\underline{q}')\gamma^\mu\gamma^5 u_{\Lambda^*_2}(-\underline{q})$$

$$\times\;\frac{g^\beta_\nu\underline{\Delta}_\mu - g^\beta_\mu\underline{\Delta}_\nu}{\underline{\Delta}^2 + m_\rho^{\;2}}\;\bar{u}_{\Lambda'_1}(\underline{q}')[g_{NN\rho}\gamma_\beta + \frac{f_{NN\rho}}{2M}\;i\;\sigma_{\beta\alpha}\;\Delta^\alpha]u_{\Lambda_1}(\underline{q}) \qquad (4)$$

$g_{NN\rho}$, $f_{NN\rho}$, $f_{N\Delta\rho}$ are appropriate coupling constants: $\sigma^{\mu\nu}=\frac{i}{2}[\gamma^\mu,\gamma^\nu]; F_\rho=(\Lambda_\rho^2-m_\rho^2)^3/(\Lambda_\rho^2+\underline{\Delta}^2)^3$. The propagator $\underline{\Delta}^2+m^2_{(\rho)}$ is a simple choice based on arguments given by Smith and Pandharipande[5]. However, it is pointed out in ref.[4] that a better choice can be made by replacing $V^{\pi(\rho)}_{N\Delta}$ of eqs. (3,4) by

$$\bar{V}^{\pi(\rho)}_{N\Delta} = 1/2\;V^{\pi(\rho)}_{N\Delta} + 1/2\;V^{\pi(\rho)}_{N\Delta}{}' \qquad (5)$$

where $V^{\pi(\rho)}_{N\Delta}$ is the same as before and $V^{\pi(\rho)}_{N\Delta}{}'$ differs from $V^{\pi(\rho)}_{N\Delta}$ in that, in the propagator, $m_{\pi(\rho)}$ is replaced by $m'_{\pi(\rho)} = [m^2_{\pi(\rho)}+m_{\pi(\rho)}(M_\Delta-M)]^{\frac{1}{2}}\approx\sqrt{3}\;m_\pi(1.17\;m_\rho)$. This modification is obtained from estimates made for certain second-order, time-ordered processes in old-fashioned perturbation theory.

RESULTS

Fig. 1 shows ^1S_o-phase shifts as function of E_{lab}. The error bars denote the empirical Livermore analysis[6]. HM2 gives the result predicted by our OBEP[2]. NO SIGMA represents the case where the whole σ-contribution, V_σ, is taken away from HM2. The other curves show the effect of adding the twice-iterated potentials to HM2-V_σ: PI uses the transition potential (3), whereas PI+RHO is obtained using (3)+(4). PI'+RHO' uses the modified prescription (5); PI'+RHO', ST is obtained if the static limit is used at the vertices of the transition potential. The parameters for the corresponding transition potentials are shown in table I. We see that the modified range reduces the isobar contribution by as much as 50%; (this is true also for higher partial waves). Furthermore, the static approximation is surprisingly good. This is very important in view of the complexity of the expressions containing the full relativistic $N\Delta$-vertex, and justifies to some extent the procedure of other authors.

In order to study whether the inclusion of these effects improves the fit of the NN-phase shifts, we have made the following calculation: In a first step, for PI', PI+RHO, PI'+RHO', we enlarge Λ_π such that these three versions predict the same $^1S_0$-phase shifts as PI in fig. 1. The resulting values for Λ_π are shown in table I in parentheses. Note that we keep Λ_ρ the same as before. This shows, by the way, that the transition potentials including ρ-exchange remain cutoff-dependent. However, this dependence is very much reduced; moreover, the resulting values for Λ_π are now in a much more reasonable range than before. In a second step, we add the σ-contribution of HM2 again, reducing the coupling constant such that all four versions predict low-energy $^1S_0$-phase shifts which are in agreement with experiment. Some of the resulting phase shifts are shown in figs. 2-4. We first notice that ρ-exchange and shorter range seem to be required in order to get the high-energy $^1S_0$-phase shifts, too. Furthermore, it is clearly demonstrated that ρ-exchange is absolutely necessary to get a good description of the data.

Table I

Parameters for the different transition potentials. The first and second value for $m_{\pi(\rho)}$ (in MeV) in the case of PI' and PI'+RHO' belong to $V_{N\Delta}^{\pi(\rho)}$ and $V_{N\Delta}^{\pi(\rho)'}$, respectively. The first value for Λ_π (in MeV) belongs to the curve in Fig. 1, the value in parentheses to the curves in figs. 2-4.

	PI	PI'	PI+RHO	PI'+RHO'
$f_{N\Delta\pi}^2$	0.36	0.36	0.36	0.36
m_π	138	138;240	138	138;240
Λ_π	850 (850)	850 (1200)	850 (980)	850 (1600)
$f_{N\Delta\rho}^2$	--	--	9	9
m_ρ	--	--	711	711;832
Λ_ρ	--	--	1500	1500

194

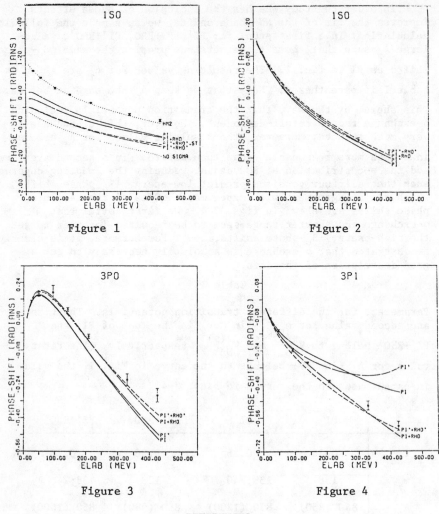

Figure 1

Figure 2

Figure 3

Figure 4

REFERENCES

1. M. Lacombe et al., Phys. Rev. D12, 1495 (1975).
2. e.g. K. Holinde and R. Machleidt, Nucl. Phys. A256, 479 (1976).
3. K. Holinde and R. Machleidt, Nucl. Phys. A280, 429 (1977).
4. J.W. Durso et al., Nucl. Phys. A278, 445 (1977).
5. R.A. Smith and V.R. Pandharipande, Nucl. Phys. A256, 327 (1976).
6. M. MacGregor et al., Phys. Rev. A82, 1714 (1969).

PHASE-EQUIVALENT NONLOCAL POTENTIALS

Peter U. Sauer
Theoretical Physics, Technical University
3ooo Hannover, Germany

ABSTRACT

The construction and application of nonlocal potentials equivalent with respect to the experimental two-body data and to the theoretical knowledge on the exterior of the nuclear force are discussed. Constraints on the nonlocality are given.

INTRODUCTION

The successful use of the Sussex interaction[1] in nuclear structure, i.e., of interaction matrix elements in oscillator representation derived from the phase shifts alone with no apparent additional assumption on the two-nucleon potential, seemed to imply that for the purposes of nuclear-structure calculations a correct account of the experimental two-body data is all what matters. This implication, e.g., illustrated in Fig. 1 by the ^{18}O spectra

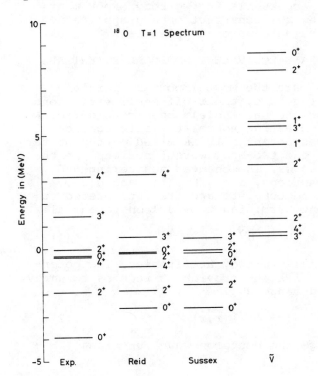

Fig. 1. Shell-model spectrum of ^{18}O. The results represent a calculation with the bare effective interaction without core-polarization correction. The spectrum of the Sussex interaction is taken from Ref.1, that of the potential $\tilde{V}$ from Ref. 3.

of the Sussex interaction and of the soft-core Reid potential, which agree well, motivated strongly the search for possible off-shell effects or for their absence. It created interest in the nuclear-structure application of potentials, which generated from an original one by nonlocal variations[2] at small relative distances, are <u>equivalent</u> among each other and to the original one with respect
(i) to the experimental two-body data and
(ii) to the accepted theoretical knowledge on the exterior
     of the nuclear force.
These phase-equivalent potentials told us, however, that the dependence of nuclear structure on the nuclear force is not simply through on-shell data only. There are nonlocal potentials, equivalent to the Reid potential in the above sense, whose spectra are entirely out of focus. An example is given in Fig. 1. Thus, the necessity of a theoretically well-founded potential model of the two-nucleon interaction for nuclear-structure calculations was proven.

However, even the best theoretically available potential, the Paris potential[4], is phenomenologically chosen at relative distances smaller than 0.8 fm. The equivalent nonlocal potentials are naive attempts to study the importance of that short-range part of the potential for nuclear structure, which is theoretically not yet known and inaccessable in two-body experiments. I was asked to review the construction and application of these phase-equivalent nonlocal potentials.

CONSTRUCTION OF PHASE-EQUIVALENT NONLOCAL POTENTIALS

$^3S_1-^3D_1$ and $^1S_0$ are the most important partial waves for nuclear structure, their off-shell variations are most interesting and have solely been considered. A possible method for creating nonrelativistic nonlocal variations of a given potential V is based on the unitary transformation U of the two-body wave functions[5] in the respective partial waves. The changed scattering and bound-state wave functions, i.e., $U|\phi(k)>$ and $U|D>$, remain orthonormal and complete and are interpreted to belong to a different hermitian two-nucleon potential

$$\tilde{V} = U[H^O + V] U^+ - H^O , \qquad (1)$$

$H^O + V$ is the original two-body hamiltonian. The potentials $\tilde{V}$ and V are equivalent, provided U reduces to unity outside a prescribed range R

$$< r| U| r'> = \delta(r-r') \quad (r,r' > R) \qquad (2)$$

and U does not change the deuteron wave function

$$U|D> = |D> . \tag{3}$$

Then, all potentials $\tilde{V}$ and V differ only at relative distances smaller than R by nonlocalities,

$$<r|\tilde{V}|r'> = <r|V|r'> \quad (r,r'>R) . \tag{4}$$

R can be chosen such that the core region alone is modified, but that all what is presently considered unquestionable fieldtheoretic knowledge on the exterior part remains untouched. U also preserves the phase-shift fit automatically, since the asymptotic behavior of the scattering wavefuntions of all potentials $\tilde{V}$ and V is the same,

$$<r|U\phi(k)> = <r|\phi(k)> \quad (r>R) . \tag{5}$$

The condition (3) on the deuteron wave function is sufficient but not necessary to leave the deuteron properties unaltered. It is powerful in reducing the size of off-shell effects due to the $^3S_1-^3D_1$ partial wave.

The method of two-body unitary transformations can be extended to many-nucleon systems[6,7]. Inverse scattering theories can provide an alternative to unitary transformations. If they employ so general potential models that the two-body data are insufficient to determine the potential in full, assumptions on the off-shell behavior have to be added. Varying these assumptions yields phase-equivalent potentials. Examples are Fiedeldey's method[8] based on two-term separable potentials or the inverse scattering theory based on the general transition matrix[9-11]. Results obtained with different techniques will be freely mixed together in Section 3.

The method of off-shell variations by unitary transformations is obviously flexible. It would be without objection provided the employed unitary transformations could at least be selected with convincing physical intuition. Instead the freedom in U has always been chosen according to mathematical convenience, e.g., the simple separable form

$$U = 1-2|g><g| , \tag{6a}$$

$$<r|g> = C r (1-\beta r)e^{-\alpha r} , \quad <g|g> = 1 \tag{6b}$$

has been a popular choice for $^1S_0$ potentials[12]. α^{-1} controls the range, β the shape of nonlocality by which the potentials $\tilde{V}$ and V differ. In examples I shall refer to those parameters in order to distinguish potential models. Especially in $^1S_0$, for which no constraint similar to the deuteron (3) existed for a long time, potentials with rather peculiar features resulted as is born out in Fig. 2

198

by the shape of sample ^1S_O wave functions. These unusual features, bumps and additional nodes, offend our sense for sound and simple physics and have understandably

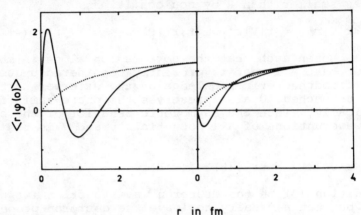

r in fm

Fig. 2. Zero-energy ^1S_O wave functions $\langle r|\phi(o)\rangle$ in the effective-range normalization of Figs. 11 and 12 of Ref. 3 without Coulomb. The dotted curve refers to the Yamaguchi potential, the solid curves refer to wave functions transformed according to Equs. (6). The range parameter α is $3.0\ \mathrm{fm}^{-1}$. $\beta = 2.00\ \mathrm{fm}^{-1}$ for the left diagram 2a, $\beta = 0.90\ \mathrm{fm}^{-1}$ (wave function with additional node on right hand side) and $\beta = 1.26\ \mathrm{fm}^{-1}$ for the right diagram 2b. The ratio of Fig. 5 is 0.05 for the two wave functions of Fig. 2b.

created a widespread allergy against these potentials. Some of these exotic features are indications that crimes not only against aesthetics, but against subtle physical requirements have been committed, which were not always realized. What are the possible pitfalls ?

(1) Pitfall no. 1:
Many ^1S_O potentials do not fit the experimental proton-proton (pp) phase shifts any more. In many applications the purely hadronic hamiltonian $H^O + V = K + V$, K being the kinetic energy operator, is varied, i.e., off-shell changes with respect to the only theoretically calculated Coulomb subtracted hadronic phase shifts are performed. When adding the electromagnetic (e.m.) interaction W back to the hadronic hamiltonian, the phase shift fit is spoilt[3]. The range of possible values for the ^1S_O pp scattering length, shown in Fig. 3, becomes even larger when in addition to static Coulomb as used for the diagram the complete one-photon exchange with magnetic and momentum-dependent contributions is added in [13]. The potentials are not equivalent as desired. They become equivalent, when generated from a hamiltonian which also includes the e.m. pp interaction W, i.e., $H^O + V = K + W + V$.

However, pitfall no. 1 is not yet avoided. For, many ^1S_O potentials $\widetilde{V}$, which then do fit the experimental pp data

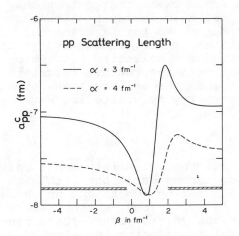

Fig. 3. Scattering length a^c_{pp} of the pp ^1S_O phase shifts. The hamiltonian used is $K + W + \widetilde{V}$ with $H^O = K$ in Equ. (1) and W being the finite-size Coulomb potential. The original potential V is the Reid soft-core potential. The unitary transformation (6) is employed with two different range parameters α, the results are shown as funtion of the parameter β. The experimental value of $- 7.823 \pm 0.010$ fm is indicated by the cross-hatched strip.

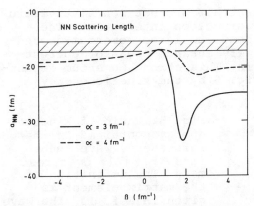

Fig. 4. nn scattering length resulting from the hadronic potentials $\widetilde{V}$ of Fig. 3 with $H^O = K + W$. The cross-hatched strip indicates the experimental value of the nn scattering length.

exactly the same as the original potential V, are unable to account for the ^1S_O neutron-neutron (nn) scattering length as the original potential V does[14]. This is demonstrated in Fig. 4. The spread of possible values becomes much larger when all one-body and two-body e.m. differences between the pp and nn systems are considered[13]. Thus, many ^1S_O potentials $\widetilde{V}$ are not equivalent when interfering with different e.m. potentials. Since charge symmetry should hold as a theoretical principle with high precision, the combined ^1S_O pp data and the nn scattering length provide some off-shell information on the two-nucleon interaction, at least in the negative sense that those parametrizations of the pp potential predicting an nn scattering length in grave disagreement with experiment

have to be rejected. However, in practice this approximate
off-shell condition should even be strengthened. The potentials are usually employed in nuclear-structure calculations, for which one assumes exact charge independence. The $^1S_0$ potentials acceptable for such calculations
should be equivalent with respect to the experimental
data in all three charge states of $^1S_0$. Charge independence
becomes an off-shell criterion for selecting the appropriate potentials for those nuclear-structure calculations. The constraint is met when the unitary transformations do not change the zero-energy pp scattering wave
function

$$U|\phi(o)> = |\phi(o)> .\qquad(7)$$

For then it can be shown[3] that the potentials $\tilde{V}$ not only
have exactly the same pp scattering data, but with high
accuracy also the same nn scattering length and pn phase
shifts as V. Condition (7) is sufficient, not necessary.
It is the $^1S_0$ equivalent to the $^3S_1-^3D_1$ bound-state condition (3) and is as powerful in reducing the size of
off-shell effects in $^1S_0$.
    (2) Pitfall no. 2:
Many phase-equivalent potentials are not nonrelativistic
parametrizations of the two-nucleon interaction. The
strong nonlocality corresponds to large momentum-dependent components, which unduely enhance operator corrections of relativistic origin. This strong momentum dependence is demonstrated in Fig. 5 by the kinetic energy

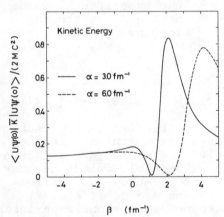

Fig. 5. Ratio of kinetic
energy contained in the
zero-energy wave funct-
ions $<r|\phi(o)>$ over rest
mass. The definition of
the matrix element is
given in Ref. 13. The wave
functions are identical
with those of Fig. 2.

contained in the zero-energy wave function[13], which is
for most potentials not small as compared to the rest
mass of the system. The same ratio for the corresponding
Reid wave function is with 1% minute on the scale of
Fig. 5. Most of these potentials should therfore be rejected for nonrelativistic nuclear-structure calculations.

If the kinetic energy is constrained to stay nonrelativistically small, the allowed shapes of the wave functions of the acceptable potentials become less terrifying as shown in Fig. 2b.

But even, if the nonlocal potentials avoid all these recognized pitfalls, their parametrization at small relative distances remains arbitrary. Thus, we should not expect those potentials to yield physically very probable results when applied to actual nuclear-physics problems. In fact, one should never try them in a precious calculation in which only one or two potentials could be tested. It is understood that these potentials should be used only with special care for quite specific questions. With these reservations in mind have there been useful applications of phase-equivalent nonlocal potentials?

APPLICATION OF PHASE-EQUIVALENT NONLOCAL POTENTIALS

The phase-equivalent nonlocal potentials have helped to clarify the interplay between the two-nucleon interaction on one side and nuclear structure and nuclear experiments on the other side in three respects:

(1) The phase-equivalent nonlocal potentials established the off-shell dependence of nuclear-structure results, but to our great disappointment improved agreement with experiment has never emerged from off-shell variations.

Examples for this type of results are the ^{18}O spectrum of the Fig. 1 which got messed up by off-shell variations, the saturation property of nuclear matter in lowest order Brueckner theory and the three-nucleon bound state properties in a comparatively exact many-body calculation. Saturation density and binding energy of nuclear matter, as shown in Fig. 6, are correlated such that a correct account of both by off-shell variations appears impossible in lowest order Brueckner theory. With respect to the three-nucleon bound state no one has yet succeeded in finding a hamiltonian, which consists only of two-body potentials and accounts simultaneously for the binding energy and the charge form factors in impulse approximation. Any improvement on the binding energy increases the disagreement between the theoretical and experimental ^{3}He charge form factor, and any improvement on the form factor worsens the binding energy. This information illustrated in Fig. 7 is valuable empirical evidence for the importance of three-body forces for the binding energy and of three-body forces and exchange-current corrections for the three-nucleon charge form factors. On the quatitative side the off-shell effects due to variations of the $^{1}S_{0}$ partial wave are gross overestimates, since some potentials being variations of hadronic

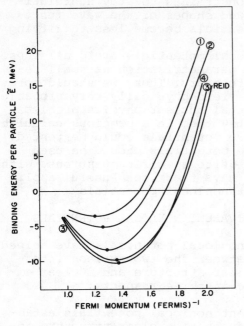

Fig. 6. Binding energy and saturation density of nuclear matter in lowest order Brueckner theory for phase-equivalent non-local potentials. The results of the Reid soft-core potential and of nonlocal variations in the ^1S_O partial wave are shown. They are taken from Ref. 12.

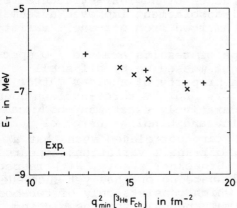

Fig. 7. ^{3}H binding energy E_T and position q^2_{min} of the ^{3}He charge form factor minimum for phase-equivalent nonlocal potentials. Results are taken from Ref. 11 (+) and Ref. 15 (x).

hamiltonians do not fit[3] the pp data and have scattering wave functions whose kinetic energy is excessive for such a nonrelativistic calculation. E.g., among the nuclear-matter off-shell variations shown in Fig. 6 the dramatic ones of 5.0–6.5 MeV due to potentials 1 and 2 are not acceptable. Results of the two non-acceptable potentials are not shown for the three-nucleon bound state in Fig. 7.

(2) The phase-equivalent nonlocal potentials estab-lished model-independent relations between nuclear-struct-ure quantities, at least illustrated known relations in a

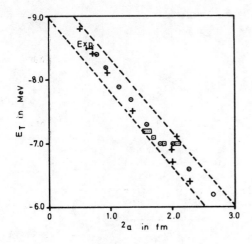

Fig. 8. The Phillips' plot (triton binding energy versus doublet neutron-deuteron scattering length) for phase-equivalent nonlocal potentials. The results are taken from Ref. 16.

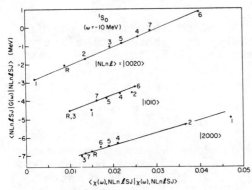

Fig. 9. ^1S_O reaction matrix for phase-equivalent nonlocal potentials. Diagonal matrix elements are plotted versus the corresponding wound integral. The results are taken from Ref. 17.

convincing manner.

Examples are the Phillips' relation, the band of Fig. 8 between doublet neutron-deuteron scattering length and the triton binding energy, or the linear relationship of Fig. 9 between the effective interaction in the nuclear medium, here diagonal oscillator matrix elements, and the corresponding defect integral, the integral over the difference between the correlated and uncorrelated wavefunctions squared. The latter linear relation translates itself to a similar dependence of the total binding energy of nuclear matter or a finite nucleus on the total wound, as shown in Fig. 10. It suggests that the relevant off-shell parameter for nuclear-structure calculations is not the individual form of the short-range correlations, which can differ grossly in character and size, but the defect integral. Since the short-range correlations differ grossly, the relations appear model-independent. Though the defect wave functions of Fig. 11 differ grossly, they are always very similar in nuclear

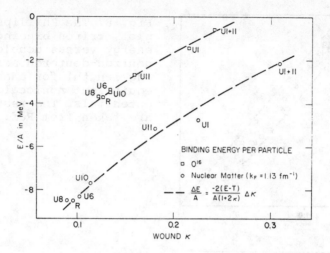

Fig. 10. Dependence of the ^{16}O and nuclear-matter binding energy on the wound. The results for phase-equivalent nonlocal potentials are taken from Ref. 18.

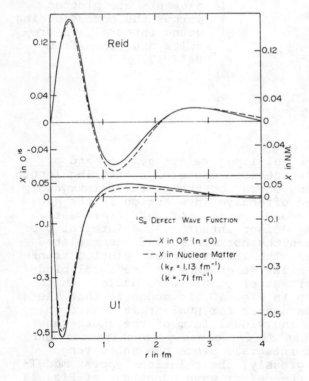

Fig. 11. The $^1S_0$ defect wave functions in ^{16}O and nuclear matter. The results of the Reid soft-core potential and a phase-equivalent potential (U 1) are taken from Ref. 18.

matter and finite nuclei. Since the defect wave functions
are the only medium-dependent ingredients of the effective
interaction in the nuclear medium, this fact supports in
a picturesque way other evidence, that it is not an
accident of the Reid soft-core potential or of similar
interactions that the local-density approximation works[19]
and that the effective interaction in finite nuclei may
be taken over from nuclear matter.

(3) The phase-equivalent nonlocal potentials corrected the traditional belief in the predictive power of
two important two-nucleon experiments.

Low-energy nucleon-nucleon scattering has been believed to provide direct and reliable evidence on the
isospin properties of the nuclear interaction. However,
charge symmetry solely applies to the hadronic part of
the nucleon-nucleon interaction. Even in the advent of
accurate and abundant neutron-neutron data, nucleon-
nucleon scattering will experimentally prove charge symmetry only to the extent that the charge-dependent one-
body and direct e.m. two-body effects can theoretically
be removed from the data in an unambiguous manner. The
traditional belief is that this can be done practically
in an almost model-independent way. Phase-equivalent potentials have contradicted this belief. The purely hadronic pp $^1S_0$ scattering length to be compared with the nn
scattering length depends strongly on the parametrization
of the nucleon-nucleon interaction used. The variation is
large when the static Coulomb is subtracted[14], it is
dramatic when all e.m. contributions are subtracted[13]. The
model-dependence is illustrated in Fig. 12, which demon-

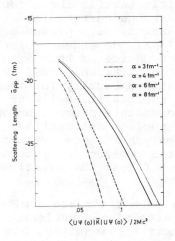

Fig. 12. Model-dependence of
the hadronic pp scattering
length. All values above the
curves up to the solid line are
possible due to model dependence.
The phase-equivalent potentials
allowed for the analysis have
constrained nonlocalities consistent with the kinetic energy
values of the abscissa. The results are given for various
range parameters α. The zero-
energy wave functions are not
changed outside 1.0 fm for
$\alpha = 6.0$ fm^{-1} and $\alpha = 8.0$ fm^{-1} and
$<\overline{K}> < 5\%$. The results are taken
from Ref. 13.

strates that the strength of nonlocalities rather than
the range of nonlocalities controls the size of model-

dependence in the hadronic pp ^1S_O scattering length. But
even when the ^1S_O constraint arising from the nonrelativistic analysis is enforced,   the value of the hadronic pp scattering length is so nonunique that an unambiguous comparison with the corresponding nn scattering
length for the purpose of establishing the amount of
nuclear charge symmetry fails. There is still hope that
bremsstrahlung or reliable calculations in few-nucleon
systems will eventually eliminate some ^1S_O potentials
employed in the analysis of Fig. 12. For the time being
the ^1S_O pp and nn scattering lengths can only establish
within the framework of sound theoretical models model-
dependent bounds on the violation of charge symmetry.
I note in passing, that I have just reinterpreted the
results discussed for pitfall no. 1 in the present
changed context.

It had also been suggested that nucleon-nucleon
bremsstrahlung experiments could be used to extract information on the off-shell behavior of the two-nucleon
interaction, e.g., to extract properties of the half-
shell T matrix of the strong interaction in an unambiguous way. Implicit in this notion is the assumption that
one knows the e.m. charge and current density operators.
However, in the presence of nonlocalities the ordinary
one-body operators of charge and current are not conserved, exchange-current corrections must be added. Knowing the fieldtheoretic source of the nonlocality, these
operator corrections can be calculated. If the nonlocality is phenomenological, one does not know how to choose
them. This ambiguity may turn out to be disastrous, if
one allows oneself strong phenomenological nonlocalities
at small relative distances. There are only two calculations with phase-equivalent potentials testing the effect
of different assumptions on the exchange current for
bremsstrahlung. Ref. 20 using potentials with presumably
weak nonlocalities finds a rather small dependence of the
bremsstrahlung cross section on the choice of the exchange current, whereas the model calculation of Ref. 21
shown in Fig. 13 indicates a dramatic ambiguity for phase-
equivalent potentials with strong nonlocalities. At any
rate, the extraction of model-independent information on
the off-shell behavior of the two-nucleon interaction
from bremsstrahlung appears extremely hard. One presumably
has to resign to pick models for the two-nucleon interaction and check, if they are after a fit to the elastic
data consistent with bremsstrahlung.

The phase-equivalent nonlocal potentials are admittedly tools from the stone-age physics of the nuclear
force. They are very limited in their applicablity, but
I hope that the set of examples, which is not exhaustive,
illustrates their usefulness for particular questions,

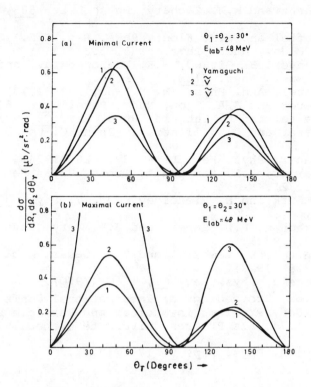

Fig. 13. Bremsstrahlung cross section in the Harvard geometry for three phase-equivalent model potentials. Two extreme assumptions are made on the exchange current, (a) uses the minimal, (b) the maximal current. The non-localities, by which the potentials differ are in potential 3 of smaller range than in 2. The results are taken from Ref. 21.

which might be hard to answer otherwise.

REFERENCES

1. J.P. Elliott, Proceedings of International Conference on Properties of Nuclear States, Montréal 1969 (Les Presses Université Montréal 1969), p. 277.
2. M.K. Srivastava and D.W.L. Sprung, Advances in Nuclear Physics Vol. 8 (Plenum Press New York 1975), p. 121, and references there.
3. P.U. Sauer, Phys. Rev. C11, 1786 (1975).
4. M. Lacombe, B. Loiseau, J.M. Richard, R. Vin Mau, P. Pires, and R. de Tourreil, Phys. Rev. D12, 1495 (1975).
5. H. Ekstein, Phys. Rev. 117, 1590 (1960).

208

6. A.W. Sáenz and W.W. Zachary, Phys. Lett. $\underline{58B}$, 13 (1975).
7. M.I. Haftel and W.M. Kloet, Phys. Rev. $\underline{C15}$, 404 (1977).
8. H. Fiedeldey, Nucl. Phys. $\underline{A135}$, 353 (1969).
9. M. Baranger, B. Giraud, S.K. Mukhopadhyay, and P.U. Sauer, Nucl. Phys. $\underline{A138}$, 1 (1969).
10. P.U. Sauer, Ann. Phys. (N. Y.) $\underline{80}$, 242 (1973).
11. P.U. Sauer and J.A. Tjon, Nucl. Phys. $\underline{A216}$, 541 (1973).
12. M.I. Haftel and F. Tabakin, Phys. Rev. $\underline{C3}$, 921 (1971).
13. P.U. Sauer and H. Walliser, J. Phys. G, to be published.
14. P.U. Sauer, Phys. Rev. Lett. $\underline{32}$, 626 (1974).
15. E.P. Harper, Y.E. Kim and A. Tubis, Phys. Rev. $\underline{C6}$, 1601 (1972).
16. N.J. McGurk and H. Fiedeldey, Can. J. Phys. $\underline{53}$, 1749 (1975).
17. H.C. Pradhan, P.U. Sauer, and J.P. Vary, Phys. Rev. $\underline{C6}$, 407 (1972).
18. M.I. Haftel, E. Lambert, and P.U. Sauer, Nucl. Phys. $\underline{A192}$, 225 (1972).
19. J.W. Negele, Phys. Rev. $\underline{C1}$, 1260 (1970).
20. L. Heller, Proceedings of International Conference on Few Body Problems in Nuclear and Particle Physics, Québec 1974 (Les Presses Université Québec 1975), p. 206.
21. E.M. Nyman, Phys. Rep. $\underline{9}$, 179 (1974).

SUPER SOFT CORE MODELS OF
THE N-N INTERACTION

D.W.L. Sprung

Physics Dept., McMaster University, Hamilton, Ontario L8S 4M1

ABSTRACT

Weak local phenomenological potential models of the N-N interaction developed by the McMaster-Montreal-Orsay collaboration
are described along with some applications.

A super soft core model of the nucleon-nucleon interaction
is a local potential model in which the usual infinite repulsive
core has been replaced by a finite repulsion, in such a way that
the perturbation series in nuclear matter appears to converge at
least in the lowest orders. This name was coined to distinguish
such a model from a hard core (infinite repulsion) or soft core
(Yukawa, as in Reid's potential[1]; or square core as in Bressel-
Kerman-Rouben[2]). The discovery of such forces was quite surprising in 1969. Since the time of Stapp's phase shift analysis in
1956[3], and the Gammel Thaler[4] potential fitted to it, it had become an article of faith that the two body data required a strong
repulsive core. The Brueckner theory was elaborated largely as a
tool to handle this short range repulsion. The Hamada Johnston[5]
and Yale[6] potentials assumed infinite repulsion, but after C. W.
Wong[7] pointed out the sensitivity of nuclear matter binding energy
to this feature, Reid developed his soft core potential.

Despite the popularity of strong repulsive cores, there were
some people who entertained the old hope that a "weak" two body
force might be found, such that perturbation theory could be used
to understand nuclear structure. One example of this effort was
the Bressel-Kerman-Rouben potential, which was developed from the
Hamada Johnston force. The infinite or hard core was replaced by
a finite square core of somewhat larger radius. The resulting
force however was still strong enough to require Brueckner theory.
More successful was the Tabakin force[8], which was something of a
curiosity because it was a non-local separable model. Tabakin had
the right idea however: the two body data determine only the on-
shell T-matrix elements, so if one wishes to explore the degree of
latitude this allows for nuclear structure, it is best to choose
a form of two body potential very different from the usual ones.
Also at MIT, Kerman, Rouben and others replaced the square local
core of the BKR force by a non local separable core.

One consequence of introducing finite repulsion at short
distances is that this region becomes interesting for the dynamics.
Professor Vinh Mau has described the Paris potential including one
and two pion exchange mechanisms, in which the N-N force is related to πN and $\pi\pi$ scattering. The three pion exchange is also
now being studied. However, as one includes more complicated ex-

ISSN: 0094-243X/78/209/$1.50 Copyright 1978 American Institute of Physics

changes, these predominate over a narrower and narrower region: one pion beyond 2.5 fm, two pion beyond 1.2 fm, etc. so it is not clear that this procedure can ever terminate. The optimistic view is that inside some distance, perhaps 0.7 fm, the NN force will become infinitely repulsive, the wave function will vanish, and all remaining unknowns become irrelevant. If one is less optimistic, the interior region requires a different approach. In our case it is phenomenological.

Our first SSC force was discovered accidentally by M.K. Srivastava. We wanted to calculate off-shell matrix elements directly from the phase shifts, without actually constructing a two body potential. Of course, this problem was resolved in an elegant manner by Baranger, Giraud, Mukhopadhyay and Sauer shortly thereafter[8]. As a step along the way we were studying the inverse scattering problem. The Marchenko procedure constructs a local potential equivalent to a given two body $^1S_0$ S-matrix S(k). This was taken from the phase-shift analysis of MacGregor, Arndt and Wright[9,10]. We fitted S(k) to a rational function, which gives a very good fit, and then analytically constructed the local potential[11]. It was a great surprise to find a repulsive core only 90 MeV high, and about 0.8 fm wide, as shown in Fig. 1. Subsequently

(Fig. 2) we modified V(r) to have an OPEP tail joined on continuously at 1.6 fm, without greatly changing the repulsive core, and also added in the coulomb force[12]. The two body phase shifts at high energy (Fig. 3) were of course much different from those of Reid or Hamada Johnston, but since they are still very poorly known this is not a serious concern. The two body relative wave function at 360 MeV shown in Fig. 4 shows that there is very little suppression of $\psi(r)$ at short distance.

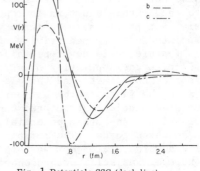

Fig. 1 Potentials SSC (dash line), SSC-NP-1 (solid line), and Reid soft core (dash-dot line), in MeV, plotted as functions of r (fm).

That the off-shell matrix elements are quite different from those of strong core forces, is shown in Fig. 5. These show the Kowalski-Noyes[13,14] ratio function of half shell t-matrix elements,

$$f(p,k) = t_0(p,k,k^2)/t_0(k,k,k^2)$$

for the laboratory energies $E = \dfrac{\hbar^2}{2m} k^2$ (E = 20 MeV => k≈0.5 fm^{-1}). At the on shell point (p = k), f is unity. The change of sign near p = 2 fm^{-1} is related to the repulsive core radius. These two

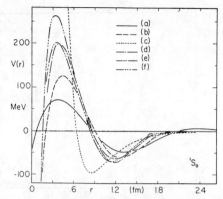

Fig. 2 A number of $^1S_0$ potentials fitted to the phase shifts of MacGregor, Arndt and Wright. Curve (a) is SSC; (b) SSC-NP-2; both fitted to the n-p phases. Curve (c) is Reid soft-core potential; (d) SSC-PP-1; (e) SSC-PP-3; (f) SSC-PP-2. All except SSC agree with the one-pion-exchange potential at large distance.

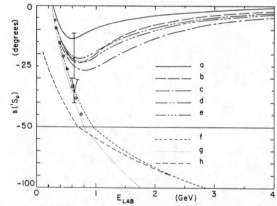

Fig. 3. High-energies phase shifts for several potentials: (a) SSC-NP-2, (b) SSC-PP-1, (c) SSC-PP-2, (d) SSC-PP-3, (e) SSC-PP-4, (f) Reid soft core, (g) Hamada-Johnston, (h) a pure hard core of radius 0.3 fm. Small circles are points on the energy dependent phase shift analysis (ref. [6]) while the error bars indicate three of their single energy analyses at 630 MeV.

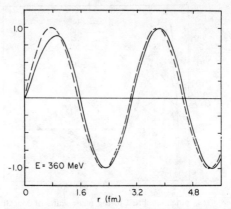

Fig. 4 Radial wave functions compared to the unperturbed sin *kr* plotted for the potential SSC-NP-2 at a number of energies. The wave functions for the pp potentials compare to the Coulomb wave function $F_0(r)$ in an almost identical manner, so were omitted from the figure.

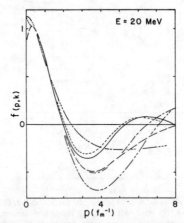

Fig. 5 Ratio of half-off-energy-shell to on-energy-shell matrix elements as a function of the final state momentum p. Several potentials are shown as follows: (a) SSC-PP-1 (full curve), (b) SSC-NP-2 (dotted curve), (c) Reid soft core (dashed curve), (d) Bressel, Kerman and Rouben (dot-and-dash curve), (e) Hamada-Johnston (double-dot-and-dash curve), (f) the separable potential, case II, of Mongan [15]) (triple-dot-and-dash curve).

points dominate the behaviour of $f(p,k)$ for $p \gtrsim 2$ fm^{-1}. For larger p, the hard core potential of Hamada Johnston shows the largest negative excursion, "soft" core less, and SSC forces the least. Also shown is a separable potential which has a smoother behaviour at large p.

Having demonstrated that a comparatively weak force can reproduce the change in sign of the $^1S_0$ phase shift, the next step was to fit a complete force of this type. In Orsay, Gogny, Pires and de Tourreil[15] had been constructing a soft local potential using as radial shape three gaussians in each S,T state. It was designed to give reasonable spherical nuclei in the HF approximation (essentially first order perturbation theory) and so a compromise had to be made in reproducing the phase shifts. R. de Tourreil and I set about to fit a complete force which gave a good fit to the phase shifts, using a rather square core [exp $-r^4$], the OPEP tail and an intermediate range regularized Yukawa[16].

In the singlet even subspace, it is well known that the $^1D_2$ state requires a weaker local force than does the $^1S_0$. This can be accounted for by writing $V(r) = V_C(r) + V_{L2}(r)L^2$, with $V_{L2}(r)$ repulsive. Fortunately, not enough is known about the $^1G_4$ state to require an L^4 potential. In the singlet odd states, the situation is even better: the $^1P_1$ and $^1F_3$ states both have a centrifugal barrier, so the potential is determined only at larger distances. The same combination of V_C and V_{L2} suffices.

In going to the triplet odd states, we have in addition to V_C and V_{L2} (which are determined by the J-weighted average of the L = 1 and L = 3 phases), spin-orbital and tensor forces. These determine the splittings of the ^3P_J phases, while $V_C + 2V_{L2}$ determines their average value. These splittings are defined as

$$\delta_C = \sum_{J=0}^{2} (2J+1)\delta(^3P_J)/\sum_j (2J+1)$$

$$\delta_{LS} = \frac{1}{12} \{5\delta(^3P_0) - 3\delta(^3P_1) - 2\delta(^3P_2)\}$$

$$\delta_T = \frac{5}{12} \{1\delta(^2P_0) - 3\delta(^3P_1) + 2\delta(^3P_2)\}$$

Of course, there is not a linear relationship between potential and phase shift, but nonetheless these combinations give a valuable qualitative measure of the importance of Central, LS and Tensor effects[17]. Now, unfortunately, the tensor force also acts to couple the $^3P_2$ and $^3F_2$ states. In principle, there is no reason why a V_T, determined to reproduce the splittings of the ^3P_J states, should also correctly describe the coupling parameter in the J = 2 state (ϵ_2). Thus, we require an additional independent interaction . It is convenient to adopt the form

$$Q_{12} = \frac{3}{2} \{\sigma^1 \cdot L\ \sigma^2 \cdot L + \sigma^2 \cdot L\ \sigma^1 \cdot L\} - \sigma^1 \cdot \sigma^2 L^2 \quad (\text{"quadratic spin-orbit}$$

force") because
(i)   this vanishes in singlet states;
(ii)   its average value is zero so it does <u>not</u> contribute to the average phase shift δ_C (in the linear approximation);
(iii)   its matrix elements are proportional to those of S_{12}; <u>except</u>
(iv)   it is diagonal to LSJ representation, so it does not contribute (linearly) to ε_2.

Thus, V_T can be chosen to describe the coupling parameter ε_2 leaving $(V_T + V_Q)$ to describe δ_T.

Other forms can be selected for the fifth operator, but they don't separate the various effects as clearly as Q_{12}. For example L_{12} is frequently used, where

$$Q_{\tilde{1}2} = 3L_{12} - \underset{\sim}{\sigma}^1 \cdot \underset{\sim}{\sigma}^2 \underset{\sim}{L}^2 .$$

It contributes to the <u>average</u> phase shift δ_C so in the potential fitting process it is difficult to disentangle L_{12} from $\underset{\sim}{L}^2$.

$$H_{12} = \underset{\sim}{\sigma}^1 \cdot \underset{\sim}{\sigma}^2 \underset{\sim}{L}^2 - L_{12}$$

$$Y_{12} = L_{12} - (\underset{\sim}{S} \cdot \underset{\sim}{L})^2$$

have also been employed but are subject to the same criticism.

A similar argument can be carried out for the $^3S_1 - {}^3D_1$, $^3D_2$ and $^3D_3$ states, leading to the conclusion that at least five operators are likely to be required. In fact, in the dTS potential, a satisfactory fit was obtained omitting V_Q except in the triplet even states. The centrifugal barrier makes the ε_2 less sensitive to the V_T, V_Q distinction than is ε_1. The resulting potential and phase shifts are seen in Figs. 6, 7.

In the beginning we had hoped to construct the potential using only Yukawa functions, taking regularized combinations of Yukawa's. This proved to be difficult, so the attempt was abandoned. For a second tack at this problem, OBEP forms were employed representing π, ρ and ω exchange. In addition a purely phenomenological intermediate range force was employed, as a proxy for the continuum 2π force. To regularize the force, a very sharp cut-off $F(r) = (\alpha r)^N / \{1 + (\alpha r)^N\}$, with $\alpha = 1.2$ fm^{-1}, $N = 20$ was employed, and essentially square cores $(1 - F(R))$ were used. This force (TRS) was developed with the aid of Ben Rouben in Montreal, now at AECL Power Projects[18]. The force is illustrated in Fig. 8. Unfortunately, my coauthors differed with me and introduced the operator L_{12} rather than Q_{12}, which I think makes it difficult to separate the various effects. It turned out to have the defect that the LS, T and L_{12} forces were each separately rather strong in the triplet odd subspace. This shows up already in second order perturbation theory and it becomes obvious in third order that the force is too strong. Recently Jean Coté, of University of Montreal, has completed his thesis in which a revised triplet

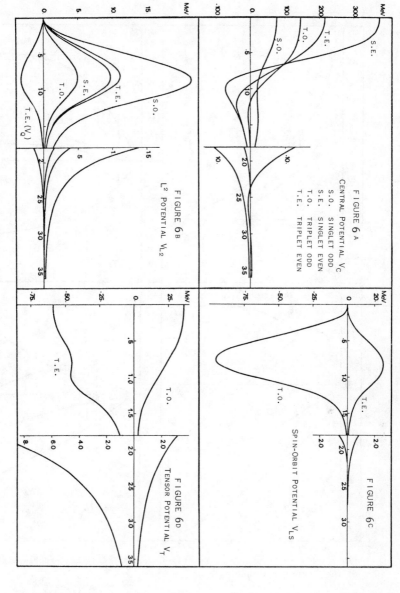

FIG.6 a-d - $V_C(r)$, $V_{L2}(r)$, $V_{LS}(r)$, $V_T(r)$ and $V_Q(r)$ for potential C.

FIGURE 6 A

CENTRAL POTENTIAL V_C

S.O.   SINGLET ODD
S.E.   SINGLET EVEN
T.O.   TRIPLET ODD
T.E.   TRIPLET EVEN

FIGURE 6 B

L^2 POTENTIAL V_{L2}

FIGURE 6 C

SPIN-ORBIT POTENTIAL V_{LS}

FIGURE 6 D

TENSOR POTENTIAL V_T

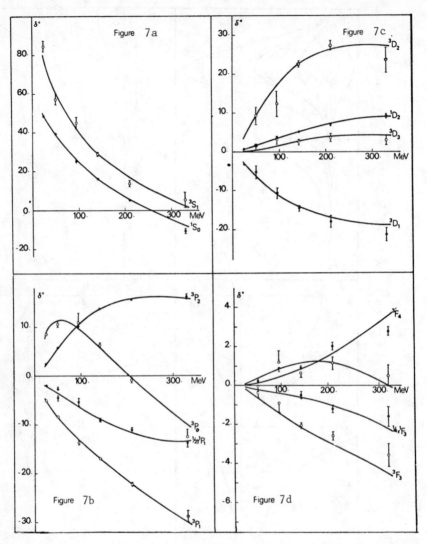

FIG.7 a-h - Phase shifts of potential C compared to the experimental values of the single energy analysis of MAW

odd states force is presented, which goes some distance towards diminishing these potential heights. Cotés thesis was directed mainly at a third order perturbation theory (PT) calculation using the revised TRS force, and a comparison with a G matrix calculation. His work shows many improvements over our earlier PT calculations. For one thing, he used the same single particle spectrum for both G matrix and PT calculations; purely kinetic energies for the states outside the Fermi sphere and self consistent parabolic spectrum for states below k_F. In the third order diagrams

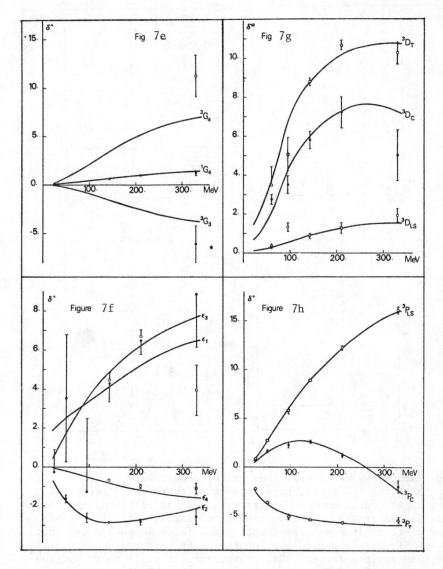

he considers only those with particle-particle ladders, again to correspond to the G matrix calculation. In earlier papers we went only to second order PT, and used only a simple effective mass spectrum for all s.p. states. This work has appeared recently[19].

For the old dTS force, it is gratifying that in each S,T subspace one sees a good convergence of PT, and very close agreement between PT (1+2+3 order) and the G-matrix (Fig. 9). Another small refinement is that for J>2 states, the G matrix calculation uses the PT theory value rather than one taken from the phase shift approximation (PSA). The difference between these two is about 1 MeV/A, and simply indicates that for high J states a potential model implies extrapolated values that differ slightly from

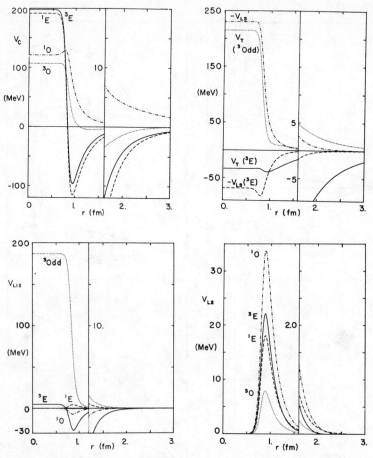

Fig. 8 The radial forms of the various components of the potential in each spin-isospin (S, T) space.

experimental values, themselves rather uncertain.

    For the TRS force, Coté found that there was a large dif-
ference between second and third order PT calculations, and that
this is due entirely to the triplet odd states (Fig. 10). This
lead us to refit our triplet odd states force in such a way that
V_T and V_{L12} are reduced in strength in the core region (Fig. 11).
For this revised force, TRSB, convergence is still poor, but the
G-matrix and PT (1+2+3) agree within a tolerable 1 MeV, as shown
in Fig. 12. I believe that replacing the operator L_{12} by Q_{12} would
allow one to find a better solution still.

    The dTS force was used by Laverne and Gignoux[20] in triton
calculations, as well as the Reid potential. They found both
forces to give very similar binding energy: E_T = -7.50, -7.12 MeV.
The contributions of various states to the potential energy are
however quite different:

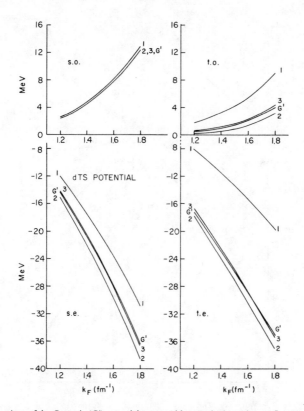

Fig. 9 Comparison of the G-matrix (G') potential energy with perturbation values to first (1), second (2) and third (3) order as a function of Fermi momentum in each spin-isospin subspace, for the dTS interaction.

Table I Potential energy in the triton

State	$^1S_0$	$^3S_1$	$^3S_1-^3D_1$	$^3D_1$	Total
Reid SC	−13.03	−0.38	−50.95	+8.24	−56.11
dTS (Force C)	−11.30	−7.74	−27.11	0.53	−45.62

The strong repulsion of the Reid force induces a greater kinetic energy. The triton form factors are also rather similar, and show the well known discrepancies with experiment.

To the extent that strong and soft repulsive core forces fit the same two body phase shifts at $E_{LAB} < 300$ MeV, these forces differ primarily in their predictions for higher energy data, where meson production and relativistic effects invalidate the Schroe-

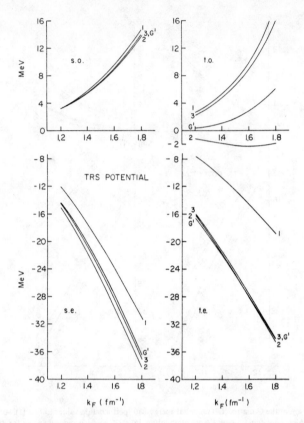

Fig.10 Comparison of the *G*-matrix potential energy with perturbation values to first, second and third order as a function of Fermi momentum in each spin-isospin subspace, for the TRS interaction

dinger theory employed. For triton properties, this high energy region appears to be fairly unimportant. In nuclear matter, it would be interesting to see whether the SSC force gives different results than the Reid force in a Pandharipande type calculation; I have not seen any such work. The principal claim we would make is that the SSC force provides an alternative model, with a different balance of force components, to some of the more standard two body forces.

In conclusion I would like to thank my collaborators R. de Tourreil, B. Rouben and Jean Coté who did so much of the work in developing these forces, and M.K. Srivastava who first showed us how it could be done. I also thank NRC for continued support under operating grant A-3198.

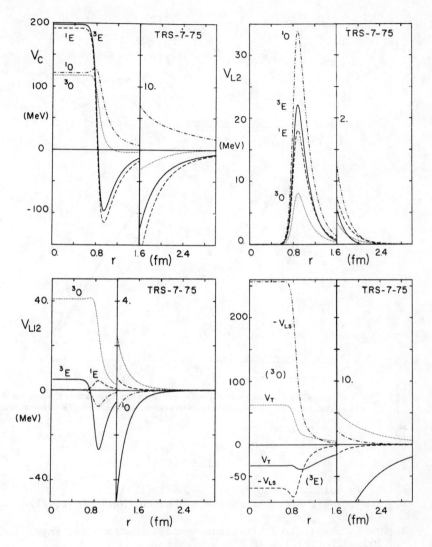

Fig. 11 The revised TRS potential, (TRS-B) which gives
better convergence in the triplet odd states

222

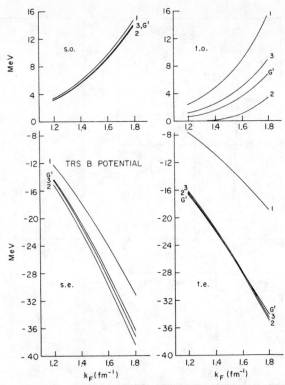

Fig. 12 Comparison of the *G*-matrix potential energy with perturbation values to first, second and third order as a function of Fermi momentum in each spin-isospin subspace, for the **TRS B** interaction.

REFERENCES

1. R. V. Reid Jr., Ann. Phys. 50, 411 (1968).
2. C. Bressel, A.K. Kerman and B. Rouben, Nucl. Phys. A124, 624 (1969).
3. H. P. Stapp, T. J. Ypsilantis and N. Metropolis, Phys. Rev. 105, 302 (1957).
4. J. L. Gammel and R. M. Thaler, Phys. Rev. 107, 291, 1337 (1957).
5. T. Hamada and I. D. Johnston, Nucl. Phys. 34, 382 (1962).
6. K. E. Lassila, M. H. Hull Jr., F. A. McDonald and G. Breit, Phys. Rev. 126, 881 (1962).
7. C. W. Wong, Nucl. Phys. 56, 213 (1964).
8. F. Tabakin, Ann. Phys. 30, 51 (1964).
9. M. H. MacGregor, R. A. Arndt, R. M. Wright, Phys. Rev. 173, 1272 (1968).

10. M. H. MacGregor, R. A. Arndt, R. M. Wright, Phys. Rev. <u>182</u>, 1714 (1969).
11. M. K. Srivastava, P. K. Banerjee and D. W. L. Sprung, Phys. Lett. <u>29B</u>, 635 (1969).
12. D. W. L. Sprung and M. K. Srivastava, Nucl. Phys. <u>A139</u>, 605 (1969).
13. K. L. Kowalski, Phys. Rev. Lett. <u>15</u>, 798 (1965).
14. H. P. Noyes, Phys. Rev. Lett. <u>15</u>, 598 (1965).
15. D. Gogny, D. Pires and R. de Tourreil, Phys. Lett. <u>32B</u>, 591 (1970).
16. R. de Tourreil and D. W. L. Sprung, Nucl. Phys. <u>A201</u>, 193 (1973).
17. D. W. L. Sprung, Nucl. Phys. <u>A242</u>, 141 (1975).
18. R. de Tourreil, B. Rouben and D. W. L. Sprung, Nucl. Phys. <u>A242</u>, 445 (1975).
19. J. Coté, B. Rouben, R. de Tourreil and D. W. L. Sprung, Nucl. Phys. <u>A273</u>, 269 (1976).
20. A. Laverne and C. Gignoux, Nucl. Phys. <u>A203</u>, 597 (1973).

N-N GOBEP MESON FIELD THEORY

A.E.S. Green, T. Ueda[†], F.E. Riewe
University of Florida, Gainesville, Fl. 32611

ABSTRACT

We update our non-relativistic N-N GOBEP models using phase parameters which incorporate new experimental data. We find that the fits of these models are substantially improved, giving increased creditability to the physics underlying our GOBEP models.

THE ROOTS OF GOBEP

It is impossible in a short paper to summarize the experimental and phenomenological studies underlying this work[1]. The theoretical roots, however, go back to the beginning of meson theory which was first proposed in scalar form by Yukawa[2], in vector form by Proca[3], and in axial vector and pseudoscalar forms by Kemmer[4], who also used isoscalar and isovector fields. Breit[5] independently arrived at vector and scalar field models from approximate relativistic considerations.

The discovery of the π in 1947 and its classification as a pseudoscalar isovector particle focused attention on Kemmer's pseudoscalar field and the singular tensor force arising from this field. This singularity problem could be overcome by the use of generalized meson fields as proposed by Green[6] in 1948 or, equivalently, the use of regulators as proposed by Pauli and Villars[7] in 1949. About the same time Green[8] proposed using pseudoscalar and scalar-vector generalized fields which leads to purely relativistic interaction components. Thus the cancellation of the attractive component mediated by the scalar field by the static repulsive component from the vector field leaves the relativistic spin-spin, spin-orbit, tensor and velocity dependent interactions as major residual interactions.

The pursuit of meson theory waned in the fifties but following the discovery of the ω, ρ and η mesons in 1961, a number of works[9-14] appeared using various one-boson-exchange (OBE) formalisms which were quite successful in describing the intermediate and outer regions of the N-N interactions. Noting these works, Green et al[15-16] renewed work on generalized pseudoscalar and scalar-vector meson fields[6],[8]. Soon variations of this early approach[17,18] led to models in quantitative correspondence with phenomenological N-N phase shifts including S and P waves as well as good deuteron properties[19]. For these waves, the use of generalized or regularized fields or the equivalent form factors as reinterpreted by Ueda and Green[18] play a critical role. These models lead to smooth velocity dependent repulsive cores in place of the hard static core of phenomenological potentials. This feature is not only important for the N-N interaction[20], but also for the N-nuclear interaction[20].

ISSN: 0094-243X/78/224/$1.50 Copyright 1978 American Institute of Physic

UPDATED GOBEP

Considerable progress in N-N scattering measurements has been made since the first N-N Conference[1]. Recently, Arndt et al. (AHR)[21] have presented new phase parameters based upon the currently available accumulation of data up to 325 MeV region. In the present work we update our earlier GOBEP by fitting the AHR phase parameters. We investigate four nonrelativistic models. Model A uses σ and ρ mesons with mass distributions related to $\pi\pi$ scattering phase shifts[22-24]. Models B, C, and D represent the σ by two poles and the ρ by one pole. In Model D the mass of the σ and ρ equivalent poles are adjusted and very low energy phase shifts and the deuteron properties are also used in the data set.

Figure 1 illustrates the accuracy of our fits and shows how well the complicated state dependencies of the phenomenological phase shifts are encompassed by our models. In general, the relative χ^2 obtained from our GOBEP fits to the new phase shifts are about one-half of those of our fits to phase shifts available about the time of the first N-N conference. The deuteron parameters for Model D, as calculated with the Gersten-Green code[19], are also quite reasonable although not as precise as we have achieved in the past.[22,18] Several of the problems with the n-p phases and the deuteron properties should be removed by virtue of the resolution of ambiguities in the $^1P_1$ and ρ_1 phase parameters made at the second conference.

DISCUSSION

It should be noted that apart from g_η^2 which has a relatively small effect on the phase shifts the coupling constants are reasonable from the viewpoint of particle physics. The two regulators which control the inner region of the potential are somewhat anomalous but many physical effects probably influence these parameters. If our models were elaborated to include additional relativistic effects such as $(p/M)^4$, L^2 and $\underset{\sim}{\sigma}_1 \cdot \underset{\sim}{L} \, \underset{\sim}{\sigma}_2 \cdot \underset{\sim}{L}$ terms the small residual deviations undoubtedly would be reduced. We might, however, note that differences in phase shifts below 325 MeV between nonrelativistic GOBEP models and relativistic GOBEP models[25,26] using momentum space are not large and that the major part of the higher order terms are absorbed in minor parameter adjustments. However, in relativistic momentum space versions, the small problem with the $^3D_2$ phase is alleviated. Holinde and Machleidt[27] have achieved some improvements in fitting old Livermore phase shifts by introducing an energy dependent form factor. Vinh Mau et al.[28] also achieves good fits to the Livermore phases by using a 14 parameter parametrization of the short range part of their dispersion theoretic potentials. The significance of these innovations, however, cannot be assessed since the phase shifts have now changed, nor are their advantages over the more physical GOBEP models with a wide σ, apparent.

Fig.1. Phase parameters versus incident energy (MeV) in laboratory system (upper scale) and incident momentum (fm^{-1}) in c.m.s. (lower scale). Solid and broken curves represent calculations of Models D and C respectively. The symbol x is for Model A. Phase parameters of Models A, B, and C close to those of Model D are not plotted. Fits of our models to energy-independent phase shifts give similar results. Solid circles represent AHR energy dependent phase parameters and open circles Breit's[29] phase parameters. The error bar of ρ at 25 MeV represents the Signell's[30] value. The adjusted parameters for Model D are $g_\pi^2 = 13.25$, $g_\omega^2 = 7.63$, $g_\omega = 8.67$, $g_\delta^2 = 2.51$ $g_\rho^2 = 0.66$, $g_\sigma^2 = 3.07$, $\eta_2^2(S^*) = 4.62$, $f_\rho/g_\rho = 4.80$, $m_\rho = 708$, $m_\sigma = 484$; monopole regulators $\Lambda_\pi = 2339$, all other $\Lambda = 1121$. The deuteron parameters obtained are B=2.2 MeV, $a(^3S_1) = 5.4$ fm, $r(^3S_1) = 1.8$ fm, $a(^1S_0) = 27.9$ fm, $r(^1S_0) = 2.40$ fm, D state probability 4.6%, $Q_D = 2.6 \times 10^{-27}$ cm^2, $\mu = 0.86$ μ_N. The n-p phases at 50 MeV and 325 MeV discussed by Arndt, Edgington, Fitzgerald, Rijken et al at the 2nd N-N conference,particularly ρ_1,move towards those shown for Model D.

In these models we interpret the σ as the I=J=0 part of the sum of both the uncorrelated and correlated two-pion exchange contributions. The latter can be related with the $\pi\pi$ data, while the former the πN data[31].

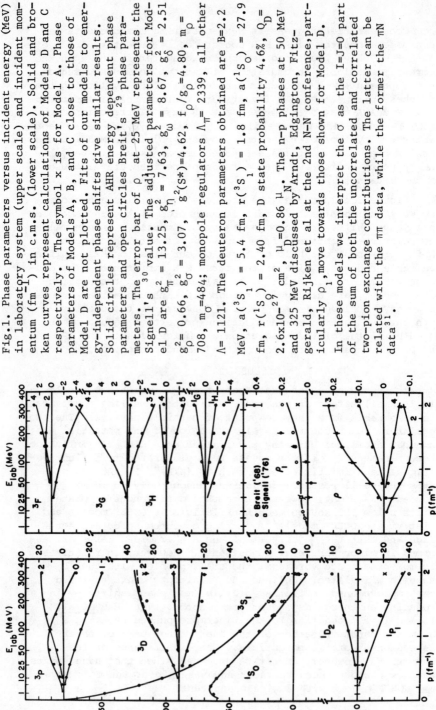

REFERENCES

1. A. Green, M. MacGregor and R. Wilson, Editors, First International-
   al Conference on N-N Interaction, Gainesville, Florida, "Rev. Mod.
   Phys. 39, 495 (1967).
2. H. Yukawa, Proc. Phys. Math. Soc. Jap. 3, 17 (1935).
3. A. Proca, J. Phys. Radium 7 (VII), 347 (1936).
4. N. Kemmer, Proc. Roy. Soc. London Ser. A 166, 127 (1938).
5. G. Breit, Phys. Rev. 51, 248 (1937).
6. A. Green, Phys. Rev. 73, 519 (1948); 75, 1926 (1949).
7. W. Pauli and F. Villars, Rev. Mod. Phys. 21, 434 (1949).
8. A. Green, Phys. Rev. 76, A 460, L 870 (1949) and unpublished.
9. S. Sawada, T. Ueda, W. Watari and M. Yonezawa, Prog. Theor. Phys.
   28, 991 (1962); 32, 380 (1964).
10. R. McKean, Jr., Phys. Rev. 125, 1399 (1962)
11. D. Lichtenberg, Nuovo Cimento 25, 1106 (1962).
12. J. Ball, A. Scotti and D. Wong, Phys. Rev. Lett. 10, 142 (1963).
13. R. Bryan, C. Dismukes, W. Ramsey, Nucl. Phys. 45, 353 (1963);
    R. Bryan and B. Scott, Phys. Rev. 135, B 434 (1964); 164, 1215
    (1967).
14. S. Ogawa, S. Sawada, T. Ueda, W. Watari and M. Yonezawa, Prog.
    Theor. Phys. Suppl. 39, 140 (1967).
15. A. Green and R. Sharma, Phys. Rev. Lett., 14, 380 (1965).
16. A. Green, T. Sawada and R. Sharma, Isobaric Spin in Nuclear
    Physics, (New York, 1966).
17. A. Green and T. Sawada, Nucl. Phys. B2, 276 (1967) and Rev. Mod.
    Phys. 39, 594 (1967).
18. T. Ueda and A. Green, Phys. Rev. 174, 1304 (1968) and Nucl. Phys.
    B10, 289 (1969).
19. A. Gersten and A. Green, Phys. Rev. 176, 1199 (1968).
20. A. Green, F. Riewe, M. Nack and L. Miller, International Sympo-
    sium on Present Status and Novel Developments in the Nuclear Many-
    body Problem, Rome, 1972. T. Ueda in same proceedings.
21. R. Arndt, R. Hackman and L. Roper, Phys. Rev. C15, 1002 (1977).
22. R. Stagat, F. Riewe and A. Green, Phys. Rev. Lett. 24, 631 (1970).
23. M. Nack, T. Ueda and A. Green Phys. Rev. D10, 3617 (1974).
24. F. Riewe, M. Nack and A. Green, Phys. Rev. C10, 2210, (1974).
25. A. Gersten, R. Thompson, and A. Green, Phys. Rev. D3 2076, 2069
    (1971).
26. T. Ueda, M. Nack and A. Green, Phys. Rev. C8, 2061 (1973).
27. K. Hollinde and R. Machleidt, Nucl. Phys. A256, 479 (1976).
28. R. Vinh Mau, The Paris Potentials, this conference.
29. R. Seamon, K. Friedman, G. Breit, R. Harcz, J. Holt, and A. Prak-
    ash, Phys. Rev. 165, 1579 (1968).
30. G. Bohannon, T. Burt and P. Signell, Phys. Rev. C13, 1816 (1976).
31. S. Furuichi, Prog. Theor. Phys. Suppl. 39, 190 (1967).
    † On leave, Osaka University, Osaka, Japan.

228

DISPERSION TREATMENT OF THE NN AND N̄N FORWARD SCATTERING AMPLITUDES*

Olgierd Dumbrajs
Research Institute for Theoretical Physics
University of Helsinki, Finland

ABSTRACT

A review is given of the recent work on analysis of the nucleon-nucleon and antinucleon-nucleon forward scattering amplitudes by means of dispersion relations.

INTRODUCTION

Elastic nucleon-nucleon and antinucleon-nucleon scattering seems to be one of the oldest and most studied problems in particle physics. The analyticity in both the energy and the momentum transfer plane, mainly in the form of dispersion relations, plays the major role in elucidating this interaction.

When studying the NN and N̄N interactions by means of dispersion relations one faces at the very beginning three important difficulties. Firstly, for this process even the forward dispersion relations have not yet been proven from the axiomatic field theory, and thus one has to believe their validity. Secondly, there is a huge unphysical cut inaccessible to direct physical measurements, which gives a large contribution to dispersion integrals. One has either to use some models for the amplitude or to rely on some very indirect methods of obtaining information about the amplitude in this region. Thirdly, elastic NN, N̄N scattering is far more complicated than meson-nucleon scattering, simply by the fact that it involves one more spinning particle which means a larger number of amplitudes even in the forward direction. Due to the lack of adequate polarization measurements one has to make certain approximations concerning the spin dependence.

Despite these difficulties many attempts have been undertaken to understand the NN and N̄N forward scattering in terms of analyticity. Of course, it is impossible to give in one talk a complete review of all existing investigations. Therefore, I will concentrate on the most recent work in this field[1,2] mentioning only briefly earlier results. For a detailed survey of earlier research the reader is referred to review articles by Dumbrajs and Staszel[3] and by Kroll.[4]

KINEMATICS AND ANALYTICITY

General form of the scattering amplitude of two 1/2 spin particles looks like:

---

*Work supported by the National Research Council of Canada and by the University of British Columbia.

$$F(s,t) = a(s,t) + b(s,t)(\sigma_1\vec{n}) \cdot (\sigma_2\vec{n}) + ic(s,t)(\sigma_1+\sigma_2)\vec{n}$$
$$+ d(s,t)(\sigma_1\vec{m}) \cdot (\sigma_2\vec{m}) + e(s,t)(\sigma_1\vec{\ell}) \cdot (\sigma_2\vec{\ell}), \tag{1}$$

where a, b, c, d and e are scalar functions of usual Mandelstam variables s and t, σ_1 and σ_2 are the Pauli spin operators for the two nucleons and $\vec{\ell}$, $\vec{m}$ and $\vec{n}$ are unit vectors in the directions $\vec{k} + \vec{k}'$, $\vec{k} - \vec{k}'$ and $\vec{k} \times \vec{k}'$, respectively ($\vec{k}$ and $\vec{k}'$ are the momenta before and after collision in centre-of-mass system: $|\vec{k}| = |\vec{k}'| = k$). From kinematics only, as a consequence of angular momentum conservation, it follows that at $t = 0$

$$c(s,0) = 0, \quad d(s,0) = e(s,0), \tag{2}$$

which leaves us with

$$F(s,0) + a(s,0) + b(s,0)\sigma_1 \cdot \sigma_2 + (d-b)(\sigma_1\vec{\ell}) \cdot (\sigma_2\vec{\ell}). \tag{3}$$

The first two terms can be expressed as the usual singlet and triplet scattering amplitudes, while the third is responsible for the double spin-flip process in the forward direction.

The optical theorem for scattering from the unpolarized initial state gives the imaginary part of the forward spin-independent amplitude

$$\sigma\text{tot}(s) = \frac{4\pi}{k} \text{Im}a(s,0). \tag{4}$$

Usually it is assumed that the double spin-flip amplitude is negligibly small and that the singlet and triplet amplitudes have the same behaviour as complex functions of s and t. In other words, we will presume that the forward scattering amplitude can be represented as a single complex function of s

$$F(s) = D(s) + iA(s). \tag{5}$$

The imaginary part of this effective amplitude is simply given by the optical theorem (4). On the other hand, the relation between the real part and the measured quantities is, in principle, much more complicated and is discussed in detail in, e.g., Ref. 3.

It is postulated that the scattering amplitude F is an analytic function in the nucleon lab energy plane ω ($\omega = \sqrt{p^2 + m^2}$, p is nucleon lab momentum m, is nucleon mass) with the analytic structure shown in Fig. 1.

230

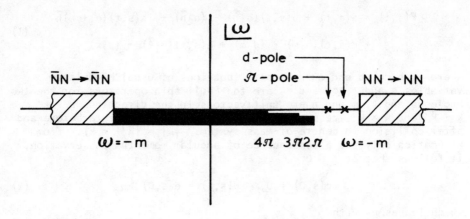

Fig. 1. Analytic structure of the NN,$\overline{\text{N}}$N forward scattering amplitude in the nucleon lab energy plane.

Then the dispersion relation (subtracted, for example, at the threshold of NN scattering[2]) holds:

$$D_{pn}(\omega) = D_{pn}(m) - f^2 \frac{\omega - m}{(\omega_\pi + \omega)(\omega_\pi + m)} + 2F(0)\sqrt{\frac{\omega_d + m}{2m}} \cdot \frac{\omega - m}{(\omega_d - m)(\omega_d - \omega)}$$

$$- \frac{\omega - m}{\pi} \int_{\omega_{2\pi}}^{m} d\omega' \frac{A_{\overline{p}n}(\omega')}{(\omega' + \omega)(\omega' + m)} + \frac{\omega - m}{\pi} P\int_{m}^{\infty} d\omega' \left\{ \frac{A_{pn}(\omega')}{(\omega' - \omega)(\omega' - m)} \right.$$

$$\left. - \frac{A_{\overline{p}n}(\omega')}{(\omega' + \omega)(\omega' + m)} \right\} , \tag{6a}$$

$$D_{\overline{p}n}(\omega) = D_{pn}(m) + f^2 \frac{\omega + m}{(\omega_\pi - \omega)(\omega_\pi + m)} - 2F(0)\sqrt{\frac{\omega_d + m}{2m}} \cdot \frac{\omega + m}{(\omega_d - m)(\omega_d + \omega)}$$

$$+ \frac{\omega + m}{\pi} \int_{\omega_{2\pi}}^{m} d\omega' \frac{A_{\overline{p}n}(\omega')}{(\omega' - \omega)(\omega' + m)} + \frac{\omega + m}{\pi} P\int_{m}^{\infty} d\omega' \left\{ \frac{A_{\overline{p}n}(\omega')}{(\omega' - \omega)(\omega' + m)} \right.$$

$$\left. - \frac{A_{pn}(\omega')}{(\omega' + \omega)(\omega' - m)} \right\} , \tag{6b}$$

where $f^2 = (m_\pi^2/4m^2) \cdot g^2/4\pi \sim 0.080$ is the πNN coupling constant, m_π is the pion mass, $\omega_\pi = -m + m_\pi^2/2m$, $\omega_{2\pi} = -m + 2m_\pi^2/m$, $\omega_d = m - B + B^2/2m$, $F(0) = 9H/[m(1-Hr)]$, $H = \sqrt{mB}$, $B = 2.22$ MeV is the deuteron binding energy and r is the $^3S_1$ effective range.

For the pp,$\overline{p}$p scattering the same relations hold with minor changes. Firstly, the deuteron pole is absent. Secondly, the pion pole term has to be divided by 2 (isospin factor).

The subtraction constant $D_{NN}(m)$ depends on the NN scattering lengths. The single subtraction of the dispersion relation is sufficient provided the Pomeranchuk theorem holds.

There are many problems which can be tackled with the aid of NN,$\overline{\text{N}}$N forward dispersion relations. One of the directions are attempts to obtain some information about the NN,$\overline{\text{N}}$N scattering amplitude at low energies and in the unphysical region. As shown in Fig. 1, 2π-exchange begins the huge unphysical region which causes the main difficulties in practical applications of the NN,$\overline{\text{N}}$N forward dispersion relations at low energies. First, it should be noted that there exist no models in a strict sense of this word for the amplitude in this region. Of course, the simplest way of obtaining the amplitude in the unphysical region would be the analytic continuation of the low-energy parametrization of the $\overline{\text{N}}$N scattering below the elastic threshold (as is done, for example, in the case of K$\overline{\text{p}}$ forward scattering). However, there are no parametrizations of this kind and no data on low-energy $\overline{\text{N}}$N scattering. Therefore, there is nothing to continue into the unphysical region. The most widespread way of handling this problem is to replace the unphysical cut by the sum of poles and to identify them with the few known physical resonances (ρ, ω, σ ...). If the problem is posed in this way, then it is immediately clear that one can, in principle, determine the values of the meson-nucleon coupling constants or, in general, to obtain some information about the unphysical region by fitting these parameters to experimental values of total cross-sections and real parts of the NN,$\overline{\text{N}}$N forward scattering amplitudes. Much work has been done in this direction (see review[3]).

However, historically this was not the first aim of applications of the NN,$\overline{\text{N}}$N forward dispersion relations. Most works were devoted to calculations of the real part of the amplitude on the basis of the approximated unphysical region and the experimental and theoretical (model) knowledge of the imaginary parts in the physical regions (see reviews[3,4]).

We now discuss the most recent works in both of these directions.

THE REAL PART OF THE FORWARD PROTON-PROTON SCATTERING AMPLITUDE AT VERY HIGH (ISR) ENERGIES[1]

In the energy range where one is far from the physical region the dispersion relation for the pp,$\overline{\text{p}}$p scattering can be written as:

$$D_{pp}(\omega) = C + \frac{m\omega}{8\pi^2} \int_m^\infty d\omega' \left\{ \frac{\sigma_{pp}(\omega')}{\omega' - \omega} - \frac{\sigma_{\overline{p}p}(\omega')}{\omega' + \omega} \right\} , \qquad (7)$$

where C is a constant incorporating the subtraction constant and the contributions from the pole term and the unphysical region. The proton-proton and antiproton-proton total cross-sections have been parametrized as:

$$\sigma_{pp} = C_1 \omega^{-\nu_1} - C_2 \omega^{-\nu_2} + \sigma_\infty , \qquad (8a)$$

$$\sigma_{\overline{p}p} = C_1 \omega^{-\nu_1} + C_2 \omega^{-\nu_2} + \sigma_\infty , \qquad (8b)$$

$$\sigma_\infty = B_1 + B_2 (\ell ns)^\gamma . \qquad (9)$$

At high energies the assumed cross-sections become equal and behave like a power of the logarithm s. The parameters in Eqs. (7), (8a), (8b) and (9) have been determined by fitting existing total cross-section data and at the same time the data for ρ from the experiment[1] and from some of the earlier experiments. The following values of the parameters have been found (cross-sections in mb and energies in GeV):

$$C_1 = 41.9 \pm 1.1,$$
$$C_2 = 24.2 \pm 1.1,$$
$$\nu_1 = 0.37 \pm 0.03,$$
$$\nu_2 = 0.55 \pm 0.02,$$
$$B_1 = 27.0 \pm 1.0,$$
$$B_2 = 0.17 \pm 0.08,$$
$$\gamma = 2.10 \pm 0.10.$$

The result of the fit is shown in Fig. 2(a) and 2(b).

The authors of the paper[1] conclude that the results of their experiment, together with previously measured data, indicate that the proton-proton and the antiproton-proton total cross-sections should rise at least up to equivalent laboratory energies of 40 TeV, where they should reach a value of about 55 mb. This, of course, is true only as long as validity of parametrizations (8a), (8b) and (9) is believed. Dispersion relations by no means should be considered as "a crystal ball that can view the road to Asymptopia".[5]

ANALYSIS OF THE NN AND $\overline{\text{NN}}$ FORWARD SCATTERING AMPLITUDES INCLUDING INFORMATION FROM $\pi\pi$ AND πN SCATTERING[2]

As already mentioned in the section on Kinematics and Analyticity, the contribution from the unphysical region was represented usually by a sum of poles. On the other hand, the information about the 2π-exchange contribution to the imaginary part of the $\overline{\text{NN}}$ scattering amplitude in the unphysical region can be obtained on the basis of the knowledge of the $\overline{\text{NN}} \to \pi\pi$ amplitude. The formalism necessary for including this information has been given a long time ago.[6] However, only recently it became possible to obtain $\pi\pi\overline{\text{NN}}$ partial wave amplitudes by numerical analytic continuation of the elastic πN scattering amplitudes using the $\pi\pi$ elastic phases.[7,8]

The main idea of Grein[2] was to insert the known total cross-sections, the real parts and the known part of the two-pion exchange contribution into the forward dispersion relations (6a), (6b). One can hope that this procedure will give some new results about our knowledge of the behaviour of the $\overline{\text{NN}}$ amplitude in the low-energy and unphysical regions, because for the first time the extra input information has been used which comes from an independent source, namely from πN and $\pi\pi$ elastic scattering.

Grein[2] uses the results for $A_{\overline{\text{NN}}}$ calculated from three different sets of $\pi\pi\overline{\text{NN}}$ amplitudes as the input in the dispersion integral in the unphysical region (Fig. 3).

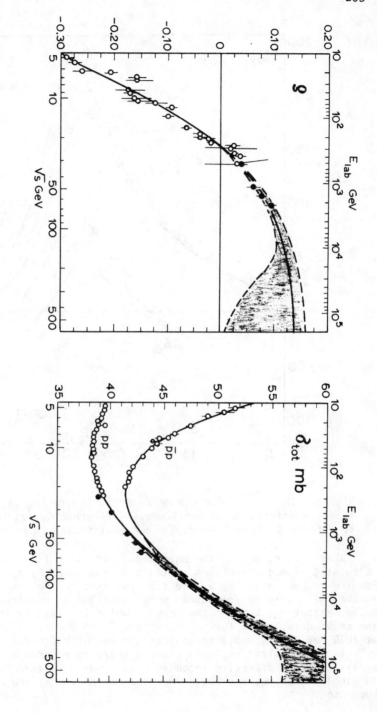

Fig. 2(a,b). The data obtained for $\rho_{pp} = D_{pp}/A_{pp}$ in Ref. 1 (full circles) together with data at lower energies. The full curves are the result of a fit, simultaneously performed on total cross-section data and on the data for ρ. The shaded areas represent the one standard deviation regions for ρ and the cross-sections. The boundaries of these regions were obtained by changing the high-energy behaviour of the cross-sections, Eq. (9), in such a way that the Chi squared of the fit increased by one.

234

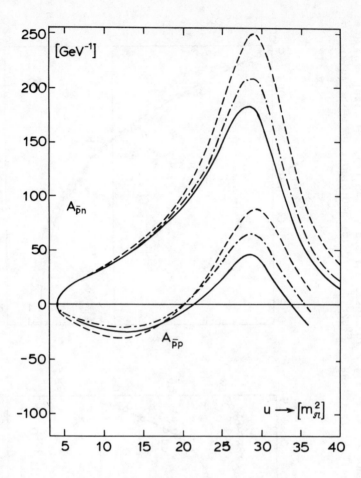

Fig. 3. The two-pion exchange contribution to $A_{\overline{NN}}$ calculated from $\pi\pi\overline{NN}$ partial wave amplitudes. Different curves correspond to different sets of $\pi N \to \pi N$ phase shifts used in calculations.

That part of $A_{\overline{NN}}$ on the unphysical cut which is not explicitly calculated from the two-pion state is approximated by a sum of poles. For the $pn,\overline{p}n$ scattering a pole term is fixed to the ρ-mass and a further pole term is introduced whose position is not fixed. For the $pp,\overline{p}p$ scattering, besides the ρ-pole, which is fixed by the fit to the $pn,\overline{p}n$ data, three further terms are introduced: at the ω-mass, at 1020 MeV for S^* and ϕ resonances and one with freely varying mass.

In the physical region A_{pp} and A_{pn} are found from a phase-shift analysis and an effective range formula is used to extrapolate to threshold. For $\sigma_{\overline{NN}}$ the following parametrization at low energies has been used:

$$\sigma_{\overline{NN}} = \frac{a}{p} + b, \tag{10}$$

with a = 43 mb GeV/c, b = 68 mb for $\overline{p}p$, and a = 34 mb GeV/c, b = 75 mb for $\overline{p}n$. At medium and high energies certain appropriate Regge-pole parametrizations of total cross-sections have been chosen. This input is shown in Fig. 4, Fig. 5(a) and Fig. 5(b).

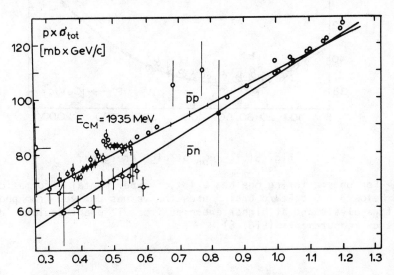

Fig. 4. $\sigma_{\overline{NN}}$ at low energies multiplied by the lab momentum. The solid lines show the extrapolation to threshold.

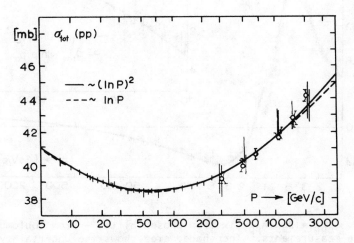

Fig. 5(a). σ_{pp} at high energies.

236

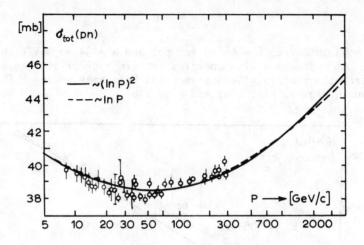

Fig. 5(b). σ_{pn} at high energies.

For pp scattering one has a lot of experimental information on D. Below $p = 0.8$ GeV/c Grein[2] uses the values derived from phase-shift analysis and at higher energies from different Coulomb inter-ference measurements (Fig. 6).

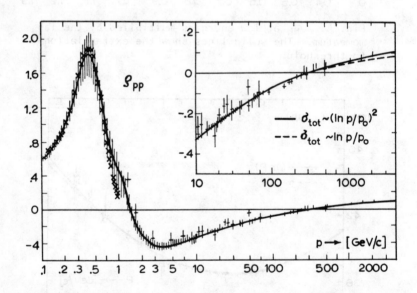

Fig. 6. $\rho_{pp} = D_{pp}/A_{pp}$; x from phase shifts; $\dagger$ from Coulomb inter-ference measurements. The shaded area shows the uncertainty due to the error of σ_{tot} when performing the fit.

There are very few real part data for p̄p scattering and no reliable phase-shift analyses (Fig. 7).

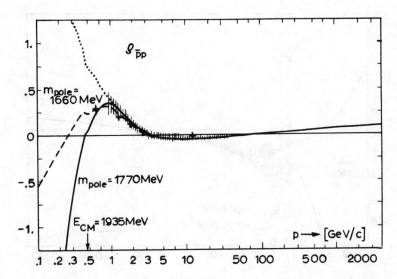

Fig. 7. $\rho_{\bar{p}p} = D_{\bar{p}p}/A_{\bar{p}p}$. Fit with a pole at m = 1770 MeV, fit with a pole at m = 1660 MeV, fit without a pole above m = 1500 MeV. Other conventions are the same as in Fig. 6.

For pn scattering real parts are used derived from phase shifts and those obtained from an analysis of pd data in the Coulomb interference region according to the Glauber model (Fig. 8).

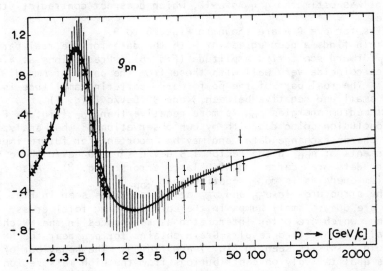

Fig. 8. $\rho_{pn} = D_{pn}/A_{pn}$. Conventions are the same as in Fig. 6.

For $D_{\overline{p}n}$ there exists no direct experimental information (Fig. 9).

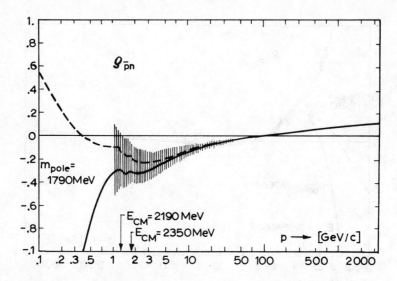

Fig. 9. $\rho_{\overline{p}n} = D_{\overline{p}n}/A_{\overline{p}n}$. Fit with a pole at m = 1790 MeV, fit
without a pole above m = 1500 MeV. Other conventions are the
same as in Fig. 6.

From the overall fits it was found that the dominant contribu-
tion to $A_{\overline{p}n}$ comes from the $\rho(763)$ resonance. Separating the isospin
zero part it turned out that there is a strong contribution from the
$\epsilon(670)$ resonance. In addition, the value of the ωNN tensor coupling
constant was estimated $g_{\omega NN}^2/4\pi \sim 12$, which does not contradict other
analyses.

Fits for $\rho = D/A$ are shown in Figs. 6 to 9.

Grein finds a good agreement with the data for the real part of
the pp forward scattering amplitude (Fig. 6). Predictions at high
energies coincide very well with those from the paper of Amaldi
et al.[1] The real part of the $\overline{p}p$ forward scattering amplitude is
rather small and negative between 3 and 90 GeV/c (Fig. 7).

At medium energies ρ_{pn} is more negative than ρ_{pp} (Fig. 6, Fig. 8).
This conclusion coincides with my own observations,[9] when analyzing
the np charge exchange data, and may be important for future improved
measurements of ρ_{pn} in the region 1.5 GeV/c $\lesssim p \lesssim$ 50 GeV/c where the
existing data are scarce and with large errors (Fig. 8). Also $\rho_{\overline{p}n}$
at medium energies is more negative than $\rho_{\overline{p}p}$.

The structures in $\rho_{\overline{p}p}$ and $\rho_{\overline{p}n}$ at low energies seen in Fig. 7 and
Fig. 9 are due to small bumps in the corresponding total cross-
sections, which are often interpreted as resonances in the $\overline{N}N$ channel.
Especially interesting results Grein obtains for $\rho_{\overline{N}N}$ near the $\overline{N}N$
threshold. This is seen in Fig. 7 and Fig. 9. The behaviour of $D_{\overline{p}p}$
depends qualitatively on contributions from the high mass region of

the unphysical cut. If one allows for important contributions from this range by including a pole near m = 1700 MeV, $\rho_{\bar{p}p}$ is negative near threshold and has a zero between 200 and 500 MeV/c. The nearer to threshold the pole is situated the more negative $\rho_{\bar{p}p}$ becomes at $\omega = -m$. The same qualitative statements hold for $\rho_{\bar{p}n}$.

The question about the near threshold behaviour of $\rho_{\overline{NN}}$ is very important, because this behaviour is predicted by different potentials and by measurements of the x-ray spectra of antiprotonic atoms. Unfortunately, these predictions are rather model-dependent, though most give the negative sign for $\rho_{\overline{NN}}$ at the threshold. Although Grein's analysis cannot exclude positive values of $\rho_{\overline{NN}}$ at the threshold, it favours the negative sign. The sign of the real part of the $\overline{NN}$ scattering amplitude at threshold is of interest in cosmological models[10] and in the context of the problem of existence of $\overline{NN}$ bound states.

In terms of analyticity the inequality $D_{\overline{NN}}(m) < 0$ would mean that there are strong contributions to $A_{\overline{NN}}$ in the high mass region of the unphysical cut. The nature of these contributions is to be clarified.

CONCLUSIONS

Noticeable progress in the dispersion treatment of the NN and $\overline{NN}$ forward amplitudes has been achieved since the appearance of the review.[3]

At high energies a very nice mutual consistency of measurements of σ_{pp} and ρ_{pp} has been demonstrated and certain predictions concerning the very high-energy behaviour of total cross-sections of pp and $\bar{p}p$-scattering have been made.[1]

At intermediate energies the inequality $|\rho_{pn}| > |\rho_{pp}|$ has been confirmed.[2,9]

For the first time new and independent information has been used as input in NN, $\overline{NN}$ dispersion relations: the $\pi\pi\overline{NN}$ partial wave amplitudes were used to obtain the contribution to $A_{\overline{NN}}$ from the two pion intermediate states in the unphysical region.[2]

Interesting connection between the contributions from the high mass region of the unphysical cut and the sign of $D_{\overline{NN}}$ near the $\overline{NN}$ threshold has been discovered and support for the inequality $D_{\overline{NN}}(m) < 0$ has been found.

Measurements of the pp scattering cross-sections with polarized beams and targets performed at Argonne[11] enable the carrying out of dispersion analysis for individual spin amplitudes. Such analysis, aside from testing the possible existence of resonances in NN scattering, would allow to disentangle the contributions from different pole terms representing the unphysical region and thus would be an improvement in determining values of different coupling constants.[12]

Finally we repeat our plea[3] for more measurements of $\bar{p}p$ scattering in the low-energy range which would clarify the very interesting and important picture of the behaviour of the $\overline{NN}$ scattering amplitude near the $\overline{NN}$ threshold and in the unphysical region.

REFERENCES

1. U. Amaldi *et al.*, Phys. Lett. <u>66B</u>, 390 (1977).
2. W. Grein, Karlsruhe University preprint TKP 77-4, March 1977.
3. O. Dumbrajs and M. Staszel, Fortschr. Phys. <u>23</u>, 399 (1975).
4. P. Kroll, Fortschr. Phys. <u>24</u>, 565 (1976).
5. S.J. Lindenbaum, in "Pion-Nucleon Scattering" (Wiley Interscience, New York, 1969).
6. M.L. Goldberger, R. Oehme, Ann. Phys. <u>10</u>, 153 (1960).
7. G. Höhler and E. Pietarinen, Nucl. Phys. B<u>95</u>, 210 (1975).
8. G. Gustafson, H. Nielsen and G.C. Oades, in "Compilation of Coupling Constants and Low Energy Parameters".
9. O. Dumbrajs, Acta Physica Polonica B (in print).
10. R. Omnes, Phys. Reports <u>3C</u>, 1 (1972).
11. A. Yokosawa, invited talk given at this Conference.
12. W. Grein and P. Kroll, in preparation.

NUCLEON-NUCLEON DYNAMICS FROM VARIATIONAL
MATRIX PADE APPROXIMANTS

D. Bessis, P. Mery[*], G. Turchetti[**], and J. Gammel[***]
DPh-T, CEN-Saclay, BP n°2, 91190 Gif-sur-Yvette, France

1. INTRODUCTION

First we summarize and analyze the operator Padé approximant (O.P.A.) method in potential scattering. These approximants are derived from a variational principle for the time evolution operator. The lowest approximant, constructed from the first two terms of the perturbation expansion of the off-shell scattering matrix, is shown to be the exact solution of the corresponding Schrödinger equation. The actual computation of the O.P.A. requires the inversion of an operator which is achieved by replacing the O.P.A. by a M.P.A. (matrix Padé approximant). The mesh points on which the calculation is performed are variational parameters for each energy, angular momentum, parity, and internal quantum numbers and are determined by the Schwinger variational principle for phase shifts. For potentials so strong that the phase shifts vary over hundreds of degrees in the energy range of interest, we reproduce the exact phase shifts within 0.1% at any energy with no more than three mesh points. Rigorous results supporting this procedure are given.

Next we discuss the extension of this procedure to Lagrangian field theories for strong interactions at low energy and give the essentials of the techniques used. We analyze the Yukawa interaction and then take into account chiral effects and vector mesons through the non-linear σ model and a renormalizable ω interaction. We report here the results obtained with only one mesh point corresponding to the physical momentum.

It is well knwon from work on the Bethe-Salpeter equation that using only one mesh point gives poor results for the S and P waves except near threshold (zero kinetic energy). Higher waves are reproduced well, however. Only by using a larger number of mesh points chosen variationally can better results be achieved for the S and P waves. Such a calculation is in progress.

2. THE VARIATIONAL APPROXIMATION

Given a renormalizable Lagrangian one can construct a renormalized pertrubation expansion for the four point Green's function (off-shell T-matrix),

[*]Permanent address : Centre Universitaire de Luminy and Centre de Physique Théorique, Marseille, France.

[**]Permanent address : Instituto di Fisica, Universita di Bologna, I.N.F.N. Sezione di Bologna, Italy.

[***]Permanent address : Saint-Louis University, Saint-Louis, Missouri (U.S.A.)

ISSN: 0094-243X/78/241/$1.50 Copyright 1978 American Institute of Physics

$$T(g) = g\, T_B + g^2\, T_{B2} + g^3\, T_{B3} + \cdots \quad , \tag{1}$$

where g is the expansion parameter and T_B and T_{B2} the first and second Born approximations, respectively. We want to construct a quasi-potential such that on-shell and off-shell the first n terms of Eq. (1) and the first n terms of the analogous expansion generated by the quasi-potential be the same. A systematic method of solving this problem is provided by the operator Padé approximant which makes stationary the Lipmann-Schwinger functional [1]

$$
F(V, V^+, t, t_o) = I - i \int_{t_o}^{t} dt'\, [V^+(t', t)\, H_I(t') + H_I(t')\, V(t', t_o)]
$$
$$
+ i \int_{t_o}^{t} dt'\, V^+(t', t)\, H_I(t')\, V(t', t_o)
$$
$$
+ i^2 \int_{t_o}^{t} \int_{t_o}^{t} dt'\, dt''\, V^+(t'', t)\, \theta(t''-t')\, H_I(t'')\, V(t', t_o) \; ;
$$
$$\tag{2}$$

where

$$\delta F(V, V^+) = 0 \quad \Rightarrow \quad F_{\text{stationary}} = V_I(t, t_o) \quad . \tag{3}$$

In these equations, $H_I(t)$ is the Hamiltonian in the interaction picture and V and V^+ are arbitrary trial operators to be varied until F is stationary. When $t_o \to -\infty$ and $t \to +\infty$, F becomes the S matrix operator.

The operator Padé approximant results from a trial operator (the Cini-Fubini ansatz) [2]

$$
V^{(N)}(t, t_o) = \Lambda_o + \Lambda_1\, V_B(t, t_o) + \Lambda_2\, V_{B2}(t, t_o) + \cdots + \Lambda_N\, V_{BN}(t, t_o) \quad ,
$$
$$\tag{4}$$

where the Λ_i are arbitrary constants determined by the variational principle. [1] The N/N O.P.A. can be shown to be exactly unitary,

$$[N/N]^+ \, [N/N] = I \quad . \tag{5}$$

Using the S and T operator formalism

$$S = I + iT \quad , \tag{6}$$

$$T(g) = T_B + g\, T_{B2} + g^2\, T_{B3} + \cdots \quad , \tag{7}$$

the first or $[1/1]$ approximant is

$$[1/1] = g\, T_B\, (T_B - g\, T_{B2})^{-1}\, T_B \quad . \tag{8}$$

We conclude this section with a discussion of some of the properties of the lowest order O.P.A. Eq.(8) defines an operator constructed from the two lowest order terms in the Born series expansion of the T-operator. It requires an inversion of the operator $T_B - g T_{B2}$. We shall see how to do this inversion in a practical way later.

We shall now prove that Eq.(8) is rigorously equivalent to the
Schrödinger equation in the case of potential scattering and therefore is an excellent candidate for generalizing the Schrödinger equation to any Lagrangian field theory.

We can replace the two-body Schrödinger equation by the integral
equation of Lipmann and Schwinger

$$T = gV + gVG_oT \quad , \tag{9a}$$

which has the exact solution

$$T = g(I - gVG_o)^{-1} V \quad . \tag{9b}$$

By iterating Eq.(9a) or expanding Eq.(9b) through second order, we
get

$$T = gV + g^2 VG_o V + g^3 VG_o VG_o V + \ldots \quad , \tag{10}$$

where, by identification, we have $T_B = gV$, $T_{B2} = g^2 VG_o V$.

Constructing the $[1/1]$ O.P.A. we find

$$[1/1] = gV [V - gVG_o V]^{-1} V = g(I - gVG_o)^{-1} V \quad , \tag{11}$$

which is identical to the exact solution of Eq.(9a). This result extends immediately to semi-relativistic equations such as the
Blankenbecler-Sugar equation and to the Bethe-Salpeter equation.

The advantage of writing the solution of these equations in the
form of an O.P.A. is that these approximants are constructed from the
Born series only and provide automatically a unitary T operator as
well as an automatic off-shell extension of the scattering amplitude
even in cases in which the Born series is not given by a Schrödinger
equation or a Bethe-Salpeter equation in ladder approximation. If
more than two Born terms can be calculated they can be incorporated
into higher order O.P.A.

We summarize here the advantages which the O.P.A. have compared
to the usual perturbation series.

Properties	Perturbation Series	O.P.A.
Analytic structure in the energy plane	No poles ; therefore, no resonances or bound states	Poles (due to operator inversion) leading to resonances and bound states
Unitarity	Badly violated for strong interactions *	Exact
Crossing symmetry	Exact	Approximately cross-ing symmetric **
Based on a varia-tional principle	No	Yes
Convergence	Not convergent ; badly divergent for strong interactions	Known to converge for large classes of functions

* A truncated perturbation series for S, $S = 1 + gS_B + g^2 S_{B2} + \cdots + g^N S_{BN}$, satisfies unitarity through order N ; that is, $SS^+ = 1 + g^{N+1} C_{N+1} + \cdots + g^{2N} C_{2N}$, where $C_{N+1}, \ldots, C_{2N}$ are obvious functions of $S_B, \ldots, S_{BN}$. For a strong interaction, unitarity is badly violated.

** Remarks similar to those just made for the unitarity of a truncated perturbation series apply. The [N/N] O.P.A. expanded in a power series is crossing symmetric through order 2N only. However, because the O.P.A. is a rational fraction, the orders beyond 2N are not so overwhelming, and the O.P.A. remains approximately crossing symmetric. Calculations verifying this assertion have been performed.

3. CONSTRUCTION OF THE [1/1] O.P.A.

Given a renormalizable Lagrangian the Feynman graphs give the first two terms of the perturbation expansion unambiguously on-shell and off-shell. Furthermore, there is no doubt that the off-shell parts are relevant to (say) p-p bremstrahlung. The graphical representation is

To simplify the notation we consider identical scalar particles of mass m . In the center of mass system we have the definitions

$$
\begin{aligned}
P_1 &= \left(\tfrac{1}{2}\sqrt{s} + \omega, \ p\hat{n}\right) , & P_2 &= \left(\tfrac{1}{2}\sqrt{s} - \omega, \ -p\hat{n}\right) , \\
P_1' &= \left(\tfrac{1}{2}\sqrt{s} + \omega', \ p'\hat{n}'\right) , & P_2' &= \left(\tfrac{1}{2}\sqrt{s} - \omega', \ -p'\hat{n}'\right) , \\
& & s &= (p_1 + p_2)^2 .
\end{aligned}
\tag{12}
$$

Relativistic invariance implies that all amplitudes are functions only of s, ω, ω', p, p' and $\hat{n}.\hat{n}' = \cos\theta$. The physical amplitude is given by

$$
\omega = \omega' = 0 \qquad , \qquad p = p' = \sqrt{\tfrac{1}{4} s - m^2} \qquad .
$$

The operator T can be partially diagonalized using various well known symmetries into irreducible blocks labelled by $T^{J,P,\alpha}(s)$, where J is the angular momentum, P the parity, α the internal quantum numbers and s the square of the energy in the center of mass system. The

[N/N] O.P.A. undergoes the same decomposition at T itself so that it is necessary to consider only the construction of the [1/1] O.P.A. to $T^{J,P,\alpha}(s)$. The Feynman graphs make it possible for us to compute the matrix elements $\langle \omega,p | T_B^{J,P,\alpha}(s) | \omega',p' \rangle$ and $\langle \omega,p | T_{B2}^{J,P,\alpha}(s) | \omega',p' \rangle$ The problem is to construct from the matrix elements an approximate phase shift $\delta_{[1/1]}^{J,P,\alpha}(s)$ given by

$$\exp\left(i\delta_{[1/1]}^{J,P,\alpha}(s)\right) \sin \delta_{[1/1]}^{J,P,\alpha}(s) \tag{13}$$

$$= \langle 0, \sqrt{\frac{s}{4}-m^2} | T_B^{J,P,\alpha}(s) \; (T_B^{J,P,\alpha}(s) - gT_{B2}^{J,P,\alpha}(s))^{-1} \; T_B^{J,P,\alpha}(s) | 0, \sqrt{\frac{s}{4}-m^2} \rangle$$

To make the operator inversion in Eq.(13), we introduce a finite mesh of points (ω_1,p_1) , (ω_2,p_2) , $\ldots$, (ω_n,p_n) and replace the operators by n×n matrices. This procedure gives rise to the matrix Padé approximant.

4. THE CHOICE OF MESH POINTS

After discretization, the approximate phase shifts $\delta_{[1/1]}^{J,P,\alpha}(s)$ will depend on the set of variables (ω_1,p_1) , (ω_2,p_2) , $\ldots$, (ω_n,p_n) while the exact phase shift is clearly independent of these variables. We now discuss the problem of the choice of the mesh points in potential scattering and in Lagrangian field theory.

In potential theory, the off-shell scattering amplitude does not depend on the relative energy ω . One can prove [3] for a positive potential that

$$\delta_{[1/1]}^{J,P,\alpha}(s,p_1,\ldots,p_n) \geq \delta_{[1/1]}^{J,P,\alpha}(s,p_1,\ldots,p_n,p_{n+1}) \geq \delta_{exact}^{J,P,\alpha}(s) \quad .(14)$$

Consequently,

$$\underset{p_1,\ldots,p_2}{\text{Inf}} \; \delta_{[1/1]}^{J,P,\alpha}(s,p_1,\ldots,p_n) \geq \delta_{exact}^{J,P,\alpha}(s) \tag{15}$$

This is the practical form of the variational principle for phase shifts in our scheme. Opposite inequalities hold for a negative potential. The same results can be applied to a singular potential provided one uses the cut-off needed for the regularization of the Born series as a variational parameter.

In the case of a potential which changes sign, there exist more than one extremum, and it is conjectured [4] that the proper one to use is the one nearest the on-shell momentum. In field theory we have emphasized already that the [1/1] O.P.A. is the exact solution of the Bethe-Salpeter equation in the ladder approximation. In the general case, that is, when we include in the second Born approximation all of the contributing diagrams (vertex graph, pion self energy graph, crossed ladder graph, and the nucleon self energy graphs which do not vanish off-shell), the [1/1] O.P.A. is still the exact solution of a Bethe-Salpeter equation with a modified two nucleon propagator given by

$$G = T_1^{-1} T_2 T_1 \qquad (16)$$

The problem of the choice of the mesh points has been investigated in the ladder approximation. The same variational properties as in potential scattering can be proved to hold in the spinless case in the ladder approximation [5]. In the general case, on the basis of numerical experience [6], one conjectures that one can set $\omega_1 = \omega_2 = \ldots = \omega_n = 0$, the p_i being chosen in a variational way exactly as in potential scattering. The choice of $\omega_i = 0$ is justified if the amplitudes do not depend too much on ω and ω'. The choice $\omega_i = 0$ in the ladder case leads to semi-relativistic equations such as the Blankenbecler-Sugar equation.

5. NUMERICAL TESTS OF THE METHOD

In the following table and figures we show the results obtained for an exponential potential $V(r) = -V_o \exp(-\mu r)$ in both attractive and repulsive cases.

	$-i\nu$	Ordinary Padé	Variational 2×2 Matrix Padé	Variational 3×3 Matrix Padé	Exact solution
Case g=-8	.1	− 7.610	− 9.237		− 9.242
	.5	−34.962	− 43.734		− 43.771
	1.	−56.975	− 76.726		− 76.861
	2.	−71.029	−110.868		−111.801
Case g=+8	.1	168.600	232.556	292.297	302.143
	.5	128.840	217.251	232.775	236.268
	1.	98.532	183.027	193.863	196.297
	2.	74.290	135.709	144.243	146.328
Case g=+10	.1	169.139	335.595	339.285	341.624
	.5	130.947	264.812	273.828	279.900
	1.	101.311	216.637	226.235	231.011
	2.	77.660	161.248	171.353	174.073

Table I — ^1S_o phase shift for an exponential potential $V(r) = -V_o e^{-\mu r}$

(phase shifts are given in degrees)

Notation : $g = 2 \dfrac{m}{\mu} \dfrac{V_o}{\mu}$; $= 2i \dfrac{\sqrt{2mE}}{\mu}$

g and ν are dimensionless.

m = 938 MeV = nucleon mass

μ = 140 MeV = pion mass

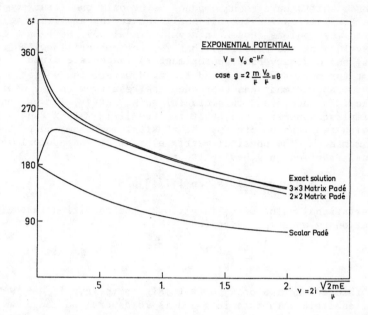

Figure 1 - The phase-shifts computed with the [1/1]P.A. in the ordinary (scalar P.A.) and in the Matrix case (V.M.P.A.) with one and two variational off-shell momenta are compared with the exact solution for an exponential potential exhibiting two bound states.

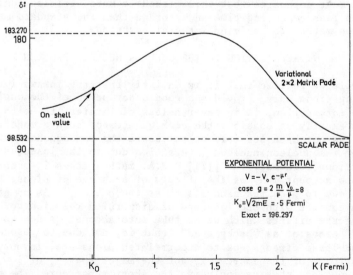

Figure 2 - The variational properties in the off-shell momentum are shown for the phase-shift computed from the [1/1] M.P.A. at a fixed energy for the previous potential.

The phase shifts have been computed using only the first two terms of the Born series for the off-shell scattering amplitude with two or three mesh points chosen in a variational way. We remark that for large values of the coupling, that is, for potentials with two bound states, the ordinary Padé approximant is inadequate since it cannot account for more than one bound state. However, the variational matrix Padé approximant can reproduce the results with any desired accuracy. To understand this crucial point better, let us consider the simplest approximation, that is, the [1/1] matrix Padé approximant with two mesh points, p_o the physical momentum and p_1 the off-shell momentum. The physical matrix element of the approximant is a rational fraction in g having two poles

$$R(g,p_o,p_1) \; = \; <p_o|\,[1/1]\,|p_o>$$

The variational value of p_1 is given by the condition that $R(g,p_o,p_1)$ be stationary when p_1 is varied,

$$\frac{\partial R(g,p_o,p_1)}{\partial p_1} \; = \; 0$$

p_1 is then a function of g so that $R(g_o,p_o,p_1(g,p_o))$ has now a much richer analytic structure in g and is in general no longer a rational fraction.

It has been proved [7] that in the case of the square well the [1/1] variational matrix Padé approximant with two mesh points is exact. For field theory numerical tests [6] on the Bethe-Salpeter equation in ladder approximation show that accurate results are obtained with a very limited number of mesh points variationally chosen.

We must emphasize that this method is more efficient and requires much less computing time and storage than the standard matrix inversion method for solving integral equations.

6. APPLICATION TO THE NUCLEON-NUCLEON PROBLEM

While we realize that it is unlikely that the Yukawa interaction including pions only provides a good description of the nucleon-nucleon interaction, it is nevertheless of interest to us to calculate as exactly as possible the consequences of such a model. Up to now it has not been possible to calculate the consequences of such a model of the nucleon-nucleon interaction due to the inadequacy of computational schemes. The [1/1] O.P.A. method makes it possible to take into account many of the effects of exchange of pions, but not effects due to pion resonances such as the ρ and ω. As an attempt to make contact with the language of dispersion relation work, we speak of the effects which we do take into account as due to "uncorrelated" pions or as "background" (that is, not due to resonances) effects. These effects due to uncorrelated pions are, in our view, at least part of the source, if not the major source, of low energy nucleon-nucleon physics including the short range core. The correlated pion exchanges which are usually introduced phenomenologically via vector meson exchange will play a role only for higher energies. They can be included in our scheme through Yang-Mills type interactions. The current algebra constraints which make it possible to describe the low energy $\pi\pi$ and πN interaction can be introduced by using a non-linear σ model.

The computational program is difficult because off-shell, the scattering amplitude has many more invariants than the traditional Fermi invariants S,V,T,A,P . It has been necessary to simplify it by taking into account only negative energy states in going off-shell, the magnitude of the momentum being kept equal to the physical value.

Before going to the results, we would like to explain why the negative energy states have been the first to be taken into account. It is because of a peculiarity of the pion ; namely, the intrinsic parity is negative.

At very low energy, it is well known that the "elementary" (that is, first Born contribution) NN force vanishes because of the negative intrinsic parity of the pion ; that is, the matrix elements of γ_5 are proportional to v/c . Therefore, the scattering lengths, which are very large experimentally, are found to be zero. However, one notices that the off-shell matrix element of γ_5 is not zero. Therefore, an approximation such as the M.P.A., which mixes the on-shell (positive energy) and off-shell (negative energy) elements, will automatically improve the scattering length [8]. We give the results for the scattering length in Tables II and III.

$g_{\pi NN}$	$\alpha_{\pi NN} = \dfrac{g_{\pi NN}^2}{4\pi}$	$a(^1S_0)$	$a(^3S_1)$
10	7.958	$-$ 1.111	$-$ 4.407
11	9.629	$-$ 1.605	$-$ 8.484
12	11.46	$-$ 2.307	$-$23.31
13	13.45	$-$ 3.345	$+$93.49
14	15.60	$-$ 4.994	$+$19.89
15	17.90	$-$ 7.951	12.45
16	20.37	$-$14.66	9.648
17	23.00	$-$43.90	8.187

Table II - Scattering length for the $^1S_0$ and $^3S_1$ waves computed from the [1/1] matrix Padé approximant to the on-shell scattering amplitude for the Yukawa model. The scattering lengths are in fermis.

$\bar{g} = \dfrac{m}{f_\pi}$	$g_{\pi NN}$	$a(^1S_0)$	$a(^3S_1)$
8	13.7	$-$ 1.408	$-$ 5.751
8.5	13.7	$-$ 2.029	$-$12.63
9	13.7	$-$ 3.083	$-$154.43
9.5	13.7	$-$ 5.214	$+$19.89
10	13.7	$-$11.59	$+$10.49
10.5	13.7	$-$856.9	$+$ 7.606
11.0	13.7	$+$14.65	$+$ 6.211

Table III - The same as Table II but for the non linear σ model.

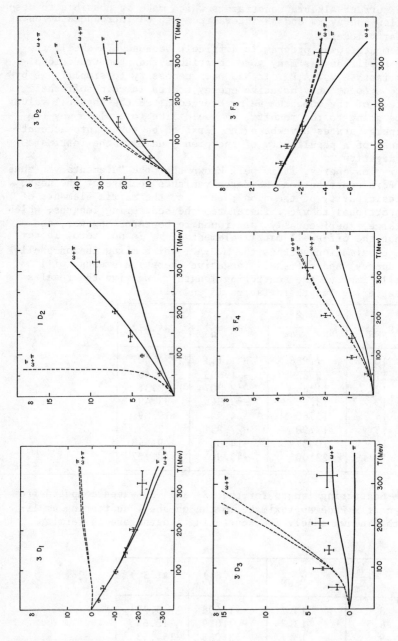

Figure 3 — The phase shifts obtained from the [1/1] matrix P.A. to the on-shell scattering amplitude for the Yukawa and ω model (continuous lines labelled by π and π+ω) are compared with the experimental data. The results obtained with [1/1] P.A. to the S matrix (dotted lines) are also shown. The masses are the physical ones (given in the text), $\alpha_\pi = 15$, $\alpha_\omega = 3.75$.

We notice that, even though the result is only qualitative for the currently accepted value of $g_{\pi NN}$, there exist two objects near threshold having the quantum numbers of the deuteron and di-neutron and that the deuteron is always the more strongly bound of the two.

For the P waves the lack of inclusion of the off-shell states limits the validity of the results to the very low energy region and we notice that the results are improved with the non-linear σ model. This reflects the role of chirality at low energy which is nevertheless far less important than for $\pi\pi$ and πN [9] .

For higher waves D,F,... the results obtained using only negative energy off-shell states, that is, without off-shell momenta, describe correctly the experimental results. Calculations performed with the Bethe-Salpeter equation show that no differences arise from the inclusion of off-shell momenta [6].

The ω-meson is usually considered to give an important contribution to the repulsive core. We have introduced such a particle within a renormalizable Lagrangian [10]. Our results show that the ω contribution to the three pion exchange force improves considerably the D waves (see Figure 3. Coupling strong enough to give correctly the hard core necessary in the S states destroys completely the correct behavior of the higher partial waves.

7. CONCLUSION

It is generally believed that scattering amplitudes in a not too large physical region can be obtained approximately by taking into account "the nearest singularities". Therefore one is tempted to classify these singularities in terms of "particles" producing the nucleon-nucleon force, the pions, the σ , the ρ and ω , etc..., the importance of the various singularities being best expressed by the product of the "distance" to the physical region by their coupling to the nucleons. However, there exists, in addition to this obvious classification of particles, another important structure which also strongly affects the physical scattering amplitude ; namely, the nearby off-shell surroundings of the physical region under inspection. Clearly this surrounding region is model dependent, and in a Lagrangian formalism it depends explicitly on the choice of the Lagrangian. For such reasons, and because up to now there has been no reliable approach to strong coupling Lagrangian field theory, the contribution of such surrounding regions has been omitted. The method of operator Padé approximants in Lagrangian field theory, which is nothing but a variational approximation method with the same inspiration as the Rayleigh-Ritz method for bound states in ordinary quantum mechanics, includes all the usual approximations such as the Bethe-Salpeter equation or the Blankenbecler-Sugar equation, but starts from the Feynman amplitudes given by the Lagrangian in the physical region and its off-shell surroundings. Being a non-linear approximation it mixes the on-shell and off-shell parts and so takes account of the surroundings of the on-shell part. To sum up, we state that we believe that a large part of the low energy phenomenology of the nucleon-nucleon interaction, including the hard core and spin-orbit forces necessary to explain the data, is provided by the off-shell surroundings of the

252

on-shell amplitudes. Calculations now in progress will show whether
we are right or wrong and whether the many years which we have devoted
to this project are justified in the way we have supposed they might
be.

ACKNOWLEDGEMENTS

We are grateful to Professor P. Cattillon for providing funds
which have made possible the conversations leading to this publication.

REFERENCES

1. D. Bessis and J. Talman, Rocky Mountain Journal of Mathematics,
   $\underline{4}$, 151 (1974) ;
   D. Bessis, in "Padé Approximants", P.R. Graves Morris Editor,
   Institute of Physics, London (1973).
2. M. Cini, S. Fubini, Nuovo Cimento $\underline{11}$, 142 (1954).
3. D. Bessis, P. Mery, G. Turchetti, to be published in Phys. Rev. D
4. L.P. Benofy, J. Gammel, P. Mery, Phys. Rev. $\underline{D13}$, 3111 (1973).
5. C. Alabiso, P. Butera, G.M. Prosperi, Nucl. Phys. $\underline{46B}$, 593 (1972).
6. J. Gammel, M. Mentzel, Phys. Rev. $\underline{D11}$, 963 (1975) ;
   J. Fleisher, J.A. Tjon, to be published in Phys. Rev. D ;
   J. Fleisher, J.A. Tjon, contributed paper to the Tampa Conference
   on "Rational Fractions with Emphasis on Padé Approximants".
7. G. Fratamico, F. Ortolani, G. Turchetti, Lett. Nuovo Cimento $\underline{17}$,
   582 (1976).
8. R.H. Barlow, M.C. Bergère, Nuovo Cimento, $\underline{11A}$, 557 (1972).
9. D. Bessis, G. Turchetti, W. Wortman, Nuovo Cimento $\underline{22A}$, 1957
   (1974) ;
   J. Fleisher, J. Gammel, M.T. Mentzel, Phys. Rev. $\underline{D8}$, 1545 (1973) ;
   A. Gersten, Z. Solov, Phys. Rev. $\underline{D10}$, 1031 (1974) ;
   D. Bessis, P. Mery, G. Turchetti, Phys. Rev. $\underline{D10}$, 1932 (1974) ;
   P. Mery, G. Turchetti, Phys. Rev. $\underline{D11}$, 2000 (1975).
10. D. Bessis, P. Mery, G. Turchetti, to be published in Nuovo Cimento.

HIGH ENERGY PROTON-PROTON ELASTIC SCATTERING
AND MULTIPLE SCATTERING MODEL

S. Wakaizumi

Department of Physics, Hiroshima University, Hiroshima 730, Japan

M. Tanimoto

Faculty of Education, Ehime University, Ehime 790, Japan

ABSTRACT

The small slope beyond the second peak in p-p elastic scattering differential cross section measured at $\sqrt{s}$=53 GeV is shown, in the constituent multiple scattering model, to indicate that the proton consists of a small number of valence constituents. The number turns out to be none other than three by imposing the requirement of translational invariance over the whole composite system (i.e. the proton).

The recent measurement of the differential cross section for elastic proton-proton scattering at center of mass energy of $\sqrt{s}$=53 GeV [1], has revealed a structureless exponential with a considerably small slope ($b_2 \approx 1.8$ $(GeV/c)^{-2}$) beyond the second maximum at a momentum transfer squared of $-t=2$ $(GeV/c)^2$ along with the well-known marked dip at $t \approx -1.3$ $(GeV/c)^2$ [2].

The diffraction-like structure of the sequential dip-bump has been investigated by many authors in the Chou-Yang model [3], but most of those calculations have given a much steeper slope of the second peak such as $b_2 \approx 5 \sim 6$ $(GeV/c)^{-2}$ which is about a half of the slope of the forward peak ($b_1 \approx 10$ $(GeV/c)^{-2}$), just coming from the convolution integrals of the eikonal involved in the model [4]. The model, moreover, predicts a second minimum in the range of $4 \lesssim |t| \lesssim 6 (GeV/c)^2$, contrary to the smooth behavior up to $|t| \approx 6.5$ $(GeV/c)^2$ obtained in the above experiment.

On the other hand, the same structure in the angular distribution has been explained as due to the multiple scattering effect in the constituent model for hadrons [5], where an interference of the single and the double scattering of constituents causes the dip at $|t| \approx 1.3$ $(GeV/c)^2$. Here we will show in this model that the small slope of the second peak just indicates a three-quark composite structure of the proton.

At high energy, scattering amplitude for the elastic proton-proton scattering as a collision between the two composite systems of constituents can be written down in terms of constituent scattering amplitude $t_{qq}(\vec{q})$ under the eikonal approximation in the multiple scattering formalism [6] as follows (and, moreover, in accord with the "Geometrical scaling"),

$$T_{pp}(t) = \frac{1}{R_o^2(s)} \binom{N}{1}^2 t_{qq}(t) [f_1(t)]^2 + \binom{N}{2}^2 \times 2! \frac{i}{2\pi} \int d^{(2)}\vec{q}_1 d^{(2)}\vec{q}_2 t_{qq}(\vec{q}_1) t_{qq}(\vec{q}_2)$$

$$\times \delta(\vec{q}_1+\vec{q}_2-\vec{q})\,[f_2(\vec{q}_1,\vec{q}_2)]^2 - {N \choose 3}^2 \times 3! \frac{1}{(2\pi)^2} \int d^{(2)}\vec{q}_1 d^{(2)}\vec{q}_2 d^{(2)}\vec{q}_3 t_{qq}(\vec{q}_1) t_{qq}(\vec{q}_2)$$

$$\times t_{qq}(\vec{q}_3)\,\delta(\vec{q}_1+\vec{q}_2+\vec{q}_3-\vec{q})\,[f_3(\vec{q}_1,\vec{q}_2,\vec{q}_3)]^2 + \cdots . \tag{1}$$

The terms in the expansion correspond to single, double, triple $\cdots$ scattering, and $f_1(t)$, $f_2(q_1,q_2)$ and $f_3(q_1,q_2,q_3)$ are the hadronic form factors related to each multiple scattering. $-t$ is the momentum transfer squared $-t=q^2$. It is assumed that a proton is composed of N constituents. $R_0^2(s) = \sigma_{inel}(s)/\pi$ and $q_1, q_2, \cdots$ are the non-dimensionalized momentum.

Hadronic form factors are calculated in accordance with the requirement of the translational invariance for the whole composite system and ,for example, $f_2(\vec{q}_1,\vec{q}_2)$ is given as

$$f_2(\vec{q}_1,\vec{q}_2) = \int \exp\{-i(\vec{q}_1\cdot\vec{r}_1+\vec{q}_2\cdot\vec{r}_2)\}|\psi(\{\vec{r}_i\})|^2 \delta(\sum_{j=1}^{N}\vec{r}_j/N)\prod_{k=1}^{N}d\vec{r}_k , \tag{2}$$

where $\psi(\{\vec{r}_i\})$ is wave-function of a proton constructed from N constituents.

At first, we take the Gaussian wave function ,for simplicity, according to the symmetric constituent model in order to investigate the relation between the slopes of the first and the second peak by calculating $T_{pp}(t)$ up to the triple scattering term. The occurrence of the dip at $|t|\simeq 1.3$ (GeV/c)^2 is reproduced by the interference of the single and the double scatterings, as is well known, if we take a dominantly imaginary constituent scattering amplitude exponentially damping in t with no structure itself and with a small real part as $t_{qq}(t) = (i+\alpha)\frac{1}{N^2}\frac{A}{4\pi}\exp(\frac{1}{2}at)$. And, the t-dependence of the peaks generated from single, double and triple scatterings proves to come both from that of the constituent scattering and from that of the form factors as follows,

$$b_1 = a + \frac{N-1}{N}\langle r^2\rangle, \quad b_2 = \frac{1}{2}(a + \frac{N-2}{N}\langle r^2\rangle), \quad b_3 = \frac{1}{3}(a + \frac{N-3}{N}\langle r^2\rangle), \tag{3}$$

where b_1, b_2 and b_3 is the slope of the first, second and third peak, respectively, and $\langle r^2\rangle$ is the mean square radius of a proton calculated as follows,

$$\langle r^2\rangle = \int r_k^2 |\psi(\{\vec{r}_i\})|^2 \delta(\sum_{j=1}^{N}\vec{r}_j/N)\prod_{\ell=1}^{N}d\vec{r}_\ell ,$$

where r_k is the distance between the k-th constituent and the center-of-mass of proton.

As is easily seen from eq. (3), if t-dependence from the constituent scattering dominates that from the form factors, the ratio R of the slopes of the second peak to the first one is $R\simeq\frac{1}{2}a/a=\frac{1}{2}$, which is quite in disagreement with the experimental value of R, $R^{exp}\simeq 1.8$ (GeV/c)$^{-2}/10$(GeV/c)$^{-2}\simeq 1/5$. On the contrary, if t-dependence from the form factors dominates, the ratio is

$$R \simeq \frac{1}{2}(\frac{N-2}{N}\langle r^2\rangle)/\frac{N-1}{N}\langle r^2\rangle = \frac{N-2}{2(N-1)} . \tag{4}$$

we can obtain the number of constituents in a proton by comparing eq. (4) with R^{exp}! That is, R is 1/4 for N=3, which is very close to $R^{exp} \approx 1/5$, but in the other cases the values of R=0 for N=2 and $R \geq 1/3$ for N>4 are in disagreement with the measurement as clearly seen in Fig. 1 where a=0.8 $(GeV/c)^{-2}$ and $\sqrt{<r^2>}$=3.3 GeV^{-1} and data is

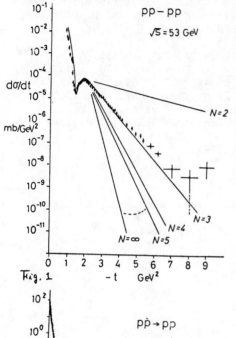

Fig. 1

taken from ref. 1. And, R approaches 1/2 if the number of constituents becomes infinite and this is the reason why the Chou-Yang model is not in accord with the elastic proton-proton scattering in the large t region. These results lead to an important conclusion that proton is an object with an extended distribution of three small constituents (quarks).

Another notable consequence of this model is that a second dip is predicted to occur. In Fig. 2 is shown the result of a calculation of $d\sigma/dt$ at $\sqrt{s}$=53 GeV, in which is used the wave function of a three-constituent system leading to a modified dipole electromagnetic form factor [7], $f_1(t)=1/(1-t/\mu_1^2)(1-t/\mu_2^2)$, μ_1^2=1.25 GeV^2 and $\mu_2^2=\mu_1^2/3$, which reproduces the measured forward peak with a curvature at small t. The slope beyond the third peak proves to have a t-dependence coming only from the constituent scattering, i.e. $b_3=\frac{1}{3}a$, as the result of the requirement of translational invariance over the whole composite system in the same way as for the Gaussian wave function as easily seen if N=3 is introduced into b_3 in eq. (3). As a is obtained to be about 2.7 $(GeV/c)^{-2}$ from the fit to the t region up to $|t|$=6 $(GeV/c)^{-2}$, a change of slope to a more gentle one of $b_3 \approx 0.9$ $(GeV/c)^{-2}$ is expected to occur from a triple scattering effect of the constituents, preceded by

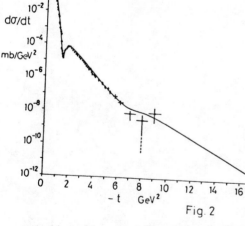

Fig. 2

a shallow dip or at least by a shoulder as seen in Fig. 2.

The authors would like to express their gratitude to Prof. M. Yonezawa for useful discussions.

REFERENCES

1. H. De Kerret et al., Phys. Letters 62B, 363 (1976).
2. A. Böhm et al., Phys. Letters 49B, 491 (1974); N. Kwak et al.,
   Phys. Letters 58B, 233 (1975).
3. T. T. Chou and C. N. Yang, Phys. Rev. 170 1591 (1968); H. Moreno
   and R. Suaya, SLAC-PUB-1161 (1972); M. Kuroda and H. Miyazawa,
   Prog. Theor. Phys. 50, 569 (1973); 51, 205 (1974); Y. Susuki,
   Prog. Theor. Phys. 54, 438 (1975).
4. U. P. Sukhatme, Phys. Rev. Letters 38, 124 (1977).
5. D. R. Harrington and A. Pagnamenta, Phys. Rev. 173, 1599 (1968);
   S. Wakaizumi, Prog. Theor. Phys. 42, 903 (1969); 54, 447 (1975);
   M. Tanimoto and S. Wakaizumi, Prog. Theor. Phys. 52, 1000 (1974).
6. R. J. Glauber, Lectures in Theoretical Physics, ed. W. E.
   Brittin et al., Vol. 1 (Interscience Publishers, Inc., New York,
   1959) p315.
7. M. KAC, Nucl. Phys. B62, 402 (1973).

FIGURE CAPTIONS

Fig. 1. The calculated differential cross section of elastic proton-proton scattering at $\sqrt{s}=53$ GeV.

Fig. 2. The calculated differential cross section at $\sqrt{s}=53$ GeV.

NUCLEON-NUCLEON PROBLEMS AT INTERMEDIATE ENERGIES

S. Furuichi

Rikkyo University, Tokyo 171, Japan

ABSTRACT

Nucleon-nucleon scattering in the energy interval between 400 MeV and a few GeV are discussed. Firstly we mention a transition character of such energy between the low and the high energy regions. Then usefulness of the method of phase shift analysis is examined. Evidences of the di-baryon resonance and similar structures are summarized, and models are presented for systematic understanding of these di-baryon structures. It is suggested that many di-nucleon resonances with broad width and high inelasticity are widely distributed in GeV-energy region.

INTRODUCTION

At the time of the first International Conference on the Nucleon-Nucleon Interaction at Florida we have reached common understanding that the inter-nucleon force in the low energy is fundamentally one-boson-exchange (OBE) type[1]. Since that time, more refined works[2] of this problem have confirmed the validity of extended OBE picture which includes the two-pion exchange force in the energy below $T_{lab} = 400$ MeV. This fact has also been stated as that the nuclear force has OBE character in the outer region than 1 fm with respect to the impact parameter:

$$b_\ell = \hbar \, (\ell(\ell+1))^{\frac{1}{2}}/p_{cm} > 1 \text{ fm}, \qquad (1)$$

where ℓ is the orbital angular momentum.

On the other hand, at sufficiently high energy, exchange of the Reggeized boson trajectories provides a good picture for the representation of N-N scattering,[3] as well for various hadron reactions. The representation is characterized by approximately linear rising trajectories and a universal slope of the leading trajectories.

The intermediate energy region between these low and high energy regions would form a clear transition region. The lower boundary of the intermediate region is 400 MeV and it corresponds to an upper bound for the energy where inelastic processes can be neglected. As for the upper boundary, however, we have not so clear standard. Sometimes we have used vaguely "a few GeV". If we tentatively take the energy of the meson-baryon scattering where individual contributions of direct channel resonances are smeared out as a measure of such boundary, the intermediate region of N-N scattering may extend to rather high energy, say $T_{lab} = 5 \sim 6$ GeV.

In this talk I would like to discuss the problems of the inter-

ISSN: 0094-243X/78/257/$1.50

mediate region from the low energy side and sketch their characteristic features.

PHASE SHIFT ANALYSIS WITH OBE TAIL

The method of phase shift analysis has been established in the
low energy N-N scattering, and sufficiently reliable set of phase
shifts has been obtained.[4] In principle, the phase shift analysis
should be a model independent procedure which naively reflects the
experimental data. At the intermediate energies, however, it seems
difficult to obtain unique set of phase shifts without any appropriate models. In order to overcome this dilemma, we built up a
program of phase shift analysis[5] in view of the result established
in the low energy region. The program is model independent except
for the OBE approximation to high partial waves. We introduce a
boundary value of angular momentum ℓ_{max} and real phase shifts are
evaluated by the OBE model for $\ell > \ell_{max}$. The value of ℓ_{max} is
chosen so that inelasticities with $\ell > \ell_{max}$ are sufficiently small
at the final result.

As for the OBE model we take contributions from the pion, the
ρ-meson, the ω-meson and an isoscalar scalar meson, which constitute
a minimal set of bosons for explaining low energy N-N scattering.
Then χ^2 minimizing search is carried out by treating both phase
shift parameters of low partial waves and boson-nucleon coupling
constants as variables. In the above sense ℓ_{max} is also a free
parameter determined a posteriori.

Utilizing this program, we analyzed the p-p and n-p scattering
data between T_{lab} = 300 MeV and 1000 MeV.[5] For p-p scattering we
obtained reasonable solutions with ℓ_{max} = 1 at 330 MeV and 425 MeV,
and ℓ_{max} = 2 and 3 are responsible for 640 MeV and 970 MeV, respectively. These boundary values ℓ_{max} for the OBE approximation are
consistent with the equation (1). Possible energy dependence of
boson-nucleon coupling constants was tested and it was shown that
the results are compatible with the energy independent coupling
constants, at least, below 1 GeV. Resultant phase shift parameters
are shown in Fig. 1.

Next let us consider the energy range where the method of phase
shift analysis guarantees feasible procedures. Equation (1)
predicts an upper bound of such energy range corresponding to a
given value of ℓ_{max} as follows:

$$T_{lab} < \ell_{max}(\ell_{max}+1) \times 83 \text{ MeV.} \qquad (2)$$

Since ℓ_{max} is related with the number of free parameters in the
analysis, technical feasibility of the analysis restricts the upper

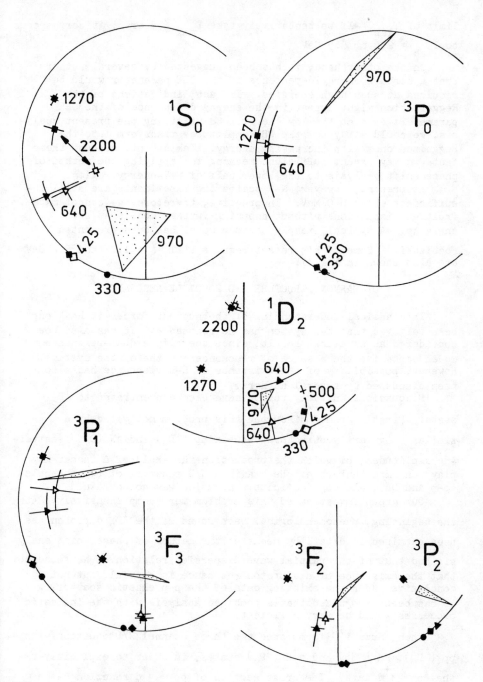

Fig. 1. Argand plot for
p-p phase shift parameters.

limit of T_{lab}. If we tentatively take $\ell_{max} \lesssim 5$ or 6, it corresponds to $T_{lab} = 2.5$ or 3.5 GeV.

On the other hand, it has been suggested by several authors[6] that a certain energy dependence of the OBE parameter would be required at such high energies. In fact, indications of the Reggeization might appear in the energy dependence of the OBE parameters even at the low energy. So utilizing the present analysis, we could study how the OBE amplitudes transform into the Reggeized ones with increasing energy. Taking into account these facts we may expect sufficient meaning for applying the method of phase shift analysis to the lower half of GeV-energy region.

At present, however, N-N scattering experiments are still insufficient above 640 MeV. In practive, therefore, we cannot made fruitful discussions without assuming further models at such energies. A typical example of such model has been presented by Hoshizaki.[7] Some of his recent results at $T_{lab} = 1.27$ and 2.2 GeV are also shown in Fig. 1.

DI-BARYON STRUCTURE AND POSSIBLE RESONANCES

In classical understanding of the nuclear force, it has long been believed that N-N system has no resonance. It has also been considered as a natural result, since the simple dual-resonance model emphasize the absence of resonances in the exotic systems. However, possibility of the existence of N-N resonance has also been discussed by several authors.

Discussions in early period have been concentrated to $^1D_2$ state. Firstly Arndt[8] discussed this problem motivated by a similar singularity structure between pp ($^1D_2$) and πN (D_{13}) scattering amplitudes, of which the production threshold of Δ isobar may play an important role. He examined N/D equation of two channels (p-p and NΔ), although definite conclusion was not obtained.

Our group investigated this problem more comprehensively. At the beginning, phenomenological background of the $^1D_2$ diproton has been examined in detail by means of the method of phase shift analysis and that of the partial-wave dispersion relation.[9] We found out that the existence of a diproton resonance is hopeful, but not conclusive under the existing data of the p-p elastic scattering. It has been recognized that a combined analysis with the inelastic processes would be very efficient.

Then, Suzuki[10] formulated the three channel N/D equation with pp ($^1D_2$), NΔ ($^5S_2$) and π^+d ($^3P_2$) states, in order to explicitly use the pp $\rightarrow \pi^+d$ data. The cross section of pp $\rightarrow \pi^+d$ reaction has a bump structure peaked at 600 MeV, and its differential cross section shows a dominance of the $^3P_2$ state at these energies. So, inclusion of the π^+d channel seems necessary. Moreover, the three channel treatment satisfies sufficient condition too, since the two pion production is still negligible small below 1 GeV. Effects of the anomalous thresholds in reaction amplitudes were fully included.

Similar approximation with that of Arndt were applied to the left hand cut and the N/D equation was solved numerically, on the basis of data set of p-p phase shifts at low energy and the cross sections of pp → NΔ and pp → π^+d reactions. Reasonable solutions have been obtained for every case, which correspond to some variety of the treatment of pp → NΔ data. As a result, three channel treatment predicted a diproton resonance with high probability. It was also shown that the anomalous threshold involved in the πd → NΔ amplitude plays an important role in fitting the inelastic reactions. His typical solutions are shown in Fig. 2 and compared with the results of phase shift analysis[4, 5].

Of course, further confirmation of the existence of $^1D_2$ resonance may leave to experiment. Most regular way to detect resonance phenomena is to successively determine the energy dependence of

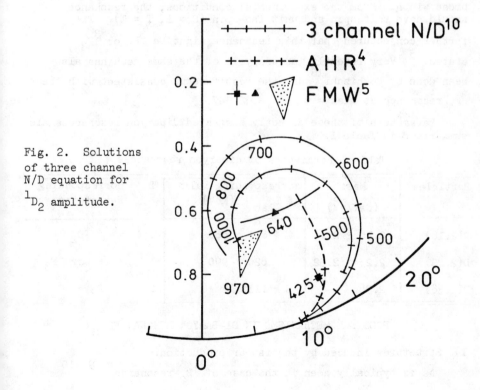

Fig. 2. Solutions of three channel N/D equation for $^1D_2$ amplitude.

amplitudes by complete experiments with a small energy interval for both elastic and inelastic processes. In this respect, recent experiments on $\Delta\sigma_T$ for p-p scattering in ANL is very interesting.[11]

Recently, new experimental evidences have been reported about the dibaryon resonance or similar structures. Kamae et al.[12] observed a resonance-like behavior of the proton polarization in $\gamma d \to pn$ reaction and proposed possible existence of a deeply bound $\Delta\Delta$ state. Based on the non-relativistic OBE potential model, they[13] conjectured that this $\Delta\Delta$ state may be (T = 0, J = 3) state with mass = 2.38 GeV. If this is the case, this dibaryon state will generate a resonance phenomenon is n-p scattering with (T = 0, $^3D_3 + {}^3G_3$) states at the energy T_{lab} = 1.14 GeV. This result seems very important, because it is the information on the n-p scattering.

Another evidence comes from the observation of p-p scattering with longitudinally polarized pure spin states by Yokosawa et al.[14] They obtained a strong energy dependence of spin parallel cross section $\sigma(\rightrightarrows)$ at $P_L \sim 1.7$ GeV/c and discussed a possible existence of direct channel p-p resonance with mass = 2.25 GeV and rather broad width. From the experimental conditions, the resonance should have unit spin and unit iso-spin (S = 1, T = 1). They further conjectured that this resonance might be $^3P_1$ or $^3F_3$ states.[15] Very recently, an analysis of the same data has also been done by Hoshizaki[16] and the result seems consistent with the $^3F_3$ resonance at mass = 2.26 ~ 2.32 GeV.

Parameters of these directly suggested dibaryon resonances are summarized in Table I.

Table I Parameters of dibaryon resonances

Particles	Mass (GeV/c^2)	corresponding value of T_{lab} (MeV)	T	corresponding N-N states
D(2.16)	2.16 ~ 2.18	610 ~ 660	1	$^1D_2$
D(2.25)	2.25 ~ 2.32	820 ~ 990	1	$^3P_1$ or $^3F_3$
D(2.38)	2.38	1138	0	$^3D_3 + {}^3G_3$

POSSIBLE MECHANISMS IN DI-BARYON RESONANCES

i) Structures induced by the isobar production.

As is typically seen in the case of $^1D_2$ resonance[8,9,10] a kind of resonance could be generated by the threshold effect of newly openned coupled channel. Of course, whether a certain threshold effect really generates resonance or not is rather sensitive to the dynamical conditions.[17] We could only expect that resonance or resonance-like structures are widely observed even for N-N scattering, although they might be a broad and highly inelastic ones.

Most natural candidate for such channels are those of a baryon isobar production in N-N scattering. At the threshold, the reaction could be dominated by the S-wave of the final system and corresponding N-N states are kinematically limited. For example, we can classify the NN → NN* type reactions, referring to their kinematical conditions as in Table II .

As is easily seen dinucleons D(2.16) and D(2.25) are reasonably understood by NΔ(1232) with S and P wave states respectively. Mass difference of these dinucleons is also reasonable in the present scheme.

Table II Kinematical classification of nucleon
isobar effects to N-N scattering

Class	isobar N^*	threshold in N-N(T_{lab})	p-p states coupled to N-N*	
			S wave N-N*	P wave N-N*
O-I	Δ(1232) P'_{33}	633 MeV	$^1D_2$	$^3P_J, ^3F_2, ^3F_3$
O-II	N(1470) P'_{11}	1210	} $^1S_0$	} $^3P_J, ^3F_2$
	N(1780) P''_{11}	2060		
E-I	N(1520) D'_{13}	1340	$^3P_1, ^3P_2, ^3F_2$	$^1S_0, ^1D_2$
	N(1535) S'_{11}	1380	} $^3P_0, ^3P_1$	} $^1S_0, ^1D_2$
	N(1700) S''_{11}	1830		
	N(1670) D_{15}	1750	$^3P_2, ^3F_2, ^3F_3$	$^1D_2, ^1G_4$
E-II	Δ(1650) S_{31}	1690	$^3P_0, ^3P_1$	} $^1S_0, ^1D_2$
	Δ(1670) D_{33}	1830	$^3P_1, ^3P_2, ^3F_2$	
O-III	Δ(1890) F_{35}	2390	$^1D_2$	$^3P_1, ^3P_2, ^3F_J, ^3H_4$
	Δ(1950) F_{37}	2570	$^1G_4$	$^3P_2, ^3F_J, ^3H_4, ^3H_5$

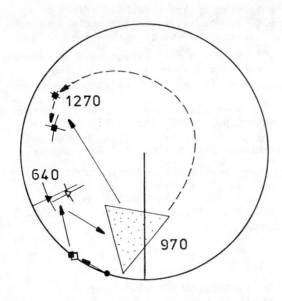

Following Table II, I would like to further conjecture that N^* isobars in O-II class and E-I class may induce some structures for $^1S_0$ at T_{lab} = 1.1 ~ 1.3 GeV and ^3P_J at T_{lab} = 1.3 ~ 1.4 GeV, respectively. In this respect, $^1S_0$ phase parameters at T_{lab} = 970 MeV in Fig. 1 is very suggestive. The phase parameter between 1.0 and 1.27 GeV involves two different possibilities as schematically shown in Fig. 3. Various experimental tests of these possibility are awaited.

Fig. 3.   Illustration of $^1S_0$ amplitudes.

ii) Dibaryon states and SU(6) symmetry.

A familiar method of considering the resonance states systematically is the symmetry group approach. For example, Dyson and Xuong[18] proposed a lowlying dibaryon multiplet in the frame work of SU(6) theory. They started from the two body system of 56 plet baryons and decompose it into representations of the dibaryon. Because baryons are Fermions, only two of the representations are permissible. These are the 490 plet and 1050 plet. They assume the 490 plet is the lowlying and extract the non-strange members as shown in Table III. Taking the threshold behavior of p-p scattering as the candidate of D_{10} and assuming this D_{10} nearly degenerates to D_{01} (deuteron), they obtained the mass formula:

$$M = A + B[T(T+1) + J(J+1) - 2] \qquad (3)$$

where A is fixed by the deuteron mass. If we take B = 50 MeV, masses of D(2.16) and D(2.38) are well reproduced by Eq. (3).

On the other hand, we have no room for D(2.25) resonance in this scheme. Even if we choose the 1050 plet is the lowlying, because this multiplet involves a T = J = 1 state, we cannot have T = 0, J = 3 state simultaneously. Moreover, this T = J = 1 state should be NΔ with $^3S_1$ state and cannot couple to p-p system. Summarizing these facts, D(2.25) could be best understood by the NΔ with

Table III Dibaryons without strangeness in 490 plet

Particle	T	J	Comment	Coupled N-N states	Possible candidate
D_{01}	0	1	Deuteron	$^3S_1 + {}^3D_1$	d(1.867)
D_{10}	1	0	Deuteron singlet state	$^1S_0$	?
D_{12}	1	2	NΔ resonance ($^5S_2$)	$^1D_2$	D(2.16)
D_{21}	2	1	charge 3 resonance	—	—
D_{03}	0	3	$\Delta\Delta$ resonance ($^7S_3$)	$^3D_3 + {}^3G_3$	D(2.38)
D_{30}	3	0	charge 4 resonance	—	—

$^5P_1$ of $^5P_3$ states, and it seems difficult to include this resonance in a lowlying multiplet, in spite of its relatively light mass.

Furthermore, the validity of the SU(6) symmetry for dibaryon system should be tested by the observation of multi charged particles (D_{21}, D_{30}) and strange partners in related SU(3) multiplets in many complex processes.

iii) Conjectured dinucleon Regge trajectories with universal slope.

Apart from the symmetry group approach, a Regge pole description is another powerful method to systematize the resonance phenomena. Several years ago, some conjectures have been made for T = 1 dinucleon Regge trajectory as a Regge recurrence of the $^1S_0$ threshold behavior.[19]

Recently a remarkable relation has been pointed out[20] among the mass values of deuteron, D(2.16) and D(2.38) resonances. These N-N states are on a linearly rising trajectory:

$$\alpha(s) = -2.28 + 0.932\, s. \tag{4}$$

Value of the slope well coincides with the universal one given by the leading trajectories of meson-meson and meson-baryon systems. This is shown in Fig. 4, where J-α(0) are plotted for resonances belonging to all leading trajectories with zero strangeness.

Based on this fact, we have proposed a degenerate set of dinucleon Regge trajectories[20], of which the one corresponding to family of (T = 1, singlet-even) states and the other corresponding to family of (T = 0, triplet-even) states. Both trajectories satisfy Eq. (4).

Other than these trajectories, D(2.25) resonance may belongs to (T = 1, triplet-odd) trajectory. If we assign $^3F_3$ state to the

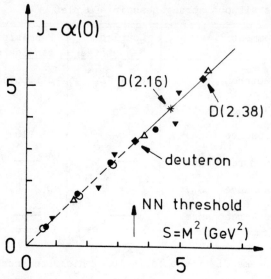

Fig. 4. J-α(0) vs. s plot; ●o are meson-meson, ▼Δ are meson-baryon, and ◆* are baryon-baryon resonances.

Intercept α(0) are taken as:
meson-meson
    0.47 (iso-spin 1)
    0.43 (iso-spin 0)
meson-baryon
    0.10 (iso-spin 3/2)
    −0.36 (iso-spin 1/2)
baryon-baryon
    −2.28.

resonance[15, 16] such odd parity trajectory slightly ahead the leading deuteron trajectory.

In the approach by the dibaryon Regge trajectory, observation of high mass states is the crucial problem. Until now, some indications have been seen in the pp → πd at forward scattering. For the purpose, various missing mass experiments involving baryon-baryon pair in final states might be useful. Although it may be possible to detect the dinucleon exchange at the backward baryon-antibaryon scattering, this is not so easy task due to very small value of the intercept α(0).

CONCLUDING REMARKS

In this talk I have considered problems to be fixed for the intermediate energy region of N-N scattering. Phenomenological situations were reviewed by the method of phase shift analysis. Then possible resonance behaviors were summarized. These dibaryon resonances seem very interesting, since some informations on the exotic resonances in the simple quark model have been accumulated

in other reactions than N-N scattering. For example, Chew[21] argued possible existence of planer trajectories which couple predominantly to baryon-antibaryon channels, the baryonium. His leading baryonium trajectories have also unit slope and different intercept. Various leading trajectories without strangeness are shown schematically in Fig. 5. This figure suggests a deep interrelation between the exoticity and the intercept of leading trajectory of a certain system.

At present, however, existing di-baryon resonances gathered in

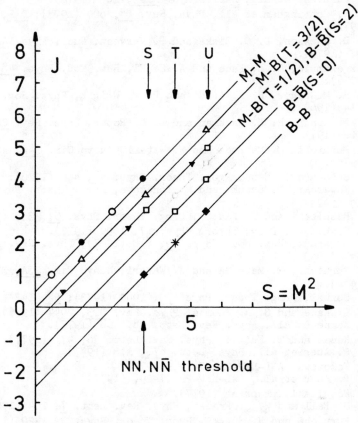

Fig. 5. Leading trajectories for various systems. Every slope is normalized to unity.

Table I are most simply understood by the combination of N-N, NΔ (S- and P-states) and ΔΔ (S-state). In the scheme, P-state NΔ could couple to $^3P_1$, $^3P_2$, $^3F_2$ and $^3F_3$ states of N-N system. Determination of D(2.25) state is important for the investigation of dynamical features of the P-state NΔ. Many theoretical works should be done in order to clarify the composite character of exotic systems at the intermediate energies.

REFERENCES

1. S. Ogawa et al., Prog. Theor. Phys. No. 39, 140 (1967).
   Rev. Mod. Phys. 39, No. 3 (1967).
   M. J. Moravcsik, Rep. Prog. Phys. 35, 587 (1972).
2. S. Furuichi, H. Kanada and K. Watanabe, Prog. Theor. Phys. 44, 711 (1970).

S. Furuichi et al., Nucl. Phys. $\underline{A193}$, 285 (1972).

W. N. Cottingham et al., Phys. Rev. $\underline{D8}$, 800 (1973); $\underline{D12}$, 1495 (1975).

A. D. Jackson, D. O. Riska and B. Verwest, Nucl. Phys. $\underline{A249}$, 397 (1975).

3. For example, G. L. Kane and A. Seidl, Rev. Mod. Phys. $\underline{48}$, 309 (1976).

4. M. H. McGregor, R. A. Arndt and R. M. Wright, Phys. Rev. $\underline{182}$, 1714 (1969).

    R. A. Arndt, R. H. Hackman and L. D. Roper, Phys. Rev. $\underline{C15}$, 1002 (1977).

5. S. Furuichi, M. Matsuda and W. Watari, Nuovo Cim. $\underline{34A}$, 467 (1976).

6. S. Sato and T. Ueda, Prog. Theor. Phys. $\underline{44}$, 409 (1970).
    M. Kawasaki, Y. Susuki and M. Yonezawa, Prog. Theor. Phys. $\underline{47}$, 589 (1972).

7. N. Hoshizaki and T. Kadota, Prog. Theor. Phys. $\underline{57}$, 335 (1977).
    N. Hoshizaki, Prog. Theor. Phys. $\underline{57}$, 1099 (1977).

8. R. A. Arndt, Rev. Mod. Phys. $\underline{39}$, 710 (1967); Phys. Rev. $\underline{165}$, 1834 (1968).

9. S. Furuichi, M. Matsuda and W. Watari, Nuovo Cim. $\underline{23A}$, 375 (1974).

10. H. Suzuki, Prog. Theor. Phys. $\underline{50}$, 1080 (1973); $\underline{54}$, 143 (1975).

11. G. L. Kane and G. H. Thomas, Phys. Rev. D13, 2944 (1976).

12. T. Kamae et al., Phys. Rev. Lett. $\underline{38}$, 468 (1977).

13. T. Kamae and T. Fujita, Phys. Rev. Lett. $\underline{38}$, 471 (1977).

14. I. P. Auer et al., Phys. Lett. $\underline{67B}$, 113 (1977).

15. A. Yokosawa, ANL-HEP-CP-77-07, (1977).
    I. P. Auer et al., ANL-HEP-PR-77-29, (1977).

16. N. Hoshizaki, preprint (1977).

17. J. S. Ball and W. R. Frazer, Phys. Rev. Lett. $\underline{7}$, 204 (1961).
    S. Furuichi and H. Suzuki, Prog. Theor. Phys. $\underline{39}$, 420 (1968).
    D. D. Brayshow, Phys. Rev. Lett. $\underline{37}$, 1329 (1976).

18. F. J. Dyson and Nguyen-Huu Xuong, Phys. Rev. Lett. $\underline{13}$, 815 (1964).

19. S. Graffi, V. Grecchi and G. Turchetti, Lett. Nuovo Cim. $\underline{2}$, 311 (1969).

    L. M. Libby and E. Predazzi, Lett. Nuovo Cim. $\underline{2}$, 881 (1969).

20. S. Furuichi and H. Suzuki, Prog. Theor. Phys. $\underline{57}$, 1803 (1977).

21. G. F. Chew, LBL-5391 (1976).

INELASTICITY AND STRUCTURE IN
pp → pp AT MEDIUM ENERGIES*

Gerald H. Thomas
Argonne National Laboratory, Argonne, Illinois 60439

ABSTRACT

Above the inelastic threshold in pp scattering, the partial
wave amplitudes get important inelastic contributions. This may
lead to observable structure in pp spin measurements. We review
some of the new pure spin total cross section data, and discuss the
possible implications. It is suggested that a systematic study of
the medium energy region might clarify some of the issues raised by
the new data, in particular that of the existence of dibaryon resonances. Elastic and inelastic polarized pp measurements at ZGS
energies are also reviewed as they may provide a clue about which
medium energy measurements should be done.

INTRODUCTION

The total cross section data[1-3] for pure spin states recently
obtained at the ZGS reveals interesting and as yet unexplained
structure. To set the stage for later discussion, let us look again
at this new data (Fig.1). The horizontal scale on the top of Fig.1,
for Medium energy Physicists, is the lab. kinetic energy in MeV; the
scale on the bottom, for particle physicists, is the lab momentum in
GeV/c; the middle horizontal scale is the invariant c.m. energy in
GeV. A noteworthy feature of Fig.1 is the vertical scale which

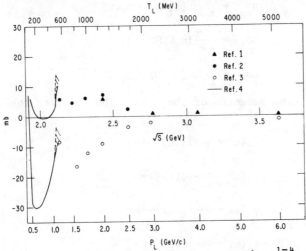

Fig. 1. Pure spin total cross sections[1-4]

*Work performed under the auspices of the United States Energy
Research and Development Administration.

extends from –30mb to +30mb. Plotted are the differences between total cross sections for pure spin states (the average being the usual spin averaged total cross section), so differences such as –30mb correspond to very large asymmetries.

Let us now look more closely at some of the distinctive features of this Figure. The upper data points[1,2] are from the ZGS and are for beam and target in pure spin states transverse to the beam direction: specifically

$$\Delta\sigma_T = \sigma(\uparrow\downarrow) - \sigma(\uparrow\uparrow) \quad . \tag{1.1}$$

The upper curve shows the prediction for $\Delta\sigma_T$ based on the Arndt, MacGregor and Wright (AMW) phase shifts.[4,5] Evidently the phase shifts do not tie on smoothly to the ZGS data. This could be an experimental problem, or indicate narrow structure around 430 MeV kinetic energy ($p_L \sim 1$ GeV/c).

The data points on the lower half of the figure come also from the ZGS, and are for beam and target in pure spin states longitudinal to the beam direction[3]: specifically

$$\Delta\sigma_L = \sigma(\rightleftharpoons) - \sigma(\rightrightarrows) \quad . \tag{1.2}$$

The lower curve shows the predictions for $\Delta\sigma_L$ based on the AMW phase shifts. In this case one solution of the phase shifts ties on nicely to the ZGS data. Structure is apparent in $\Delta\sigma_L$ at $p_L \cong 1.2$ and 1.5 GeV/c, corresponding to a c.m. energy $\sqrt{s} = 2150$ and 2250 MeV respectively.

If we imagine the structure to be caused by activity in a single partial wave, then we can try to verify that $\Delta\sigma_T$ and $\Delta\sigma_L$ behave in the proper manner. In terms of s-channel helicity amplitudes[6], the three optical theorems for pp elastic scattering are[7]

$$\sigma_{tot} = \frac{2\pi}{k} \operatorname{Im} [\phi_1(0) + \phi_3(0)] \tag{1.3}$$

$$\Delta\sigma_T = -\frac{4\pi}{k} \operatorname{Im} \phi_2(0) \tag{1.4}$$

$$\Delta\sigma_L = \frac{4\pi}{k} \operatorname{Im} [\phi_1(0) - \phi_3(0)] \tag{1.5}$$

where k is the c.m. momentum of the beam. The notation for the helicity amplitudes is[8]

$$\phi_1(t) = \langle++|\phi(t)|++\rangle$$

$$\phi_2(t) = \langle--|\phi(t)|++\rangle$$

$$\phi_3(t) = \langle+-|\phi(t)|+-\rangle \tag{1.6}$$

$$\phi_4(t) = \langle+-|\phi(t)|-+\rangle$$

$$\phi_5(t) = \langle++|\phi(t)|+-\rangle$$

as functions of the invariant momentum transfer t ; the differential cross section is

$$\frac{d\sigma}{dt} = \frac{\pi}{2k^2} \left[|\phi_1|^2 + |\phi_2|^2 + |\phi_3|^2 + |\phi_4|^2 + 4|\phi_5|^2 \right] \quad . \quad (1.7)$$

These helicity amplitudes can be related to the spin partial waves,

singlet $\qquad R_J \qquad = \cos\rho_J e^{2i\delta_J} - 1$

triplet $\begin{cases} R_{JJ} \quad = \cos\rho_{JJ} e^{2i\delta_{JJ}} - 1 \\[2mm] R_{J\pm1,J} = \cos\rho_{\pm J}\cos 2\varepsilon_J e^{2i\delta_{\pm J}} - 1 \\[2mm] R^J \quad = i\sin 2\varepsilon_J e^{i(\delta_{+J}+\delta_{-J}+\alpha_J)} \end{cases}$ $\qquad (1.8)$

of AMW.[4] [Note we allow for inelastic effects through ρ and α .] The experimental phase shifts reflect a parametrization of the scattering amplitude in the spin basis. There is a rotation involved in going from one basis to another. Taking this rotation into account, the (s-channel) helicity partial waves, defined for each c.m. scattering angle Θ by

$$\langle\mu\lambda|\phi(\Theta)|\rho\sigma\rangle = \frac{1}{2ik} \sum_J (2J+1) T^J_{\mu\lambda\rho\sigma} d^J_{\rho-\sigma,\mu-\lambda}(\Theta) \qquad (1.9)$$

are given in terms of the spin partial waves by the expressions[9]

$$T^J_1 = R_J + \frac{J+1}{2J+1} R_{J+1,J} + \frac{J}{2J+1} R_{J-1,J} + \frac{2\sqrt{J(J+1)}}{2J+1} R^J$$

$$T^J_2 = -R_J + \frac{J+1}{2J+1} R_{J+1,J} + \frac{J}{2J+1} R_{J-1,J} + \frac{2\sqrt{J(J+1)}}{2J+1} R^J$$

$$T^J_3 = R_{JJ} + \frac{J}{2J+1} R_{J+1,J} + \frac{J+1}{2J+1} R_{J-1,J} - \frac{2\sqrt{J(J+1)}}{2J+1} R^J \qquad (1.10)$$

$$T^J_4 = -R_{JJ} + \frac{J}{2J+1} R_{J+1,J} + \frac{J+1}{2J+1} R_{J-1,J} - \frac{2\sqrt{J(J+1)}}{2J+1} R^J$$

$$T^J_5 = -\frac{\sqrt{J(J+1)}}{2J+1} \left[R_{J+1,J} - R_{J-1,J} \right] + \frac{R^J}{2J+1} \quad .$$

The notation T_i is analogous to ϕ_i in (1.6).

We can now envision several possibilities. For example, suppose there were $^1D_2$, $^3P_2$ and $^3F_3$ resonances, along with a $^3P_1$ background. These contributions to $\Delta\sigma_T$ and $\Delta\sigma_L$ would be as follows

$$(\Delta\sigma_T) = \text{Im}\left\{\left(\frac{4\pi}{k}\ \frac{1}{2ik}\right)\left(5R_2 - 2R_{12}\right)\right\}$$

$$(\Delta\sigma_L) = \text{Im}\left\{\left(\frac{4\pi}{k}\ \frac{1}{2ik}\right)\left(5R_2 - R_{12} - 7R_{33} - 3R_{11}\right)\right\} . \qquad (1.11)$$

In other words the $^1D_2$ structure must show up with equal magnitude in both $\Delta\sigma_T$ and $\Delta\sigma_L$. More generally, any singlet wave contributes equal-ly, and with positive sign.* Based on the data in Fig.1 it is evident, assuming the data are correct, that in the region of the NΔ threshold, no strong $^1D_2$ structure exists. This is contrary to what has long been believed.[10] I should note that the $\Delta\sigma_T$ data are preliminary, so the situation could in principle change.

Next, let us consider the triplet contributions to (1.11). All triplet $L=J$ waves are absent from $\Delta\sigma_T$ whereas they each give negative contributions to $\Delta\sigma_L$. At $p_L = 1$ GeV/c ($T_L = 431$ MeV), the $^3P_1$ wave has phase $\delta_{11} \sim -40°$; assuming no inelasticity the $^3P_1$ contribution alone is

$$-\left(\frac{2\pi}{k^2}\right)(3)(1 - \cos 80°) = -30.5 \text{ mb} \qquad (1.12)$$

which shows the phase shift prediction in Fig.1 gets a large contri-bution from this wave. In addition to this partial wave, other waves such as the $^3F_3$ could also contribute structure to $\Delta\sigma_L$; such structure should not be in $\Delta\sigma_T$. At $p_L = 1.5$ GeV/c ($T_L = 830$ MeV) there is evid-ence for an effect in $\Delta\sigma_L$ but not $\Delta\sigma_T$. Hidaka et al.[11] argue that this could be due to a dibaryon resonance in the $^3F_3$ channel, based on a moment analysis of polarization data.

Finally, there are triplet waves with $L=J\pm 1$ which contribute to both $\Delta\sigma_T$ and $\Delta\sigma_L$. For $J > 0$ their contribution to $\Delta\sigma_T$ is larger than to $\Delta\sigma_L$. Moreover their contribution to $\Delta\sigma_T$ is always negative, while $L=J+1$ adds to and $L=J-1$ subtracts from $\Delta\sigma_L$. [For this dis-cussion, the mixing term R^J is ignored.] In the medium energy region under consideration, $1 \lesssim p_L \lesssim 3$ GeV/c [$430 \lesssim T_L \lesssim 2200$ MeV] there is no strong evidence at present that the structure is due to these waves with $L=J\pm 1$, since structure in $\Delta\sigma_L$ is not matched with corresponding structure in $\Delta\sigma_T$, and vice versa.

To sum up, the new ZGS data shows structure in the medium energy region, although the exact nature of this structure is unclear. It would be nice to separate inelastic mechanisms from those which are primarily elastic. On the one hand we saw that the $^3P_1$ wave can

---

*

$$\text{Im}\ \left\{\left(\frac{4\pi}{k}\right)\frac{1}{2ik}(2J+1)R_J\right\}$$

$$= \frac{2\pi}{k^2}(2J+1)(1-\cos\rho_J\cos 2\delta_J) \geq 0 .$$

Similar arguments hold for each of the other partial wave contribu-tions.

cause a large negative $\Delta\sigma_L$ due only to a large real phase shift. The understanding of this comes from treating the elastic forces in a unitary theory [e.g. potential models]. On the other hand the S-wave $n\Delta^{++}$ threshold is supposed to give a large positive $^1D_2$ contribution to both $\Delta\sigma_T$ and $\Delta\sigma_L$. In addition, we may imagine that the opening of more and more inelastic channels not only builds a total spin averaged cross section (usual diffraction mechanism), but also plays some role in the behavior of the total cross section in pure spin states, and hence for $\Delta\sigma_T$ and $\Delta\sigma_L$.

The most interesting possibility to me is that one source for the structure is dibaryon resonances. To establish such resonances, care will have to be paid to the elastic and inelastic effects mentioned above. This is especially true since current lore does not expect **nor** require their existence.

Therefore one would like to have a systematic study of the medium energy region for the purpose of establishing the nature of the observed structure. Such a study needs both theoretical input, as well as a more exhaustive set of experiments than currently exists.

In the rest of my talk, I will try to indicate some of the elements which should go into a systematic study. One important element is the role inelasticity plays in the medium energy structure. In Sec. II, calculations of the elastic and single π contributions to $\Delta\sigma_T$ and $\Delta\sigma_L$ are described, and their shortcomings noted. Some information at large t scattering at 12 GeV/c is shown to possibly reflect on the issue of inelasticity, and its possible spin dependence. Related to this question of spin dependence is the energy dependence of $\Delta\sigma_T$ and $\Delta\sigma_L$.

Another element in a systematic study is a list of experiments which still need to be done. This does not exist, but amplitude measurements at higher energy may give a clue about what experiments should be considered. In Sec. III a summary is given of the types of experimental information now available at ZGS energies, and some of the future plans. This includes some comments about single and double pion production, as well as elastic scattering. Special attention is paid to the problems of amplitude analysis.

INELASTIC CHANNELS

A. $pp \rightarrow pn\pi^+$ Threshold

After the first measurements of the pure spin total cross sections, the situation looked something like Fig. 2. The phase shift solutions below single π production are (claimed to be) unambiguous; above π production threshold there is admitted uncertainty. Nevertheless the ZGS data for $\Delta\sigma_T$ appeared to connect nicely with the low energy information. Enough so that the data were used to resurrect the idea of a $^1D_2$ resonance phenomena.[12] Such a resonance, even if it existed, would be strongly influenced by the prominent $pp \rightarrow pn\pi^+$ channel.

To study the contribution of this channel to $\Delta\sigma_T$ and $\Delta\sigma_L$, a model was constructed[13] assuming pion exchange dominance. The amplitude used is shown graphically in Fig. 3. The off shell

274

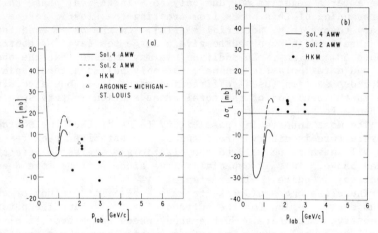

Fig. 2. Pure spin total cross sections before the recent data were known.[13]

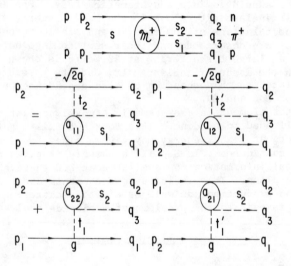

Fig. 3. Deck amplitude used in the calculation of Ref. 13.

$\pi_{off} N \to \pi_{on} N$ scattering is taken to be the same as on shell for equal values of πN c.m. energy and scattering angles. For on shell, the usual c.m. scattering angle is used; for off shell, the angle is between the incident and final nucleon in the final πN rest frame (i.e. the Jackson angle). The results of the calculation for the spin averaged single π cross sections are shown in Fig. 4. Agreement is adequate for $pp \to pn\pi^+$, but less good for $pp \to pp\pi^0$. Fortunately the latter is much less important for $\Delta\sigma_T$ and $\Delta\sigma_L$ so the observed discrepancy plays no important role in what is done.

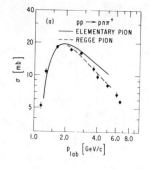

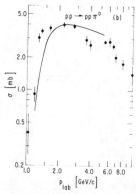

Fig. 4. Calculation of spin averaged cross section for $pp \to pn\pi^+$ and $pp \to pn\pi^0$ (Ref.13).

The results of this 'Deck' model for the single π production contribution to $\Delta\sigma_T$ and $\Delta\sigma_L$ are shown in Fig.5, compared with the latest data. I believe you will readily agree that the structure in the data and in the model do not agree, contrary to our expectation from the earlier information of Fig.2.

B.   Elastic Contribution

Attempts to understand this discrepancy have been made by Berger and Farmelo.[14] They estimate the elastic contribution to $\Delta\sigma_T$ and $\Delta\sigma_L$ with a Regge model. The t channel amplitudes are asymptotically related to the s channel amplitudes of Eq.(1.6) by[15]

$$N_0 = \frac{1}{2} (\phi_1 + \phi_3)$$

$$N_1 = \phi_5$$

$$N_2 = \frac{1}{2} (\phi_4 - \phi_2) \qquad (2.1)$$

$$U_2 = \frac{1}{2} (\phi_4 + \phi_2)$$

$$U_0 = \frac{1}{2} (\phi_1 - \phi_3)$$

where N_k is a natural parity amplitude with k units of helicity flip; U_k is an unnatural parity amplitude with k units of helicity flip. Asymptotically, exchanges with π-like quantum numbers contribute to U_2 while those with A_1-like quantum numbers contribute to U_0. Natural parity exchanges contribute in principle to N_0, N_1 and N_2. It is found that Pomeron, f and ω exchanges have large non-flip couplings and so are dominantly helicity conserving (N_0), whereas ρ and A_2 exchanges have large flip couplings so are dominantly in N_2. All exchanges (ρ,ω,f,A_2 and Pomeron) could be in N_1 though high energy analysis tends to exclude a large Pomeron flip coupling.

To attempt a Regge description in the medium energy region, one needs also lower lying trajectories $\tilde{f}$ and $\tilde{\omega}$ which have quantum numbers like the f and ω, but intercept one unit lower ($\alpha_{\tilde{f},\tilde{\omega}} \approx -\frac{1}{2}$). These trajectories govern for example the change of sign of $\rho = \mathrm{Re}N_0/\mathrm{Im}N_0$, the real to imaginary ratio of the spin averaged amplitude.

With these assumptions, the elastic asymmetries C_{ss}, C_{nn} and $C_{\ell\ell}$ can be obtained (C_{ij} is the parameter obtained when the beam is polarized in direction i and the target in direction j). From these asymmetries, the elastic contributions to $\Delta\sigma_L$ and $\Delta\sigma_T$ can then be found.

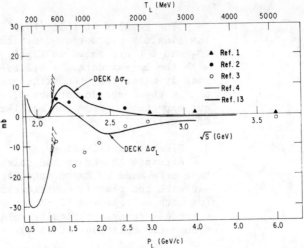

Fig. 5. Calculation of single pion contribution to the pure spin total cross sections.[13]

$$\Delta\sigma_T = -\int dt\frac{d\sigma}{dt}\left[C_{ss} + C_{nn}\right]$$

$$\Delta\sigma_L = -2\int dt\frac{d\sigma}{dt}\,C_{\ell\ell}\quad. \tag{2.2}$$

The preliminary results from the calculation allow a qualitative description of the structure of $\Delta\sigma_T$ and $\Delta\sigma_L$, though the results are sensitive to the parameterization of the t dependence of the $\tilde{f}$ and $\tilde{\omega}$ trajectories.

Even with this partial success, because of the uncertainties in the calculation I believe the total cross section differences are not satisfactorily explained. However the above calculation shows the importance of elastic unitarity: One can think of (2.2) as a way of generating imaginary parts for $\phi_2(0)$ and $\phi_1(0) - \phi_3(0)$ (which in the original model were absent). If the procedure is iterated using the generalization of (2.2) for $\text{Im}\phi_2(t)$ and $\text{Im}[\phi_1(t) - \phi_3(t)]$ then a set of amplitudes satisfying elastic unitarity results. This procedure can be modified to include inelastic effects as well.

C. Spin Independence of Inelasticity

As one goes to higher energies above the $N\Delta$ threshold, more channels open in addition to the S-wave. Not only are there more partial waves in single π production, but gradually multipion processes become important. This sounds very complicated, unless there are some simple approximate rules for the behavior of the inelasticity.

The eikonal picture of very high energy scattering suggests that the elastic amplitude results from the shadow of the inelastic processes. If the scattering is between spinless particles, the scattering amplitude

$$f(q) = \frac{1}{2ik}\sum_{\ell}(2\ell+1)(\eta_{\ell}e^{2i\delta_{\ell}}-1)\,P_{\ell}(\cos\theta) \tag{2.3}$$

in the eikonal approximation becomes

$$f(q) = ik \int_0^\infty bdb \, (\, 1-\eta(b) \,) \, J_0(bq) \qquad (2.4)$$

where $q = \sqrt{-t}$ and $b = (\ell + \frac{1}{2}) / k$ is the impact parameter. Because of the properties of the Bessel transform, we see that the small impact parameter behavior governs the large-t behavior. The forward scattering behavior is just the area of the profile function $1 - \eta(b)$, and is directly measured by the total cross section

$$\sigma_{tot} = 4\pi \int_0^\infty bdb \, (\, 1 - \eta(b) \,) \quad . \qquad (2.5)$$

All these ideas are of course well known, and are summarized here just to introduce the notation.

Our interest is in the corresponding behavior of the inelasticities which enter pp elastic scattering. Our assertion is that the singlet $\eta_J = \cos \rho_J$, and triplet $\eta_{JJ} = \cos \rho_{JJ}$, $\eta_{JJ\pm 1} = \cos \rho_{JJ\pm 1}$ [cf.1.8] inelasticities are approximately equal at high energy and functions only of impact parameter $b = (J + \frac{1}{2}) / k$:

$$\eta_J(b) \approx \eta_{JJ}(b) \approx \eta_{JJ\pm 1}(b) \quad . \qquad (2.6)$$

Moreover, the possible mixing between different L states (R^J) is presumed small.

To ascertain the reasonableness of (2.6) one can ignore δ_ℓ phase and use (1.10) and (1.9) to compute the helicity amplitudes. Using the eikonal approximation, the result is

$$\phi_1(t) = \phi_3(t) \cong ik \int_0^\infty bdb \, (\, 1-\eta(b) \,) \, J_0(b\sqrt{-t})$$

$$\phi_2(t) = \phi_4(t) \cong \phi_5(t) \approx 0 \qquad (2.7)$$

where $\eta(b)$ is the common value of (2.6). This expresses the idea that the shadowing due to inelastic processes conserves s-channel helicity and agrees with the phenomenological picture of diffraction obtained for pp elastic scattering.

There is some theoretical justification for (2.6) [or equivalently (2.7)] by Low.[16] He uses the M.I.T. bag model to compute the (bare) Pomeron. Because bags interact via color vector gluon exchange, it is plausible that the imaginary part of the scattering amplitude M reflects a structure similar to one photon exchange:

$$M = i \, p_L \, f(q) \bar{u}(\gamma_\mu + \frac{\lambda}{2m} \, \sigma_{\mu\nu} q^\nu) u$$

$$\times \, \bar{u}(\gamma^\mu + \frac{\lambda}{2m} \, \sigma^{\mu\sigma} q_\sigma) u \quad . \qquad (2.8)$$

Where m is the proton mass. It should be noted that (2.8) is not the Born term (which would be real) but the contribution to Im M

that a Born term would make through unitarity. In terms of the helicity amplitudes, (2.8) implies

$$\phi_1 = \phi_3 = f(q)$$

$$\phi_4 = - \phi_2 = \lambda^2 \frac{q^2}{4m^2} f(q) \tag{2.9}$$

$$\phi_5 = \lambda \frac{q}{2m} f(q) \ ,$$

so that the strength λ governs the flip to non flip coupling. In the bag model, this strength is predicted to be the anomolous isoscalar magnetic moment for the nucleon, and hence

$$\lambda \approx 0 \ . \tag{2.10}$$

One then derives (2.7) and (2.6).

As an aside, it is interesting to note the implications if λ were not small. At ZGS energies, there are observables such as $R = (osos)$ [cf.Sec.III] which should behave as $-\cos \Theta_R$ if $\lambda \approx 0$; for $\lambda \approx 1$ the behavior would be substantially different. This is not observed to the case up to Serpukhov energies. At much higher energies such as those available at FNAL and ISR, an eikonalized bare Pomeron of the type (2.8) could generate small but measureable unnatural parity contributions at $-t = 0$. Estimates of these yield

$$\Delta\sigma_T \sim 0 \left(\frac{\lambda}{2mR} \right)^6 \sim +5\mu b$$

$$\Delta\sigma_L \sim 0 \left(\frac{\lambda}{2mR} \right)^4 \sim +50\mu b \tag{2.11}$$

where the numerical values correspond to $\lambda \sim 1$ and the hadron size $R \sim 1$ fm. At even higher energies, a spin flip part to diffraction could conceivably lead to measurable ℓns type growth of $\Delta\sigma_T$ and $\Delta\sigma_L$ analogous to the observed growth of the spin averaged total cross section. Such results indeed can come out of eikonalized versions of Low's model if (2.8) is taken to be the eikonal and p_L is replaced by p_L^γ with $\gamma > 1$.

Returning now to our main point, the inelasticities we see are approximately spin independent. Since this is based on the small t behavior of the amplitudes, we may expect deviations from spin independence for small impact parameters, and hence large t.

Such deviations can be due to a breakdown of (2.6), or due to an important real phase contribution. Recall that our implicit assumption in using the eikonal approximation is that in the partial wave amplitude

$$\frac{\eta e^{2i\delta} - 1}{2i} = \frac{i}{2}(1-\eta) + \eta e^{i\delta}\sin \delta \ , \tag{2.12}$$

we can ignore the second term on the right hand side. Under certain circumstances, the second term might dominate the large t behavior.

To illustrate these ideas, let us consider the following simple model.[17] Imagine all triplet partial wave amplitudes are equal and sum to give $f_t(q)$ in the ϕ_1 amplitude; the singlet amplitudes sum to give $f_s(q)$ in ϕ_1. Consultation with (1.10) implies that in the eikonal approximation,

$$\phi_1 = f_s + f_t$$

$$\phi_2 = -f_s + f_t$$

$$\phi_3 = 2f_t \qquad\qquad (2.13)$$

$$\phi_4 \approx 0$$

$$\phi_5 \approx 0$$

Note the eikonal limit is t fixed, $s \to \infty$ so the pp symmetry properties will not be evident in (2.13); in particular $\phi_3 = -\phi_4$ at 90° is not apparent, but would automatically hold for any model for the partial waves using the exact expression (1.9) - (1.10).

The model consists of the assertions that 1) at low t, $f_s \sim f_t$ which is the statement that $1 - \eta(b)$ dominates the forward scattering behavior; 2) at large t, due to a small b effect (perhaps due to a core?) in the triplet partial waves,

$$|f_s| \ll |f_t| \quad . \qquad\qquad (2.14)$$

Of course without prior knowledge of data, I would not construct such a crazy model. This model predicts that the elastic polarized beam, polarized target asymmetries will be

$$C_{nn} \sim C_{ss} \sim C_{\ell\ell} \sim 1/3 \qquad\qquad (2.15)$$

at large values of t, whereas at small values of t the asymmetries will be zero (i.e. small). The symmetric result (2.15) can be seen in a direct way. One can show generally that

$$\frac{d\sigma}{dt}(1 - C_{nn} - C_{ss} - C_{\ell\ell}) \propto |\phi_1 - \phi_2|^2 \qquad\qquad (2.16)$$

is due purely to the singlet contributions. So if the singlet contributions are negligible in a model which treats the triplet amplitudes symmetrically, the three asymmetries in (2.16) should be equal to 1/3. Measurement of C_{nn}, C_{ss} and $C_{\ell\ell}$ is a good way of isolating the singlet and triplet contributions.

The simple model may suggest some interpretation of the new ZGS data[18] taken by the Michigan group for C_{nn} and the polarization asymmetry A shown in Fig.6. At small t ($p_\perp^2$) C_{nn} is indeed small; the surprizing result is that at large t, C_{nn} is in fact rather close to 1/3. I think this illustrates that when spin independence is

280

violated, it can lead to dramatic effects. The evidence from the
Michigan data indicates that large violations may occur only at small
impact parameters.

D. Real Parts and Spin Independence

We have assumed spin independence for the inelasticities Eq.(2.6),
but not for the real parts. Each of the partial wave amplitudes have
the structure

$$f_\ell = \frac{1}{2i}(\eta e^{2i\delta} - 1) \tag{2.17}$$

which can be decomposed as

$$f_\ell = f_\ell^{(R)} + f_\ell^{(I)} \tag{2.18}$$

where

$$f_\ell^{(R)} = \eta e^{i\delta} \sin\delta$$
$$f_\ell^{(I)} = \frac{i}{2}(1-\eta) \quad . \tag{2.19}$$

The assumption of spin independence for $f_\ell^{(R)}$ clearly fails: in the
phase shift region where $\eta = 1$, the phases δ are observed to be
strongly spin dependent ; in the high energy region ($p_L \gtrsim 3$ GeV/c)
Reggeized boson exchanges are found to determine spin dependent
phases δ .

One can expect to see the effects of δ in the total cross
section only when δ is large:

$$\text{Im} f^{(R)} = \eta \sin^2\delta$$

governs the contribution to total cross sections.

For pp scattering with $p_L \lesssim 3$ GeV/c, $\text{Im}\phi_3$ shows structure whereas $\text{Im}\phi_1$ does not [cf. Eq.(1.3) and (1.4)]. Figure 7 shows the energy

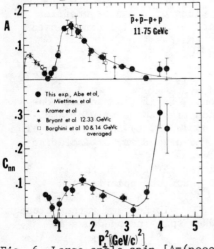

Fig. 6. Large angle spin [A=(nooo)]
and spin-spin [C_{nn}=(nnoo)] asymmetries.[18]

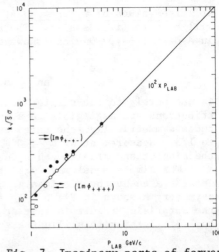

Fig. 7. Imaginary parts of forward
helicity amplitudes as obtained from
pure spin total cross sections.[3]

dependence of these amplitudes. As we have argued previously a structure in $Im\phi_3$ suggests that it originates from $L = J$ triplet partial wave amplitudes R_{JJ}[cf. Eq.(1.10)]. To keep the notion of spin independence for η, we would have to show that the structure in Fig.7 results from structure in some or one of the phase shifts δ_{JJ}. At present this is an open question.*

E. Summary

A systematic study of the medium energy region evidently needs to interpolate from the elastic phase shift region where inelastic effects are absent, to the high energy domain where inelastic shadowing determines the gross features of the scattering.

The task of obtaining amplitudes for $p_L \gtrsim 3$ GeV/c has been investigated by high energy phenomenologists. Notions of Diffraction, Regge exchanges, etc. may suffice for a good qualitative explanation. The difficulty will be interpolating between the Regge region and the elastic phase shift region ($1 \lesssim p_L \lesssim 3$ GeV/c). Possible constraints on such an interpolation are: 1) That it should fit to the low energy phase shifts; 2) Inelastic threshold effects of $pp \rightarrow p\pi^+n$ should be properly treated; 3) A sensible treatment of inelasticity should be given, which ties on to the high energy notion of diffraction; 4) At low energies one should of course be consistent with elastic unitarity; 5) Finally, some amount of (partial wave) analyticity should be imposed. At various times attempts have been made to carry out such a program, or parts of it.[19] Because of the potential interest in finding dibaryon resonances, and other anomolous structures in the pp system, such a program should be attempted.

MEDIUM ENERGY DATA AT THE ZGS

It is almost certain that without more data, the structures in the present medium energy data will remain unexplained. The question is then what experiments should be done next. As yet there are no clear choices, though the problems are evident. There are five complex pp amplitudes at each s and each t to be determined. A direct amplitude reconstruction can be obtained from 9 measurements (plus a few extra to remove discrete ambiguities) at each s and t. Everything but the overall phase is determined. Above the inelastic region, elastic unitarity no longer determines this phase. Thus there are intrinsic ambiguities associated with this unknown phase.

One would therefore expect to need more experiments for a partial wave analysis in the inelastic region than in the elastic region. There appear in fact to be fewer experiments done. One would also expect that spin information about the dominant inelastic processes (e.g. $pp \rightarrow pn\pi^+$) would be crucial; little, if any information exists of this type. Since the NN structure will depend on the isospin, one would hope to have as much np data as pp data. The information on np scattering is quite meager.

---

*Similarly the argument of Sec. II C does not necessarily imply that spin independence for inelasticities breaks down; it may be that real parts become important at small b.

282

In the absence of such data in the $1 \leq p_L \leq 3$ GeV/c region, I thought it would be helpful to summarize just a smattering of higher energy data, which might be useful to have at lower energies.

A. $pp \to p\pi^+n$, $pp \to \Delta^{++}\pi^-p$

The EMS group at Argonne[20] has collected large numbers of events of the type

$$p\uparrow p \to p\pi^+ n \qquad\qquad (3.1)$$

$$p\uparrow n \to p\pi^- p \qquad\qquad (3.2)$$

$$p\uparrow p \to \Delta^{++}\pi^- p \qquad\qquad (3.3)$$

where the incident beam is transversely polarized. In Fig.8, we see that the behavior of some observables follows roughly that expected from one pion exchange. The statistics of the experiment are 140,000 events of type (3.2) and 800,000 events of the type (3.1). A great deal of information can in principle be extracted from data of this precision. One interesting systematic of the data is that the pro-

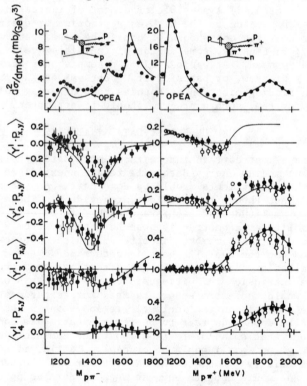

Fig. 8. EMS data[20] for $p\uparrow p\to p\pi^+n$ and $p\uparrow n\to p\pi^-p$ at 6 GeV/c. The curves[20] result from an absorbed one pion exchange model.

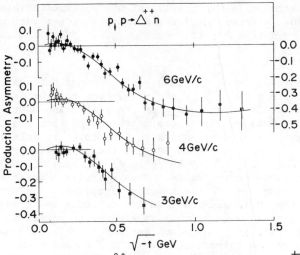

Fig. 9. EMS data[20] for the subprocess p↑p→Δ⁺⁺n .

duction asymmetry for the subprocess p↑p → Δ⁺⁺n [Fig. 9] is large and roughly independent of energy between p_L of 3–12 GeV/c. I have no particular understanding of this phenomena. A simple exchange degeneracy argument for π and B exchange would predict no asymmetry whereas the observed asymmetry is as large as 40% in magnitude. It would be interesting if this effect remains large at lower energies.

An important subprocess for 2 pion production is (3.3). The EMS group has found some interesting systematics associated possibly with diffractive production of N* resonances which then decay into Δ⁺⁺π⁻. Figure 10 shows the decay cosine for helicity $\frac{1}{2}$ and $\frac{3}{2}$ Δ's.

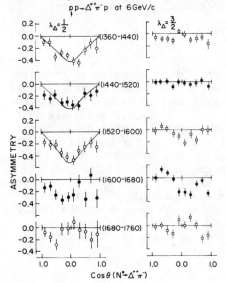

Fig. 10. EMS[20] data for p↑p→Δ⁺⁺π⁻p at 6 GeV/c.

The strong structure in the $\frac{1}{2}$ helicity state suggests that the $J^P = \frac{1}{2}^+ N^*$ (1470) is produced diffractively, together with a large nonresonant $J^P = \frac{3}{2}^-$ Deck contribution.

B. pp → pp: Amplitude Analysis

As more partial waves contribute, it becomes impossible to perform an effective partial wave analysis. One has recourse to a fixed s and t amplitude analysis for the scattering matrix up to one overall phase. This requires more measurements than are needed for the low energy partial wave analysis. One point to make is that the present plans at the ZGS are to measure the pp amplitudes at $p_L = 6$ and 12 GeV/c for $0.1 \lesssim -t \lesssim 1.5$ GeV/c^2. As you know the ZGS is scheduled to be shutdown in two years. Thus there is a question of priorities to be answered for the remaining program. One possibility is to extend the amplitude measurements to lower energies; another possibility is to do some of the experiments at higher t. Still another possibility is to use combinations of polarized neutron beams and polarized proton targets to carry out the amplitude measurements for np scattering. All of these possibilities and more are feasible, but only a small subset of them can be done before the scheduled shutdown.

There is a systematic program[21] for extracting the amplitudes which has been applied at $p_L = 6$ GeV/c for $-t \lesssim 1.0$ GeV/c^2.[22] The idea of the strategy is to first note the expectation that one amplitude N_0 [Eq.2.1] should dominate and then test this hypothesis. If one measures the parameters I(onon) and I(olos),

$$I(\text{onon}) = |N_0|^2 + 2|N_1|^2 + |N_2|^2 - |U_0|^2 - |U_2|^2 \quad (3.4)$$

$$I(\text{olos}) \tilde{=} |N_0|^2 - |N_2|^2 + |U_0|^2 - |U_2|^2 \quad (3.5)$$

and finds them nearly equal to the differential cross section

$$I(\text{oooo}) = |N_0|^2 + 2|N_1|^2 + |N_2|^2 + |U_0|^2 + |U_2|^2 , \quad (3.6)$$

then N_0 is indeed dominant.

The notation we are using is that I(abcd) is obtained from the experiment with a beam polarized along "a", a target polarized along "b", a scattered particle's polarization analyzed along "c" and the recoil's polarization analyzed along "d". The a,b,c and d are 3-vectors in the rest frame of the beam, target, scattered and recoil frames, respectively. For each particle we use the convention that "n" is a unit vector along the normal to the scattering plane, and "ℓ" is a unit vector along the lab momentum vector of the particle [by convention, "ℓ" of the target is along the beam direction]. The third independent spin direction for each particle $s = n \times \ell$ lies in the scattering plane perpendicular to the lab momenta of the particle. An unpolarized particle is represented by a "0". The quantity (abcd) = $\frac{I(\text{abcd})}{I(\text{oooo})}$ is then the corresponding Wolfenstein parameter.

Given that N_0 is the largest amplitude [(onon) ≈ 1 and (oℓos)≈ 1], 8 experiments determine the 4 remaining complex amplitudes including their phases relative to N_0:[23]

$$(onoo) \cong -2N_1^\perp \big/ |N_0| \qquad (3.7)$$

$$(osos) \cong -\cos\Theta_R - 2N_1^{||} \big/ |N_0| \qquad (3.8)$$

$$(nnoo) \cong -2N_2^{||} \big/ |N_0| \qquad (3.9)$$

$$(nsos) \cong 2N_2^\perp \big/ |N_0| \qquad (3.10)$$

$$(\ell\ell oo) \cong -2U_0^{||} \big/ |N_0| \qquad (3.11)$$

$$(\ell son) \cong -2U_0^\perp \big/ |N_0| \qquad (3.12)$$

$$(ssoo) \cong 2U_2^{||} \big/ |N_0| \qquad (3.13)$$

$$(snos) \cong 2U_2^\perp \big/ |N_0| \quad , \qquad (3.14)$$

where

$$X^\perp = \text{Im}(N_0 X^*) \big/ |N_0|$$

$$X^{||} = \text{Re}(N_0 X^*) \big/ |N_0| \quad , \qquad (3.15)$$

and Θ_R is the lab recoil scattering angle. These equations are linear in the unknown amplitudes, with corrections which are quadratic in the small quantities $X \big/ |N_0|$, $X = N_1$, N_2, U_0, U_2. Given the exact expressions for the above parameters, one has a quickly convergent scheme for determining amplitudes without discrete ambiguities. Of course the convergence depends upon the assumption of N_0 dominance, and hence on how close (3.4) and (3.5) are to I(oooo). At 6 GeV/c, the assumption of dominance appears to be justified for $-t \lesssim 0.8$ GeV/c^2, and the convergence of the above scheme is rapid.

The 11 experiments (3.4) - (3.14) have been carried out and are either published, or in the process of being analyzed. When the analysis is complete, one will have, to some accuracy, the pp amplitudes at 6 GeV/c and small t. What do we hope to learn from such amplitudes?

Of course, given the amplitudes, we can always test our favorite model. More than that, we want to test certain features that many models have in common. One feature is that Regge exchanges obey a duality requirement. The imaginary part of the scattering amplitude at low energies, builds up in the sense of a dispersion integral, the high energy Regge behavior. This means that for an exotic channel like pp where no resonances are expected, there is no imaginary part of the amplitude, and hence the high energy Regge exchanges must add (called Exchange Degeneracy) to produce purely real amplitudes.[24] Let us see what the implications are for the 6 GeV/c data.

We expect the Regge contributions to the scattering matrix to decrease like a power of p_L, whereas the diffraction contribution to

286

be roughly energy independent. It is observed that the differential cross section is indeed roughly energy independent; since $I(oooo) \cong |N_0|^2$, we expect N_0 to have a large diffractive component. The polarization decreases like a power of the energy, so we believe N_1 will have a large Regge component. Exact exchange degeneracy would have N_1 real (assuming only Regge exchanges contribute). In Fig.11 we have a picture of the amplitudes N_0 and N_1 in the complex plane. The assumptions are that the phase of N_0 at $-t \lesssim 0.3$ GeV/c^2 is roughly what it is at $-t = 0$ (for the purpose of illustration) and that N_1 is purely real.

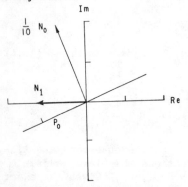

Fig. 11. Construction of N_1 at $-t \approx 0.3$ GeV/c^2 assuming N_1 purely real using P_0 data.

We see that the polarization determines only the component of N_1 perpendicular to N_0; the reality condition determines the projection of N_1 along N_0. This expectation will be tested to some degree when the R=(osos) parameter data become available. For the picture here

$$R + \cos\theta_R \cong -0.05. \qquad (3.16)$$

The value reflects the small (effective) flip coupling of the Reggeon. In the picture

$$|N_1|/|N_0| \cong 0.2\sqrt{\frac{-t}{4m^2}} \qquad (3.17)$$

for $-t \simeq 0.3$ GeV/c^2 (here m=nucleon mass). If the picture is correct, it will be difficult to measure the single flip coupling accurately.

The double flip Regge contributions to N_2 may be quite large.

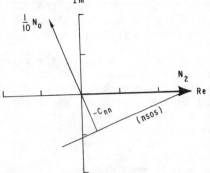

Fig. 12. Construction of N_2 at $-t \approx 0.3$ GeV/c^2 assuming N_2 purely real using C_{nn} data.

From meson baryon scattering one learns that the ρ and A_2 flip couplings are big; by factorization one expects these trajectories to contribute to NN scattering. In Fig.12 is shown a picture of the N_2 and N_0 amplitudes. As before we take a representative phase for N_0 and presume N_2 is purely real. We use the data on C_{nn} = (nnoo) to determine the projection of N_2 along N_0. To justify the assumption that N_2 is mainly Regge we note that C_{nn} decreases like a power of p_L at fixed t. We therefore predict a quite large perpendicular component for N_2, This component is given by (nsos), which is predicted to be $\approx +0.25$. Again this parameter has been measured, and when the data are analyzed, we will be able to learn about the phase of N_2. We note that if the above picture is roughly correct,

$$|N_2|/|N_0| \stackrel{\sim}{=} 1.6 \left(\frac{-t}{4m^2}\right) \tag{3.18}$$

showing that N_2 has a large coupling. Indeed for $-t \approx 1.0$ GeV/c^2 the above coupling would give (olos) ≈ 0.6; in other words as $-t$ increases, one would begin to move away from N_0 dominance.

The above arguments concern the natural parity trajectories only. With somewhat less reliability, one can make similar duality arguments for the unnatural parity exchanges; these exchanges contribute to U_0 and U_2 (though their cuts contribute to the other amplitudes).

Rather than giving such arguments, I would like to describe one new feature to come out of the 6 GeV/c amplitude analysis, namely information on the exchanges with A_1-like quantum numbers, which contribute asymptotically only to U_0. As determined by Berger[25], the size of the __axial__ __vector__ A_1 like coupling is comparable in pp scattering to that of the __vector__ exchanges ω and ρ. Figure 13 shows what is known about U_0 at $-t = 0$. Here the phase of N_0 is determined by experiment. The imaginary part of U_0 is known from $\Delta\sigma_L$. The real part of U_0 could in principle be obtained from a $C_{\ell\ell} = (\ell\ell oo)$ measurement at $-t = 0$. At $-t \sim 0.1$, $C_{\ell\ell} \approx 0$; if this remained true then U_0 would be as drawn. We note however that a negative real part for U_0 is in contradiction with a forward dispersion relations calculation (ReU$_0$ comes out positive).[26] Thus an experimental determination of ReU$_0$ could provide some useful checks on dispersion relations as applied to non spin averaged amplitudes.

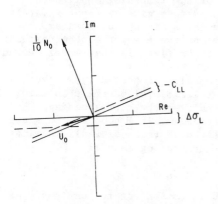

Fig. 13. Construction of U_0 at $-t=0$. GeV/c^2 using $\Delta\sigma_L$ data and approximating $C_{\ell\ell}(0)$ by $C_{\ell\ell}(-t=0.2$ GeV/c$^2)$.

One final comment about Figure 13. For U_0 as drawn, the perpendicular component is quite large. This component is measured by (ℓson) which for the picture is ≈ 0.10. Away from the forward direction, this parameter has been measured, though the data are not analyzed. From this data and $C_{\ell\ell}$ we will get knowledge of U_0 at least for non-forward t.

Analogous arguments can also be presented for determining U_2 which receives contributions from π-like exchanges. Here $C_{ss} = (ssoo)$ determines the projection of U_2 along N_0 and (snos) determines the perpendicular component. Both have been measured, and agree roughly with ones expectations for π exchange.[27] Also $\mathrm{Im}U_2$ as measured by $\Delta\sigma_T$ at 6 GeV/c[1] is quite small, in contrast to its behavior at lower energies.[1,2]

In summary we see that there soon will be enough experiments to determine the amplitudes. For some sensible choice of the phase of N_0, questions about the reality of the Regge contributions can be answered. We have emphasized this point only to be able to give an idea of the potential value of the data. Many other physics questions will also be answered by the data. For example the t dependence will test detailed models of the cut corrections to the elementary Regge exchanges. The actual values of the couplings as obtained from the data can be compared with their counter parts in other two body reactions.

D. Formal Amplitude Analysis

The previous discussion about amplitudes presupposed that the N_0 amplitude was dominant. There is no guarantee that this will remain true at large -t, nor will it evidently be true for np→np (it is certainly false for np charge exchange around the OPE peak). Nevertheless one would like to know how to decide which experiments are important to do, and which are not. This is a long standing problem, for which no universal strategy yet exists. In the last few minutes, I would like to describe one attempt[28] at a partial solution: the case where one amplitude dominates, though not necessarily N_0. In fact the discussion will be given quite generally for any spin½ spin½ scattering process, with an obvious generalization to arbitrary spin scattering.

The main assumption is that one knows ahead of time which amplitude dominates. The strategy always consists of choosing a short list of experiments which verify the dominance, and a long list of experiments which measure the interference of the remaining amplitudes against the dominant one. The equations then give an (approximately) linear relationship between the observables and the unknown amplitudes. The differential cross section determines the modulus of the dominant amplitude, and its phase is presumed to be fixed in some model dependent manner.

To get a rough idea how the mathematics of the analysis goes, we introduce the following notation. Let I(a,b,c,d) be the symbol for the experiment under consideration, and also the corresponding Wolfenstein parameter (times cross section). If M denotes the scattering matrix, I(abcd) can be represented in terms of Pauli matrices $\sigma_a = 1, \sigma_x, \sigma_y, \sigma_z$ for a = 0,x,y,z and the scattering matrix:

$$I(abcd) = \mathrm{Tr}(\sigma_c \sigma_d M \sigma_a \sigma_b M^\dagger) . \qquad (3.19)$$

This is the usual representation. It is convenient to consider M as a column vector ψ (by relabeling the elements) in which case

$$I(abcd) = \psi^\dagger V_{abcd} \psi \qquad (3.20)$$

where

$$V_{abcd} = \sigma_a^t \otimes \sigma_b^t \otimes \sigma_c \otimes \sigma_d \qquad (3.21)$$

is just the tensor product of Pauli matrices. The matrices V satisfy simple algebraic properties

$$V^2 = I \qquad (3.22)$$

$$VV' = \pm V'V \qquad (3.23)$$

$$TrV = 0 \text{ unless } V = I . \qquad (3.24)$$

The first is true since $\sigma^2 = 1$ for each term in the tensor product. The second is true since each pair of Pauli matrices in V and V' either commute or anticommute. The minus sign in (3.23) results wherever an odd number of the Pauli matrices in V and V' anticommute. Finally the third relation is true since $Tr\sigma = 0$ for at least one factor in V.

With the above algebraic properties one can characterize both the <u>short list</u> and <u>long list</u> of experiments necessary to determine the amplitude analysis. When N_0 was dominant, we saw that two observables (onon) and (oℓos) were approximately unity. Let us consider the converse for the general situation and suppose some observable I(abcd) = I(oooo) $\equiv 1$ (in the following, we measure everything in units of the differential cross section). This means

$$\psi^\dagger V\psi = 1 \qquad (3.25)$$

for the corresponding operator V. But note V has eigenvalues ± 1 since $V^2 = I$, and there are as many +1 as -1 eigenvalues. The only way (3.25) can be true is if ψ is an eigenvector of V. Obviously the argument remains true if $\psi^\dagger V\psi = -1$; we shall consider for illustration only +1 eigenvalues. For spin 1/2 scattering ψ has 16 complex components, and $\psi^\dagger V\psi = 1$ effectively removes half. In general if 4 matrices V are found which commute (they can be simultaneously diagonalized), are independent, and satisfy $\psi^\dagger V\psi = 1$, then ψ is determined. The corresponding 4 measurements then comprise the short list.

For pp elastic scattering, we required only two because the symmetries of the system provided the remaining constraints. For example (nnnn) = 1 by parity conservation. Conversely when (nnnn) = 1 all parity violating amplitudes are zero. Thus a single albeit complex spin experiment can decide whether all parity violating amplitudes are small. Note such an experiment would look for deviations of (nnnn) from 1 which would be quadratically small in the parity violating components.

The long list of measurements results from the observation that those matrices U which anticommute with a V on the short list must give a zero parameter

$$\psi^\dagger U\psi = 0 . \qquad (3.26)$$

To see this note UV = -VU implies $\psi^\dagger U\psi = \psi^\dagger UV\psi = -\psi^\dagger VU\psi = -\psi^\dagger U\psi$ since $V\psi = \psi$ by assumption. As soon as ψ is not precisely an eigenvector, $\psi^\dagger U\psi$ measures the interference between the eigenvector direction and and the deviation. The only point to check for a set of anticommuting

U's (each with respect to some V in the short list) is that no two determine the same quantity. Calling two experiments equivalent if both fix the same quantity, the anticommuting matrices split up into disjoint equivalence classes. Thus for pp elastic scattering, the 8 measurements on the long list are only 1 possible choice of experiments. Many other equivalent choices are possible.[29]

The number of experiments needed to determine dominance we saw was 4. If there are N amplitudes (for spin 1/2 scattering, N = 16) only 2(N-1) parameters remain to be determined, and we tried to make plausible that for each unknown quantity only one measurement need be performed. Thus the long list will have 2(N-1) measurements. A complete set (including the differential cross section) consists of 2N + 3 measurements, which can be further reduced by strong interaction symmetries.

CONCLUDING REMARKS

I have tried to communicate some of the potential interest pp scattering experiments at ZGS energies may hold for both medium energy and high energy physicists. We saw that the pure spin total cross sections display interesting and yet unexplained structure, possibly due to dibaryon resonances. We also saw that the inelasticities at high energy may have interesting systematics. Use of these systematics in phase shift analysis may prove useful. Finally we suggested that medium energy experiments of the type done to measure amplitudes could be of some use to continuing progress in understanding the nucleon nucleon interaction in the inelastic region.

REFERENCES

1. W. DeBoer et al., Phys. Rev. Letters $\underline{34}$, 558 (1975).
2. J. Roberts, private communication.
3. I. P. Auer et al., Phys. Lett. $\underline{67B}$, 113 (1977); I. P. Auer et al., to be published.
4. R. A. Arndt, M. H. MacGregor and R. M. Wright, Phys. Rev. $\underline{182}$, $\underline{1714}$ (1969); $\underline{173}$, 1272 (1968); $\underline{169}$, 1149 (1968). These references will be referred to as AMW. See also R. A. Arndt, R. H. Hackman, and L. D. Roper, Phys. Rev. $\underline{C9}$, 555 (1974).
5. G. H. Thomas, "Behavior of Low and Medium energy NN scattering and amplitude Reconstruction", in Physics With Polarized Beams: Report of the ANL Technical Advisory Panel (Chairman G. Kane), ANL-HEP-CP-75-73, p.83.
6. M. Jacob and G. C. Wick, Ann. Phys. $\underline{7}$, 404 (1959).
7. F. Halzen and G. H. Thomas, Phys. Rev. $\underline{D10}$, 344 (1974).
8. M. L. Goldberger, M. T. Grisaru, S. N. MacDowell and D. Y. Wong, Phys. Rev. $\underline{120}$, 2250 (1960).
9. A. Scotti and D. Y. Wong, Phys. Rev. $\underline{138}$, B145 (1965).
10. There are many refs. to the possibility of having a $^1D_2$ resonance, some of which are F. J. Dyson and N.-H. Xuong, Phys. Rev. Letters $\underline{13}$, 815 (1964); $\underline{14}$, 339 (1965). R. A. Arndt, Phys. Rev.

$\underline{165}$, 1834 (1968). S. Graffi, V. Grecchi and G. Turchetti, Nuovo Cimento Letters $\underline{2}$, 311 (1969). H. Suzuki, Prog. Theor. Phys. $\underline{50}$, 1080 (1973); $\underline{54}$, 143 (1975). D. D. Brayshaw, Phys. Rev. Letters $\underline{37}$, 1329 (1976).

11. K. Hidaka et al., "Evidence for a Dibaryon resonance in the pp system", Argonne preprint ANL-HEP-PR-77-48.

12. G. L. Kane and G. H. Thomas, Phys. Rev. $\underline{D13}$, 2944 (1976).

13. E. L. Berger, P. Pirila and G. H. Thomas, ANL-HEP-75-72 (unpublished).

14. E. Berger, private communication.

15. E. Leader and R. C. Slansky, Phys. Rev. $\underline{148}$, 1491 (1966).

16. F. E. Low, Phys. Rev. $\underline{D12}$, 163 (1975).

17. The model gives a picture of the scattering similar to that of A. D. Krisch, Phys. Rev. Letters $\underline{11}$, 217 (1963).

18. J. R. O'Fallon et al., Univ. of Michigan preprint UM-HE-77-28.

19. See for example Ref. 10. Presently, P. Johnson, R. Warnock and myself are thinking about this program.

20. See e.g. the review of the polarization studies using the Argonne Effective Mass Spectrometer (EMS) by R. Diebold, in the proceedings of the conference High Energy Physics with Polarized Beams and Targets; ed. M. L. Marshak (AIP, N.Y.1976) p. 92. Figures 8-10 are courtesy of A. B. Wicklund, private communication.

21. The experimental program to measure the NN amplitudes has been reviewed at this conference by A. Yokosawa, this proceedings; also ANL-HEP-PR-CP-47.

22. The present status of the amplitudes as determined by ZGS experiments is given by P. Johnson, R. C. Miller and G. H. Thomas, Phys. Rev. $\underline{D15}$, 1895 (1977).

23. Of course in practice one always uses the exact expressions (cf. Refs. 7 and 22).

24. Eg. H. Harari, "Recent Developments, Duality and Hadron Dynamics", Summer School in Elementary Particle Physics, July 22-August 29, 1969, ed. R. F. Peierls [BNL 50212(C-58) Brookhaven 1970], p. 385; J. L. Rosner, Physics Reports $\underline{C11}$, 189 (1975).

25. E. Berger, private communication and ANL-HEP-77-08.

26. W. Grein and P. Kroll, Wuppertal report no. WU B 77-6 and contribution to this conference.

27. For reviews see C. Quigg and G. C. Fox, Ann. Rev. Nucl. Science $\underline{23}$, 219 (1973); G. L. Kane and A. Seidl, Rev. Mod. Phys. $\underline{48}$, 309 (1976).

28. P. W. Johnson and G. H. Thomas, Argonne preprint ANL-HEP-PR-77-26 (to be published in Phys. Rev. D.)

29. Indeed, a set of experiments equivalent to Eq.(3.4)-(3.15) has been proposed by D. Besset et al., "A Complete Experiment for pp scattering at Medium Energies", Karlsruhe preprint (1977). These experiments are specifically tailored to the low and medium energy regime where the scattered particles' spin can be analyzed.

pp→dπ+ NEAR THRESHOLD

G. Jones
University of British Columbia, Vancouver, B.C. Canada

ABSTRACT

The p+p→d+π+ reaction has been studied over a range of proton energies from 305 to 425 MeV. The pions were detected (in a single arm experiment) using a 50 cm broad-range Browne-Buechner magnetic spectrograph. Measurements of the differential cross-section and analysing power for the reaction (determined by measurements of the pion azimuthal asymmetry resulting from the use of a polarized beam) enabled the contribution of d-wave pion production in this energy range to be clearly discerned.

INTRODUCTION

In 1955, Gell-Mann and Watson[1] published their now-famous pheno-menological analysis of the near-threshold production of pions in the nucleon-nucleon reaction. Their analysis was oriented to the near-threshold region in order to limit the number of amplitudes in-volved (by centrifugal barrier effects). In the following, I hope to show that the initial expectations have failed to be fulfilled and nature has turned out to be much less co-operative than origin-ally anticipated.

In Table I, we show the dependence of the differential cross-section for the pp→dπ+ reaction (for polarized protons) on the ampli-tudes for each of the states associated with a partial wave decompo-sition of the reaction.[2]

Table I   (a) Angular momentum decomposition of pp→dπ+ reaction.

Initial pp state	(deuteron state · ℓ_π)$_j$	Amplitude
$^1S_0$	$(^3S_1\ p)_0$	a_0
$^3P_1$	$(^3S_1\ s)_1$	a_1
$^1D_2$	$(^3S_1\ p)_2$	a_2
$^3P_1$	$(^3S_1\ d)_1$	a_3
$^3P_2$	$(^3S_1\ d)_2$	a_4
$^3F_2$	$(^3S_1\ d)_2$	a_5
$^3F_3$	$(^3S_1\ d)_3$	a_6

ISSN: 0094-243X/78/292/$1.50

(b) Expansion of γ_i in terms of amplitudes.

Term	Type	γ_0	γ_1	γ_4
$|a_0|^2$	$|p|^2$	2	0	0
$|a_1|^2$	$|s|^2$	2	0	0
$|a_2|^2$	$|p|^2$	1	3	0
$|a_3|^2$	$|d|^2$	$5/2$	$-3/2$	0
$|a_4|^2$	$|d|^2$	$5/2$	$5/2$	0
$|a_5|^2$	$|d|^2$	$5/7$	$30/7$	$-25/7$
$|a_6|^2$	$|d|^2$	$5/4$	$3/2$	$5/4$
Re $a_0^*a_2$	p-p	$2\sqrt{2}$	$-6\sqrt{2}$	0
Re $a_1^*a_3$	s-d	$-\sqrt{2}$	$3\sqrt{2}$	0

etc.

(c) Expansion of λ_i in terms of amplitudes.

Term	Type	λ_0	λ_1	λ_2	λ_3
Im $a_0^*a_1$	s-p	$2\sqrt{2}$	0	0	0
Im $a_0^*a_3$	p-d	-4	0	0	0
Im $a_1^*a_2$	s-p	-2	0	0	0
Im $a_1^*a_4$	s-d	0	$4\sqrt{5/2}$	0	0
Im $a_2^*a_3$	p-d	$-2\sqrt{2}$	0	$9\sqrt{2}$	0
Im $a_2^*a_4$	p-d	0	0	$6\sqrt{5/2}$	0
Im $a_3^*a_4$	d-d	0	$2/5$	0	0
Im $a_0^*a_6$	p-d	3	0	15	0
Im $a_1^*a_5$	s-d	0	$-6\sqrt{5/7}$	0	0
Im $a_2^*a_5$	p-d	0	0	$6\sqrt{5/7}$	0
Im $a_2^*a_6$	p-d	$-3/2\sqrt{2}$	0	$-9/2\sqrt{2}$	0
Im $a_4^*a_5$	d-d	0	$50\sqrt{1/14}$	0	$-100\sqrt{1/14}$

etc.

The differential cross-section for polarized protons is given by:

$$\frac{d\sigma}{d\Omega} = \gamma_0 + \gamma_2 \cos^2\theta + \gamma_4 \cos^4\theta$$
$$+ \vec{p} \cdot \vec{n} \sin\theta (\lambda_0 + \lambda_1 \cos\theta + \lambda_2 \cos^2\theta + \lambda_3 \cos^3\theta)$$

(1)

where $\vec{n}$ is a unit vector in direction $\vec{k}_p \times \vec{k}_\pi$. If the reaction is investigated sufficiently near threshold, only s- and p-wave pion production need be considered, and only three complex amplitudes are required to determine the reaction. Thus, there are <u>five</u> quantities that need to be defined experimentally (one phase is arbitrary). Such measurements have included:

1. Unpolarized angular distribution:

$$\frac{d\sigma}{d\Omega} (\theta) = \gamma_0 + \gamma_2 \cos^2\theta = K(A_o + \cos^2\theta) \tag{2}$$

2. 90° asymmetry. In terms of analysing power:

$$A_\pi = \frac{1}{P} \frac{\sigma_L - \sigma_R}{\sigma_L + \sigma_R} \qquad \underline{Q} = A_\pi (\frac{\pi}{2}) = \frac{\lambda_0}{\gamma_0} \tag{3}$$

3,4. Total cross-section measurements. <u>Assuming</u> that the near-threshold energy dependence is determined by the centrifugal barrier, one expects[1]:

$$\sigma_t = 4\pi K(A_o + \frac{1}{3}) = \underline{\alpha}\eta + \underline{\beta}\eta^3 \tag{4}$$

where $\eta = (\frac{P_\pi}{\mu c})$ c.m.

5. <u>Polarization</u> of the outgoing deuteron resulting from use of un-polarized protons.

This is the program that occupied much experimental effort (primarily by groups at the Lawrence Lab., Carnegie Institute of Technology, and the USSR) during the decade following 1955. As a result of this effort the experimental situation can be summarized as follows:

1. As shown in Fig. 1, the shape of the angular distribution was quite well known (primarily from measurements of the inverse reaction $\pi^+d\rightarrow2p$), but only for values of $\eta > 0.5$. For values of $\eta < 1$, there was no evidence for terms higher than $\cos^4(\theta)$.

2. Fig. 2 illustrates the asymmetry in pion production at 90° when using polarized protons. Measurements at two values of η, viz 0.41, 0.97 are shown, with the two values consistent with the η dependence expected from the phenomenological theory.[1]

$$Q = \frac{\sqrt{2} \, \eta \, \eta_c \, \sin(\psi-\tau_1)}{\eta^2 + \eta_c^2} \tag{5}$$

3,4. The total cross-section was known over the largest energy range of the experiments quoted (again based primarily on results of measurements of the inverse reaction). The results were quite consistent with a dependence of the form (Eq. (4))[4]:

$$\sigma_t = 0.138 \, \eta + 1.01 \, \eta^3 \text{ mb for values of } \eta \leqslant 1.0 \tag{6}$$

This expression is little different from the more recent fits[3] to the cross-section data shown in Fig. 3.

5.   Finally, in 1956, Tripp[5] (of the Lawrence Lab.) published data on the polarization of the deuterons produced by the pp→dπ$^+$ reaction at a proton energy of 340 MeV (η = 0.59).   And thus the three amplitudes were uniquely determined.

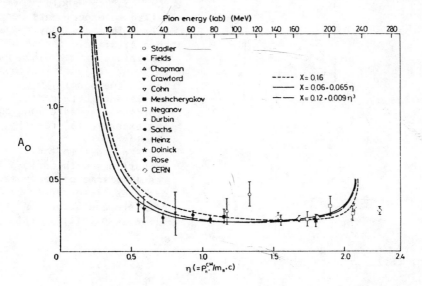

The parameter A from the expression for the differential cross-section of the form $d\sigma/d\Omega = K(A_0 + \cos^2\theta - B\cos^4\theta)$.

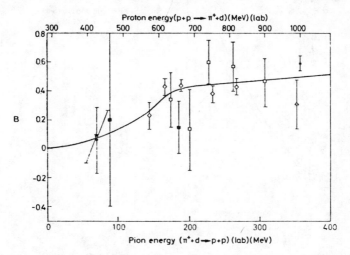

The parameter B from the expression for the differential cross-section of the form $d\sigma/d\Omega = K(A_0 + \cos^2\theta - B\cos^4\theta)$
The solid line is simply a guide for the eye.

Fig. 1. Angular Distribution Parameters for π+d→pp.

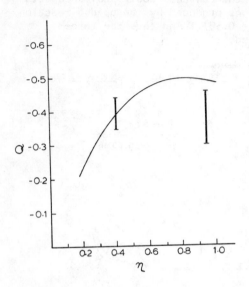

The underlying question concerning all this work was, of course, the relevance of the "near-threshold" formulation to experiments performed in the energy range then attainable (principally for values of $\eta \gtrsim 0.5$). Basically, the justification lay in the quality of the consistency between such "threshold" predictions and the experimental results obtained.

In order to facilitate comparisons with our own results, some of the numerical results obtained by Crawford and Stevenson[4] are listed explicitly:

Fig. 2. $Q = \frac{1}{P} \epsilon \left(\frac{\pi}{2}\right)$ plotted against η.

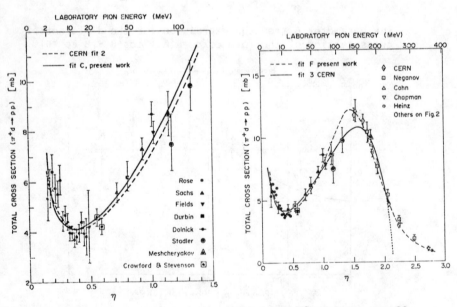

CERN 1 and CERN 2 yielded: $\sigma_t = G_0 \alpha \eta + G_1 \beta \eta^3$ with $\alpha = \underline{0.188}$, $\beta = \underline{0.90}$ where G_0 and G_1 are Coulomb correction factors.

Fig. 3. Total Cross-section for $p+p \rightarrow \pi^+ + d$.

σ_t given by Eq. (6)

$$A_o = X(1 + \frac{\eta_c^2}{\eta^2}), \text{ from Gell-Mann & Watson where}$$

(7)

$$\eta_c^2 \propto \frac{\alpha}{\beta} = .082 + \frac{.056}{\eta^2}$$

$Q = -0.39$ at $\eta = \underline{0.41}$

(8)

With the passage of time, inconsistencies became apparent. A principal area of concern was the value of α (Eqs. 4 and 6) deduced by Crawford and Stevenson (as well as other workers in the field) as compared to the value deduced from measurements of threshold pion photo-production $\gamma p \rightarrow n\pi^+$ and low energy $\pi^+ p$ elastic scattering. From the latter measurements a value of $\alpha = 0.25$ was expected.[6] Following the suggestion of Thomas and Afnan that s-wave pion production can deviate significantly from such a simple η dependence even in the near-threshold region, Measday and Spuller[6] fitted the total cross-section data shown in Fig. 3 to a function of the form of (6) but with $\alpha \rightarrow \alpha_o - 0.20 \eta$ yielding:

$$\sigma_t = (0.247 \pm .017) \eta - 0.2 \eta^2 + (0.6 \pm 0.3) \eta^3$$

(9)

$$+ (1.0 \pm 0.5) \eta^4 - (0.6 \pm 0.2) \eta^5$$

shown by the dashed line in Fig. 3. The threshold limit of the s-wave term is now in good agreement with the other "threshold" data. There is clearly not much to choose in the quality of the σ_t fits themselves. The new parameterization does, however, lead one to question the validity of the amplitudes deduced by Crawford and Stevenson, and Tripp.

EXPERIMENT

Our experiment was performed by the following group: G. Jones, E.G. Auld, R.R. Johnson, T. Masterson, P. Walden, E. Mathie, A. Haynes, D. Ottewell, C. Winter, M. Sivertz, R. Feenstra and B. Tatischeff (Orsay). The differential cross-section $\frac{d\sigma}{d\Omega}$ (θ) and the analysing power A_π (θ) for the pions from both $pp \rightarrow \frac{d\pi^+}{pn\pi^+}$ were measured simultaneously over the angular range of 35° to 145° (or the kinematic limit imposed by the pion detection threshold of 12 MeV). Both polythene (with carbon difference) and liquid hydrogen targets were used at proton energies of 305, 310, 320, 330, 350, 375, 400 and 425 MeV. We used a proton beam from the TRIUMF cyclotron with a beam current of about 1 nA, and polarizations of at least 60%. The pion detector was a broad-range Browne-Buechner 50 cm magnetic spectrograph, shown diagramatically in Fig. 4. It has a 24 element scintillation counter hodoscope on the focal plane.

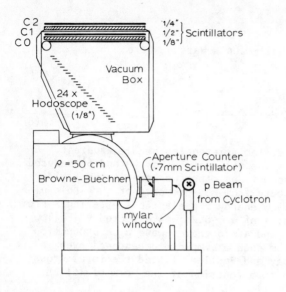

Fig. 4. Magnetic spectrograph.

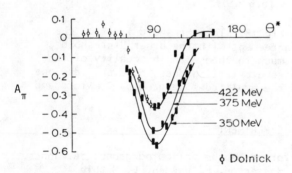

Fig. 6. Analysing power for pp→dπ⁺.

RESULTS

A sample of the data we obtained at a bombarding energy of 350 MeV is shown in Fig. 5. On the abscissa is shown a point 2 MeV below the peak energy in order to give an indication of the energy resolution of the system. The analysing power A_π for the pions from the break-up reaction is strikingly similar to that for the 2-body reaction. As this data is still being analysed, the remainder of this paper is restricted to a discussion of the preliminary analysis of the two-body reaction only. In the next figure (6), the analysing power as a function of angle is shown for several bombarding energies. Our data at 425 MeV is in very good agreement with that of Dolnick.[2] The spin-averaged angular distribution at 350 MeV proton energy is shown in Fig. 7. The values of A_o obtained from such analyses are plotted as a function of η (in the manner of Crawford and Stevenson) in Fig. 8a.

Our result for X, (X = 0.13), is not very different from that determined by Crawford and Stevenson (Eq. 7): 0.082 ± 0.034. When the total cross-section expansion of Measday and Spuller is used, the results appear rather similar, and are shown in Fig. 8b. In this case, X is obtained from:

$$A_o = X + X(X + 1/3) \ f \ (\eta) \tag{10}$$

Note that the value of X (used in the phenomenological expression for the dependence of Q on η) is quite different for the two fits to the data (0.21 for Measday-Spuller compared to 0.13 above).

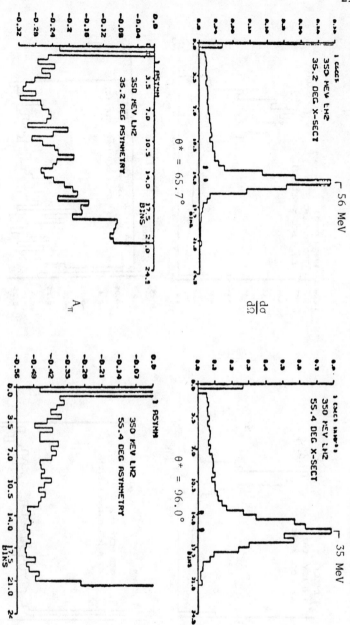

Fig. 5. Sample $pp \to \frac{d\pi^+}{pn\pi^+}$ spectra T_p = 350 MeV η = 0.64.

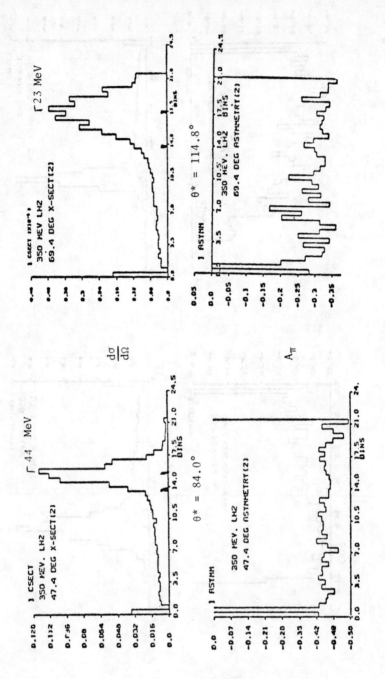

Fig. 5. Sample $pp \to d\pi^+_{pn\pi^+}$ spectra $T_P = 350$ MeV $\eta = 0.64$.

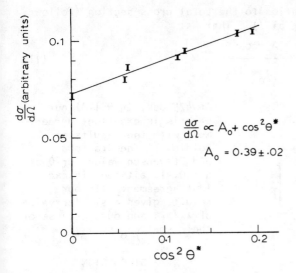

Fig. 7. $\frac{d\sigma}{d\Omega}$ ($\theta*$) at 350 MeV, $\eta = 0.64$.

$$\frac{d\sigma}{d\Omega} \propto A_o + \cos^2\theta^*$$

$$A_o = 0.39 \pm .02$$

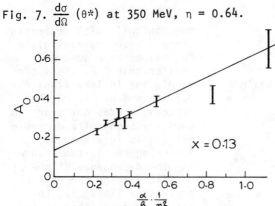

Fig. 8a. A_o vs. $\frac{1}{\eta^2}$.

$X = 0.13$

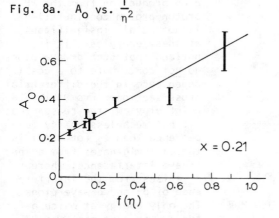

Fig. 8b. A_o vs. $f(\eta)$.

$X = 0.21$

In Fig. 9 the values of Q obtained from our data (together with those of the earlier workers) is shown plotted as a function of η. The dashed line is the prediction of the phenomenological theory using the simple expansion for the total cross-section (Eq. 5). The solid line shows the result when the more reasonable expansion of Measday and Spuller is used. The latter, using a value of $X = 0.21$, and a value of $\sin(\psi - \tau_1) = 0.70$ is obviously a much better representation of this data.

Fig. 10 is a composite of all the analysing power data we have taken for a bombarding energy of 350 MeV. The results for both polythene and liquid hydrogen targets are included. The solid curve is the expectation if only s- and p-wave pion production is involved. The clear non-symmetry about 90° is direct evidence of the presence of odd powers of $\cos(\theta)$ in the $\lambda(\theta)$ of Eq. (1). Such terms can only occur if d-wave pion production contributes.

The next two figures (11 and 12) illustrate the energy dependence of the coefficients of a power series expansion of the $\lambda(\theta)$ (in $\cos(\theta)$). Older, higher energy data is included for comparison. In order to remove the gross (3,3) resonance energy dependence from these values, the λ_j co-

302

efficients are plotted as ratios to the total cross-section (following the example of Akimov et al.[7]). That is:

$$\frac{\lambda(\theta)}{\gamma_0 + \frac{1}{3}\gamma_2} = \frac{A_\pi(\theta)}{\sin\theta} \times \frac{A_0 + \cos^2\theta}{A_0 + \frac{1}{3}} \qquad (11)$$

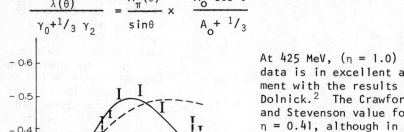

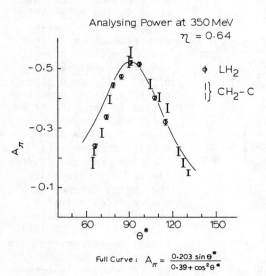

Fig. 9. $Q = \frac{1}{P} \varepsilon \left(\frac{\pi}{2}\right)$ plotted against a function of η.

Analysing Power at 350 MeV
$\eta = 0.64$

ϕ LH$_2$

$\left.\begin{matrix} | \\ | \end{matrix}\right\}$ CH$_2$-C

Full Curve: $A_\pi = \dfrac{0.203 \sin\theta^*}{0.39 + \cos^2\theta^*}$

Fig. 10. Analysing power at 350 MeV $\eta = 0.64$.

At 425 MeV, ($\eta = 1.0$) our data is in excellent agreement with the results of Dolnick.[2] The Crawford and Stevenson value for Q at $\eta = 0.41$, although in excellent agreement with our result, gives a smaller value of λ_0/σ_t than ours because of their small A_0.

DISCUSSION

Although λ_2 decreases monotonically with decreasing energy, becoming negligible for values of η somewhat smaller than 0.5, it is certainly not insignificant for $0.5 < \eta < 1.0$. λ_1 on the other hand does not start to show any signs of a decrease until η is less than 0.5. Referring to the table showing the dependence of the λ_j coefficients on the various amplitudes (Table I), the results suggest that d-wave pion production from the ^3P proton-proton configuration is not at all insignificant at these energies. It is noticed that such d-wave pions do <u>not</u> contribute to a $\cos^4(\theta)$ dependence in the differential cross-section. From this point of view, the reason that λ_1 has much less energy dependence than λ_2 for $\eta < 1$ is that λ_1 originates from s- and d-wave interference, whereas λ_2 arises from the interference of p- and d-wave pions. The only energy at which a fit for λ_3 was also made was 425

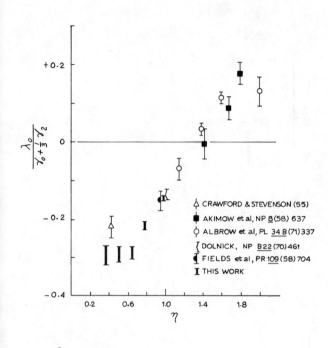

$$\frac{\lambda_0}{\gamma_0 + \frac{1}{3}\gamma_2} = \frac{Q}{1 + \frac{1}{3A_0}}$$

Fig. 11. Dependence of λ_0 on η.

MeV, ($\eta = 1.0$) for which the composite of our data and that of Dolnick spanned a very large angular range, thus helping to define the coefficients more precisely than possible with either data set by itself. The results for the composite set, as well as our data alone were consistent in preferring a small negative value for λ_3. Our best estimate for λ_3 is $-.075 \pm 0.11$. Since the separate terms in the power series expansion are not orthogonal functions, the errors in the separate terms are strongly correlated. Since the value for λ_3 was arbitrarily taken to be zero in the higher energy fits ($\eta > 1$) made by earlier workers[7,8], their errors for λ_1 are probably significantly underestimated.

CONCLUSION

In summary, $\eta = 0.5$ is clearly <u>above</u> "threshold" with d-wave pion production quite significant. Thus, experiments concerned with threshold pion production in the $pp \rightarrow d\pi^+$ reaction, are not, as has usually been thought, about to celebrate their silver anniversary, but are just now at the threshold of being begun!

REFERENCES

1. M. Gell-Mann and Watson, Ann. Rev. of Nucl. Sci. <u>4</u>, 219 (1954).
2. C.L. Dolnick, Nucl. Phys. <u>B22</u>, 461 (1970).
    F. Mandl and T. Regge, Phys. Rev. <u>99</u>, 1478 (1955).
    The amplitudes of these authors have been divided by $\sqrt{2\ell+1}$ where ℓ is the relative angular momentum of the incident protons, in order to arrive at amplitudes as usually defined.

304

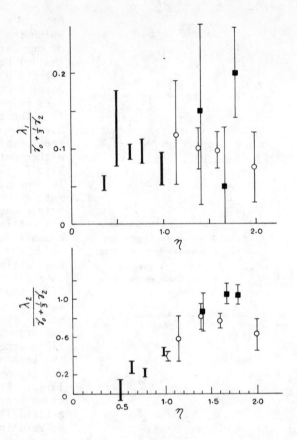

Fig. 12.   Dependence of λ_1 and λ_2 on η.

3.   C. Richard-Serre, W. Hirt, D.F. Measday, E.G. Michaelis, M.H.J. Saltmarsh and P. Skarek, Nucl. Phys. B20, 413 (1970).
4.   F.S. Crawford, Jr. and M.L. Stevenson, Phys. Rev. 97, 1305 (1955).
5.   R.D. Tripp, Phys. Rev. 102, 862 (1956).
6.   J. Spuller and D.F. Measday, Phys. Rev. D12, 3550 (1975).
7.   Yu. K. Akimov, O.V. Savchenko and L.M. Soroko, Nucl. Phys. 8, 637 (1958).
8.   M.G. Albrow, S. Andersson-Almehed, B. Bosnjakovic, F.C. Erne, Y. Kimura, J.P. Lagnaux, J.C. Sens and F. Udo, Phys. Lett. 34B, 337 (1971).
9.   T.H. Fields, J.G. Fox, J.A. Kane, R.A. Stallwood and R.B. Sutton, Phys. Rev. 109, 1704 (1958).

SINGLE AND DOUBLE PION PRODUCTION
IN PROTON-PROTON COLLISIONS AT MEDIUM ENERGIES*

P. R. Bevington
Case Western Reserve University, Cleveland, Ohio 44106

ABSTRACT

The current status of our knowledge of pion production from proton-proton collisions at medium energies is reviewed briefly. New measurements at 800 MeV are reported for both single- and double-pion production. The momentum spectra for single-pion production are compared to existing models. The magnitude of the integrated cross section for double-pion production indicates that there is no anomalously high yield near threshold as there is in neutron-proton collisions.

INTRODUCTION

This discussion of pion production will be limited to proton-proton collisions at medium energies, with emphasis on incident kinetic energies near 800 MeV. The status of our knowledge of proton-proton interactions at these energies can be gauged from agreement with predictions of phenomenological phase shifts. The most widely quoted set of p-p phase shifts is that of MacGregor, Arndt, and Wright[1] (MAW), which is fairly reliable up to 350 MeV with dubious extrapolations to 750 MeV.

Measurements of Willard et al.[2] made by the Case Western Reserve University group (with collaborators from the University of Idaho and LASL) at LAMPF of the differential cross section for proton-proton elastic scattering at 800 MeV indicate that the MAW phase shifts underestimate the inelasticity in the $^1D_2$ phase shift, which would be characteristic of pion production enhanced by the 3/2-3/2 resonance. The polarization $P(\theta)$ in elastic scattering should be the parameter most accurately reproduced by the phase-shift calculations; yet very recent (unpublished) measurements by the CWRU/U. Idaho/LASL group at LAMPF at 800 MeV indicate that the analyzing power of hydrogen is about 15% lower than predicted, suggesting that the polarization peaks near 700 MeV.

Measurements of most of the observables from elastic proton-proton scattering are in progress at a number of laboratories and should improve the picture considerably in the near future, but since the major discrepancy appears in the inelastic channel, we must also investigate the production of pions in proton-proton collisions.

TWO-BODY FINAL STATE

The greatest effort in pion production in recent years has been toward an understanding of the two-body final state reaction $pp \rightarrow \pi^+d$, including the inverse reaction. Experimental data at CERN of the reaction $\pi^+d \rightarrow pp$ were analyzed thoroughly by Richard-Serre et al.[3] The dominance of the $^1D_2$ state is illustrated by the sharp peaking of the excitation function (Fig. 1) around a center-of-mass energy equivalent to the mass of a nucleon-delta pair. Brayshaw[4] has recently fit the integrated cross section over a wide energy range (Fig. 1) with consistent covariant dynamics of NN scattering, pion production, and πD scattering. His

---

*Work supported in part by the U.S. Energy Research and Development Administration.

phase shifts differ from those of MAW (Fig. 2) in both the real and imaginary parts of the $^1D_2$ phase, but also indicate greater involvement of the $^3P_1$ phase as well, with a sharp break near the threshold for single-pion production. Suzuki[5] arrived at a similar set of phase shifts starting with a different model for pion production.

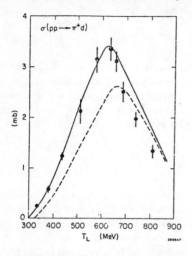

Fig. 1. Data[3] and theory[4] for yield of $pp \rightarrow \pi^+ d$.

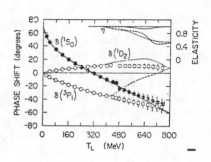

Fig. 2. Phase shifts of Brayshaw[4] vs MAW[6] for $pp \rightarrow \pi^+ d$.

The enhancement of the $^1D_2$ phase is well established; the exact behavior of the $^3P_1$ phase shift is not so well determined, but there are indications that it should vary markedly in the region of the 3/2-3/2 resonance. The latest energy-dependent phenomenological phase shifts of Arndt, Hackman, and Roper,[6] which extend reliable p-p phase shifts to 500 MeV, include an enhancement of the $^1D_2$ phase shift above that of MAW, but ignore any rapid energy dependence in the $^3P_1$ phase shift.

The angular distribution of pions was shown by Richard-Serre et al.[3] to be of the form

$$\sigma_\pi(\theta) = K(A + \cos^2\theta - B\cos^4\theta) \tag{1}$$

where A ranges from 0.2 to 0.3 and B increases from 0.0 to 0.5 with increasing energy. More recent determinations of Aebischer et al.[7] at CERN at 398, 455, and 572 MeV and by Felder[8] at LAMPF at 800 MeV are in substantial agreement.

Aebischer et al.[7] found asymmetries $P(\theta^*)$ consistent with a negative asymmetry (pions scattered to the **left**) rising from zero near 400 MeV increasing with energy to about −0.2 at 600 MeV, consistent with previous measurements.

THREE-BODY FINAL STATE

Less attention has been paid to the three-body final state reaction $pp \rightarrow \pi^+ pn$. The integrated cross section rises rapidly from threshold to peak at an incident proton kinetic energy of 1 GeV and then decreases slowly with energy (e.g., see Lock and Measday,[9] Ch. 8). The single-isobar model of Mandelstam[10] predicts a sharp peak near 800 MeV, but the data are better represented by the extended-isobar model of Sternheimer and Lindenbaum.[11] We might therefore expect the Δ(1232) isobar to dominate, with small contributions from heavier isobars.

The Rice/Houston group of Hudomalj-Gabitzsch *et al.*[12] has performed a kinematically complete experiment at LAMPF at 800 MeV in which they detected the emitted proton in coincidence with the pion. Their data (Fig. 3) show peaks in the correlated pion momenta corresponding to strong final-state interactions between the proton and pion indicative of the 3/2-3/2 resonance. They find cross sections at these peaks several hundreds of times as great as three-body phase-space predictions. These enhancements can be fit fairly well with a simple isobar model similar to that of Mandelstam,[10] which assumes that the momentum dependence is due entirely to the Δ^{++} resonance.

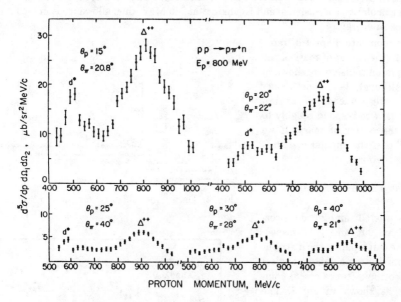

Fig. 3. Momentum spectra of protons in coincidence with π^+ (Ref. 12).

Cochran *et al.*[13] measured the pion spectra at Berkeley at 730 MeV. Their data are shown in Fig. 4 together with theoretical curves calculated by Suslenko and Kochkin[14] using a Ferrari-Selleri OPE Model with a form factor

$$G(k^2) = g\mu^2/(k^2 + 10\mu^2) \qquad (2)$$

and off-mass-shell corrections of the Selleri type solution.[15]

Glass *et al.*[16] have measured at LAMPF at 647, 771, and 805 MeV the momentum spectra of neutrons emitted from the reaction pp → npπ⁺. These spectra have been fit successfully with a modified version of the calculation developed by Stephenson, Gibbs, and Gibson.[17] Pratt *et al.*[18] have reported measuring neutron spectra at 0°, 14°, and 27° at LAMPF at 800 MeV.

Gugelot *et al.*[19] have observed neutral pion production at CERN at 600 MeV. They find angular distributions which contain strong $\cos^2\theta$ terms and some evidence for $\cos^4\theta$ dependence. The momentum spectra tend to rise fairly linearly with pion momentum. These data have been analyzed further by Borie, Drechsel, and Weber,[20] who fit the angular distributions from π^0 and π^+ production with a model of (3,3) resonance and nucleon pole, which predicts $\cos^2\theta$ terms with similar magnitudes for both charge states.

The CWRU/LASL group measured the yield of pions from proton-proton collisions at LAMPF at 800 MeV as part of a survey of all particles emitted in such collisions. The data are unpublished and preliminary from the Ph.D. research of F. Cverna; experimental details will be discussed in the next section. The pion momentum spectra in the laboratory shown in Fig. 5 exhibit much the same peaking at the mass of the Δ^{++} as was seen in the data correlated with proton detection, and are in considerable disagreement with three-body phase-space predictions which peak at higher momenta. The data are in good agreement with the calculations of Suslenko and Kochkin[14] at 740 MeV when allowance is made for the difference in energy.

The momentum spectra transformed to the center-of-mass of the proton-proton collision are shown in Fig. 6. Although there is not a one-to-one correspondence of angles, the 50° data correspond roughly to 90° in the center of mass. If we assume[9] that the angular distribution has the form

$$\sigma_\pi(\theta) = a(1/3 + b\cos^2\theta) \qquad (3)$$

and that the 50° and 20° data correspond to 90° and 38°, respectively, we can separate the distributions into isotropic (at all angles) and anisotropic (times $\cos^2\theta$) components as shown in Fig. 7. The data are consistent with values of a = 2.0 and b = 0.8. The isotropic portion peaks near the pion momentum (230 MeV/c) corresponding to decay of a stationary Δ^{++}. The anisotropic portion has a shape similar to that of the statistical model for three-body phase space.[21] Neither compares favorably with the predictions of the isobar model of Lindenbaum and Sternheimer.[22]

Vovchenko et al.[23] compared somewhat less precise measurements made at Dubna at 660 MeV by Batusov et al.[24] with the resonance model of Mandelstam[10] and the isobar model of Lindenbaum and Sternheimer[22] and found their

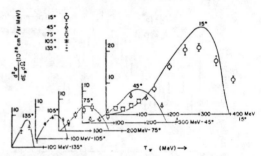

Fig. 4. Data[13] and theory[14] for pp → π^+pn at 740 MeV.

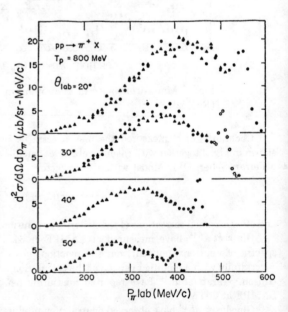

Fig. 5. Momentum spectra (lab) of π^+ from pp at 800 MeV. Symbols denote spectrometer magnet currents.

data in agreement with the isotropic part, but not the anisotropic part, which predicts peaking at high momenta.

DOUBLE-PION PRODUCTION

The status of double-pion production from proton-proton collisions near 800 MeV has been a point of controversy. The excitation function for the reaction pp → $\pi^+\pi^-$pp is shown in Fig. 8, where the solid line is adapted from calculations of Lindenbaum and Sternheimer[22] and the data are from refs. 13, 25-29, respectively, from 730 MeV to high energy. The point at 730 MeV is the work of Cochran *et al.*[13] at Berkeley measured with a scintillator hodoscope and may overestimate the yield because of background contamination. The points at 970 MeV are the work of Bugg *et al.*[25] and Barnes *et al.*[26] at Birmingham with a bubble chamber and represent detection of one event. The magnitude is somewhat corroborated by observation of similar yields for two other double-pion production reactions but is subject to the questionable statistics of single events.

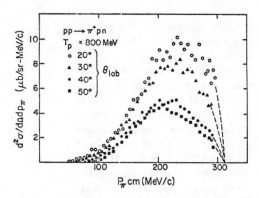

Fig. 6. Momentum spectra (cm) for pp → π^+pn at 800 MeV.

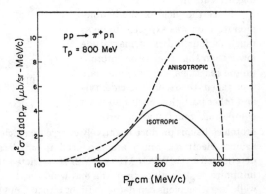

Fig. 7. Constant and angle-dependent portions of Fig. 6.

The measurements of Pickup *et al.*[28] at Brookhaven at 2 GeV indicate that the momentum spectra of pions peak nearly symmetrically between the kinematic limits with isotropic angular distributions (though nucleons have a strong $\cos^2\theta$ dependence) and indicate a dominant contribution from excitation of both protons to $\Delta(1232)$ isobars. These observations are also predicted by the isobar model of Lindenbaum and Sternheimer.[22]

Measurements of double-pion production from neutron-proton collisions by Rushbrooke *et al.*[30] at Birmingham at 970 MeV indicate similar pion distributions, although the magnitude of the integrated cross section is anomalously high up to 1 GeV (e.g., see Lock and Measday,[9] Ch. 8) until it settles down near a ratio of 2:1 over that in proton-proton collisions.

Our group, consisting of F. Cverna, H. B. Willard, P. R. Bevington, H. W. Baer, and M. W. McNaughton from Case Western Reserve University, and N.S.P. King from the University of California at Davis, has recently measured the yield of negative pions from the reaction pp → $\pi^-\pi^+$pp at LAMPF at 800 MeV. The data presented here are unpublished and preliminary from the Ph.D. research of F. Cverna.

310

The experimental arrangement is shown in Fig. 9. The external proton beam of LAMPF, which is extremely clean and well collimated, passes through a 6.5-cm-long LH$_2$ target. The trajectories of emitted π^- are traced through three pairs (horizontal and vertical) of multiwire proportional counters (MWPC)[31] M1, M2, M3, with a spectrometer magnet between the last two pairs of chambers. The geometry provides a wide range of momentum acceptance (from 80 to over 300 MeV/c) at one magnet current.

Since the yield of the double-pion reaction was expected to be as much as four orders of magnitude lower than that from other proton-proton reactions, considerable care was taken to exclude accidentals and other particles. Fortunately no competing reactions in proton-proton collisions produce negatively charged particles. A gas Cerenkov counter discriminates against electrons, which are produced in π^0 decay. The origin of accepted events, as determined by MWPCs M1 and M2, is restricted to the middle 5 cm of the LH$_2$ target to minimize contributions from the end windows, but target-empty subtractions are made as well. Negative pions can decay in flight into μ^-, but most of these are rejected by the vertical trajectory analysis of M1-M3 or the horizontal trajectory between M1 and M2. The data are corrected for loss of pions from decay.

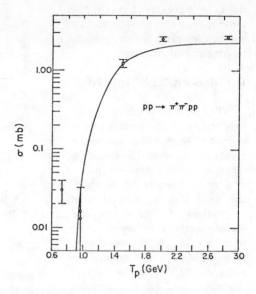

Fig. 8. Yield of double-pion production from pp.

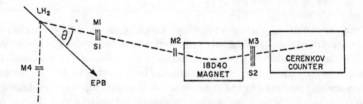

Fig. 9. Experimental arrangement.

Time-of-flight information from a pair of scintillators reduces the number of accidentals, but a significant number of protons, which trace acceptable trajectories through the first two MWPCs and are swept to the right by the spectrometer magnet, scatter back into the third MWPC. A fourth MWPC M4 identifies elastically scattered protons, and the contributions are found to be concentrated at one end of MWPC M3, corresponding to high pion momentum.

The observed momentum distributions of π^- in the laboratory system for 800 MeV are shown in Fig. 10. Note that the cross sections are of the order of nb/sr-MeV/c. The data are binned in 20-MeV/c intervals, with statistical uncertainties only shown as error bars. The contribution of accidentals at momenta above the kinematically allowed limit is quite noticeable, especially at back angles, such as 60° where the kinematic limit is below 200 MeV/c.

The distributions transformed to the center-of-mass of the proton-proton collision are shown in Fig. 11 along with statistical three-body phase-space calculations.[21] Contamination from accidentals appearing at the high-momentum end of the spectra are quite consistent in character at various angles. Contributions from particles identified by MWPC M4 as elastically scattered protons have the same shape as this contamination and a slightly larger magnitude. We believe these residual contributions to be from inelastically scattered protons.

The data are too crude to discriminate between various models insofar as the shape of the distribution is concerned, especially since the models predict similar peaking this close to threshold. The angular distribution does seem to be isotropic as expected.

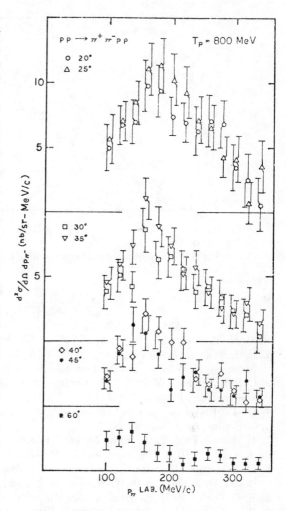

Fig. 10. Momentum spectra (lab) of π^- from pp at 800 MeV.

The three phase-space curves are normalized to integrated cross sections of 2, 3, and 4 μb/sr, respectively. The measured integrated cross section is thus 3.0 ± 1.0 μb/sr, which is in good agreement with the extrapolated curve and all but the data at 730 MeV of Fig. 8. Hence it appears that there is no anomalously large contribution near threshold to double-pion production from proton-proton collisions as there is in neutron-proton collisions.

312

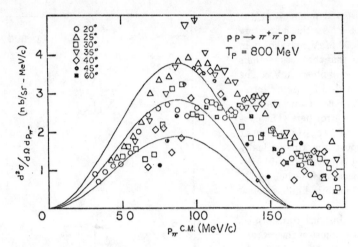

Fig. 11. Momentum spectra (cm) and phase space for pp →
$\pi^-\pi^+$pp at 800 MeV.

REFERENCES

1. M. H. MacGregor, R. A. Arndt, and R. M. Wright, Phys. Rev. **169**, 1149 (1968).
2. H. B. Willard, B. D. Anderson, H. W. Baer, R. J. Barrett, P. R. Bevington, A. N. Anderson, H. Willmes, and N. Jarmie, Phys. Rev. **C14**, 1545 (1976).
3. C. Richard-Serre, W. Hirt, D. F. Measday, E. G. Michaels, M.J.M. Saltmarsh, and P. Skarek, Nucl. Phys. **B20**, 413 (1970).
4. D. D. Brayshaw, private communication: "Pion Production and Covariant NN Dynamics" (unpublished).
5. H. Suzuki, Prog. Theor. Phys. **54**, 143 (1975).
6. R. A. Arndt, R. H. Hackman, and L. D. Roper, Phys. Rev. **C15**, 1002 (1977).
7. D. Aebischer, B. Favier, L. G. Greeniaus, R. Hess, A. Junod, C. Lechanoine, J.-C. Nikles, D. Rapin, and D. W. Werren, Nucl, Phys. **B106**, 214 (1976).
8. R. D. Felder, Ph.D. dissertation, Rice University (1976), unpublished.
9. W. O. Lock and D. F. Measday, "Intermediate Energy Nuclear Physics," Methuen & Co., Ltd. (1970).
10. S. Mandelstam, Proc. Roy. Soc. **A244**, 491 (1958).
11. R. M. Sternheimer and S. J. Lindenbaum, Phys. Rev. **123**, 333 (1961).
12. J. Hudomalj-Gabitzsch, T. Witten, N. D. Gabitzsch, T. Williams, G. S. Mutchler, J. Clement, G. C. Phillips, E. V. Hungerford, L. Y. Lee, M. Warneke, B. W. Mayes, and J. C. Allred, Phys. Letters **60B**, 215 (1976); J. Hudomalj-Gabitzsch, private communication.
13. D.R.F. Cochran, P. N. Dean, P.A.M. Gram, E. A. Knapp, E. R. Martin, D. E. Nagle, R. B. Perkins, W. J. Shlaer, H. A. Thiessen, and E. D. Theriot, Phys. Rev. **D6**, 3085 (1972).
14. V. C. Suslenko and V. I. Kochkin, JINR report P2-5572, Dubna (1971), unpublished.
15. F. Selleri, Nuovo Cimento **40A**, 236 (1965).
16. G. Glass, M. Jain, M. L. Evans, J. C. Hiebert, L. C. Northcliffe, B. E. Bonner, J. E. Simmons, C. Bjork, P. Riley, and C. Cassapakis, Phys. Rev. **D15**, 36 (1977).

17. W. R. Gibbs, B. F. Gibson, and G. J. Stephenson, Jr., cont. to VI Int. Conf. on High Energy Physics and Nuclear Structure, Santa Fe (1975) (unpublished); R. C. Slansky, G. J. Stephenson, W. R. Gibbs, and B. F. Gibson, Bull. Am. Phys. Soc. II **20**, 83 (1975).

18. J. Pratt, R. Bentley, H. Bryant, R. Carlini, C. Cassapakis, B. Dieterle, C. Leavitt, T. Rupp, and D. Wolfe, cont. to VI Int. Conf. on High Energy Physics and Nuclear Structure, Santa Fe (1975) (unpublished).

19. P. C. Gugelot, S. Kullander, G. Landau, F. Lemeilleur, and J. Yonnet, Nucl. Phys. **B37**, 93 (1972).

20. E. Borie, D. Drechsel, and H. J. Weber, Z. Physik **267**, 393 (1974).

21. M. M. Block, Phys. Rev. **101**, 796 (1956).

22. S. J. Lindenbaum and R. M. Sternheimer, Phys. Rev. **105**, 1874 (1957).

23. V. G. Vovchenko, N. I. Kostanashvili, and V. A. Yarba, Yad. Fiz. **11**, 810 (1970); Sov. Jrnl. Nucl. Phys. **11**, 453 (1970).

24. Yu. A. Batusov, N. I. Kostanashvili, G. I. Lebedevich, D. S. Nabichvrishvili, and V. A. Yarba, JINR Preprint R1-4491 (1969), unpublished.

25. D. V. Bugg, A. J. Oxley, J. A. Zoll, J. G. Rushbrooke, V. E. Barnes, J. B. Kinson, W. B. Dodd, G. A. Doran, and L. Riddiford, Phys. Rev. **B133**, 1017 (1964).

26. V. E. Barnes, D. V. Bugg, W. P. Dodd, J. B. Kinson, and L. Riddiford, Phys. Rev. Letters **7**, 288 (1961).

27. A. M. Eisner, E. L. Hart, R. I. Louttit, and T. W. Morris, Phys. Rev. **B138**, 670 (1965).

28. E. Pickup, D. K. Robinson, and E. O. Salant, Phys. Rev. **125**, 2091 (1962).

29. E. L. Hart, R. I. Louttit, D. Luers, T. W. Morris, W. J. Willis, and S. S. Yamamoto, Phys. Rev. **126**, 747 (1962).

30. J. G. Rushbrooke, D. V. Bugg, A. J. Oxley, J. A. Zoll, M. Jobes, J. Kinson, L. Riddiford, and B. Tallini, Nuovo Cimento **33**, 1509 (1964).

31. P. R. Bevington and R. A. Leskovec, Nucl. Instr. & Meth., to be published.

THE PHYSICS OF LOW ENERGY PION PRODUCTION[*]

W. R. Gibbs
Theoretical Division, Los Alamos Scientific Laboratory
Los Alamos, New Mexico   87545

INTRODUCTION

I wish to start by reviewing the fundamental physics that goes into pion production. For this reason I may say some things that seem obvious to some people. However I wish to lay out a pedagogical foundation from the start which emphasizes the very important and direct connection between pion production and the sizes of the systems involved.

Let us consider the basic prototype reaction

$$p \rightarrow n + \pi^{+}.$$

Of course we know that this reaction is forbidden in free space by momentum-energy conservation but it is interesting to contemplate the degree of forbiddeness. If we consider the threshold case (the neutron and π^{+} have zero momentum) and conserve energy then the proton must have (on-shell) 533 MeV/c. Thus momentum fails to be conserved by that amount. If we now place an additional object into the picture such that all (or any one) of

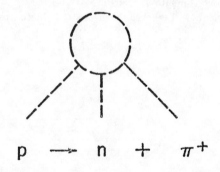

$$p \longrightarrow n + \pi^{+}$$

the particles can interact with it we can, of course, conserve momentum as well. To see how this works consider first the initial proton interacting with the additional object. The interaction will have some range and inside of this range the proton will not have a definite momentum but a distribution of momenta, that is to say, its wave function near to the object is not a single plane wave but a collection of many plane waves of different momenta. The width of the spread of this distribution about the central (on-shell) momentum is given by the uncertainty principle as the

[*]Work performed under the auspices of the U.S. Energy Research and Development Administration.

inverse of the range of the interaction. Thus if the additional object is a nucleon and we take the range of the nucleon-nucleon interaction to be characterized by one pion exchange then the width of the momentum distribution will be given by ~ 140 MeV/c. To conserve momentum the initial proton must have none, so that it is 533 MeV/c away from its central value and very unlikely to be found in a distribution with a width of only 140 MeV/c.

Continuing with the example of the additional object being a nucleon we see that the neutron in the final state suffers the same problems since its range is essentially the same as the proton's. In order for the pion to conserve momentum alone it also must find 533 MeV/c but the spacial range of the pion nucleon system is much shorter leading to a typical range in momentum of 350-700 MeV/c. Some, in fact, would take larger values for this number but most modern analyses of pion nucleon data lie in this range. Clearly the pion is the favored candidate for supplying the missing momentum and suggests that "pion rescattering" may be expected to be the dominant mechanism in pion production.

Another way of looking at the problem is to consider three graphs for the three different types of interaction (Fig. 1). Of course in most realistic calculations all of the interactions will be calculated at once, but it is useful to consider them one at a time to discover which is most important. In each case all lines are on-shell except one, as I have labeled. In order to reach the (off-shell) momentum required for each line labeled "off", support must be found in the half-off-shell t-matrix which leads to it.

Figure 2 shows a comparison of two nucleon-nucleon off-shell t-matrices[1] compared with the separable π-N form of Londergan, McVoy and Moniz.[2] Here is illustrated that the π-N t-matrix has much stronger support for high momentum components (>750 MeV/c) than the nucleon-nucleon t-matrix. One may also estimate the degree to which the off-shell N-N t-matrix

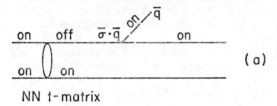

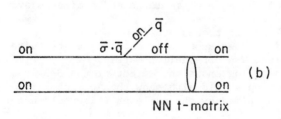

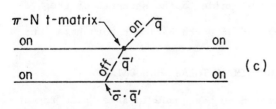

Fig. 1. Three different possibilities for the off-shell momentum.

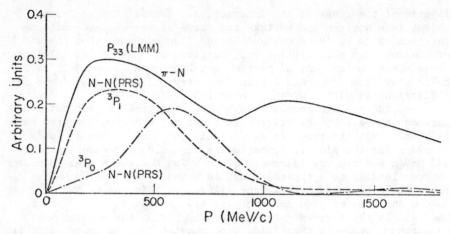

Fig. 2. Comparison of pion-nucleon and nucleon-nucleon t-matrices.

will be probed by looking at the width of the peak involved. It
appears that distances of the order of 1-2 f^{-1} are involved.

Returning to Fig. 1 for a moment we can see another reason for
the dominance of graph c. The fundamental production amplitude is
itself proportional to $\bar{\sigma} \cdot \bar{q}'$ and thus a large q' is favored. Also
note that since it is graph c that is the dominant contributor at
threshold the production amplitude is <u>not</u> proportional to $\bar{\sigma} \cdot \bar{q}$ but
rather to the pion-nucleon (off-shell) scattering amplitude.

It is also interesting to note that the "Galilean Invariance
Ambiguity"

$$[\sigma \cdot q' \rightarrow \sigma \cdot (q' - \lambda \frac{\mu}{m} \bar{p})]$$

is less serious at threshold because q' is large and dominates over
the other term.

There are a number of points to be born in mind

1) The pion production cross section is expected to be quite
sensitive to the range of the π-N off-shell t-matrix.

2) Since the π-N scattering shows the effect of the 3-3
resonance, the pion production should also show this effect. If
the missing momentum were supplied by the nucleons this would not
be the case.

3) Because the "rescattering" takes place far off-shell, the
on-shell t-matrix may be a poor guide to the strength of the "rescattering".

4) There is no suggestion of perturbation theory here, but
rather the ideas more resemble DWBA.

COMPARISON OF SIMPLE IDEAS WITH DATA

A number of calculations have been made in this "pion rescattering" picture. A comparison with pionic atom data was made by
Koltun and Reitan[3] in 1966. They used only s-wave rescattering
and showed that this dominated over the direct capture terms.

Lazard, Ballot and Becker[4] made a rescattering calculation in the resonance region for the p+p → d+π+ reaction. They did not worry about the range of the system and the p-wave off-shell t-matrix was taken to have an s-wave character as concerns the dependence on the magnitude of the off-shell momentum. These two approximations compensated such that they found reasonable agreement with data.

In 1974 Goplen et al.[5] extended Koltun and Reitan's calculation for the p+p → d+π+ reaction to finite energy. It was found that a rescattering t-matrix of the pure Kisslinger form (zero spacial range) gave a cross section much too large. If one assumes that there is no form factor on the absorption vertex, the range on the scattering vertex required to produce agreement with data is ~ 350 MeV/c. It was later pointed out[6] that if equal ranges are taken for the rescattering and absorption form factors then the data requires about 600 MeV/c for this π-N range parameter. The truth doubtless lies in between. Note that this model is somewhat crude in that it uses the same range on all p-waves.

It was pointed out by Preedom et al.[7] (using this same pion rescattering calculation) that the angular distribution of the inverse reaction (π++d → p+p) was sensitive to the N-N interaction and possibly could be used to test nucleon-nucleon interaction models. Thus we seem to have the situation in which the N-N

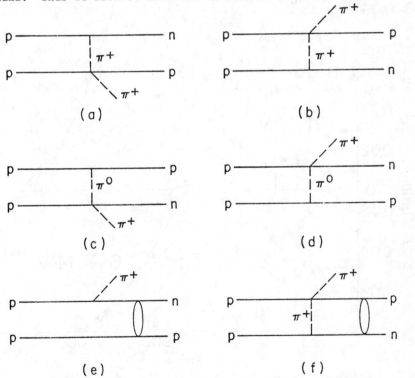

Fig. 3. Graphs considered by Glass et al.

318

interaction cannot supply the required off-shell momenta, but after these are supplied by the π-N rescattering the reaction is sensitive to "details" of the nucleon-nucleon system.

Glass et al.[8] compared their data for p+p → n+p+π⁺ where the neutron is observed at 0° with such a pion rescattering calculation. The graphs considered are shown in Fig. 3. It is interesting to note that graph e was found to be negligible although f was quite important. In fact it is f which makes the calculation difficult. There exists a point in the momentum spectrum where the two nucleons may be found with zero relative kinetic energy. This strong final state interaction is difficult to calculate, and in fact a model

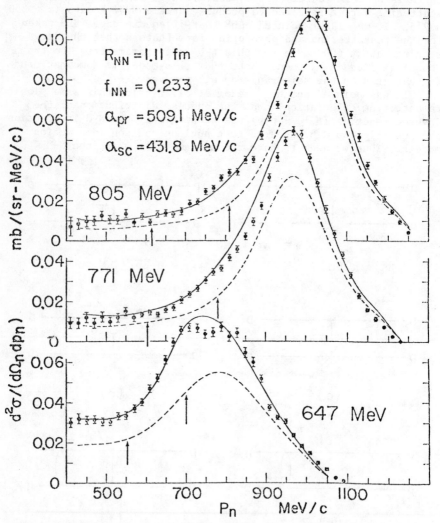

Fig. 4. Results of Glass et al.

was used which involved some parameterization to include 3-body final state effects. However the fit (Fig. 4) was made at the highest incident energy, and the other two energies were calculated without changing the values of any of the parameters. The reproduction of the data is remarkable although there are still some evident difficulties in the representation of the final state N-N interaction. This seems to be clearly a case in which a three body calculation is needed. Never-the-less, we see that the range parameters are of the same general size as those found in $\pi^+ + d \to p+p$ and π-N elastic scattering and that the shape and position of the effect of the 3-3 resonance is well reproduced. This same group is currently applying these same ideas to the calculation of inclusive reactions (paper C 20).

Effects of the nucleon-nucleon final state interactions are also being looked for in more complete (kinematically) experiments by the Rice-Houston collaboration (papers C 27 and C 37).

I would like to digress slightly from the nucleon-nucleon problem for a moment to point out that this kind of picture also works for pion production on (slightly) heavier nuclei. Hess and I recently have been calculating[9] in a model similar to one originally proposed by Bashin and Duck.[10] This model involves pion rescattering from the core nucleons coherently. It also involves a somewhat crude approximation to the off-shell momentum integral which is always present. In view of the approximations made this should be regarded as an estimate, and factors of two discrepancy should not be taken seriously although factors of ten will tend to indicate that some physics is missing. The calculation in its simple form provides a factor with which to multiply the pion elastic scattering cross section to obtain the production cross section. It also provides a slight angle shift between the production angle and the elastic scattering angle.

$$\sigma_{p\pi}(\theta) = F(\theta)\sigma_{\pi}[\theta'(\theta)] \quad .$$

For $A \to \infty$, $F(\theta)$ becomes a constant and $\theta' = \theta$, but for small A, $F(\theta)$ increases at back angles. We see in Fig. 5 a comparison with recent data from LAMPF by Källne et al. showing agreement perhaps better than might be expected. Agreement is by no means always that good and, in fact, there is sufficient disagreement at lower energies and forward angles to say that something is lacking in the model in that region. A reason for part of the discrepancy is known but there are still more problems.

Let me hasten to point out that this type of picture is not contradictory to the type of pionic stripping calculation done with optical models (such as practiced by Jerry Miller e.g.). It was shown early on by Jones and Eisenberg[11] that calculation of pion production with rescattering by a Kisslinger optical model gave a cross section 50 times too large. Miller and Phatak[12] showed that if finite range pion-nucleon t-matrices are used reasonable magnitudes and shapes are obtained. There is no real difference in the physics put into these two types of calculation, only in the order

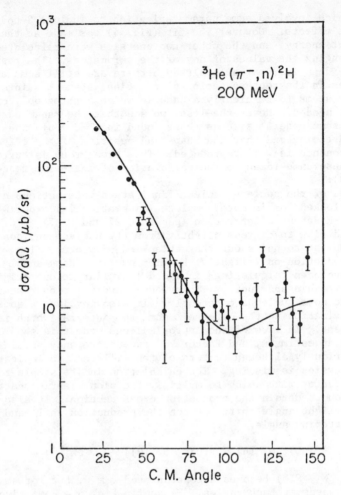

Fig. 5. Comparison of the rescattering model
with Källne et al.

in which things are discussed. The same central point emerges, that
pion rescattering (distortion) must be included and must be included
correctly. A simple rule of thumb seems to be that pionic stripping
models without any pion rescattering are about a factor of five
low.[5,11,13]

There are other different types of models for pion production
on nuclei. One of the more successful of these is the one considered
by Fearing.[14] In this case the cross section is related to the
experimental p+p → d+π⁺ cross section. I don't believe the basic
philosophy is different, just a different approximation in evaluating
the expressions.

MORE REFINED CALCULATIONS

A number of refinements are possible at this point. The first obvious one is that the pion-nucleon range should be different for each TJ component. In fact LMM[2] give a set of functions and these can be used. Recently Goplen's code has been modified by Dave Giebink to include these effects. The result of a calculation of this type is shown in Fig. 6. Note that these calculations have no

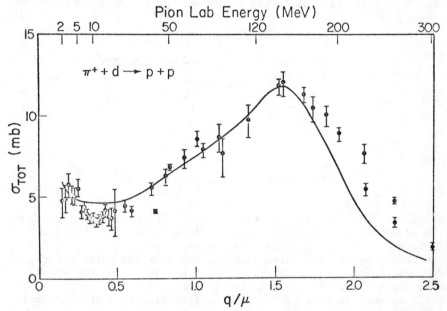

Fig. 6. Total cross section for pion absorption on deuterium.

adjustable parameters and are done with numbers taken from the literature.

In the case of pion production from nuclei some improvement can also be made by removing the restriction of the approximation to the momentum integral used by Bhasin and Duck and by Hess and myself. In fact, it seems imperative to do so, since this approximation automatically predicts no polarization asymmetry and it is known that large asymmetries exist (at least at low energies and forward angles where the agreement with the cross section data is poor). Recently Steve Young and I have been calculating this process by actually doing the integral and using phase shifts from pion-nucleus elastic scattering to get the pion-nucleus amplitude which is now needed. Large effects can be obtained although asymmetries as large as observed have not, as yet, been obtained. The results are very sensitive to the input quantities (nuclear structure, pion-nucleus on- and off-shell amplitudes, etc.).

There is another approach to pion production in the two nucleon system with a long history. The most recent practicioners are Green and coworkers,[15] Riska, Brack and Weise[16] and Borie, Drechsel and Weber,[17] Woloshyn, Moniz and Aaron,[18] Weber, Eisenberg and Shuster,[19] and Brayshaw.[20] In the language we have been using, this approach consists of noticing that the pion-nucleon rescattering will be dominated by the Δ_{33}. Thus the calculation consists of producing a Δ and letting it decay into a pion and a nucleon. In this form this reaction is the prototype for production of Δ's in

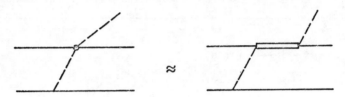

$$P + P \rightarrow N + \Delta \rightarrow N + N + \pi^+$$

nuclei, and can provide a reasonable model for a three nucleon force (the pion from the decay of the delta is absorbed by a third nucleon) which may be important for nucleon energies around 1 GeV.[21]

This type of model also supplies the starting point for three body calculations in which two incident nucleons are treated as one nucleon plus a nucleon-pion system in a 11 state with total energy of a nucleon. In this way nucleon-nucleon elastic scattering and pi production are computed together and inelasticity is incorporated into the phase shifts in an integral fashion. Such a calculation is presently being undertaken by Silbar, Kloet, Aaron and Amado and also independently by Brayshaw.

Recent calculations[16] have shown that inclusion of ρ-exchange as well as π-exchange will replace a large amount of the potential range, at least in the case of production. This approach is more fundamental in the sense that it represents processes in terms of particle exchanges instead of potentials but it has one serious drawback. The ρ coupling constants are not well enough to make a prediction. Since there is necessarily a large calcellation between π- and ρ-exchange, the cross section is very sensitive to these coupling constants. Unfortunately pion production cannot be used to obtain these numbers because the remaining form factors (or potentials) are also largely unknown. Perhaps a consistent treatment of production and elastic scattering could provide everything at once.

A recent calculation of this type (paper C 14 of this conference) by Epstein and Riska seems to be successful in explaining the p-p polarized elastic scattering data.

Of course there are considerably more constraints to be applied to the theories than just the total cross sections. One of these is the angular distribution already mentioned. Another is the polarization asymmetry measured with the use of polarized proton beams.

The standard way of expressing the differential cross section for pion production with a polarized beam is

$$\sigma_p(\theta) = \frac{1}{32\pi} \left\{ (\gamma_0 + \gamma_2\cos^2\theta + \gamma_4\cos^4\theta) \right.$$

$$\left. + \vec{P}\cdot\hat{n} \ \sin\theta[\lambda_0 + \lambda_2\cos^2\theta + \cos\theta(\lambda_1 + \lambda_3\cos^2\theta)] \right\} .$$

There has been data[22] in the literature for several years on λ_0, λ_1 and λ_2 and new data[23] has just become available from TRIUMF. In spite of this there is available only a single calculation by Niskanen[24] which is still in preprint form. A comparison of this data with Niskanen's calculation is shown in Fig. 7.

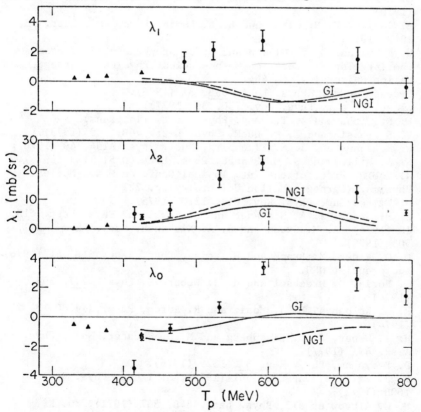

Fig. 7. Comparison of polarization asymmetry data with calculations by Niskanen.[24]

As you see the agreement is very poor with the forward-backward asymmetry parameter (λ_1) even of the wrong sign. I doubt that this failure is a characteristic of the particular model being used. It is possible that straightforward refinements of these calculations will produce agreement but it is also very possible that these are early indications that some critical piece of physics is still missing in our understanding of pion production.

REFERENCES

1. H. S. Picker, E. F. Redish, and G. J. Stephenson, Jr., Phys. Rev. C4, 287 (1971), Phys. Rev. C8, 2495 (1973), G. J. Stephenson, Jr., private communication.
2. J. T. Londergan, K. W. McVoy and E. J. Moniz, Ann. Phys. 86, 147 (1974).
3. D. S. Koltun and A. Reitan, Nucl. Phys. B4, 629 (1968) and Phys. Rev. 141, 1413 (1966), and 115, 1139 (1967).
4. C. Lazard, J. L. Ballot, and F. Becker, Nuovo Cimento 65, 117 (1970).
5. B. Goplen, W. R. Gibbs and E. L. Lomon, Phys. Rev. Lett. 32, 1012 (1974).
6. W. R. Gibbs, B. F. Gibson and G. J. Stephenson, Jr., Proc. of the Int. Topical Conf. on Meson-Nuclear Physics, (Carnegie-Mellon University) p. 464.
7. Preedom et al., Phys. Lett. 65B, 31 (1976).
8. Glass, et al., Phys. Rev. 150, 36 (1977).
9. W. R. Gibbs and A. T. Hess, Phys. Lett. (in press).
10. V. S. Bhasin and I. M. Duck, Phys. Lett. 46B, 309 (1973).
11. W. R. Jones and J. M. Eisenberg, Nucl. Phys. A154, 49 (1970).
12. G. A. Miller and S. C. Phatak, Phys. Lett. B 51, 129 (1974).
13. J. Noble, Proc. of the Int. Topical Conf. on Meson-Nuclear Physics, (Carnegie-Mellon University) p. 221.
14. H. W. Fearing, Phys. Rev. C11, 1210 (1975).
15. A. M. Green, J. A. Niskanen and S. Hakkinen, Phys. Lett. 61B, 18 (1976); A. M. Green and J. A. Niskanen, Nucl. Phys. A271, 503 (1976).
16. D. O. Riska, M. Brack and W. Weise, Phys. Lett. 61B, 41 (1976) and preprint MSU.
17. E. Borie, D. Drechsel and H. J. Weber, Z. Physik 267, 393 (1974).
18. R. M. Woloshyn, E. J. Moniz and R. Aaron, Phys. Rev. C 13, 286 (1974).
19. H. J. Weber, J. M. Eisenberg and M. D. Shuster, Nucl. Phys. A278, 491 (1977).
20. D. Brayshaw, Phys. Rev. 37, 1329 (1976).
21. S. J. Wallace, BAPS 22, 561 (1977), M. Ikeda, Phys. Rev. C6, 1608 (1972).
22. M. G. Albrow et al., Phys. Lett. 34B, 337 (1971); Yu. K. Akimov, O. V. Savchenko and L. M. Soroko, Nucl. Phys. 8, 637 (1958); Carl L. Dolnick, Nucl. Phys. B22, 461 (1970); D.

325

Aebischer et al., Nucl. Phys. <u>B106</u>, 214 (1976).

23. T. Masterson et al., BAPS <u>22</u>, 591 (1977).

24. J. A. Niskanen, University of Helsinki, preprint HU-TTT-7-77.

RESULTS ON ππ-SCATTERING[*]

Wolfgang Ochs

Max-Planck-Institut für Physik und Astrophysik, D-8000 München 40,
Fed. Rep. of Germany

ABSTRACT

We review the main facts which have led to our present understanding of single pion production and the ππ-scattering amplitude.

1. INTRODUCTION

There has been considerable progress in the field of ππ-phase
shift analysis within the last five years mainly due to high stati-
stics experiments on single pion production ($\pi N \to \pi\pi N$ or $\pi N \to \pi\pi\Delta$) at
beam momenta $\gtrsim 5$ GeV. In particular the high statistics CERN-Munich
experiment [1] (~300 k $\pi^+\pi^-$ pairs at 17 GeV/c) has allowed a detailed
investigation of the basic one pion exchange mechanism for single
pion production, and "amplitude analysis" has proven to be a powerful method to isolate it from "background". The surprisingly significant polarization effects found recently [2], however, require a
scheme more complicated than thought originally (i.e. including A_1-
exchange).

The various pion production data, and the recent data from K_{e4}
decays [3] determine the partial waves with angular momentum $l \leq 3$ and
isospin I=0,1,2 from threshold up to ~1800 MeV almost uniquely. There
may be one ambiguity left related to the strength of the $\rho'(1600) \to \pi\pi$-
coupling; this missing piece of information is expected from e^+e^--
annihilation or diffractive photoproduction in the near future.

The interest in ππ-scattering comes from three sources. Firstly,
the elastic scattering of the lightest hadrons plays an important role
as input in dispersion relations to calculate the forces in more complicated reactions, such as πN,KN and, as discussed at this conference,
nucleon-nucleon scattering. Secondly, many theoretical approaches to
hadron physics yield predictions on the most simple interaction involving only the spinless pions, such as the bootstrap schemes, dual Regge
models and current algebra with PCAC. Finally, the spectroscopy of
ππ-resonance states is still very important today in the light of the
recent progress with the "new" spectroscopy of the charmed particles.

In the following we review the main facts which have led to our
present understanding of pion production and ππ-amplitudes. Further
information may be found in two very comprehensive reviews by Martin,
Morgan and Shaw [4] and by Petersen [5].

2. SINGLE PION PRODUCTION AND ππ-SCATTERING

The results on ππ-scattering considered here come from a) single
pion production in the "Regge region" (i.e. $p_{lab} \gtrsim 5$ GeV), b) K_{e4}-decay

---

* Invited talk at the IInd International Conference on Nucleon
Nucleon Interaction, Vancouver, B.C., Canada, June 1977

(see Fig.1). The latter method yields information on the threshold re-

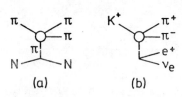

gion ($M_{\pi\pi} \lesssim 400$ MeV) using well established weak interaction theory only [6]
while the first requires a separation
of the one pion exchange (OPE) process
[7] from other hadronic background which
is more model-dependent. In this section we discuss how that can be done.

Fig.1: Information on $\pi\pi$-
scattering is obtained from
pion production and K_{e4}-decay.

21. SOME DEFINITIONS

The process in Fig.1a can be des-
cribed by helicity amplitudes $H^{\ell}_{\lambda\lambda',m}$
(λ,λ' are the helicities of the nucleon, ℓ,μ spin and helicity of
the pion pair). The spin-quantization axis is either the momentum
direction of the incoming π ("t-channel") or of the pion pair ("s-
channel"). Often the combinations

$$H^{\ell(\pm)}_{\lambda\lambda',m} = \frac{1}{\sqrt{2}} (H^{\ell}_{\lambda\lambda',m} \mp (-)^m H^{\ell}_{\lambda\lambda',-m}) \qquad (1)$$

corresponding to natural (+) or unnatural (−) parity exchange are used.
In an experiment with unpolarized target the amplitudes cannot be measured directly but instead the differential cross section $d\sigma/dt M_{\pi\pi} d\Omega$
(t is the N,N' momentum transfer). The dependence on the $\pi\pi$-decay
angles Ω is represented by the spherical harmonic moments $\langle Y^M_L \rangle$ with a
cutoff $L \leq L_{max}$. These moments are linearly related [8] to the density
matrix elements $Re\rho^{\ell\ell'}_{mm'}$ but usually the moments do not fully determine
$Re\rho$. Furthermore, ρ is related to the above amplitudes by

$$\rho^{\ell\ell'}_{mm'} = \frac{1}{2} \sum_{\lambda\lambda'} H^{\ell}_{\lambda\lambda'm} H^{\ell*}_{\lambda\lambda'm'}. \qquad (2)$$

2.2 PION EXCHANGE AND ABSORPTION

One pion exchange (Fig.1a) predicts a cross section dependence
$d\sigma/dt \sim t/(t-m^2_\pi)^2$. The data [1] indeed show a forward dip and a peak near
$t=m^2_\pi$ as predicted, but a careful analysis shows that $d\sigma/dt$ stays finite at $t=0$. Furthermore, the t-channel angular distribution is not
azimuthally isotropic ($\langle Y^M_L \rangle \neq 0$ for $M \neq 0$) as expected for spin zero exchange. This type of effect has been explained by absorption models
[9,10] but a rather good description of the low t-data is given by a
simple recipe known as "Williams model" [11] or "poor man's absorption
model" [12]. It keeps all the real OPE s-channel helicity amplitudes
except those which have non-flip n=0 where $n = |\lambda-\lambda'+m|$.

$$\frac{t}{t-m^2_\pi} \to \frac{t}{t-m^2_\pi} - C_A. \qquad (3)$$

Williams puts $C_A=1$, the ratios of moments with the same L are predicted without free parameters. Fig.2 shows that this is a good first
approach to reality, but at larger t one has to go beyond that model
and it was also found [13] that $C_A < 1$ at larger $M_{\pi\pi}$.

328

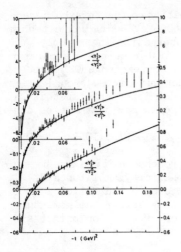

Instead of "improving" these models further one can also try to determine the H-amplitudes directly from the data. This is not possible with unpolarized target data; but we may succeed by supplementing the data with some assumptions which are abstracted from a class of theoretical models. This procedure is called today "amplitude analysis". Two simplifying assumptions or slight modifications of those have been made:

I. Phase coherence: the phase of $H^{\ell}_{\lambda\lambda',m}$ at fixed $M_{\pi\pi}$ and t does not depend on the helicities, only on ℓ.

Fig.2: The moment ratios of $\pi^- p \to \pi^+ \pi^- n$-data [1] in the ρ-region compared to the "poor man's absorption" model [11,12]

II. Spinflip dominance ($H^{\ell}_{++,m}=0$), at least for the (-)-amplitudes.

These assumptions can be implemented in various schemes. Schlein [14] in his pioneering work on amplitude analysis uses only (I), but this is not yet sufficient for a unique solution. Ochs [15] and Frogatt and Morgan [16] get an over-determined system of equations with (I) and (II) and the resulting constraints are fulfilled by the data [17]. Estabrooks and Martin [18,19] with (II) only could demonstrate that solutions with a large violation of (I) are in principle possible but unphysical. Their determination of the three t-channel $\ell=1$ amplitudes is shown in Fig.3.

The OPE-pole contributes to H_0 $H^1_{+-}(-)$ only and indeed this amplitude dominates at small t and is well described by the OPE model. The "background" amplitudes H_- and H_+ are built up by the "absorptive π-cut" and A_2-exchange. Fig.3 demonstrates how powerful the method of amplitude analysis is in separating the OPE-pole contribution. This we need to calculate the $\pi\pi$-amplitudes from the Feynman diagram in Fig.1a.

2.4 THE METHOD OF t-AVERAGED MOMENTS

In practice such a t-dependent analysis for the purpose of phase shift extraction is still rather complicated with in-

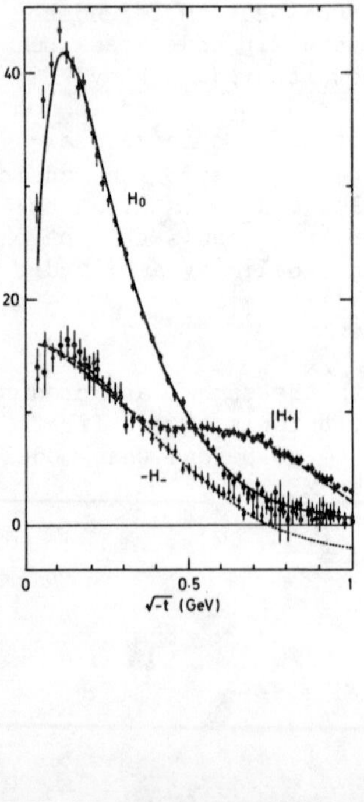

Fig.3: t-Channel helicity amplitudes [20] for $\pi^- p \to \rho n$ at 17 GeV from Estabrooks and Martin amplitude analysis [18,19]. Only H_0 has the pion exchange pole.

creasing number of partial waves. Instead one can also perform an ana-
sis as above (with I and II) but working with moments $\langle Y_L^M \rangle$ integrated over a sufficiently small t-range. This "Ochs-Wagner" method [17]
can be justified by explicit model calculations and by comparison with
the differential method [20]. It is almost as simple or complicated as
the analysis of on-shell ππ-scattering and in each ππ-mass bin only
one additional parameter C_A from eq.(3) is necessary, which describes
the strength of absorption. For illustration we show in Fig.4 how
the moments $\langle Y_L^M \rangle$ with M=0 and M=1 are fit
by a particular ππ-amplitude in the ρ,f
and g-meson region (only 5 out of 13 moments shown).

2.5 THE POLARIZED TARGET EXPERIMENT

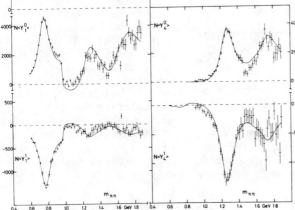

A test of the
above assumptions is
possible with experi-
ments on a polarized
target. In particul-
ar, if there are only
flip amplitudes as in
(II) no polarization
is expected. In Fig.5
we show some of the
new preliminary re-
sults from a trans-
versely polarized
target experiment by
the CERN-Munich group
[2]. Significant po-
larization effects
show up in the moments

Fig.4: ππ-Mass spectrum (N) and selection of
angular moments from $\pi^- p \to \pi^+\pi^- n$-data for $0.01 \leq$
$-t \leq 0.15$ GeV^2 and the Ochs-Wagner fit using the
ππ-amplitudes of Fig.10 [17].

$\langle \cos\psi Y_M^L \rangle$, where ψ is
the angle between
the normal to the
production plane and
the direction of po-

larization. The moment $\langle \cos\psi Y_0^1 \rangle$ gives evidence for a non-flip amplitude corresponding to unnatural exchange, such as A_1 exchange, which
has been neglected so far. Does that imply we have to revise all previous phase shift results?
    First we note that there is no change in the determination of
density matrix elements if we replace II by the weaker condition [15]

    (II') the non-flip amplitudes are proportional to the flip
           amplitudes $H_{++m}^{\ell}(\pm) = \alpha^{\pm} H_{\frac{1}{2}m}^{\ell}(\pm)$, or equivalently $\text{Rg}\rho^{(\pm)} \leq 1$.

This allows for polarization if $\alpha^{\pm}$ is complex. Indeed a previous analysis [22], exploiting the positivity constraints for ρ, has given already some evidence for $\text{Rg}\rho^{(-)}=1$ at the peak of the ρ-meson, and it
is clearly important to know how accurate this is in general. If II'

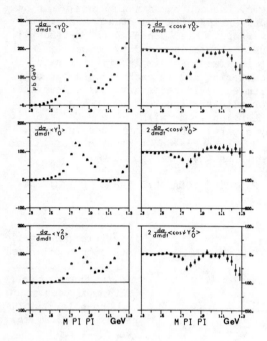

Fig.5: Angular moments from $\pi^-p \to \pi^-\pi^+n$ at 17 GeV with transversely polarized target (CERN-Munich [2]). The non-zero moment $\langle \cos\psi Y_0^1 \rangle$ gives direct evidence for A_1-exchange.

holds and second $\alpha^\pm$ is independent of the $\pi\pi$-mass, there will be exactly no change of phase shift results (for example from the method in 2.4), because all what happens is a decrease of the OPE-contribution to all $\rho_{oo}^{\ell\ell'}$ by the fraction $(1+|\alpha^-|^2)^{-1}$, which may be absorbed into the overall normalization. Furthermore, the model analysis by Kimel and Owens [23] of that data suggests that typically $H_{++o}^1 \approx$ 0.1 H_{+-o}^1 and, therefore, the A_1-contribution to ρ_{oo}^{11} is of the order of 1% only.

In summary, there is a clear OPE signal in $\pi N \to \pi\pi N$ data, which can be effectively isolated by the methods of amplitude analysis. We will have to watch the results from the polarized target experiment which requires a revision of the present theoretical schemes, but we do not expect a significant change of phase shift results.

3. EXPERIMENTAL RESULTS ON $\pi\pi$-PARTIAL WAVE AMPLITUDES

The considerable progress in the last years is due largely to the advent of electronic experiments which yield an order of magnitude more events than the previous bubble chamber experiments. In Table I we list the most relevant experiments with high statistics.

Table I   High Statistics $\pi\pi$-Experiments

Group	Reaction	P_{Lab}(GeV)	Events	Reference
Geneva Saclay	K_{e4} decay		30K	3
Saclay	$\pi^-p \to \pi^+\pi^-n$	2.77	19K	24
	$\pi^-p \to \pi^-\pi^0p$		11K	
Berkeley	$\pi^+p \to \pi^+\pi^-\Delta^{++}$	7	27K	26
CERN-Munich	$\pi^-p \to \pi^+\pi^-n$	17	300K	1
	$\pi^+p \to \pi^+\pi^+n$	12.5	45K	27
Karlsruhe, IHEP, Pisa Vienna, CERN	$\pi^-p \to \pi^0\pi^0n$	40	~100K	28

3.1 THE ρ-MESON REGION AND $K\bar{K}$-THRESHOLD

The energy region from ~500 MeV up to ~1100 MeV has been studied by many groups and Fig.6 shows a selection of results. Below the $K\bar{K}$-

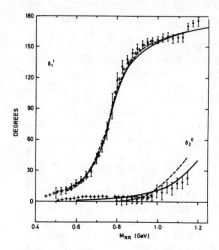

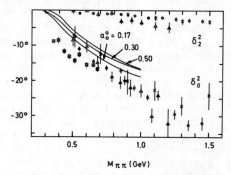

Fig.6b: δ_0^2, δ_2^2: ($\bullet$) Hoogland et al. [27]; (x) Durusoy et al. [29]; (▦) Prukop et al. [30]; (▲) Losty et al. [31]; the curves [5] are predictions from dispersion relations by Basdevant, Frogatt and Petersen [32];

Fig.6: Results on $\pi\pi$-phase shifts δ_ℓ^I from various groups. 6a: δ_1^1, δ_2^0: ($\bullet$) Baton et al. [24]; (+) Estabrooks, Martin et al. [19]; --- Protopopescu et al. [26]; —— Ochs, Wagner et al. [17] (from Ref. [4]).

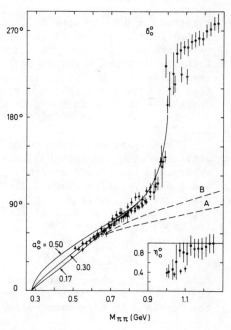

Fig.6c: δ_0^0: (▼) Protopopescu et al. [26]; ($\bullet$) Ochs, Wagner et al. [17]; (▲) Estabrooks and Martin [33]; curves are predictions by Basdevant et al. [32]; Morgan's subtraction of the S* with parameterizations A,B results in the dashed curves [34].

threshold elastic unitarity holds approximately and the $\pi\pi$-amplitude can be represented by phase shifts δ_ℓ^I for fixed isospin I and spin ℓ. The best parameters [35] for the ρ-meson which dominates δ_1^1 are $M_\rho = 773 \pm 3$ MeV, $\Gamma_\rho = 152 \pm 3$ MeV.

The I = 2 phase shifts δ_0^2 and δ_2^2 are found in a reasonably narrow band. Their negative sign has been established by the Saclay group studying $\pi^-\pi^0$ final states [24]. Much interest has been focussed on the isoscalar S-wave. Today there is a unique solution (Fig.6c) based on Berkeley and CERN-Munich data. The exclusion of other solutions ("up down" ambiguity) was possible below the ρ by unitarity constraints above the ρ by studying the analytic behaviour of the amplitude near the $K\bar{K}$-threshold [25,26], by comparing with the $\pi^0\pi^0$-mass spectrum [19] and also by unitarity [17]. In all analyses based on CERN-Munich data δ_0^0 continued rising above 1 GeV and passed through 270° in the f-region. Though the behaviour of δ_0^0 and η_0^0 is quite well established the interpretation in terms of resonance states and their classification is a controversial subject.

Many authors have fit the data by various parameterizations of the amplitudes with correct analytic properties. They all agree [5, 36] on a IInd sheet pole of the S-matrix at

$$\text{ReE} = 980\ldots1010 \text{ MeV}, \qquad \text{ImE} = 5\ldots30 \text{ MeV} \qquad (4)$$

which is called the S^*. Furthermore [34,36], $(g_{K\bar{K}}^{S^*}/g_{\pi\pi}^{S^*})^2 \approx 4$.

This does not exclude a large width $\Gamma_{\pi\pi}^{S^*} \sim 700$ MeV in a 2-channel Breit-Wigner formula [37], still the pole position will be as above. This discrepancy which is, of course, due to the near by $K\bar{K}$-threshold makes it quite difficult to draw meaningful conclusions on the classification of that state. That there is in fact a S^*-resonance and not merely a "cusp-effect" may, however, be concluded from the neat SPEAR result on the decay $\psi \rightarrow \phi\pi^+\pi^-$ (Fig.7). The OZI-quark line rule in this decay suggests that the pions come from a strange quark $s\bar{s}$-state and so the peak at ~ 900 MeV could be identified with the S^*.

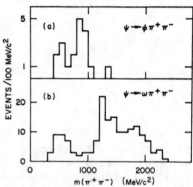

Using particular parameterizations A and B, Morgan [34] has subtracted the S^*-effect from δ_0^0 and a slowly rising phase is left over (dashed line in Fig.6c). This could correspond to an elastic* broad resonance ε with

$$M = 1000 \ldots 1300 \text{ MeV},$$
$$\Gamma = 500 \ldots 700 \text{ MeV}.$$

At least it is clear that besides the S^* there is at most one further ε-state in this mass region.

Fig.7: SPEAR results on ψ-decays [38]. The peak at 900 MeV could be due to $\psi \rightarrow \phi S^*$.

---

* A new $K\bar{K}$-analysis seems to require a considerable inelasticity in the f-region [39].

3.2 THE THRESHOLD REGION

This region is determined by the three scattering lengths a_0^O, a_0^2, a_1^1 (notation a_ℓ^I). Here strong restrictions on the behaviour of the partial wave amplitudes can be put from analyticity, crossing symmetry and unitarity with the help of dispersion relations: The gross features of δ_0^O and δ_1^1 as discussed above determine essentially all three isospin amplitudes except for one parameter, say a_0^O. In particular a_1^1 is essentially fixed and a_0^O and a_0^2 are related by an "universal curve" [40]. An extensive study of dispersion relations has been carried out by Basdevant, Frogatt and Petersen [32]. Using experimental results on δ_0^O below 1000 MeV and the ρ-meson parameters they arrive at (in units of m_π^{-1})

$$a_1^1 = 0.040 \pm 0.004; \qquad\qquad -0.05 < a_0^O < 0.6$$
$$2a_0^O - 5a_0^2 = 0.69 + 0.96(a_0^O - 0.3) + 0.7(a_0^O - 0.3)^2 \qquad (5)$$

The missing piece of information on the threshold region is provided by a very accurate measurement of K_{e4}-decays [3], which allows a clean determination of $\delta_0^O - \delta_1^1$ without that model dependent assumptions as above. The data are shown in Fig.8 together with theoretical curves as above calculated for various values of a_0^O. The result is [5]

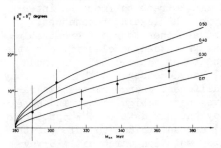

Fig.8: The phases $\delta_0^O - \delta_1^1$ from K_{e4}-decays [3]. The curves represent dispersion theory calculations [32] for various values of the scattering length a_0^O.

$$a_0^O = 0.26 \pm 0.05 \qquad (6)$$

The "universal curve" then defines a narrow strip in the (a_0^O, a_0^2) plane and $a_0^2 = -.028 \pm .014$. Earlier K_{e4}-data [41,42] agree with that within the larger errors. A value $a_0^O = 0.44 \pm 0.1$ is derived from low-mass $\pi^- p \to \pi^+ \pi^- n$ data by Männer [43] but the energy dependence of his δ_0^O and δ_1^1 violates the dispersion relation calculation of Basdevant et al. [44].

Theoretical calculations for δ_0^O and δ_0^2 with the same scheme are also shown in Fig.6 for the three values of a_0^O. It appears that the experimental phases δ_0^2 fall somewhat below the predictions with $a_0^O > 0.17$. The theoretical error at 1000 MeV is about 3°, whereas the discrepancy between the expectation from $a_0^O = 0.26$ and the most accurate CERN-Munich data is $\sim 5^O$. Only the Prukop et al. data disagree completely.

In summary quite a consistent picture of the $\pi\pi$-amplitudes below 1000 MeV emerges, and the data seem to be roughly compatible with dispersion relations.

3.3 THE f- AND g-MESON REGION $(1\,\text{GeV} \lesssim M \lesssim 1.8\,\text{GeV})$

Now one has to include S-, P-, D- and F-waves in the analysis, and the problem of phase shift ambiguities becomes more serious. A systematic treatment is possible with the method of "Barrelet zeros" [45]. A partial wave expansion of the amplitude $T_{\pi\pi}$ with $\ell \leq L$ can be written in terms of the (complex) amplitude zeros z_k

$$T_{\pi\pi} = C_k \prod_{k=1}^{L} (z-z_k), \qquad z = \cos\theta. \tag{7}$$

From the measurement of the differential cross section

$$\frac{d\sigma}{dz} \sim |T|^2 = |C|^2 \prod_k [(z-\mathrm{Re}z_k)^2 + (\mathrm{Im}z_k)^2] \tag{8}$$

one can determine $\mathrm{Re}z_k$ and $|\mathrm{Im}z_k|$. Then the overall phase of C is unknown and there are the discrete ambiguities to choose the signs of $\mathrm{Im}z_k$. In the absence of a reliable total cross section measurement one either can fix the overall phase by requiring Breit Wigner-type behaviour of the leading resonances f and g or use dispersion relations. The discrete ambiguities can be seen in Fig.9. The

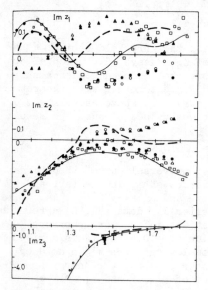

symbols (except for the open squares) are results of the energy independent analysis by Männer [46] and very similar results are obtained by Estabrooks and Martin [21]. At low masses $\mathrm{Im}z_1 > 0$ is selected by the S^*-effect and $\pi^\circ\pi^\circ$-data, $\mathrm{Im}z_2 < 0$- and $\mathrm{Im}z_3 < 0$-results from the interplay of the leading resonances. There are then four different solutions which can be classified by the sign of $\mathrm{Im}z_k$ at 1.6 GeV. Shimada [48] has demonstrated that the study of different charge states can resolve these ambiguities. Indeed already the raw data on $\pi^\circ\pi^\circ$-production [28] are sufficient to rule out the $\mathrm{Im}z_2 > 0$-solution which has a S-wave resonance under the g-meson. The ambiguity with $\mathrm{Im}z_1$ could be resolved by studying $\pi^\pm\pi^\circ$ in the f-region. Shimada prefers the solution with the signs (---) or "sol.A", which is the only one with nice straight line zeros $\mathrm{Re}z_k$ in all charge states, a feature appealing for some theoretical reasons. On the other hand,

Fig.9: Amplitude zeros from various phase shift solutions: ▲△●○ Männer [36], □—— Ochs and Wagner [17], --- Frogatt and Petersen [47].

backward dispersion relations [49] prefer (+--) or "sol.B". The corresponding partial wave amplitudes are shown in Fig. 10 (right-hand side). Whereas sol.B/+-- has a P-wave resonance (ρ') under the g-meson with elasticity $x \approx 0.25$ and a fully elastic S-wave, sol.A has no ρ' or $x < 0.02$ and a less elastic S-wave. Also shown is the analysis by Ochs and Wagner [17] which fits a K-matrix parameterization to the original moments (Fig.4). Apart from the leading resonances one pole in the S-wave and one in the P-wave (ρ') was required to obtain a good fit. Frogatt and Petersen [47] first reconstructed "on shell" moments from previous phase shift results and then used t = 0 and u = 0 dispersion relations to fix the overall phase and select the correct solution. These two results are similar to sol.

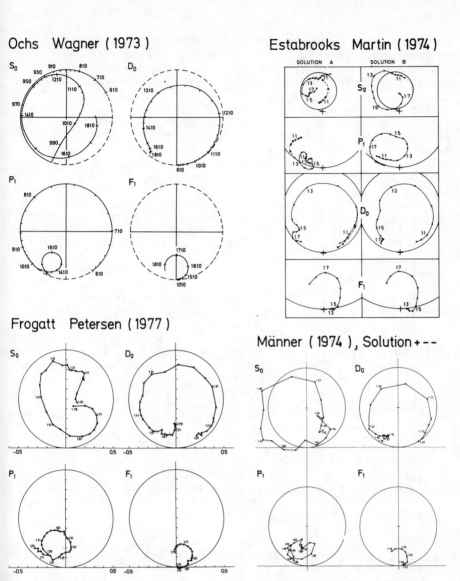

Fig.10: ππ-Phase shifts based on CERN-Munich data [1]: the K-matrix fit by Ochs and Wagner [17], the two preferred solutions by Estabrooks and Martin [21], one (out of four) solution by Männer [46] possibly preferred, and the solution by Frogatt and Petersen [47] constrained by dispersion relations.

B/+--, though there are differences in detail.

Both analyses require some energy smoothness and looking back at Fig.9 their solutions do not always follow the points from Männer's completely energy-independent bin by bin analysis. Ochs and Wagner also found in a bin by bin analysis solutions (the squares) which give a good fit to the data and are near the continuous solution. It appears that Imz_k is not that well determined by the data. Whereas Rez_k is given by the dips in $d\sigma/dz$, Imz_k is given by the cross section at the dip position and that is strongly affected by absorptive effects.

The uniqueness of solutions is questioned by Martin and Pennington [50] in their new approach. They use dispersion relations in a sophisticated way to determine the phase of the amplitude at each energy and angle going beyond a cut-off partial wave expansion. They find again two solutions α and β close to the old A and B.

Hopefully the ambiguity will be settled soon by a measurement of the $\rho'(1600) \to \pi\pi$ branching ratio with e^+e^--machines or from diffractive photoproduction. The existence of the $\rho' \to \pi\pi$-decay mode is already obvious from such an experiment at FNAL [51] (see Fig. 11).

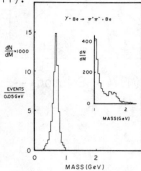

Fig.11: Evidence for $\rho'(1600) \to 2\pi$ from diffractive photoproduction at FNAL [51].

3.4 THE HIGH MASS REGION

No detailed analysis exists here. The outstanding feature is the well-established [28,52] h-meson, the fourth state on the linear ρ-trajectory. Total $\pi^+\pi^-$-cross sections have also been determined [53] up to $M_{\pi\pi} = 10$ GeV and found to be ≈ 15 mb compatible with pomeron factorization.

4. THEORETICAL IMPLICATIONS

4.1 CURRENT ALGEBRA

The results on the threshold region in 3.2 allow a test of Weinberg's [54] calculation. First a linear expansion of the amplitude is used, which results in

$$2a_o^o - 5a_o^2 = 18a_1^1 \equiv 6L. \tag{9}$$

Current algebra predicts for the universal length $L = (4\pi f_\pi^2)^{-1}$. With the pion decay constant $f_\pi = 131$ MeV one finds $L = 0.09$ or

$$a_1^1 = 0.030 \tag{10}$$

This is slightly smaller than the "experimental" value $a = 0.040 \pm 0.004$. Furthermore, a particular assumption on $SU(2) \times SU(2)$ symmetry breaking, namely the non-exoticity for the σ-terms, lead to

$$a_o^o/a_o^2 = -7/2 \qquad \text{or} \qquad a_o^o \simeq 0.17 \tag{11}$$

This is somewhat below the K_{e4}-number $a_o^o = 0.26 \pm 0.05$. Current algebra appears to be reasonably supported by the $\pi\pi$-data.

4.2 POLE AND ZERO STRUCTURE OF THE ππ-AMPLITUDE

The ρ-, f-, g- and h-mesons define the linear parent trajectory.
The first daughter trajectory is missing altogether if we put the ε
under the f (no f' under the g, no ρ' under the f). Then there is
evidence for the second daughter with ε and ρ'(1600) and evidence
against the third daughter, the ε under the g (sol. $Imz_2>0$). So
the odd daughters appear to be absent, different from the simple
Lovelace-Veneziano model [55]. Another feature of dual models [56],
the existence of straight line zeros which "kill" the double poles
in the amplitudes (where the full lines in Fig.12 cross) is approxi-

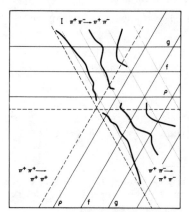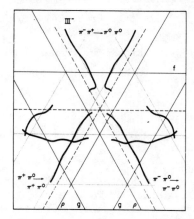

Fig.12: Zero trajectories [57] (Rez_k) reconstructed from ππ-phase
shifts [17] in the Mandelstam plane for two charge configurations.
Straight lines are resonance poles, dotted lines are zeros as expected from dual models.

mately correct (there is still an extra zero through the "Adler point").
In $\pi^+\pi^-\to\pi^0\pi^0$ we then expect with the odd daughters missing the spacing
between neighbour zero trajectories to be doubled. This is, in fact,
observed in Fig.12 (the second zero trajectory does not appear in the
upper triangle, because L = 4 waves are never included in the analysis).
If there was an ε under the ρ, we would have an ε-ρ-double pole which
would require an extra zero trajectory. So the spectra of zeros is
just consistent with the spectra of observed states in this scheme.

REFERENCES

[1] G. Grayer et al: Nucl. Phys. B75, 189 (1974);

[2] H. Becker et al.: Contribution to XVIII Int. Conf. on High Energy
     Physics, Tbilisi (1976);

[3] L. Rosselet et al.: Phys. Rev. D15, 574 (1977);

[4] B.R. Martin, D. Morgan and G. Shaw: Pion-pion Interaction in
     Particle Physics (Academic Press, New York, 1976);

[5] J.L. Petersen: The ππ-Interaction, CERN 77-04 (1977);

[6] A. Pais and S.B. Treiman: Phys. Rev. 168, 1858 (1968);

[7] C. Goebel: Phys. Rev. Letters 1, 337 (1958),
     G.F. Chew and F.E. Low: Phys. Rev. 113, 1640 (1959);

[8] see for example Ref. [1];

[9] K. Gottfried and J.D. Jackson: Nuov. Cim. 34, 735 (1964);

[10] F.S. Heney, G.L. Kane, J. Pumplin, M. Ross: Phys. Rev. 182,
     1579 (1969);

[11] P.K. Williams: Phys. Rev. D1, 1312 (1970);

[12] G.C. Fox: Proc. Caltech Conf. on Phenomenology in Particle
     Physics, Pasadena (1971);

[13] W. Ochs and F. Wagner: Phys. Letters 44B, 271 (1973);

[14] P.E. Schlein: Phys. Rev. Letters 19, 1052 (1967);

[15] W. Ochs: Nuov. Cim. 12A, 724 (1972);

[16] C.D. Frogatt and D. Morgan: Phys. Letters 40B, 655 (1972);

[17] W. Ochs: Thesis, Ludwig-Maximilians-Universität, Munich
     (1973),
     B. Hyams et al.: Nucl. Phys. B64, 134 (1973);

[18] P. Estabrooks and A.D. Martin: Phys. Letters 41B, 350 (1972),
     Proc. of Tallahassee Conf. on ππ-Scattering, A.I.P. Conf.
     Proc. 13, 357 (1973);

[19] P. Estabrooks et al.: ibid A.I.P. Conf. Proc. 13, 37 (1973);

[20] P. Estabrooks, A.D. Martin and C. Michael: Nucl. Phys. B72, 454
     (1974);

[21] P. Estabrooks and A.D. Martin: Phys. Letters 53B, 253 (1974),
     Nucl. Phys. B95, 322 (1975);

[22] G. Grayer et al.: Nucl. Phys. B50, 29 (1972);

[23] J.D. Kimel and J.F. Owens: Tallahassee preprint, FSU HEP
     76-10-21 (1976);

[24] J.P. Baton, G. Laurens and J. Reignier: Phys. Letters 33B,
     528 (1970);

[25] S.M. Flatté et al.: Phys. Letters 38B, 232 (1972);

[26] S.D. Protopopescu et al.: Phys. Rev. D7, 1279 (1973);

[27] W. Hoogland et al.: Contr. to XVIII Int. Conf. on High
     Energy Physics, Tbilisi (1976). A smaller data sample is
     published in Nucl. Phys. B69, 266 (1974);

[28] N.D. Apel et al.: Phys. Letters 57B, 398 (1975);

[29] N.B. Durusoy et al.: Phys. Letters 45B, 517 (1973);

[30] J.P. Prukop et al.: Phys. Rev. D10, 2055 (1974);

[31] M.J. Losty et al.: Nucl. Phys. B69, 185 (1974);

[32] J.L. Basdevant, C.D. Frogatt and J.L. Petersen: Nucl. Phys. B72, 413 (1974);

[33] P. Estabrooks and A.D. Martin: Nucl. Phys. B79, 301 (1974);

[34] D. Morgan: Phys. Letters 51B, 71 (1974); Proc. Summer Symposium on New Directions in Hadron Spectroscopy, ANL 1975, ANL-HEP-CP-75-58, p.45

[35] Particle Data Group: Rev. of Mod. Phys. 48 (1976);

[36] A.D. Martin, E.N. Ozmutlu and E.J. Squires: Nucl Phys. B121, 514 (1977);

[37] S.M. Flatté: Phys. Letters 63B, 228 (1976);

[38] G.J. Feldman and M.L. Perl: SLAC-PUB-1972 (1977), F. Vannucci et al.: Phys. Rev. D15, 1814 (1977);

[39] D. Cohen: K̄K-Amplitude Analysis: ANL-HEP-CP-77-37 (1977);

[40] D. Morgan and G. Shaw: Nucl. Phys. B10, 261 (1968);

[41] A. Zylbersztejn et al.: Phys. Letters 38B, 457 (1972);

[42] E.W. Beier et al.: Phys. Rev. Letters 30, 399 (1973);

[43] W. Männer: Int. Conf. on Exp. Meson Spectroscopy, Boston 1974, AIP Conf. Proc. 21, p. 22

[44] J.L. Basdevant et al.: PAR-LPTHE 75.11 (1975);

[45] E. Barrelet: Nuov. Cim 8A, 331 (1972);

[46] see Ref. [43] and B. Hyams et al.: Nucl. Phys. B100, 205 (1975);

[47] C.D. Frogatt and J.L. Petersen: see Ref. [5] and Niels Bohr Inst. Preprint NBI-HE-77-7 (1977);

[48] T. Shimada: Prog. Theor. Phys. 54, 758 (1975);

[49] R.C. Johnson, A.D. Martin and M.R. Pennington: Phys. Letters 63B, 95 (1976);

[50] A.D. Martin and M.R. Pennington: in preparation;

[51] Wonyong Lee: Proc. Int. Symp. Lepton and Photon Interaction, Stanford, 1975, p. 213;

[52] W. Blum et al.: Phys. Letters 57B, 403 (1975);

[53] J. Hanlon et al.: Phys. Rev. Letters 37, 967 (1976);

[54] S. Weinberg: Phys. Rev. Letters 17, 616 (1966);

[55] G. Veneziano: Nuov. Cim. 57A, 190 (1968), G. Lovelace: Phys. Letters 28B, 264 (1968);

[56] R. Odorico: Phys. Letters 38B, 411 (1972);

[57] T. Eguchi, T. Shimada and M. Fukugita: N.P. B74, 102 (1974).

POSSIBLE EVIDENCE FOR NARROW BOUND STATES

RELATED TO THE p̄p SYSTEM

P. Pavlopoulos[3], G. Backenstoss[1], P. Blüm[2], K. Fransson[4],
R. Guigas[2], N. Hassler[2], M. Izycki[1], H. Koch[2], A. Nilsson[4],
H. Poth[2], M. Suffert[5], L. Tauscher[1] and K. Zioutas[3]

---

1) Institut für Physik, University of Basle, Basle, Switzerland.

2) Kernforschungszentrum und Universität Karlsruhe, Inst. für Kernphysik, Karlsruhe, Germany.

3) CERN, Geneva, Switzerland.

4) Research Institute for Physics, Stockholm, Sweden.

5) Division des basses Energies, CRN Strasbourg, France.

ISSN: 0094-243X/78/340/$1.50 Copyright 1978 American Institute of Physics

1. INTRODUCTION

During the last decade, several theoretical treatments[1-27] and a plethora of experiments[28-39] [see the summary of Montanet[32]] were performed in a search for high mass, non-strange bosons strongly related to the N$\bar{\text{N}}$ system. These mesons, clustered around the nucleon-antinucleon threshold, have the exciting characteristic of being exceptionally narrow. The existence of such exotic mesons is of great interest both to nuclear physics and to elementary particle physics.

What are these mesons and why are they so important? Let me try to give an answer to these questions through a synopsis of the most representative theoretical predictions. However, this short phenomenological view of the theory should only be an introduction to the idea of our experiment. A complete review of the theoretical interpretations of nucleon-antinucleon narrow states will be presented this afternoon by Myhrer[19]. Exotic mesons coupled to the N$\bar{\text{N}}$ system were predicted mainly on the basis of N$\bar{\text{N}}$ potentials[2-19]. Already in 1949 Fermi and Yang[1] had initiated various compound models of elementary particles on the basis of nucleon-antinucleon states. In 1956 Bethe and Hamilton[2] and Afrikyan[3] tried to solve the same question in analogy to the deuteron problem using a nuclear physics potential approach. In 1958 Ball and Chew[4] constructed a N$\bar{\text{N}}$ potential from NN data by G parity[40] transformation. Ball, Scotti and Wong[5], using a one-boson exchange model, obtained a heavy-mass boson spectrum. They found five bound states of the N$\bar{\text{N}}$ system with the same quantum numbers as those of the π, ρ, σ_0, η, ω mesons. A common characteristic of these papers was that the influence of the strong annihilation on the formation of bound and resonant states was not taken into account.

A very interesting point concerning the relevance of the N$\bar{\text{N}}$ states to nuclear physics is the understanding of the relation between the NN and the N$\bar{\text{N}}$ boson exchange potential through the G-parity transformation. The difference between e^-e^- and e^-e^+ electromagnetic interactions is the sign. This is a consequence of the fact that the photon, which is exchanged, has charge-parity -1. For the strong interactions in the one-boson exchange model the difference between NN and N$\bar{\text{N}}$ interactions is described by the G parity of the exchanged meson[4,10,40]. So one may attempt to describe the elastic N$\bar{\text{N}}$ interactions through the well-known NN potentials (V_x), using a potential of the type

$$\tilde{V}_x = G_x V_x ,$$

where $\tilde{V}_x$ is the N$\bar{\text{N}}$ potential and G_x is the G parity of the meson ($x = \pi, \rho, \eta, \omega, \sigma_0, \sigma_1$). This quantum mechanical potential approach has been extensively developed by Shapiro et al. in a series of papers[9-11]. A very rich spectrum of bound and resonant states arose as a result of their work. It is very appealing that some of these states can be as narrow as a few MeV. The repulsive hard core of the NN potential becomes a deep well in the case of N$\bar{\text{N}}$ (Fig. 1) owing to the negative G parity of the ω meson. This well offers the possibility for a great number of states to be formed, contrary to the nucleon-nucleon spectrum, which has only one bound state, the deuteron. Shapiro et al.

Electromagnetic interactions:

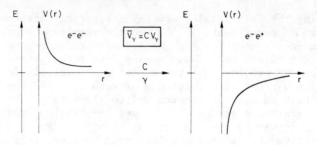

Strong interactions:

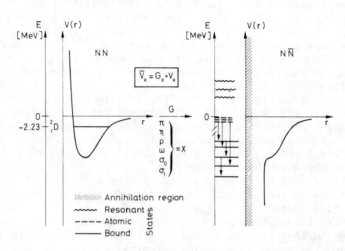

Fig. 1 Particle-antiparticle potential transformation

also considered for the first time the $\overline{N}N$ annihilation process, and under the assumption that the annihilation radius is proportional to $1/2\, m_N$ [41], they treat the annihilation in perturbation theory. In particular for the bound states, Shapiro et al.[20,21], predicted the possibility of E1 and M1 electromagnetic transitions from the lowest atomic states to the various quasinuclear bound states. The nuclear physics potential approach was extended and generalized by Dover et al.[12-15], and for the first time the possibility of pionic transitions to quasinuclear states was discussed[12]. Later, Myhrer et al.[16-18] proved that the annihilation process cannot be described in perturbation, and so the existence in general and the width in particular of these bound and resonant states depend strongly on the annihilation parameters of the optical potential. Since the theoretical predictions of $\overline{N}N$ bound states or resonances are almost independent of the assumption involved, we may not only learn something about nucleon-antinucleon forces but also about nucleon-nucleon short-range interactions, which are the most relevant ones in the $\overline{N}N$ system.

On the other hand, through the development of the quark models in recent years, many authors predicted narrow $N\bar{N}$ states[23-27] within the framework of quantum chromodynamics. In 1968 Lichtenberg[22] first adopted the idea that baryons in the SU(6) representation are composed of a quark and a diquark. This assumption led directly to the consideration of objects composed of a diquark and an antidiquark, the so-called d or exotic mesons. Owing to duality arguments, the pionic decay of these four quark structures is weakly forbidden[24-27] and so one obtained for the first time a clear statement that these states should be narrower than those of ordinary mesons.

Nevertheless, at this moment it is difficult to make a strict distinction between these two different points of view -- the dual model and the nuclear potential model -- since each model illustrates a different perspective of the physics and both have a limited degree of validity.

2. EXPERIMENT

The $N\bar{N}$ formation experiments yielded a number of well-established candidates for narrow resonances with masses above the $N\bar{N}$ threshold. As far as the table of the observed resonances is concerned, the reader is referred to the review paper of Montanet[32]. Mesonic structures below the NN threshold, or bound states, have not been detected in the $\bar{p}p$ system so far. Only a narrow $\bar{p}n$ bound state at 1794 ± 1.4 MeV has been reported by Gray et al.[39].

The method we have used is based on the observation of monoenergetic γ-rays accompanying the annihilation of stopped $\bar{p}$ in liquid hydrogen, where the $\bar{p}$ forms an antiprotonic atom with the target proton. A possible formation of these states might occur via γ-ray transition from an antiprotonic atomic state[20,21]. In this talk I will try to describe an experiment in which bound states were observed in the $\bar{p}p$ system for the first time. In this experiment the γ-spectrum, after $\bar{p}$ absorption in liquid hydrogen, was measured between 30 and 1000 MeV, where a special technique to handle the NaI was invoked to allow the observation of these structures.

2.1 Beam and $\bar{p}$ trigger

The experiment took place at the k_{19} low-momentum electrostatically separated beam of the CERN Proton Synchrotron (PS). We found the optimum conditions at 800 MeV/c, and with a momentum spread of $\pm 2\%$ we stopped in our 1.74 g/cm^2 thick hydrogen target about 250 $\bar{p}$ per burst. A configuration of scintillation and Cerenkov counters, the so-called telescope, typical for stopped particle experiments, identified the incoming particles (Fig. 2). A carbon moderator diminished the energy of the $\bar{p}$ such that 60% to 70% of them stopped in the target. A time-of-flight (TOF) counter system worked parallel to the telescope so that we finally obtained a 10^{-4} pure $\bar{p}$ trigger. The target was surrounded by seven large scintillation counters, called side counters (SC), which detected the charged $\bar{p}p$ annihilation products. We registered a $\bar{p}$ as stopped if at least two SC were triggered. Stopped

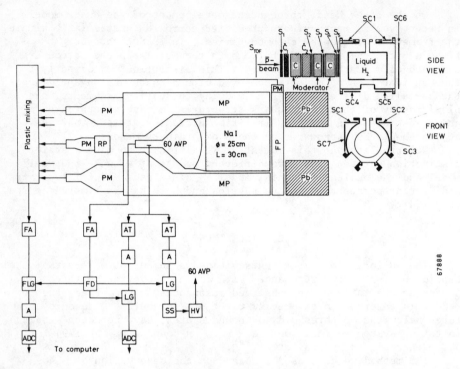

Fig. 2 Schematic drawing of the experimental set-up:

      FA = Fast amplifier
      AT = Attenuator
       A = Amplifier
    FLG = Fast linear gate
      FD = Fast discriminator
      LG = Linear gate
      SS = Stabilizer

$\bar{p}$ annihilating into neutrals (Table 1) produced no SC trigger. On the other hand, $\bar{p}$'s passing through the target were identified by requiring coincidence with SC6 in the absence of any other SC.

2.2 NaI(Tℓ) spectrometer

The 10" × 12" NaI(Tℓ) spectrometer[42] of Strasbourg was used in order to detect high-energy γ-rays. A thick plastic scintillator surrounds the NaI crystal completely and is composed of three pieces (Fig. 2): the front plastic (FP), 6 cm thick; the main plastic (MP), at least 10 cm thick; and the rear plastic (RP), 10 cm thick. The basic idea[43] of our method is to collect in the plastic scintillator the energy escaping from the crystal. Owing to the nature of the escaping shower, we can assume that the energy deposited in the plastic scintillator $E_{p\ell}$ is approximately a constant fraction of the energy missing in the crystal E_{NaI}. If, then, we add the output of the plastic scintillator, multiplied by an experimentally determined coefficient k to the output of the crystal, we obtain a pulse height corresponding to the energy of the incident photon E_γ:

$$E_\gamma = E_{NaI} + k \times E_{p\ell} \ . \tag{1}$$

The advantages of this method are first the increase of the effective crystal volume (and consequently the increase of the crystal efficiency), and secondly the improvement of the resolution (Fig. 3). In order to obtain an energy-independent efficiency for γ-ray detection, events which deposit an unreasonably large fraction f of the shower in the plastic were rejected if

$$\frac{E_{p\ell}}{E_{NaI}} \geq f \ . \tag{2}$$

This constraint condition automatically rejects γ-rays traversing the NaI crystal with bad geometry, thereby improving the resolution of the system (see Fig. 3). In order to define the factors k, f, we have measured the response of the spectrometer at the Bonn 500 MeV electron-synchrotron using tagged photon beams of energies ranging from 137 MeV to 387 MeV. The spectrometer was also tested under the running-conditions at CERN with the 129 MeV γ-ray produced by radiative capture of the π^- at rest in hydrogen. Our tests resulted in

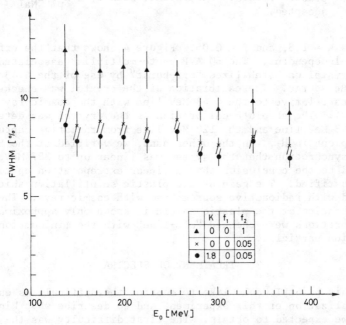

Fig. 3 Energy resolution of the NaI spectrometer for different selection conditions (see text). Only events with $f_1 \leq E_{p\ell}/E_{NaI} < f_2$ accepted.

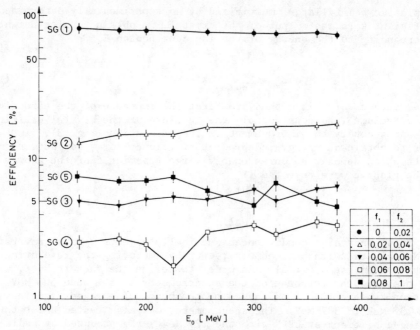

Fig. 4   Energy dependence of the NaI efficiency for different
plastic levels. Only events with $f_1 \leq E_{p\ell}/E_{NaI} < f_2$
accepted.

the values k = 1.8, and f = 0.05. Figure 4 shows that the efficiency
is energy-independent. The 60 AVP photo-multiplier associated with
the NaI crystal was stabilized "off-burst" by use of the 4.43 MeV line
of an Am-Be source. The calibration of the crystal was checked every
eight hours relative to the 4.43 MeV line with the cosmic-ray peak
($\sim$ 139 MeV). The absolute calibration of the crystal was determined
by the 4.43 MeV line and the 129 MeV line from radiative capture of
negative pions in H_2. On the other hand, we verified at the Bonn
electron-synchrotron that the system was linear up to 387 MeV, and
checks led to the conclusion that a linear extrapolation up to 1 GeV
is well justified. The gain of the plastic scintillation shield was
stabilized with radioactive sources and with cosmic rays. The absolute calibration of the plastic shield is known only approximately.
Both calibrations were checked "off-line" with the minimum ionizing
annihilation particles.

3.   TREATMENT OF SPECTRA

At this point I would like to mention the difficulties encountered
in the realization of this experiment and to describe what kind of
spectrum we expected to obtain. The first difficulty was the very
small yield of the gammas, which from previous experiments was estimated to be below the limit of 3% per stopped $\bar{p}$ [44]. On the other
hand, we expected an average of 1.76 π^0's (Table 1) per annihilation

Table 1 Contributions of pion states to $\bar{p}p$ annihilation at rest[45,46]

Final state	Resonant state	%
a) Without neutrals		
$\pi^+\pi^-$	–	0.33 ± 0.04 ⎫
$2\pi^+2\pi^-$	$\rho^0\pi^+\pi^-$	5.8 ± 0.3 ⎬ ∿ 8%
$3\pi^+3\pi^-$	–	1.9 ± 0.2 ⎭
b) With one neutral		
$\pi^+\pi^-\pi^0$	–	3.7 ⎫
	$\rho^0\pi^0$	1.4 ± 0.2
	$\rho^\pm\pi^\mp$	2.7 ± 0.4
$2\pi^+2\pi^-\pi^0$	$\omega\pi^+\pi^-$	3.8 ± 0.4
	$\omega\rho^0$	0.7 ± 0.3
	$\rho^0\pi^+\pi^-\pi^0$	7.3 ± 1.7 ⎬ ∿ 29%
	$\rho^\pm\pi^\mp\pi^+\pi^-$	6.4 ± 1.8
	$\eta\pi^+\pi^-$	1.2 ± 0.3
	$\eta\rho^0$	0.22 ± 0.17
$3\pi^+3\pi^-\pi^0$	$\omega + 2\pi^+ + 2\pi^-$	1.3 ± 0.3
	$\eta + 2\pi^+ + 2\pi^-$	0.6 ± 0.2 ⎭
c) With many neutrals		
$\pi^+\pi^-\pi^0\pi^0$	–	9.3 ± 3.0 ⎫
$\pi^+\pi^-\pi^0\pi^0\pi^0$	–	23.3 ± 3.0
$\pi^+\pi^-\pi^0\pi^0\pi^0\pi^0$	–	2.8 ± 0.7
$2\pi^+2\pi^-\pi^0\pi^0$	–	16.6 ± 1.0 ⎬ ∿ 59%
$2\pi^+2\pi^-\pi^0\pi^0\pi^0$	–	4.2 ± 1.0
$\pi^0\pi^0$	–	∿ 4 × 10^{-2}
$m\pi^0$ m > 2	–	3.2 ± 0.5 ⎭

according to the bubble chamber data[45]. Because the π^0's decay into
two gammas, we should have on the average 3.52 background gammas per
annihilation. In other words, this means that we expected a spectrum
consisting of a few weak lines on the top of a smooth but very high
background. In Fig. 5 we see the spectrum from stopped $\bar{p}$'s in hydrogen obtained under the following conditions:

- only prompt events are allowed;
- at least two of the SC had to be triggered;
- charged particles were rejected;
- optimum spectrometer parameters were used.

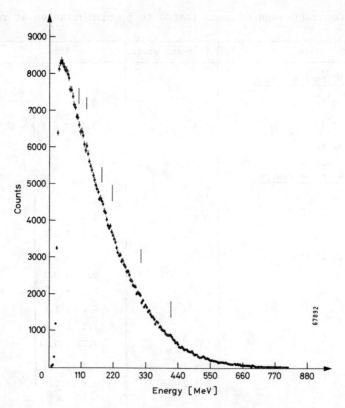

Fig. 5 Spectrum of gamma-rays accompanying p̄p annihilation at rest

The shape of this spectrum is produced by γ-rays from the annihilation π⁰'s. This π⁰ background is known to a large extent and is smooth. We may therefore construct it from Monte Carlo calculations, at least for the known annihilation channels[*), or we may approximate it by a smooth analytical function. In addition, background was present from neutrons, which are produced by charged pions from the annihilation interacting with the detector surroundings. We subtracted also this neutron-induced background, which was determined separately and extends only to about 150 MeV. Using Monte Carlo calculations of the well-known annihilation channels[*,46)], we constructed a high-statistics γ spectrum, which we have approximated through a third-order polynomial. We normalized this polynomial function to the number of p̄ in our experiment, and after subtraction of this well-known part of the background we obtained the spectrum seen in Fig. 6, which is already two-point smoothed. In order to make the apparent structures of Fig. 6 better visible, we subtract an arbitrary smooth background and thus obtain the spectrum shown in Fig. 7.

---

*) Pionic annihilation channels with only one neutral pion in the
   final state. For these channels the intermediate states are known.

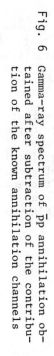

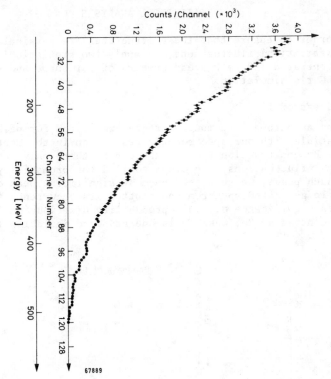

Fig. 6 Gamma-ray spectrum of p̄p annihilation obtained after subtraction of the contribution of the known annihilation channels

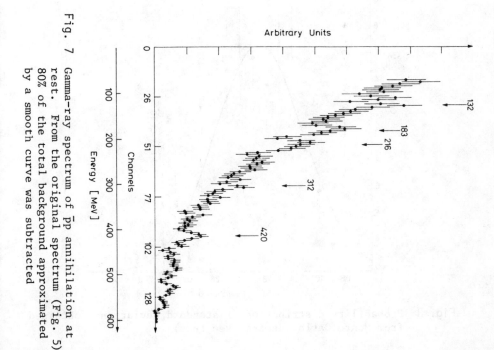

Fig. 7 Gamma-ray spectrum of p̄p annihilation at rest. From the original spectrum (Fig. 5) 80% of the total background approximated by a smooth curve was subtracted

4. STATISTICAL ANALYSIS

In order to ensure that these structures are physical effects and not statistical fluctuations, we apply two statistical tests[47]. We have generated 150 background spectra of our shape and statistics by Monte Carlo simulation.

4.1 Goodness of fit

Using an automatic computer program we search for peaks that are compatible with our spectrometer resolution in the Monte Carlo spectra. By construction these peaks can be there only because of statistical fluctuations. In order to find the probability distribution of such peaks, we selected the most significant peak of each Monte Carlo generated spectrum and plotted its occurrence frequency versus its significance σ. This probability distribution is a Gaussian centred at 2σ, where σ is one standard deviation (Fig. 8).

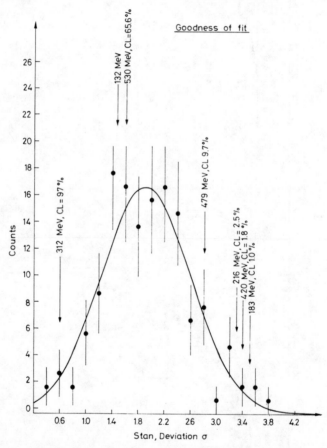

Fig. 8 Probability distribution of standard deviation σ obtained from Monte Carlo spectra (see text)

4.2 Parametric approach

The Monte Carlo spectra are fitted, once assuming only a poly-nomial background and then assuming one line in addition. The quality of the fit is characterized by χ^2 (background) and χ^2 (background + line), respectively. The difference $\Delta\chi^2 = \chi^2$ (background) $- \chi^2$ (back-ground + line) is a measure for the presence of a line. For the Monte Carlo spectra, the $\Delta\chi^2$ distribution is found to be a χ^2-distribution peaked at $\Delta\chi^2 = 2$ as expected (Fig. 9).

By applying the method 4.1 to our experimental spectrum (Fig. 5) we find different structures with different confidence levels. Struc-tures with acceptable significance are found at 183 MeV, 216 MeV, and 420 MeV. The study of the experimental spectrum with the test 4.2 leads to similar results.

The 129 MeV line is expected to be present in the spectrum, owing to the few, very slow, negative pions from annihilation, stopping in the H_2 target and being captured radiatively. For this line we applied the method 4.1 by searching for peaks at only this particular energy, and the confidence level of this line amounts to 0.7%. This confi-dence level means that in seven of a thousand cases this line is due to statistical fluctuations. It should be pointed out that the only constraint applied in the analysis of the spectra was that we required the lines to have a width corresponding to the known resolution of our spectrometer. In other words, the applied analysis is valid only for structures with FWHM compatible with our spectrometer resolution. On the other hand, since we do not know whether and where peaks should occur, these tests search for peaks of unknown position and intensity.

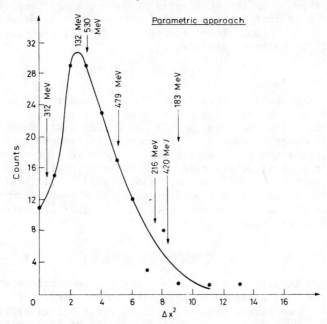

Fig. 9 Distribution of $\Delta\chi^2$ obtained from Monte Carlo spectra (see text)

5. POSSIBLE EXPLANATIONS OF THESE STRUCTURES

5.1 γ-rays from annihilation products

Owing to the kinematics of the annihilation products it is quite improbable that a γ coming from the different intermediate states of the annihilation (Table 1) can produce narrow structures of the above energies. In order to exclude such an effect, we produced a Monte Carlo spectrum with high statistics according to Table 1. The result of these calculations was that γ-rays coming from the annihilation mesons produce a smooth background.

5.2 Charged particles

Charged particles were triggered out automatically by the selection of events following condition (2). In addition to this, a coincidence between SC7 and FP (Fig. 2) vetos charged particles. We have found that both these triggers reject charged particles up to a factor of a few 10^{-5}. On the other hand, we have found that because of geometrical reasons the mixing of the pulse from the plastic with the NaI pulse (condition 1), reduces peaks produced by the charged particles to broad bumps.

5.3 Effects induced by the apparatus

Such effects are best investigated by measuring on a target which is not expected to produce any structure at all, or at least a different one. We selected ^{4}He as the target. Comparison with the H$_2$ spectrum shows that the structures of hydrogen are completely absent in ^{4}He, and in contrast we obtained a structure in ^{4}He which is absent in the $\bar{p}$-H$_2$ spectrum. The statistics of the ^{4}He spectrum correspond to 70% of the hydrogen spectrum.

6. STRUCTURE IN $\bar{p}$ ^{4}He

Applying the same methods as those applied to the $\bar{p}$-H$_2$ case, we have found a structure with 365 ± 26 MeV energy. The confidence level of this line amounts to 7.2% for a line width compatible with the spectrometer resolution. If, however, a broader structure is admitted, then the confidence of this structure improves considerably. The spectrum after subtraction of an arbitrary polynomial background is shown in Fig. 10. If the confidence level of 7.2% allows a discussion for a line, then I believe that the formation of such bound states inside the nuclear matter is another exciting aspect of the nature of these states.

7. <u>RESULTS AND DISCUSSION</u>

We conclude that we have seen three narrow structures related to the $\bar{p}p$ system at energies of 183, 216, and 420 MeV. The confidence levels of these structures are 1%, 2.5% and 1.8%, respectively (Table 2). We cannot deduce quantum numbers of these states from our data. The results are presented in Table 2. The spectrum after subtraction of a cubic polynomial background is shown in Fig. 11.

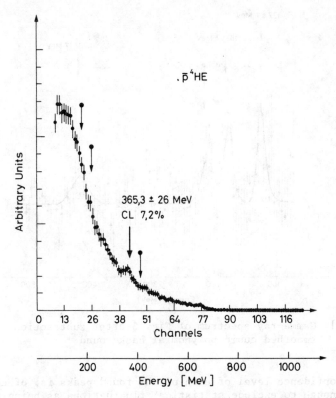

Fig. 10   Gamma-ray spectrum of $\bar{p}$ annihilation at rest in $_2^4$He.
From the original spectrum 80% of the total background
approximated by a smooth curve was subtracted.   The
arrows with black points indicate the position of structures in the $\bar{p}$-hydrogen spectrum

Table 2   Results

Energy (MeV)	Instrumental line width (MeV)	Confidence level (%)	Yield per annihilation
132 ± 6	16	0.7	$(5.1 \pm 2.7) \times 10^{-3}$
183 ± 7	19	1.0	$(7.2 \pm 1.7) \times 10^{-3}$
216 ± 9	21	2.5	$(6.0 \pm 1.9) \times 10^{-3}$
420 ± 17	34	1.8	$(8.5 \pm 2.0) \times 10^{-3}$

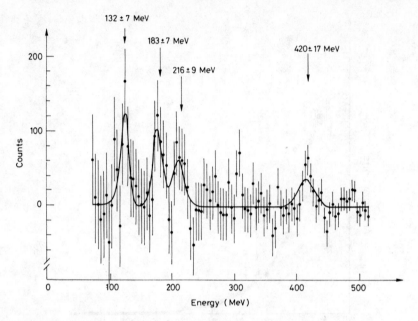

Fig. 11 Gamma-ray spectrum of Fig. 5 after subtraction of a
smoothed cubic polynomial background

The confidence level of each of the found peaks is, of course,
not high enough to exclude statistical fluctuations as being their
source. However, we should keep in mind the following two facts:
The first if the presence of the 129 MeV $\pi^- p$ radiative capture
line. We know that this γ line must exist with a yield of about 10^{-3}
per stopped antiproton. The fact that we observe this line adds to
the reliability of our analyses. The second point is that the
presence of three such significant simultaneous fluctuations in one
spectrum is statistically very unlikely. So the evidence arising from
our experiment is hopefully convincing enough to support the idea that
the nucleon-antinucleon system bears physics in it which is full of
suprises.

REFERENCES

1. E. Fermi and C.N. Yang, Phys. Rev. 76, 1739 (1949).
2. H. Bethe and R. Hamilton, Nuovo Cimento 4, 1 (1956).
3. A.M. Afrikyan, Soviet Phys. JETP 3, 503 (1956).
4. J.S. Ball and G.F. Chew, Phys. Rev. 109, 1385 (1958).
5. J.S. Ball, A. Scotti and D.Y. Wong, Phys. Rev. 142, 1000 (1966).
6. G.F. Chew and J. Koplik, Nuclear Phys. B79, 365 (1974).
7. J. Rosner, Phys. Rev. Letters 21, 950 (1968).
8. G. Schierholz and S. Wagner, Nuclear Phys. B32, 306 (1971).
9. O.D. Dalkarov, V.B. Mandelzweig and I.S. Shapiro, JETP Letters
   10, 257 (1969); Soviet J. Nuclear Phys. 11, 496 (1970);
   Soviet Phys. JETP 32, 744 (1971); Nuclear Phys. B21, 88 (1970).
10. I.S. Shapiro, Soviet Phys. Uspekhi 16, 173 (1973).
11. L.N. Bogdanova, O.D. Dalkarov and I.S. Shapiro, Ann. Physics 84,
    261 (1974).
12. Carl B. Dover, Proc. 4th Internat. Symposium on N̄N interactions,
    Syracuse University, 1975 (eds. T.E. Kalogeropoulos and
    K.C. Wali) (Syracuse Univ., Syracuse, New York, 1975), Vol. II,
    Chapter VIII, p. 37-91.
13. C.B. Dover, S.H. Kahana, Phys. Letters 62B, 293 (1976).
14. C.B. Dover and M. Goldhaber, Phys. Rev. D 15, 1997 (1977).
15. C.B. Dover, S.H. Kahana, T.L. Trueman, BNL-22542 (1977).
16. F. Myhrer and A.W. Thomas, Phys. Letters 64B, 59 (1976).
17. F. Myhrer and A. Gersten, Nuovo Cimento 37A, 21 (1977).
18. O.D. Dalkarov and F. Myhrer, Nuovo Cimento 40A, 152 (1977).
19. F. Myhrer, in 2nd Internat. Conf. on NN interactions, Vancouver,
    July 1977, to be published.
20. O.D. Dalkarov, V.M. Samoilov and I.S. Shapiro, Soviet J. Nuclear
    Phys. 17, 566 (1973).
21. L.N. Bogdanova et al., Proc. 4th Internat. Symposium on N̄N Inter-
    actions, Syracuse Univ., 1975 (Syracuse Univ., Syracuse, New
    York, 1975) Vol. 2, p. 1.
22. D.B. Lichtenberg, Phys. Rev. 178, 2197 (1968).
23. C.F. Chew, LBL-5391 (1976).
24. C.F. Chew, Proc. 3rd European Symposium on N̄N Interactions,
    Stockholm, 1976 (Pergamon Press, New York, 1977), p. 515.
25. C. Rosenzweig, Phys. Rev. Letters 36, 697 (1976).
26. G.C. Rossi and G. Veneziano, CERN-TH 2287 (1977).
27. H.G. Dosch and M.G. Schmidt, CERN-TH 2296 (1977).
28. A.S. Carroll et al., Phys. Rev. Letters 32, 247 (1974).
29. T. Kalogeropoulos et al., Phys. Rev. Letters 34, 1047 (1975).
30. V. Chaloupka et al., Phys. Letters 61B, 487 (1976).
31. W. Brückner et al., Phys. Letters 67B, 222 (1977).
32. L. Montanet, NN review in 5th Internat. Conf. on Experimental
    Meson Spectroscopy, Boston, 1977, to be published.
33. R.J. Abrams et al., Phys. Rev. D 1, 1917 (1970).
34. D.C. Peaslee et al., 4th Internat. Symposium on N̄N Interactions,
    Syracuse University, 1975 (eds. T.E. Kalogeropoulos and
    K.C. Wali) (Syracuse Univ. Syracuse, New York, 1975) Vol. I,
    p. 84-91.
35. C. Baltay et al., Phys. Rev. Letters 35, 891 (1975).

36. P. Benkheiri et al., Phys. Letters 68B, 483 (1977).
37. B.R. French et al., CERN/D.Ph.II/Phys. 75-22 Rev. (1975).
38. C. Evangelista et al., European Conf. on Particle Physics, Budapest, Hungary, 4-9 July 1977.
39. L. Gray, P. Hagerty and T. Kalogeropoulos, Phys. Rev. Letters 26, 1491 (1971).
40. T.D. Lee and C.N. Yang, Nuovo Cimento 3, 749 (1956).
41. A. Martin, Phys. Review 124, 614 (1961).
42. M. Suffert et al., J. de Phys. 31C 5b, 261 (1971).
43. M. Suffert, P. Pavlopoulos, K. Zioutas, I. Hegerath, The response of a 10" × 12" NaI (Tℓ) spectrometer to 130 MeV to 387 MeV γ-rays (to be published).
44. T. Kalogeropoulos et al., Phys. Rev. Letters 35, 824 (1975).
45. L. Montanet, private communication.
46. Table XI in "High energy physics" (E.H.S. Burhop).
47. F. James, private communication.

LOW-ENERGY NUCLEON-ANTINUCLEON INTERACTION

F. Myhrer
CERN, Geneva, Switzerland

1. INTRODUCTION

At this conference the nucleon-nucleon (NN) forces mediated by mesons are particularly under discussion. Here I will discuss a more exotic but closely related subject, nucleon-antinucleon (N$\bar{\text{N}}$) meson exchange forces. The NN pion exchange diagrams in Fig. 1 are connected to the N$\bar{\text{N}}$ ones by

$$T_{N\bar{N}} (n\pi \text{ exchange}) = (-)^n T_{NN} (n\pi \text{ exchange}) \quad (1)$$

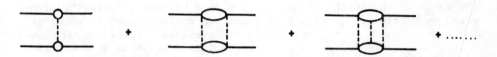

Fig. 1   One-, two-, and three-pion exchange diagrams

More generally NN and N$\bar{\text{N}}$ meson exchange diagrams are connected by G conjugation, for example, a meson of odd G-parity changes the sign of the diagram when going from NN to N$\bar{\text{N}}$. Since π and ω have odd G-parity, for example, the strong ω-repulsion in NN will lead to a strong ω-attraction in N$\bar{\text{N}}$. On the other hand, the $\pi\pi(J = 0)$ exchange (the scalar meson exchange, ε or σ_0) which has even G-parity, will remain attractive in the N$\bar{\text{N}}$ case.

Generally speaking the present NN meson exchange potential predicts a strongly attractive N$\bar{\text{N}}$ potential. This attractive N$\bar{\text{N}}$ potential will (because of the strong L·S forces involved) give many N$\bar{\text{N}}$ bound states and low-energy resonances, as was recognized early on by Shapiro et al.[1]. However, in addition, the N$\bar{\text{N}}$ can annihilate. This has no counterpart for NN scattering and a major part of this talk will be focused on the problem of annihilation.

Because of the short time available, I will concentrate on the most recent N$\bar{\text{N}}$ theoretical work. But at the same time, I shall try to present some general experimental information. All potential descriptions of low-energy N$\bar{\text{N}}$ scattering are in the spirit of the Ball-Chew model[2]. For a review of early works see Phillips[3]. Here a summary of the "predicted" potential resonances will be given. We fit existing $\bar{\text{p}}$p data very well, but there are some open questions that I shall discuss. In the last section I will give a short presentation of the dual quark model approach for B$\bar{\text{B}}$ scattering. B$\bar{\text{B}}$ resonances are also predicted in this model as discussed recently by Chew[4], and Rossi and Veneziano[5].

ISSN: 0094-243X/78/357/$1.50   Copyright 1978 American Institute of Physics

2. THE POTENTIAL MODEL

Following the arguments presented in the Introduction I should like to use information from NN scattering to predict what will happen for N$\bar{\text{N}}$. For a summary of the boson exchange forces see Brown and Jackson, and Vinh-Mau[6]. In this talk I will use the one-boson-exchange potential (OBEP) of Bryan and Scott[7], and introducing a simple model for annihilation I will show the fit the potential model gives to $\bar{\text{p}}$p data.

Bryan and Phillips[8] used the OBEP of Ref. 7 (all parameters fixed by fitting NN data), and added an absorptive potential of Wood-Saxon form,

$$W(n) = -i\, W_c \,/\, (1 + \exp(bn)) \tag{2}$$

The range of the annihilation forces should be short, following arguments by Martin[9]. In Fig. 2 at least two nucleons must be exchanged in the t-channel. Therefore, the inverse range of this graph is at least two nucleon masses (2M). Bryan and Phillips chose b = 0.2 fm^{-1} in Eq.(2), and needed $W_0 \simeq 8$ GeV in order to fit $\bar{\text{p}}$p data. With a different value for b they had to change W_0 a great deal in order to get a fit[3,8].

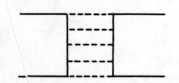

Fig. 2   Annihilation graph contributing to N$\bar{\text{N}}$ elastic scattering

Not long after the Bryan and Phillips work, Shapiro et al. (see, for example, Ref. 1) pointed out that OBEP will give rise to many N$\bar{\text{N}}$ bound states and resonances around N$\bar{\text{N}}$ threshold, because (as stated in the Introduction) the N$\bar{\text{N}}$ potential is very attractive. The N$\bar{\text{N}}$ bound states are deuteron-like objects, consisting of a nucleon and an antinucleon loosely bound together (a few MeV). The radii of these N$\bar{\text{N}}$ bound states are roughly 1 fm [1]. Because they can be understood in a Schrödinger picture (like the nucleus, which is also weakly bound, consisting mainly of protons and neutrons), Shapiro calls these N$\bar{\text{N}}$ states "quasinuclear" states. However, if the N$\bar{\text{N}}$ binding is much stronger than, say, 100 MeV relativistic aspects will be important, if not dominant, and it will no longer be meaningful to think of such an N$\bar{\text{N}}$ state as consisting of a nucleon and an antinucleon. One of the major experimental features of these predicted N$\bar{\text{N}}$ potential resonances should be that they couple mainly to the N$\bar{\text{N}}$ elastic channel.

Later Dover[10] redid the calculations of Shapiro et al.[1] using the same OBEP. He did make some comments concerning Shapiro et al.'s work, but generally speaking confirmed their results. Also Vinh Mau and co-workers[11], with their much better NN potential (built from

$\pi + 2\pi + \omega$ exchange plus a core parametrization), support the results of Shapiro et al.[1]. The results of Shapiro et al. are summarized in Figs. 3 and 4.

Before I go on to discuss annihilation, a few comments should be made regarding the accuracy of the positions predicted for the above $N\bar{N}$ resonances. Common for all OBEP (including the Paris potential) is that they need a cut-off to take care of the r^{-3} behaviour of the L·S and tensor forces at shorter distances. One way of introducing this cut-off is the following [for example, Bryan and Scott[7]]. The one-pion exchange (OPE) diagram gives a potential

$$V_{OPE}(r) = g_\pi^2 \ f(\mu r) \ \frac{e^{-\mu r}}{r} \tag{3}$$

where g_π^2 is the coupling constant, μ is the pion mass and $f(\mu r)$ is an analytic expression in μ and r containing the tensor force with the r^{-3} singularity. Instead of using this potential, Eq. (3), in the Schrödinger equation, the following cut-off potential is used

$$V(r) = \left[V_{OPE}(r) - g_\pi^2 \ f(\Lambda r) \ \frac{e^{-\Lambda r}}{r} \right] \frac{\Lambda^2}{\Lambda^2 - \mu^2} \tag{4}$$

where Λ is the cut-off parameter. This V(r) is free from the r^{-3} behaviour as $r \to 0$. When this potential is applied to NN this cut-off at short distances ($\Lambda \sim 1-2$ GeV) does not affect the NN phase shifts (apart from the S-waves) very much, because of the strong ω repulsion But, since in $N\bar{N}$ we do not have a strong boson-exchange repulsion at shorter distances, a small variation in the value of this cut-off affects the $N\bar{N}$ phase shifts very much [see Richard et al.[11] or Myhrer and Gersten[12]]. A 2% variation of Λ in the Bryan-Phillips OBEP can change the position of the $N\bar{N}$ resonances by 50-100 MeV [12].

Another comment concerns how far down (towards shorter distances) one can extrapolate the NN OBEP. From NN data the OBEP is not known beyond roughly 0.8 fm. We should also keep in mind that nucleons and mesons are not point particles but we have a radius of ~ 0.8 fm. So the present OBEP with only $\pi + 2\pi + \omega$ exchange, and constant coupling constant, cannot be expected to reproduce the actual NN and $N\bar{N}$ forces at shorter distances. A hint that all is not rosy was given by Dalkarov and myself[13]. If we take the Bryan-Phillips OBEP with their cut-off, we find that the $N\bar{N}$ potential is 1 GeV or deeper at $r \simeq 0.5$ fm in all S-, P- and D-waves. With such a deep potential what about relativity? Buck and Gross[14] claim that relativity (v^2/c^2 terms from the negative energy components) will introduce strong repulsion in NN. But this is a controversial subject on which there is still some work to be done.

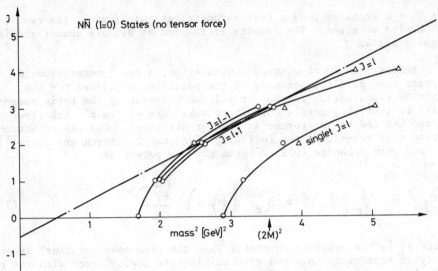

Fig. 3   NN̄ (I = 0) states from Ref. 1 and the leading trajectory in
Ref. 4.   The triangles are from my own recalculation of Ref. 1.

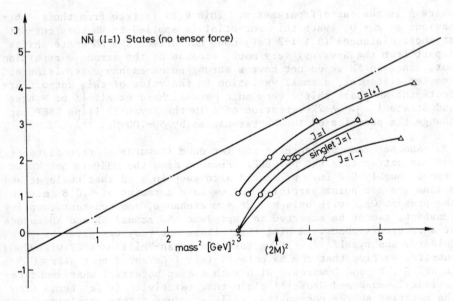

Fig. 4   Same as Fig. 3, NN̄ (I = 1) states

Despite these critical comments which have to be examined, it is evident that the NN OBE forces predict many weakly bound states and resonances (quasinuclear states). However, in the N̄N system their exact positions are not very well predicted.

We now turn to the annihilation part of the N̄N system. Here only very crude models exist. First I will briefly discuss the arguments of Shapiro et al.[1]). They suggest that because the annihilation is of short range ($\sim$ 2M), see Fig. 2, they can estimate the influence annihilation has on the bound states and resonances. In particular, since the N̄N bound state has a radius R $\approx$ 1 fm, whereas the annihilation amplitude f_a, varies rapidly only over a range $\sim$ $(2M)^{-2}$ (= 0.1 fm), the annihilation width of the N̄N states can be estimated by

$$\Gamma \approx f_a \cdot |\Psi(o)|^2 \tag{5}$$

where $\Psi(r)$ is the N̄N wave function from the OBEP.

Here one must face the basic argument of whether or not these states exist! The point is that the bound-state wave function $\Psi(r)$, and its radius R, are given by the strength and the range of OBEP (R is generally smaller than m_π^{-1} because the attraction is mainly given by σ_0 and ω exchange). To find the annihilation widths of these states one must not only know the range of the annihilation forces, but also their strengths as compared to the OBEP strength. It is a mixture of these parameters that will determine if a N̄N bound state or resonance will be wide (washed out), or not, in a particular N̄N partial wave.

To find out at what distances the annihilation is effective Gersten and myself[12]) examined the optical potential model of Bryan and Phillips more closely. If we used their OBEP we found many N̄N resonances [W_0 = 0 GeV in Eq. (2)]. Then we slowly turned on the strength of absorption ($W_0 \neq 0$). Figure 5 illustrates what happened when the absorptive strength was increased. At full strength (W_0 = 8.3 GeV) we can no longer define a resonance from this Argand diagram. The reason for this is that the Bryan and Phillips imaginary potential is so strong that it is felt in all N̄N channels, even at large distances ($\sim$ 1 fm). In a paper with Thomas[14]), it was demonstrated that the resonance pole moves away from the physical region very quickly as the absorptive strength is increased. It just becomes a pole in the scattering amplitude, with no observable consequences.

To avoid the long tail of the absorptive potential in Eq. (2), Dalkarov and the author introduced absorption through a boundary condition à la Feshbach and Weisskopf[15]) but kept the OBEP of Bryan and Phillips. Here the absorption takes place by demanding only incoming waves at the boundary. Again the same question can be asked. At what distance is the annihilation boundary effective? For small values of ℓ ($\leq$ 4) we force the logarithmic derivative of the radial wave function u_ℓ to satisfy

$$u_\ell'/u_\ell \Big|_{r=r_c} \approx -iK \tag{6}$$

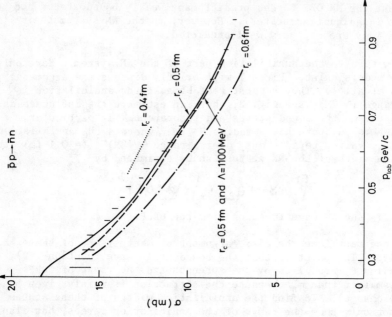

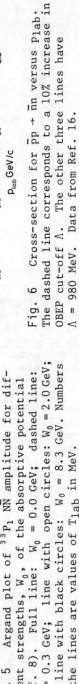

Fig. 5 Argand plot of $^{33}P_1$ N$\bar{\text{N}}$ amplitude for different strengths, W_0, of the absorptive potential (Ref. 8). Full line: $W_0 = 0.0$ GeV; dashed line: $W_0 = 0.3$ GeV; line with open circles: $W_0 = 2.0$ GeV; and line with black circles: $W_0 = 8.3$ GeV. Numbers on the lines are values of T_{lab} in MeV.

Fig. 6 Cross-section for $\bar{\text{p}}\text{p} \rightarrow \bar{\text{n}}\text{n}$ versus P_{lab}. The dashed line corresponds to a 10% increase in OBEP cut-off Λ. The other three lines have $\Lambda = 980$ MeV. Data from Ref. 16.

where $\kappa = \sqrt{M[E - V(r_c)]}$ is the effective wave number at radius r_c where $V(r)$ = OBEP, see Ref. 13. Equation (6) is like a WKB approximation, which is valid if κ does not vary too strongly with r. This is alright for $r_c \simeq 0.5$ fm. With r_c our only free parameter we reproduce $\bar{p}p$ scattering data rather well (cf. Figs. 6-11).

In Fig. 6 we plot the total σ for $\bar{p}p \rightarrow \bar{n}n$. It is a smooth function of energy as also shown by the data[16]. (The fits to data at the lowest momenta in Fig. 6 are not very good, because we have neglected the p-n mass difference, which at $p_{lab} \simeq 300$ MeV/c is 10-15% of available $\bar{p}p$ phase space). However, experimentally one finds an $N\bar{N}$ resonance in this energy region which is seen in σ_{el} and σ_{tot} [17]. In Fig. 7 we show the results from the experiment of Brückner et al.[17] ($\bar{p}p \rightarrow$ charged mesons) where S(1936) is seen. Clearly, it is a small bump on top of a large background, which will be explained when I discuss Figs. 10 and 11.

Figures 8 and 9 show the $d\sigma/d\Omega$ for $\bar{p}p \rightarrow \bar{n}n$ at two energies. The forward dip-bump is seen experimentally[3] and is explained as being due to double spin flip in the one-pion exchange according to Phillips[3]. Obviously one needs to partly shelter the other boson exchanges by a boundary at $r \geq 0.5$ fm in order to get this dip-bump with the present OBEP. Compare also Figs. 8 and 9 to see the very rapid change in $d\sigma/d\Omega$ at backward angles for increasing energy. Finally in Figs. 10 and 11 we show $(d\sigma/d\Omega)_{el}$ for different energies (for $r_c = 0.5$ fm). The data in Fig. 10 are from Eisenhandler et al.[18]. The curves in Fig. 11 also reproduce the data. In Figs. 8, 9 and 10 one can see the strong dependence of $d\sigma/d\Omega$ (backward) on the boundary radius r_c. Notice also that the forward slope of $d\sigma/d\Omega$ in Fig. 10 does not depend strongly on the annihilation radius r_c. It is evident from Fig. 11 that even at small energies, $T_{lab} = 50$ MeV, many partial waves contribute to $d\sigma/d\Omega$. Therefore, if one has a resonance in one partial wave this will only have a minor effect on the total elastic cross-section. (The other partial waves will give most of σ_{el}.)

What happens to the OBEP resonances in this model? The annihilation boundary is at a large radius ($r_c \simeq 0.5$ fm) compared to the nucleon Compton wavelength, and since the annihilation acts at $r_c = 0.5$ fm in all $N\bar{N}$ partial waves, no resonance survives. However, we[13] know that, for example, for D-waves the OBEP plus the centrifugal barrier becomes repulsive for $r \simeq 0.4$ fm. Thus, if in a more realistic model OBEP + $\ell(\ell + 1)/r^2$ can become repulsive in certain $N\bar{N}$ partial waves for $r > r_c$ (the annihilation boundary), then the annihilation region will be sheltered by a repulsive potential and the corresponding resonance (with given LSJT) will survive. Since very little is understood about annihilation I will cut short any speculations, but say that a few OBEP resonances might have a chance of surviving the annihilation. It should be noted that this last statement is not accepted by Shapiro et al. who, as mentioned earlier, think that they can treat annihilation in a perturbative sense.

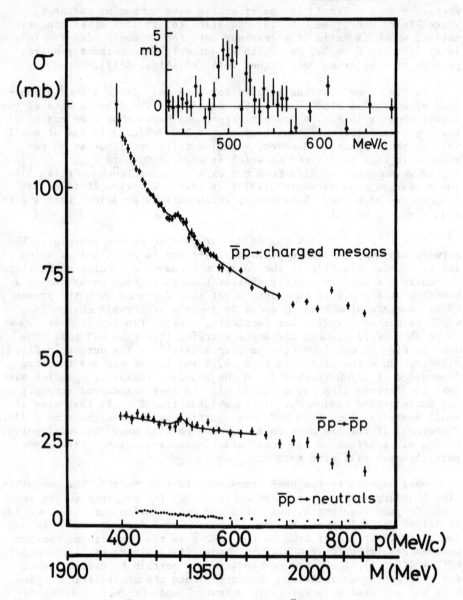

Fig. 7   Data from Ref. 17 (Brückner et al.)

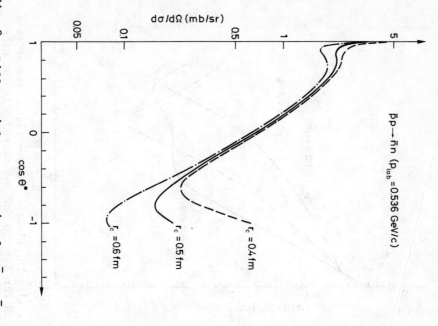

Fig. 8 Differential cross-section for $\bar{p}p \to \bar{n}n$ in c.m. versus cos θ for Plab = 0.536 GeV/c for different annihilation boundaries r_c.

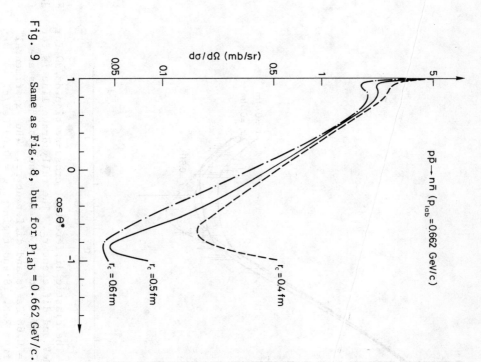

Fig. 9 Same as Fig. 8, but for Plab = 0.662 GeV/c.

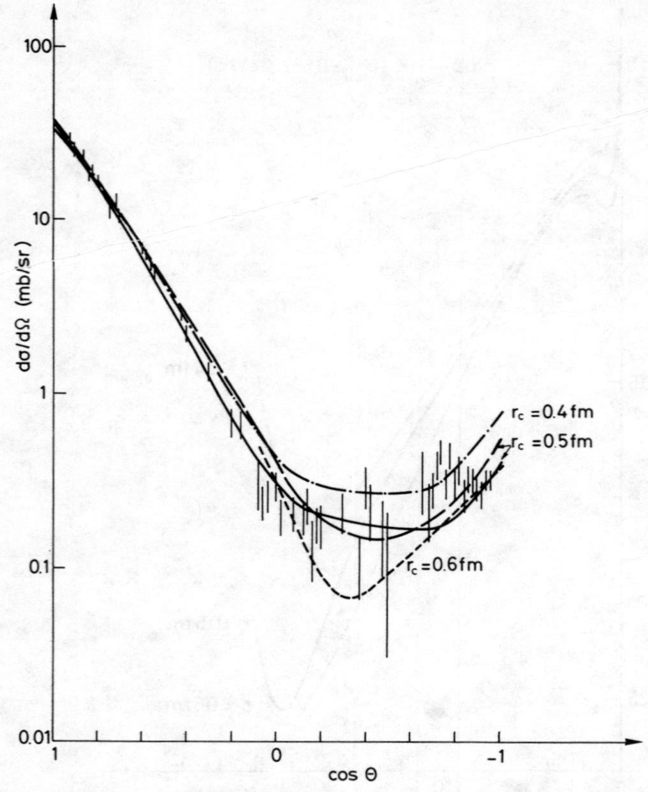

366

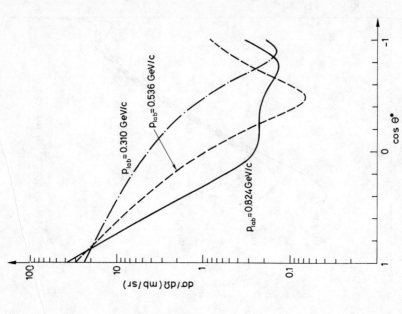

Fig. 10 The elastic differential cross-section $d\sigma/d\Omega$ for $\Lambda = 980$ MeV and different r_c. The fully drawn line is for Plab = 0.73 GeV/c and $r_c = 0.5$ fm. The other lines are for Plab = 0.66 GeV/c and different r_c. The experimental data are from Ref. 18 and their Plab = 0.69 GeV/c.

Fig. 11 The elastic differential cross-section $d\sigma/d\Omega$ for $\Lambda = 980$ MeV and $r_c = 0.5$ fm for different values of Plab

Before I conclude, let me sketch the arguments of the dual-quark model leading to predictions of low-energy p̄p resonances.

3. THE DUAL-QUARK MODEL AND BARYONIUM

The dual-quark model has had some successes with the mesons (qq̄), but as soon as baryons (qqq) were introduced there were problems[19] to face. Before I present the argument concerning baryonium, I shall briefly illustrate the dual-quark model (DQM) arguments as applied to mesons.

Usually one assumes that, for example, Φ(1040) consists mainly of strange-antistrange (ss̄) quarks and π and ρ consist mainly of up and down quarks. In addition one has the Zweig rule which says that, for example, Φ does not like to decay into ρπ . These rules can be understood with the assumption of the <u>dominance</u> of dual, planar diagrams. In Fig. 12a is shown the allowed Φ decay into KK̄ and in Fig. 12b the forbidden decay of Φ into ρπ in terms of quark lines.

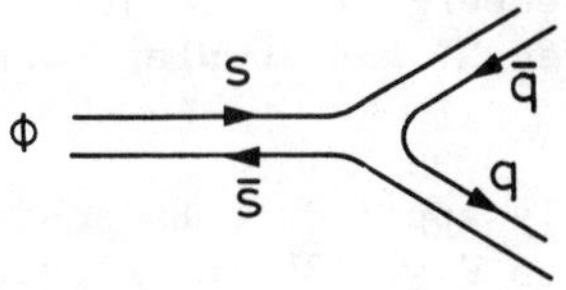

Fig. 12a   Planar diagram
for Φ decay

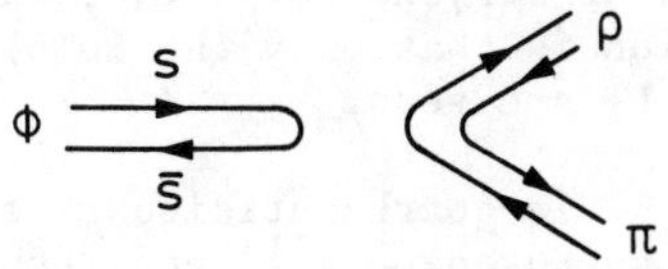

Fig. 12b   Illustration of
a non-planar diagram for
forbidden Φ decay

The diagram in Fig. 12b is disconnected and therefore not allowed according to the Zweig rule. Further we describe ππ scattering by Fig. 13a which can be looked at either in the s-channel, Fig. 13b, producing a ππ resonance, for example ρ, or in the t-channel, Fig. 13c, giving, for example, a ρ exchange. This is in the dual-quark language the same diagram looked at in two different projections.

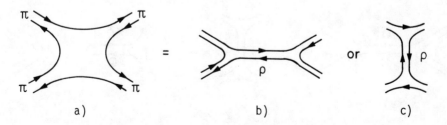

a)   b)   c)

Fig. 13   Dual-quark diagram for ππ scattering

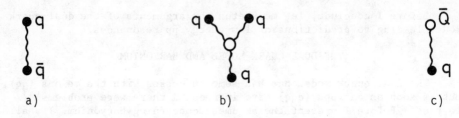

a)   b)   c)

Fig. 14   Examples of mesons and baryons from quarks as discussed in the text

The mesons themselves can be thought of as a quark and an anti-quark held together by all possible gluon exchanges, which in Fig. 14a is represented by a string between q and $\bar{q}$. A baryon has three quarks and, according to gauge-invariance arguments by Rossi and Veneziano[5], the strings representing gluon exchanges between the quarks must be connected in a junction, Fig. 14b. If one freezes one degree of freedom in the baryon, Fig. 14b, one arrives at the quark-antidiquark picture of baryons, Fig. 14c, suggested by Lichtenberg[20]. The justification is that only the SU(6)56-plet (with even L) and 70-plet (with odd L) are seen.

This quark-antidiquark system was recently applied to $B\bar{B}$ scattering by Rosenzweig[21] and Chew[4]. In analogy to Fig. 13b for $\pi\pi$ scattering, we can draw planar, dual diagrams for $B\bar{B}$ scattering -- see Fig. 15a. We find in the s-channel $Q\bar{Q}$ states, called baryonium by Chew[4]. Using the fact that Q can have spin 1 and isospin 1 and that they have relative angular momentum ℓ, Chew finds the possible quantum numbers for the QQ states as in $q\bar{q}$. He then draws Regge trajectories and the leading one is the dashed-dotted line in Figs. 3 and 4 with intercept $-\frac{1}{2}$. The $Q\bar{Q}$ states do not like to decay into mesons, since one will then have a disconnected diagram, Fig. 15c, analogous to Fig. 12b. Therefore baryonium should be narrow. Unfortunately, this is not a dynamical theory, and nothing is known about $Q\bar{Q}$ coupling to mesons other than the rule of the dominace of planar diagrams taken over from mesons.

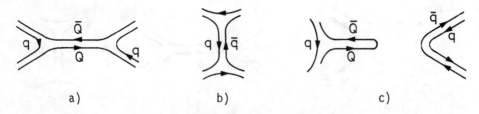

a)   b)   c)

Fig. 15   Dual diagrams of $B\bar{B}$ scattering. In (c) is shown the coupling of $Q\bar{Q}$ to ordinary mesons via a non-planar diagram.

To go one step further, I will discuss the Rossi and Veneziano approach to baryonium. In Fig. 14b one sees that the baryon is a planar object. To describe B$\bar{\text{B}}$ scattering I will draw the quark diagrams in three dimensions. There are three diagrams plus three more because of B$\bar{\text{B}}$ crossing. In Figs. 16a, b and c three s-channel B$\bar{\text{B}}$ scattering diagrams are shown in two projections, where on the r.h.s. are shown the projections along the cuts A, B and C of the l.h.s. diagram. The diagram in Fig. 16a gives the s-channel resonance M_4^J (J = junction)[5], which corresponds roughly to the $Q\bar{Q}$ states of Chew[4]. The s-channel resonance M_2^J in Fig. 16b contains qq with two junctions and M_0^J in Fig. 16c is just a gluon ball.

To find the Regge trajectories for these three objects one can look at the diagram of Fig. 16a in the t-channel projection, Fig. 16d. In Fig. 16d the s-channel intermediate state represents B$\bar{\text{B}}$ annihilation into a single $q\bar{q}$ meson plane. At high energies this ladder graph is controlled by Regge exchange of M_4^J. Using arguments from a simple multiperipheral-like model on Fig. 16d, Rossi and Veneziano find the intercept of M_4^J to be

$$\alpha_4^J(0) = 2\alpha_B(0) - 1 + (1 - \alpha_R) = -\tfrac{1}{2} \tag{7a}$$

since $\alpha_R = \tfrac{1}{2}$ and $\alpha_B = 0$. They[5] find for M_2^J and M_0^J

$$\alpha_2^J(0) = 2\alpha_B(0) - 1 + 2(1 - \alpha_R) = 0 \tag{7b}$$

$$\alpha_c^J(0) = 2\alpha_B(0) - 1 + 3(1 - \alpha_R) = \tfrac{1}{2} \tag{7c}$$

To get the slopes Rossi and Veneziano use arguments from a multiperipheral chain model. Since (for example) Fig. 16d has $q\bar{q}$ as an intermediate state (as does Fig. 13), they find that M_4^J has the same slope as the normal meson (ρ, A_2, ...) trajectory. Their results are shown in Fig. 17, where I have placed S(1936) in a possible position as a star.

Why are these resonances narrow? This is because their junction has to be conserved, for example, M_4^J will not decay easily into mesons, because they do not contain the junction. This means that they[5] propose a new Zweig rule, illustrated in Fig. 16e, where the r.h.s. does not contain a junction, and therefore is "disconnected" like Fig. 12b or Fig. 15c. Not very much is understood about Zweig rule breaking, and the new Zweig rule for baryonium is only a proposal. The dual models for baryons are based on very rough arguments, which do nevertheless result in Regge trajectories. As yet, no explicit calculation has been made to substantiate these arguments.

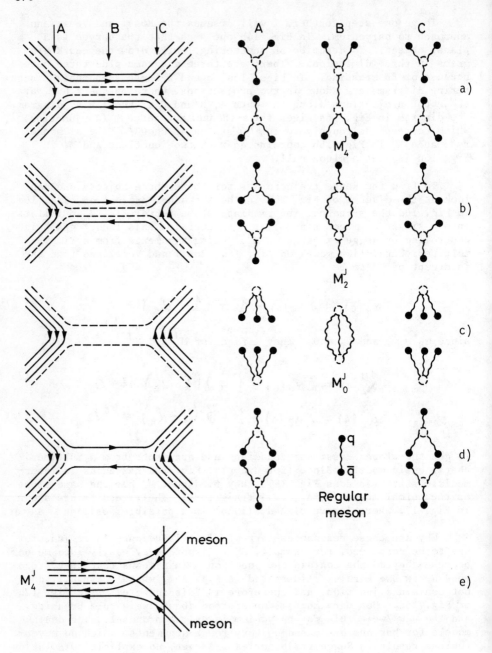

Fig. 16   Dual quark diagrams for B$\bar{\text{B}}$ scattering. The dashed line in the l.h.s. diagrams is the junction which in the r.h.s. diagrams is denoted by a dashed circle.

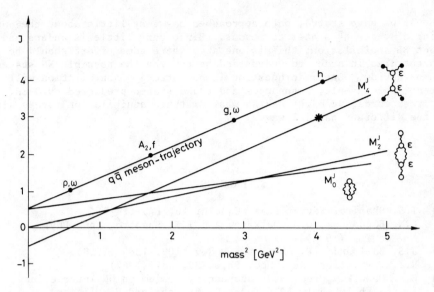

Fig. 17   Regge trajectories as drawn in Ref. 5 (Rossi and Veneziano)

4. CONCLUSIONS

To summarize, we can first compare the differences between the potential and the dual-quark model predictions. The latter gives iso-spin degenerate Regge trajectories, whereas in general the former does not. This is because (for example) pion exchange will give a splitting of the I = 0 and I = 1 $\bar{p}p$ trajectories. Further, the quark models predict I = 2 states which can only be achieved if one speculates about a $\Delta-N(\Delta\bar{N})$ potential giving narrow $\Delta N(\Delta\bar{N})$ states. It should be stressed here that the potential model describes weakly bound $N\bar{N}$ states, i.e. the states consist of a nucleon and an antinucleon (three quarks and three antiquarks in two bags). For strongly bound states the potential (Schrödinger) picture cannot be used. However, it is interesting to note that both models predict $\bar{p}p$ states of roughly the same low spins (J $\simeq$ 2 ± 1), around threshold.

Experimentally one resonance, S(1936) has been seen with a width $\Gamma \leq 4$ MeV [17] -- a number which is of the order of the normal p-n or π^0-π^+ mass difference. We also know that $\Gamma_{el}/\Gamma_{tot} > 20\%$ for this resonance[17]. Lastly, S(1936) is not seen in $\bar{p}p \rightarrow \bar{n}n$ [16]. Other possible candidates for $\bar{p}p$ resonances have been seen [see summary of Montanet[22]], and evidence for $\bar{p}p$ bound states was presented this morning by Pavlopoulos[23]. But as I showed you, many partial waves contribute to $\bar{p}p$ even at very low energies. Therefore a resonance in one of them will be a small bump on a large background, and this makes the experimental search difficult.

As we have stated, both approaches say very little about the coupling of these B$\overline{\text{B}}$ states to mesons. Since very little is understood about B$\overline{\text{B}}$ annihilation, this is one area where some effort should be concentrated in order to understand better why the possible N$\overline{\text{N}}$ states can be narrow. One main question is how strongly annihilation will affect, for example, resonances and bound states predicted by OBEP. In other words, we might ask at what distance annihilation forces act in the different partial waves?

REFERENCES

1) I.S. Shapiro, Soviet Phys. Uspekhi 16, 173 (1973).
   L.N. Bogdanova, O.D. Dalkarov and I.S. Shapiro, Ann. Phys.
      84, 261 (1974).
2) J.S. Ball and G.F. Chew, Phys. Rev. 109, 1385 (1958).
3) R.J.N. Phillips, Rev. Mod. Phys. 39, 681 (1967).
4) G.F. Chew, in Proc. 3rd European Symposium on N$\overline{\text{N}}$ Interactions,
      Stockholm, July 1976 (eds. G. Ekspong and S. Nilsson)
      (Pergamon Press, New York, 1977) p. 515.

   G.F. Chew, preprint LBL-5391 (1976).
5) G.C. Rossi and G. Veneziano, preprint CERN-TH 2287 (1977).
   H.G. Dosch and M.G. Schmidt, preprint CERN-TH 2296 (1977).
6) G.E. Brown and A.D. Jackson, The nucleon-nucleon interactions,
      NORDITA Lectures 1973-1974.
   R. Vinh-Mau, preprint IPNO/TH 77-14, to appear in Mesons in
      nuclei (eds. M. Rho and D. Wilkinson).
7) R.A. Bryan and B.L. Scott, Phys. Rev. 177, 1435 (1968).
8) R.A. Bryan and R.J.N. Phillips, Nuclear Phys. B5, 201 (1968).
9) A. Martin, Phys. Rev. 124, 614 (1961).
10) C. Dover, Proc. 4th Internat. Symposium on N$\overline{\text{N}}$ Interactions,
      Syracuse, NY, 1975 (eds. T.E. Kalogeropoulos and K.C. Wali)
      (Syracuse, NY, 1975) Vol. 2, p. VIII, 37.
11) J.M. Richard, M. Lacombe and R. Vinh-Mau, Phys. Letters 64B, 121
      (1976).
12) F. Myhrer and A. Gersten, Nuovo Cimento 37A, 21 (1977).
13) O. Dalkarov and F. Myhrer, Nuovo Cimento 40A, 152 (1977).
14) W. Buck and F. Gross, Phys. Letters 63B, 286 (1976).
   F. Myhrer and A.W. Thomas, Phys. Letters 64B, 59 (1976).
15) H. Feshbach and V.F. Weisskopf, Phys. Rev. 76, 1550 (1949).
16) M. Alston-Garnjost et al., Phys. Rev. Letters 35, 1685 (1975).
17) A.S. Carroll et al., Phys. Rev. Letters 32, 247 (1974).
   V. Chaloupka et al., Phys. Letters 61B, 487 (1976).
   W. Brückner et al., Phys. Letters 67B, 222 (1977).
18) E. Eisenhandler et al., Nuclear Phys. B113, 1 (1976).
19) L. Rosner, Phys. Rev. Letters 21, 950 (1968).
20) D.B. Lichtenberg, Phys. 178, 2197 (1968).
21) C. Rosenzweig, Phys. Rev. Letters 36, 697 (1976).
22) L. Montanet, to be published in Proc. 5th Internat. Conf. on
      Experimental Meson Spectroscopy, Boston, 1977.
23) P. Pavlopoulos, Talk at this conference.

THE DEUTERON*

A.W. Thomas

TRIUMF, Vancouver, B.C., Canada   V6T 1W5

ABSTRACT

"It is of particular interest to study the simplest nuclear system, that is the diplon [deuteron], which almost certainly consists
of a neutron and a proton."[1]

INTRODUCTION

It is more than forty years since Bethe and Peierls wrote those
words about the deuteron. This conference provides a timely opportunity to review some of the progress that has been made since then. We
shall see that in many ways this progress has been enormous; for example, our knowledge of the form factor from high-energy electron
scattering now extends to very large momentum transfer $(q^2 \sim 100 \text{ fm}^{-2})$.[2]
However, in some of the newer areas of investigation (such as the iso-
bar content of the deuteron) one might summarize the situation by
misquoting Bethe and Peierls's statement that the "[deuteron most of
the time] consists of a proton and a neutron".

Experimentally our knowledge of the deuteron is restricted to a
very few observables, namely its binding energy $(\varepsilon_d = 2.22461(7) \text{ MeV})$,[3]
spin (J=1), magnetic moment $(\mu_d = 0.857406(1) \text{ n.m.})$,[4] and quadrupole
moment $(Q = 0.2860(15) \text{ fm}^2)$.[5] In addition, electromagnetic interactions are so basic that the electric and magnetic form factors,
measured by electron scattering, almost have the status of observables.
Finally there are a number of "not-so-observable", but very important,
properties, such as the d-state probability (P_D), the asymptotic $(r\rightarrow\infty)$
d-wave to s-wave ratio η (or AD/AS), and now one might also add the
probability for $\Delta\Delta$ or NN* components. There are, of course, a number
of excellent reviews of these properties which complement the present
work—particularly that of Sprung at the Laval Conference.[6]

Non-relativistically the deuteron is described as a neutron-
proton bound state, with wave function

$$\Psi_d^{1M}(\vec{r}, \vec{\sigma}_1, \vec{\sigma}_2) = \frac{u_0(r)}{r} \, \mathcal{Y}_{01}^{1M}(\hat{\vec{r}}, \vec{\sigma}_1, \vec{\sigma}_2) + \frac{\omega(r)}{r} \, \mathcal{Y}_{21}^{1M}(\hat{\vec{r}}, \vec{\sigma}_1, \vec{\sigma}_2). \qquad (1)$$

If one knew the N-N force, and if non-relativistic quantum mechanics
were exact, the radial wave functions u_0 and ω could be calculated
exactly. Unfortunately the N-N force is still poorly known [at least
for $r \lesssim (1-2)$ fm], and much of what we have to say concerns what the
observable deuteron properties described above tell us about u_0 and
u_2. Before proceeding with this discussion, it is worth while to
remind the reader of the relevance of Ψ_d to nuclear physics.

Lovelace has shown quite generally that the fully off-shell t-
matrix, in a N-N channel which has a bound state, is dominated by

---

*Supported by a grant from the National Research Council of Canada.

that pole as $E \to -\varepsilon_d$.[7] That is, in the $^3S_1$-$^3D_1$ channel,

$$t(\vec{p},\vec{p}';E) \xrightarrow[E \to -\varepsilon_d]{} \frac{[(p^2 + \alpha^2)\Psi_d(\vec{p})][\Psi_d^*(\vec{p}')(p^2 + \alpha^2)]}{E^+ + \varepsilon_d}, \qquad (2)$$

where $\alpha^2 = m_N\varepsilon_d$. Now in order to calculate nuclear binding energies (e.g. for the triton E_T) one needs the (off-shell) N-N t-matrix, at small negative energies.[8-10] Thus one might expect the pole term [Eq.(2)] to dominate the calculation. That this is dramatically correct was shown by Afnan and Read,[10] who used the bound-state wave function for the Reid potential[11] to construct a rank one separable interaction which gives -7.15 MeV for E_T ($^1S_0$ and $^3S_1$-$^3D_1$ N-N interactions only), compared with the exact result of -7.02 MeV. A striking dependence on the properties of the deuteron (particularly P_D) has also been demonstrated in nuclear matter calculations.[12] Clearly then, a knowledge of Ψ_d is of considerable importance from the point of view of nuclear structure physics.

We now turn to a detailed analysis of the "observables" mentioned above—with the intention of extracting u_0 and ω. (In particular, as we shall discuss later, many nuclear properties depend critically on P_D, so that we place most emphasis on this parameter.) Unfortunately, it will soon become obvious that this task is not simple. For example, our efforts will sometimes be frustrated by exchange currents. (This leads to the idea, discussed briefly in the section on isobars, of including meson degrees of freedom explicitly in Ψ_d). In other cases, the reaction proposed as a probe is too complicated to be reliable. There are, however, a few examples of real progress, with (at least a promise of) a significant advance in our knowledge of Ψ_d.

SPIN AND MAGNETIC MOMENT

It is fair to say that theory and experiment are in exact agreement as to the spin of the deuteron. At first glance the situation for the magnetic moment (μ_d) seems almost as good. The experimental value (0.857 406 n.m.)[4] differs from the naive impulse approximation ($\mu_p + \mu_n$ = 0.879 696 n.m.) by only -0.0223 n.m. This can be exactly accounted for in terms of an orbital contribution from the d-state component of Ψ_d,

$$\Delta\mu_d \equiv \mu_d - \left(\mu_p + \mu_n\right) = -\frac{3}{2}\left(\mu_p + \mu_n - \frac{1}{2}\right)P_D, \qquad (3)$$

if P_D is 4%. If there were no other corrections, this would provide a very direct measure of this elusive quantity.

Unfortunately there is a host of corrections, many of which cannot be accurately calculated, but all of which are expected to be the same size as $\Delta\mu_d$. Feshbach showed that the existence of a spin-orbit force, together with minimal coupling, produced a magnetic moment-like term in the Hamiltonian [see, however, the correction in Eq.(2.6) of Ref. 6].[13] Depending on the potential model, this produces a correction $(\Delta\mu)_{LS} \in$ (0.0024, 0.0093) n.m.,[6] which corresponds to an increase in P_D of from 0.4% to 2.0%.

Some of the larger meson current corrections are shown in Fig. 1. Many of the early estimates of the $\rho\pi\gamma$ process[14-18] suffer from a lack of knowledge of the coupling constant $g_{\rho\pi\gamma}$, which has only been determined recently.[19] The overall impression is that the contribution of Fig. 1(a) is rather ill determined, but probably lies between ±0.02 n.m. (Horikawa et al.[16] found 0.013 n.m. and 0.027 n.m. for the Ueda-Green I and Bryan-Scott III potentials, respectively.) For the ω-exchange in Fig. 1(b) Blankenbecler and Gunion found ~0.017 n.m.,[17] Jackson et al.[20] found a correction from the pair [Fig. 1(a)] and recoil currents [Fig. 1(d)] of 0.05 n.m. [Note that the former (with a pion exchanged) is only equivalent to the seagull term of Fig. 1(e) in the soft pion limit.[21]]

Finally, we observe that even a very small percentage of isobars in the deuteron could produce a large correction to μ_d (of up to 6%).[22] Of course, this conclusion cannot really be quantitative since neither the $\Delta\Delta$ probability of the deuteron nor the magnetic moment of the Δ has been measured. In summary, what started out as an apparently innocuous, well understood property of the deuteron has turned into a nightmare.

ASYMPTOTIC D- TO S-WAVE RATIO (η)

Although this ratio is not usually considered an observable property of the deuteron, we shall see that there is now a very clever technique to measure it. The importance of η, which is defined by[23]

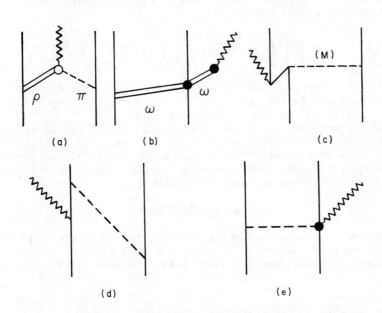

(a)   (b)   (c)

(d)   (e)

Fig. 1. Various exchange current corrections to the deuteron magnetic moment.

$$\eta = \lim_{p^2 \to -\alpha^2} \left[\Psi_2(p)/\Psi_0(p) \right] , \tag{4}*$$

as a constraint on the intermediate and long-range tensor force is seldom emphasized strongly enough.

Most potential models with a OPE tail, which reproduce Q and P_D (within reasonable limits to be described soon) give $\eta \in (0.025, 0.032)$. This is in agreement with the model-independent value of Wong (0.029),[24] obtained many years ago using only the OPE cut, and an effective range expression of the low-energy S-wave NN phase shifts, as input to a dispersion relation. It would be valuable if this work could be updated, and some reliable limits placed on the resulting value of η. (One might then hope to turn the problem around and use η to eliminate some potential models.)

An exciting technique for measuring η has recently been described by Knutson et al.[25a] The idea is that sub-Coulomb (d,p) reactions are very sensitive to the tail of the deuteron wave function. By following the classical Coulomb trajectories, and realizing that the positive deuteron quadrupole moment means that its spin and matter distribution are related, they establish that the stripping probability is greatest when the deuteron spin is aligned along the $\vec{q}$ (momentum transfer) direction. Moreover, within the semi-classical approximation they show that the tensor polarization $T_{2m}(\theta)$ is just A $Y_2^m[(\pi+\theta)/2,0]$, with θ the scattering angle, and A is determined entirely by η.

Using a more rigorous quantum mechanical description of stripping, Johnson et al.[25b] find a similar result, with the polarization determined by a single parameter D_2:

$$D_2 = \frac{1}{15} \left(\int_0^\infty dr \ r^3 \ \omega(r) \right) \Big/ \left(\int_0^\infty dr \ r \ u(r) \right) . \tag{5}$$

However, to a very good approximation one can use the asymptotic form of Ψ_d to evaluate Eq.(5), with the result that D_2 is proportional to η—thus confirming the semi-classical result.

The current best value for η, extracted from 9 MeV (d,p) tensor analysing powers on Pb is 0.023 $\pm$ 0.003.[25a] Improved statistical accuracy at a lower energy (say 7 MeV) where strong distortion is even less significant, should eventually result in an accuracy of (3-4)% for η.[25c] This would be an extremely valuable result.

THE QUADRUPOLE MOMENT

This very important property measures the departure of the matter distribution from the spherical along the z-axis $[\langle (3z^2-r^2)/4 \rangle]$ when the deuteron has spin projection $M_z = +1$. In terms of (u,ω) this is

---

*Or more usually by the asymptotic condition

$\{u(r) \to N e^{-\alpha r}, \ \omega(r) \to \eta \ N[1 + 3(\alpha r)^{-1} + 3(\alpha r)^{-2}]e^{-\alpha r}\}$, as $r \to \infty$.

$$Q = 50^{-1/2} \int_0^\infty dr \; r^2 \; \omega(r) \left(u(r) - 8^{-1/2} \; \omega(r) \right). \tag{6}$$

The "experimental" determination of Q is rather indirect. One actually obtains the electric quadrupole interaction constant $\left(\frac{e \, q \, Q}{h} \right)$ from careful measurements of transition energies for HD and D_2 molecules in strong magnetic fields.[26] One must then be able to calculate "q" (the derivative of the electric field at the deuteron along the molecular axis) very accurately, using the best variational atomic wave functions. Reid and Vaida obtained the result $Q = 0.2860 \pm 0.0015$ fm^2.[5] Although these authors were extremely careful, as in any variational calculation one obtains more accurate eigenvalues than eigenfunctions. Therefore, it is very difficult to exclude the possibility of larger deviations of Q from the quoted value than the error suggests. Certainly an independent check on Q, using say polarized electron scattering (see ELECTRON SCATTERING below), could be very useful.[27]

Even allowing for possible meson exchange corrections [which seem better understood here, and amount to about 0.009 fm^2 from Figs. 1(c) and 1(d)[20,27,28]], it has proven difficult to reproduce such a large value of Q with purely local potentials. Thus Q has served as a very important constraint on all models of the N-N interaction. Perhaps the most valuable constraint of this type that it has yielded is a model-independent lower limit on P_D (i.e. $P_D \gtrsim 3.3\%$).[29] Since this has already been discussed in detail in several reviews of deuteron properties,[6,9] we say no more.

ELECTRON SCATTERING

The theory of electron scattering from the deuteron has been very thoroughly set out by Gourdin,[30,31] and others,[32,33] including a number of relativistic corrections [$0(q^2/4M_d^2)$, with $q^2 = \vec{q}^2 - q_0^2$ the 4-momentum transfer]. Ignoring these corrections (which amount to $\lesssim 10\%$ for $q^2 < 20$ fm^{-2},[34] but are of course essential at the large momentum transfers now available[2]) for the sake of simplicity, one finds that the cross-section for unpolarized electrons is

$$d\sigma/d\Omega = \left(d\sigma/d\Omega \right)_{Mott} \left[G_{ES}^2 \left(F_0^2 + F_2^2 \right) \right.$$
$$\left. + G_{MS}^2 \; F_{mag}^2 \left(1 + 2 \; \tan^2(\theta/2) \right) \right]. \tag{7}$$

The first term is the electric scattering, which can be separated from the second (magnetic) term by measuring at fixed q^2 for different θ (by changing the electron energy). Information about the deuteron structure is contained in the monopole and quadrupole form factors

$$F_0(q^2) = \int_0^\infty (u^2 + \omega^2) \; j_0(qr/2) \; dr, \tag{8}$$

$$F_2(q^2) = 2 \int_0^\infty \omega(u - 8^{-1/2}\omega) j_2(qr/2) \; dr. \tag{9}$$

One immediate problem is that G_{ES} is the sum of the neutron and proton electric form factors, and the former can only be found from experiments on deuterium. Indeed, Galster *et al.* in their determination of G_{En} stressed the ambiguities due to a lack of knowledge of Ψ_d[34]—we refer to this work for a clear exposition of this problem. While this is a nuisance for $q^2 < 20$ fm^{-2}, in the region of the recent SLAC measurement[2] (up to 100 fm^{-2}) G_{En} is essentially unknown, and one must rely on models such as that of Iachello *et al.*[35] The ambiguity is not so bad as to hide the fact that exchange current effects [in particular Figs. 1(a) and 1(c)] are much smaller than had been suggested[18]—partially because of the neglect of form factors[36] by Chemtob *et al.* However, it seems hopeless at this stage to attempt to obtain information about Ψ_d from the high -q^2 data.

In Fig. 2 we show the contributions F_0^2 and F_2^2 to the electric deuteron form factor. The deep minimum in F_O occurs at about 16 fm^{-2} for most realistic Ψ_d (with the Paris model a notable exception), but its precise structure contains valuable information on the short-range behaviour of the deuteron wave function. Unfortunately in this most interesting region the quadrupole form factor is large, and hides the variation of F_O. As several authors have observed, F_2 itself is of interest in this region—in particular, its height is related to P_D (cf. next section).[27,37] In addition, $[F_2(q^2)/q^2]$ tends to $(2^{1/2}Q/6)$ as q^2 goes to zero, and thus one could in principle obtain an independent measure of the deuteron quadrupole moment, without the problems of calculating a molecular field gradient (as discussed in the preceding section).[9,27]

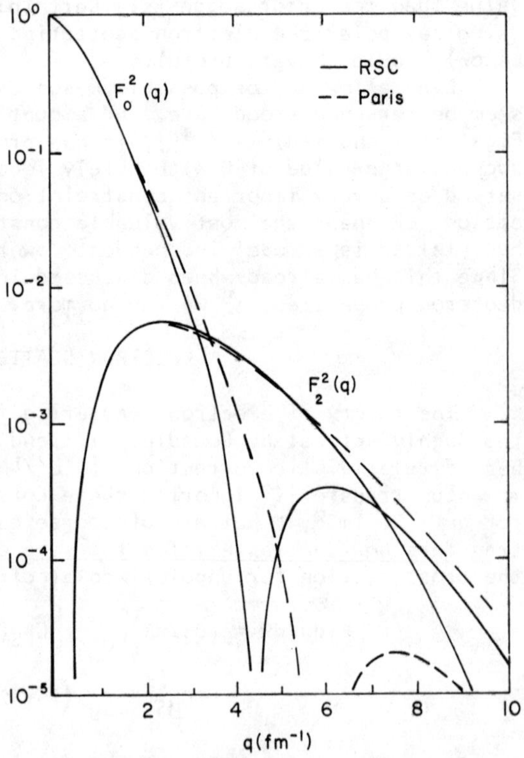

As they stand the ideas in the last paragraph for accurately measuring $F_2(q^2)$ are unrealistic, because of the lack of knowledge of G_{En}.

Fig. 2. The charge and quadrupole form factors of the deuteron for the Reid soft core[11] (RSC) and Paris (see the review by Vinh Mau at this conference) potentials.

A more promising approach which eliminates this unknown is to measure the tensor polarization of the elastically scattered electron. (This was originally suggested by

Moravcsik and Ghosh[38] as a tool for determining P_D—an idea we return
to in the next section.) Defining "x" to be $F_2(q^2)/F_0(q^2)$, and
eliminating the magnetic scattering by an appropriate extrapolation
in θ at fixed q^2,[9],[39] one finds

$$T_{20}^{(e)}(x) = \left(2x + x^2/2^{1/2}\right) \Big/ \left(1 + x^2\right) . \tag{10}$$

The superscript "e" indicates the above extrapolation has been carried out. Clearly the zero in $F_0(q^2)$ is characterized by a value of
$2^{-1/2}$ for $T_{20}^{(e)}$. Further work to exploit this approach, and to understand how exchange currents can affect matters, is certainly called
for. Finally we mention that a more systematic approach to the
problem of relating Ψ_d to the form factor, such as defining a "high-
energy component of the deuteron wave function",[37] may be useful in
this analysis.

THE D-STATE PROBABILITY

Of all the basic deuteron properties this has proven to be the
most significant and the most elusive. In 1963 Moravcsik[40] wrote
that "there is no reliable method of determining [P_D] accurately". As
we indicate very briefly below, in spite of many ingenious suggestions
this statement is still true today!
Before reviewing some of these suggestions, a few words on the
significance of P_D in nuclear physics seems appropriate. The most
reliable nuclear binding energy calculation possible with a realistic
N-N interaction is undoubtedly the evaluation of E_T—the triton binding energy.[9],[10],[41],[42] A series of model calculations by Phillips[43]
and Afnan *et al.*[12] established that separable potentials which produce
identical low-energy scattering parameters and deuteron properties
(ε_d, Q), but different values of P_D, gave rise to a large variation in
E_T. (In fact, they found $d|E_T|/dP_D$ equal to -0.4 MeV/% and -0.6 MeV/%,
respectively.) These model results seem to have been confirmed by
recent calculations using realistic N-N interactions. Indeed, the
OBE potentials which usually have lower d-state probability [$\sim(4-5)\%$]
than purely local potentials [$\sim(6-7)\%$] tend to bind the triton by
about an extra MeV (~8 MeV, compared with ~7 MeV).[42] In nuclear
matter the calculations are probably less reliable, but the tensor
force plays a greater role in determining the saturation properties,
and the dependence on P_D is very striking!
To summarize, there are other uncertainties in nuclear binding
energy calculations (e.g. possible short distance non-locality[44]), and
P_D is only one measure of the strength of the intermediate range
tensor force. Nevertheless, a good determination of P_D would impose a
very significant constraint on any potential model, and dramatically
reduce existing ambiguities in the calculation of nuclear binding
energies. Let us now consider some of the schemes which have been
suggested to determine P_D.

Tensor polarization in electron scattering
Because our ignorance of G_{En} does not affect calculations of $T_{20}^{(e)}$
(as defined earlier), this seems a more promising method to obtain P_D

than measuring the differential cross section. Several groups have observed that at low momentum transfer $[q^2 \sim (1-8)\,fm^{-2}]$[9,27,37-39] the quadrupole form factor, and hence $T_{20}^{(e)}$, is sensitive to P_D. Unfortunately, this sensitivity reaches a minimum in the most likely range, namely (4-7)%. Lomon[27] has proposed a measurement to an accuracy of order (0.1-0.2)% in $T_{20}^{(e)}$, which is apparently feasible with present technology, and is under consideration at Bates.

A very important result of this measurement would be (essentially) another measure of Q—without relying on HD and D_2 wave functions. It is, however, very difficult to believe that one could obtain a model-independent value of P_D. For example, Afnan and Read[46a] get $T_{20}^{(e)}$ at $q^2 = 4\,fm^{-2}$ to be 0.5794 for the H-J[47] potential (P_D = 7.02), 0.5083 for the B-S[48] potential (P_D = 5.47), but 0.5816 for the dT-SC[49] (P_D = 5.45). Indeed, the most thorough calculation of this type (including enough meson exchange corrections to give the experimental value of Q),[46b] has established that almost all realistic potentials (with P_D from 4.5% to 7.0%) then give the same value of $T_{20}^{(e)}$ (e.g. at $q^2 = 5\,fm^{-2}$ Lomon and Moniz quote 0.63 ± 0.03[46b]). They do, however, find a couple of isolated models outside of this region, so that an unexpected experimental result could have serious implications.

The d(γ,p)n reaction

In a recent paper Arenhövel and Fabian[50] discussed the role of the tensor force, and in particular the value of P_D, in forward photodisintegration. They observed that the forward cross section due to E1 absorption would be zero without the d-state (and E1 is the dominant process between 20 MeV [76%] and 80 MeV [66%]). Because of the E1 dominance, meson exchange currents (and isobars) are not expected to be important.

Careful measurements of this reaction in the forward direction from 20 to 120 MeV[51] have revealed a cross section systematically (30-40)% lower than that calculated with the H-J,[47] RSC[11] or BS[48] potentials.[50] The latter, with its lower value of P_D did seem the best of the three, however. Arenhövel et al. concluded that all three potentials have too much tensor force. An approximate calculation* with the Holinde-Machleidt relativistic OBE potential[52] (P_D = 4.3%) produced much better agreement, as shown in Fig. 3. (Recently Lomon has also shown that a BCM with P_D = 4.5% can reduce the discrepancy by 50%.[50])

In summary, this reaction does seem to provide some measure of the d-state wave function. It is also attractive because it involves the well-known electromagnetic interaction. However, the calculations have not yet reached the sophistication of the electron scattering work described above, and much work must be done before this can seriously be considered a tool to determine P_D.

Pion production (pp→π^+d)

It has been realized for some time that s-wave pion production in this reaction is very sensitive to cancellations between the S and D waves of the deuteron.[53] Indeed Thomas and Afnan suggested that an

---

*It was approximate because the H-M potential[52] is given in momentum space, and Arenhövel et al. use co-ordinate space wave functions.

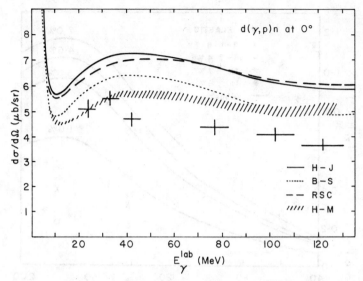

Fig. 3. Comparison of the calculations of Arenhövel *et al.*[50]
(for several realistic N-N potentials[11,47,48,52]) of forward
photodisintegration, with the data of Hughes *et al.*[51]

experimental determination of the threshold cross section could
severely constrain various potential models, and possibly help to determine P_D.[54] Unfortunately this reaction involves very high momentum
transfer, and in consequence is even more sensitive to ambiguities in
the wave functions at small internucleon separations,[55] and in the
off-shell behaviour of the πN interaction.[56]

Very recently there has been a vast improvement in the quality of
the data for $pp \leftrightarrow \pi^+d$.[57,58] Calculations of these cross sections at
40, 50 and 60 MeV pion energy, within the model of Goplen *et al.*,[56]
tend to favour a rather low value of P_D, of about 4% [as did the
$\gamma(d,p)n$ reaction above].[57] However, there have been no studies of the
dependence of this result on either the model used for the calculation
or the off-shell behaviour of the N-N and πN interactions. In view of
the discussion above for purely s-wave pion production, and the large
differences between alternative theories for the much simpler πD
elastic process (see below), it seems unlikely that one could determine
P_D in this way.

Tensor polarization (T_{20}) in elastic πD scattering

Some years ago Gibbs[59] established within his fixed scatterer
approximation (FSA) multiple scattering theory that the tensor polarization in backward πd elastic scattering, $T_{20}(180°)$, was very sensitive
to P_D. In fact, it was essentially zero over a wide range of energies
unless P_D was non-zero, and at a given energy increased monotonically
with P_D. Gibbs's most recent results[60] are shown in Fig. 4, together
with the results of a relativistic three-body (Faddeev) calculation by
Rinat and Thomas.[61] Clearly both theoretical models are in qualitative
agreement concerning the effect of increasing P_D. Unfortunately there

382

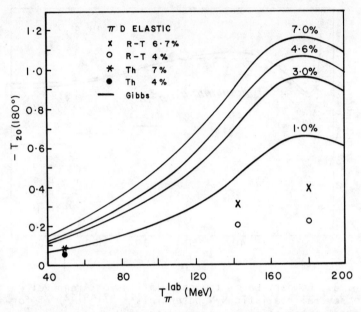

Fig. 4. A comparison of the predictions of several models [Gibbs,[59],[60] Rinat *et al.* (R-T)[61] and Thomas (Th)[61]] for the tensor polarization (T_{20}) in backward πd elastic scattering.

is a considerable quantitative difference. (Note that these calculations produce very similar differential cross sections.) While this experiment, which has been proposed at SIN,[62] is extremely interesting from the point of view of relativistic three-body scattering theory, it is clear that for some time one is not going to determine P_D this way.

Summary

We have examined several techniques for determining P_D, but considerable personal bias was involved in the choice! Amongst the many other (equally undefinitive) attempts to measure P_D we mention (i) high-energy p-d elastic scattering (favours $P_D = (7-9)$%, but could be as low as 2%),[63] (ii) tensor polarization in low-energy n-d scattering, where one can perform sophisticated Faddeev calculations,[60],[64] and (iii) a sophisticated p-d polarization experiment (3.6 GeV/c deuterons on a polarized hydrogen target) at high energy[65] (which gave $P_D = 6.5 \pm 2.0$%).

Several of the techniques which were discussed in detail favoured rather low values of P_D (~4%), which would be rather attractive from several points of view.[27] For example, this would bind the triton by over 8 MeV[9],[66]—leaving a relatively small (and believable) correction due to three-body forces and relativistic effects. Unfortunately there does not seem to be any conclusive experiment on this matter. In fact, we may eventually have to rely on the sort of theoretical arguments proposed by Holinde in a contribution to this conference.[28] That is, if we can understand the NNπ form factor, and extend our knowledge of the N-N force due to heavier meson exchanges (and for the tensor force, particularly ρ-exchange) down to 1 fm or so, it may be possible to place a reasonable theoretical constraint on P_D. This program should certainly be pursued vigorously.

ISOBARS

This is a topic which has aroused much interest and discussion recently,[67] and to which we cannot possibly do justice in the space available. Hopefully with the availability of meson factories, our ability to calculate processes involving <u>real</u> isobars will rapidly improve—with a consequent improvement in virtual processes, too. Thus far there is very little evidence for their presence in nuclei, despite many experiments (e.g. see the review by Weber[68]). The case which is canonically quoted as hard evidence is, of course, the 10% exchange current correction in threshold np $\rightarrow$ γd.[69-71] Unfortunately this correction, originally quoted as 1.45% in the amplitude, should now be reduced by at least 50% with the inclusion of ρ-exchange and a reliable d-state wave function,[72] and is probably the least definite of the three effects included in Ref. 69.

Almost all calculations of isobars in nuclei ignore their width (e.g. the recent calculation of a 100 MeV bound state of the $\Delta\Delta$ system[73]). It is possible that three-body calculations along the line proposed by Weber[74] will eventually allow this restriction to be removed. This brings us to a very important preprint by Goldhaber,[75] who stresses that not enough attention has been paid to the obvious fact that a "Δ" in a nucleus (e.g. the deuteron) is far off-mass-shell. Applying some simple corrections, he finds that experiments looking for the (usual) ~1% $\Delta\Delta$ component of the deuteron wave function need a much greater sensitivity than heretofore suggested. In any case, this is a field in which we can expect dramatic progress in the next few years.

CONCLUSION

In this paper we have mentioned several cases (e.g. sub-Coulomb (d,p) and the tensor polarization in low -q^2 electron scattering) where a presently feasible experiment should lead to significant progress in our understanding of the only bound state of the N-N system. Unfortunately there are many other examples (such as the attempts to determine P_D in a model-independent way, or to calculate such a simple property as μ_d) where George Bernard Shaw's comment that "science is always wrong; it never solves a problem without creating ten more" seems most appropriate. It is our hope that by clearly pointing out where such problems exist one is contributing to their eventual resolution.

384

ACKNOWLEDGEMENTS

It is a pleasure to thank my many colleagues at TRIUMF and else-where for their advice and encouragement. In particular, I am indebted to Drs. I.R. Afnan, W. Gibbs, B. Gibson, W. Grüebler, E. Lomon, L. Mathelitsch, H.P. Noyes, D. Riska, H. Weber and H. Zingl for their assistance (in one way or another) in the preparation of this manuscript.

REFERENCES

1. H. Bethe and R. Peierls, Proc. Roy. Soc. (London) <u>148</u>, 146 (1935).
2. R.G. Arnold, B.T. Chertok, E.B. Dally, A. Grigorian, C.L. Jordan, W.P. Schütz, R. Zdarko, F. Martin and B.A. Mecking, Phys. Rev. Lett. <u>35</u>, 776 (1975).
3. R.C. Greenwood and W.W. Blake, Phys. Lett. <u>21</u>, 702 (1966).
4. I. Lindgren, in Alpha, Beta and Gamma Ray Spectroscopy, Vol. 2, ed. K. Siegbahn (North-Holland, Amsterdam, 1968), p.1623.
5. R.V. Reid and M.L. Vaida, Phys. Rev. Lett. <u>29</u>, 494 (1972); Erratum <u>34</u>, 1064 (1975).
6. D.W.L. Sprung, in Problèmes à Petit Nombre de Corps dans la Phy-sique de Noyau et de Particules Élémentaires, ed. R.J. Slobodrian *et al.* (Les Presses de l'Université Laval, Québec, 1975), p.475.
7. C. Lovelace, Phys. Rev. <u>135B</u>, 1225 (1964).
8. M.G. Fuda, Phys. Rev. <u>166</u>, 1064 (1968).
9. J.S. Levinger, in The Two and Three Body Problem (Springer-Verlag, New York, 1974), p.88.
10. I.R. Afnan and J.M. Read, Phys. Rev. <u>C8</u>, 1294 (1973).
11. R.V. Reid, Ann. Phys. <u>50</u>, 411 (1968).
12. I.R. Afnan, D.M. Clement and F.J.D. Serduke, Nucl. Phys. <u>A170</u>, 625 (1970).
13. H. Feshbach, Phys. Rev. <u>107</u>, 1626 (1957)
14. R.J. Adler and S.D. Drell, Phys. Rev. Lett. <u>13</u>, 349 (1964).
15. R.J. Adler, Phys. Rev. <u>141</u>, 1499 (1966).
16. Y. Horikawa, T. Fujita and K. Yazaki, Phys. Lett. <u>42B</u>, 173 (1972).
17. R. Blankenbecler and R.J. Gunion, Phys. Rev. <u>D4</u>, 718 (1974).
18. M. Chemtob, E. Moniz and M. Rho, Phys. Rev. <u>C10</u>, 344 (1974).
19. B. Gobbi *et al.*, Phys. Rev. Lett. <u>33</u>, 1450 (1974).
20. A.D. Jackson, A. Lande and D.O. Riska, Phys. Lett. <u>55B</u>, 23 (1975).
21. M. Rho, Meson Fields in Nuclei, Erice Lectures (1976).
22. H. Arenhövel, M. Danos and H.T. Williams, Phys. Lett. <u>31B</u>, 109 (1970); Nucl. Phys. <u>A162</u>, 12 (1971).
23. H.F.K. Zingl, Separable Interactions, Universitat Graz preprint (1975); L. Mathelitsch (private communication).
24. D.Y. Wong, Phys. Rev. Lett. <u>2</u>, 406 (1959).
25. a) L.D. Knutson, E.J. Stephenson and W. Haeberli, Phys. Rev. Lett. <u>32</u>, 690 (1974); L.D. Knutson and W. Haeberli, *ibid.* <u>35</u>, 558 (1975).
    b) R.C. Johnson *et al.*, Nucl. Phys. <u>A208</u>, 221 (1973).
    c) L.D. Knutson and W. Haeberli (private communication).

26. R.F. Code and N.F. Ramsey, Phys. Rev. $\underline{A4}$, 1945 (1971).
27. E. Lomon, Is the D State of the Deuteron about 4%?, contribution C33 to this conference, and private communication.
28. K. Holinde, Influence of the πNN Form Factor on Two-Nucleon Data, contribution C24 to this conference.
29. S. Klarsfeld, Orsay preprint IPNO/TH 74-5 (1974).
30. M. Gourdin, Nuovo Cimento $\underline{28}$, 533 (1963); $\underline{32E}$, 493 (1964).
31. M. Gourdin, Nuovo Cimento $\underline{36}$, 129 (1965).
32. F. Gross, Phys. Rev. $\underline{142}$, 1025 (1966) and references therein; *ibid.* $\underline{152E}$, 1517 (1966).
33. J.L. Friar, Ann. Phys. $\underline{81}$, 332 (1973).
34. S. Galster, H. Klein, J. Moritz, K.H. Schmidt, D. Wegener and J. Bleckwein, Nucl. Phys. $\underline{B32}$, 221 (1971).
35. F. Iachello, A.D. Jackson and A. Landé, Phys. Lett. $\underline{43B}$, 191 (1973).
36. M. Gari and H. Hyuga, Nucl. Phys. $\underline{A264}$, 409 (1976).
37. L. Mathelitsch and F.K. Zingl, Relations between the Deuteron Form Factors and the Wave Functions, contribution C34 to this conference and Graz preprint UTP 02/77 (1977); Phys. Lett. (to be published).
38. M. Moravcsik and P. Ghosh, Phys. Rev. Lett. $\underline{32}$, 321 (1974).
39. T.J. Brady, E.L. Tomusiak and J.S. Levinger, Can. J. Phys. $\underline{52}$, 1322 (1974).
40. M. Moravcsik, The Nucleon-Nucleon Interaction (Clarendon Press, Oxford, 1963), p.40
41. C. Gignoux and A. Laverne, Phys. Rev. Lett. $\underline{29}$, 436 (1972).
42. I.R. Afnan and J.M. Read, Phys. Rev. $\underline{C12}$, 293 (1975).
43. A.C. Phillips, Nucl. Phys. $\underline{A107}$, 209 (1968).
44. A.W. Thomas and I.R. Afnan, Phys. Lett. $\underline{55B}$, 425 (1975).
45. E. Lomon, private communication.
46. a) I.R. Afnan and J.M. Read, The Deuteron Tensor Polarization in e-d Scattering, Flinders University preprint (1975).
    b) E. Lomon and E. Moniz (private communication).
47. T. Hamada and I.D. Johnston, Nucl. Phys. $\underline{34}$, 382 (1962).
48. R.A. Bryan and B.L. Scott, Phys. Rev. $\underline{177}$, 1435 (1969).
49. R. de Tourreil and D.W.L. Sprung, Nucl. Phys. $\underline{A201}$, 193 (1973).
50. H. Arenhövel and W. Fabian, Forward Proton Production in d(γ,p)n, Univ. Mainz preprint 1976; E.L. Lomon, Dependence of Forward Deuteron Photodisintegration on the D-State, MIT preprint (1977).
51. R.J. Hughes, A. Zeiger, H. Wäffler and B. Zeigler, Nucl. Phys. $\underline{A267}$, 329 (1976).
52. K. Holinde and R. Machleidt, Nucl. Phys. $\underline{A256}$, 479 (1976).
53. I.R. Afnan and A.W. Thomas, Phys. Rev. $\underline{C15}$, 2143 (1977).
54. A.W. Thomas and I.R. Afnan, Phys. Rev. Lett. $\underline{26}$, 906 (1971).
55. H.C. Pradhan and Y. Singh, Can. J. Phys. $\underline{51}$, 343 (1972).
56. B. Goplen, W.R. Gibbs and E.L. Lomon, Phys. Rev. Lett. $\underline{32}$, 1012 (1974).
57. B.M. Preedom *et al.*, Phys. Lett. $\underline{65B}$, 31 (1976).
58. G. Jones, Invited paper on pp→π⁺d at this conference.
59. W.R. Gibbs, Phys. Rev. $\underline{C3}$, 1127 (1971).
60. W.R. Gibbs, in Proc. 4th Int. Conf. on Polarization Phenomena (Birkhauser, Zurich, 1976), p.61.

386

61. A.S. Rinat and A.W. Thomas, Nucl. Phys. A282, 365 (1977), and to be published; A.W. Thomas, Nucl. Phys. A258, 417 (1976).
62. W. Gruebler, SIN proposal #R-73-01.1, and private communication.
63. E.A. Remler and R.A. Miller, Ann. Phys. 82, 189 (1974).
64. P. Doleschall, in Proc. 4th Int. Conf. on Polarization Phenomena (Birkhauser, Zürich, 1976), p.51.
65. J. Saudinos and C. Wilkin, Ann. Rev. Nucl. Sci. 24, 365 (1974).
66. I.R. Afnan and J.M. Read, Phys. Rev. C12, 293 (1975).
67. A.M. Green, Nucleon Resonances in Nuclei, University of Helsinki report ISBN 951-45-0876-9 (1976).
68. H.J. Weber, in Meson Nuclear Physics-1976 (Carnegie-Mellon Conference), American Institute of Physics Conference Proceedings No. 33 (AIP, New York, 1976), p.130.
69. D.O. Riska and G.E. Brown, Phys. Lett. 38B, 193 (1972).
70. H.P. Noyes, Nucl. Phys. 74, 508 (1965).
71. G. Stranahan, Phys. Rev. 135, B953 (1964).
72. D.O. Riska, private communication.
73. T. Kamae and T. Fujita, Phys. Rev. Lett. 38, 471 (1977).
74. H.J. Weber, Nucl. Phys. A264, 365 (1976).
75. A.S. Goldhaber, Extracting Deltas from the Deuteron, LAMPF preprint (1977).

nD SCATTERING AT 180° FOR NEUTRON ENERGIES FROM 200 to 800 MeV*

B. E. Bonner
Los Alamos Scientific Laboratory, University of California
Los Alamos, New Mexico 87545

C. L. Hollas, C. R. Newsom, and P. J. Riley
University of Texas, Austin, Texas 78712

G. Glass
Texas A & M University, College Station, Texas 77843

THEORETICAL BACKGROUND

Near the end of the last decade, Kerman and Kisslinger,[1] in attempting to understand the backward peaking that had been observed in pD elastic scattering at 1 GeV, found that simple one-nucleon exchange failed to fit the backward peak, the calculation being low by a factor of two. Taking a bold step, they postulated the existence of Isobars in the deuteron, finding that a 1% admixture of the N* (1688) sufficiently augmented the calculated cross section to bring it into agreement with the 1-GeV data of Bennett et al.[2] In intervening years, many experiments and calculations have been performed relating to backward pD elastic scattering at medium energy and Isobars in nuclei.[3] Alternative explanations of the observed peaking were proposed. One was the calculation of Craigie and Wilkin[4] in which triangle diagrams were used to relate the cross sections for pD elastic and pp → dπ+. Many authors have pursued this technique with success in fitting backward pD scattering. Another recent calculation[5] takes account of the fact that np angular distributions also are backward peaked and incorporates this into an extended Glauber Multiple Scattering calculation. With this approach, backward pD scattering is fit over a wide range of energies when a radically different form factor for the deuteron is postulated for the large momentum transfers encountered. The fact that the recent SLAC measurements[6] tend to verify the postulated shape of the deuteron form factor brings us back to the original hope that information on the high-momentum components of the deuteron wave function can be obtained from the scattering of hadrons of modest energy.

EXPERIMENTAL BACKGROUND

In the last few years, several angular distributions have been reported[2] for pD backward elastic scattering at medium energies. In each of these measurements, the backward peak is observed. Representative results between 600 and 1000 MeV are shown in Fig. 1. The data of Boschitz[7] and of Alder[8] near 600 MeV and the three measurements[2,9,10] at 1000 MeV are shown. Two observations can be made:

*Work supported by the U. S. Energy Research and Development Administration.

(1) Gross disagreement (a factor of two) between the data sets at 1000 MeV exists. (2) Most of the data on pD backward scattering do not extend to angles beyond about 160° to 170°. Yet there are several plots in the literature of the extrapolated 180° cross section as a function of energy. In a previous experiment[11] by this group that is about to be published, we measured for the first time the neutron-deuteron elastic scattering cross section at 800 MeV. That data is also included in Fig. 1. The experimental difficulty associated with measurements at 180° c.m. for proton beams does not arise for neutron beams, and we have taken the angular distribution all the way in to 179° c.m. We found in that experiment that the extrapolation to 180° is very straightforward--a simple exponential function $Ae^{B(u - u_{180})}$ fits the data at 800 MeV. The experiment to be described here has as its aim the measurement of nD backward scattering from about 120° to 180° over the incident energy range from 200 to 800 MeV. Part of the reason for the interest in such a measurement is the fact that a bump or shoulder in the (extrapolated) 180° excitation function has been hinted at by previous pD experiments at medium energy. It has been noted before that this behavior cannot be reproduced in calculations involving Isobars in the deuteron, but does follow naturally from the triangle diagram calculations. We report here only on the 180° excitation function, which was measured in nD elastic scattering for incident energies of 200 to 800 MeV. Only a small fraction of our data have been analyzed at this time.

EXPERIMENTAL TECHNIQUE

Basically, the experiment consists of using a continuum neutron beam incident on a liquid deuterium target and detecting the scattered deuterons in a multiwire proportional chamber spectrometer. The experimental layout is shown in Fig. 2. The 800-MeV proton beam at LAMPF strikes an aluminum target and is then deflected and buried some distance away. Neutrons emerging at 0° are collimated to form a neutron beam, then cleared of charged particles before encountering a liquid deuterium target. Charged particles produced in the target are momentum analyzed in a multiwire proportional chamber spectrometer. Particle identification is achieved by a simultaneous measurement of their time of flight through the spectrometer. This enables a calculation of the particle mass from the relation $M = P/\beta\gamma$. Particle identification is unambiguous for more than 99% of the events. The elastic deuterons are not the only ones that are observed-they form a rather small fraction of the total. However, the elastic deuterons can be cleanly separated from the others by measuring in addition the time of flight of the incident neutron. This is shown in Fig. 3 where the incident time of flight is plotted versus deuteron momentum for both the elastic and inelastic processes that lead to a deuteron in the final state. Therefore, in a manner analogous to identification of particle type by measurement of transit time through the spectrometer, we can determine the type of process which gave rise to the deuteron by measurement of the incident neutron time of flight. Once we have identified the elastic deuterons, then the measured deuteron momentum and angle uniquely specify the incident neutron energy to within the accuracy of the deuteron momentum determination (about 1%).

The other requirement that must be met before the cross section can be determined is that the incident neutron spectrum must be known absolutely. This was achieved in a separate measurement, using the same apparatus and techniques, with the exception that a liquid hydrogen target instead of liquid deuterium was used. The normalizing reaction is np → dπ°, the cross section for which is known, using isospin invariance, from the reaction pp → dπ+. An indication of the validity of the technique is seen in Fig. 4, which is from a separate communication[12] to this conference. This gives the variation with momentum of the np charge exchange cross section at 180° (or t = 0) as determined using the same techniques and equipment as the present experiment. Since the normalizing reaction np → dπ° is not available below about 300 MeV (800 MeV/c), another form of normalization is required. The one used for the data reported below was to extrapolate the np charge exchange cross section (shown in Fig. 4) from 300 MeV down to 180 MeV.

RESULTS OF PRESENT EXPERIMENT

In Fig. 5 is shown the preliminary results of the present experiment compared to previous pD measurements in the energy range from 150 to 1000 MeV. References to the previous experiments are given in the figure caption. Present results are indicated by the solid circles and were obtained for neutron c.m. scattering angles greater than 175°. It is apparent that the agreement with the lower energy measurements[13,14] is quite good, but the medium-energy SREL data[8] differ considerably from the present results. Agreement is found with the 425-MeV point of Booth et al.[15] The curve is merely drawn to guide the eye.

Calculations using the triangle diagram technique have been made. For our 800-MeV angular distribution, the shape is fit very well. The energy dependence of the calculated 180° cross section is not correct however when the 800-MeV normalization is used. The calculation is about a factor of two too high at 500 MeV in that case. This is not surprising since other mechanisms such as nucleon and T = 1/2 Isobar exchange are probably present.

REFERENCES

1. A. K. Kerman and L. S. Kisslinger, Phys. Rev. 180, 1483 (1969).
2. G. W. Bennett et al., Phys. Rev. Letters 19, 387 (1967). A review of the experimental situation in pD backward scattering is given in the article by J. E. Simmons, Proc. Conf. on High Energy Physics and Nuclear Structure, 1975 (AIP, New York, 1975).
3. A review of the status of Iosbars in nuclei is given in the article by H. J. Weber, Proc. Conf. on Meson Nuclear Physics, 1976 (AIP, New York, 1976).

4. N. S. Craigie and C. Wilkin, Nucl. Phys. B14, 477 (1969); V. M. Kolybasov and N. Ya. Smorodinskaya, Sov. J. Nucl. Phys. 17, 630 (1973).
5. S. A. Gurvitz and A. S. Rinat, Phys. Letters 60B, 405 (1976).
6. R. G. Arnold et al., Phys. Rev. Letters 35, 776 (1975).
7. E. T. Boschitz et al., Phys. Rev. C6, 457 (1972).
8. J. C. Alder et al., Phys. Rev. C6, 2010 (1972).
9. E. Coleman et al., Phys. Rev. 164, 1655 (1967).
10. L. Dubal et al., Phys. Rev. D9, 597 (1974).
11. B. E. Bonner et al., to be published.
12. B. E. Bonner et al., Abstract C8 at this conference.
13. K. Kuroda et al., Nucl. Phys. 88, 33 (1966).
14. G. Igo et al., Nucl Phys. A195, 33 (1972).
15. N. E. Booth et al., Phys. Rev. D4, 1261 (1971).
16. J. Banaigs et al., Nucl. Phys. B23, 596 (1970).

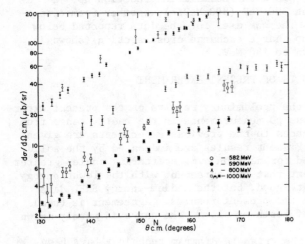

Fig. 1. Angular distributions for pD backward elastic scattering at energies of 582 MeV (Ref. 7), 590 MeV (Ref. 8), and the three measurements at 1000 MeV (Refs. 1, 9, and 10). Data at 800 MeV are from a recent experiment on nD elastic scattering (Ref. 11).

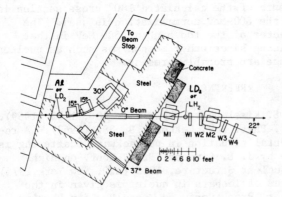

Fig. 2. Layout of neutron beam and spectrometer used in the present experiment.

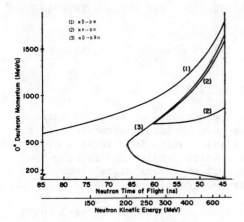

Fig. 3. Plot of the 0° deuteron momentum versus the incident neutron time of flight for the three reactions leading to deuterons in the final state: (1) nD elastic scattering, (2) quasifree deuteron production n'N' → dπ, and (3) nD → dNπ. This illustrates that deuterons from the elastic process can be separated from the other processes.

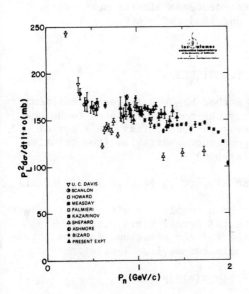

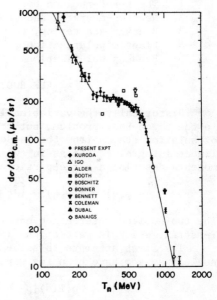

Fig. 4. Figure from Ref. 12 showing the variation with neutron momentum of the 180° cross section for np elastic scattering (multiplied by P^2). Below 800 MeV/c, this reaction was used for normalization of the results in Fig. 5.

Fig. 5. Present results on nD elastic scattering at 180° compared to extrapolation of previous pD measurements plotted versus the nucleon kinetic energy. The curve is merely drawn to guide the eye. References to previous pD measurements are, from the top, 13, 14, 8, 15, 7, 11, 2, 9, 10, and 16.

THE FEW NUCLEON PROBLEM WITH REALISTIC NUCLEON-NUCLEON INTERACTIONS[*]

W. M. Kloet

Theoretical Division, Los Alamos Scientific Laboratory
Los Alamos, New Mexico 87545

INTRODUCTION

In view of the subject of this conference it seems of interest to discuss the use of realistic nucleon-nucleon interactions in the few nucleon problem. Recently several results in this area have been reported. We address the question, which predictions for the few nucleon system do follow from two-nucleon interactions alone. At this stage three-nucleon forces are not considered, though certainly not ruled out.

Since two-nucleon interactions are only reasonably well known for low energy (0-300 MeV), we discuss only low energy few nucleon processes. The phenomena of interest are
1. The three-nucleon bound state, energy and form factors,
2. Neutron-deuteron and proton-deuteron elastic scattering,
3. Breakup scattering, nd → nnp and pd → npp,
4. Elastic polarization,
5. Breakup polarization.

FEW BODY TECHNIQUES

Historically the variational method has the longest tradition in the three-body problem, but today the most widely used technique of solution starts from the Faddeev equations. They are a set of linear integral equations in two continuous variables (two relative momenta describe a three-particle system)

$$\psi(p,q) = \phi(p,q) + \int dp' \, dq' \, K(p,q;p',q')\psi(p',q') \quad . \qquad (1)$$

The two-nucleon interaction enters as a completely off-shell two-body T-matrix, which is part of the integral kernel $K(p,q;p',q')$.

The first attempts to solve the Faddeev equations for the three-nucleon system have been for separable two-nucleon T-matrices

$$\langle p|T(s)|p' \rangle = g(p)\tau(s)g(p') \quad . \qquad (2)$$

In that case the equations reduce to linear integral equations in one continuous relative momentum

$$F(q) = f(q) + \int dq' \, H(q;q')F(q') \quad , \qquad (3)$$

where

$$\psi(p,q) = g(p)\tau(s-E_q)F(q) \quad . \qquad (4)$$

Subsequently the problem has been solved with methods suited

ISSN: 0094-243X/78/392/$1.50 Copyright 1978 American Institute of Physics

for any type of two-body interaction and in practice a local interaction was used.

There are no fundamental difficulties in doing a full three-
body calculation with a reasonably realistic two-nucleon potential,
including a substantial number of its partial waves. But due to
complications of the large set of coupled integral equations which
has to be solved when more than s-wave two-nucleon interactions are
included, progress in going to a full size realistic potential (such
as the Reid soft core potential) has been rather slow. Let us look
at some results.

THREE-BODY BOUND STATES, ^{3}H AND ^{3}He

The realistic potentials used here are Hamada-Johnston,[1] Reid
soft core,[2] Sprung de Tourreil (A,B,C),[3] some one boson exchange
potentials developed at Bonn[4], the Bressel-Kerman-Rouben[5] and the
Rouben potential.[6]

To calculate the triton binding energy usually only the $^1S_0$
and the $^3S_1$-$^3D_1$ components of the interaction are used. In variational and Faddeev calculations the effect of higher partial waves
with $J \leqslant 2$ has been tested and found to add at most 0.2 MeV to the
binding energy.[7,11,19]

In Table I some results are given for these potentials. E_T is
the triton binding energy in MeV. $F_{ch}(^3He)$-dip is the position in
fm^{-2} of the minimum in the ^{3}He charge form factor, which is
another interesting quantity. For the two-body potential the deuteron D-state probability is also given, $P_D(d)$.

Table I. Results for three-nucleon bound state properties.

Potential	method	ref.	$P_D(d)$	E_T	$F_{ch}(^3He)$-dip
Hamada-Johnston	variational	8	7.0	6.5	12.5
	variational	9		6.7	
Reid soft core	Faddeev(r-space)	10	6.5	7.0	14.0
	Faddeev(q-space)	11		7.0	13.9
	variational	12		6.7-7.7	12.6
Sprung-de Tourreil-A	Faddeev(r-space)	10	4.4	7.6	14.8
Sprung-de Tourreil-B	Faddeev(r-space)	10	4.3	7.7	15.0
Sprung-de Tourreil-C	Faddeev(r-space)	10	5.5	7.5	14.4
Sprung-de Tourreil	variational	13		6.7	
Bonn-EHM	Faddeev(q-space)	14	5.7	7.2	15.3
Bonn-HM1	Faddeev(q-space)	14	5.8	7.5	15.5
Bressel-Kerman-Rouben	Faddeev(q-space)	15	6.5	6.2±0.2	14.4
Rouben	Faddeev(q-space)	16	4.6	7.9	21.5
Experiment				8.48	11.6

The calculations for the one-boson-exchange potentials are particularly interesting, since they employ interactions which have the

best theoretical basis. None of the calculations reproduces the experimental binding energy of 8.48 MeV, nor the high momentum transfer behaviour of the ^{3}He charge form factor. In Fig. 1 the charge form factors are shown for the Reid soft core and the two one-boson-exchange potentials.

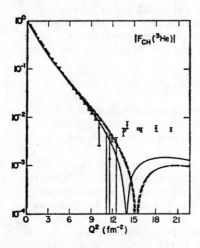

Fig. 1. ^{3}He charge form factors for Reid soft core, Bonn-EHM and Bonn-HM1. Data are from Ref. 17.

One observes in Fig. 1 as well as in Table I, the feature that for potentials with larger binding energy (closer to experiment) the ^{3}He charge form factor has its minimum at larger momentum transfer (which means larger disagreement with experiment). However the disagreement between calculated charge form factor and experiment is not too disturbing, since estimates of mesonic exchange effects have had modest success in reducing this discrepancy.[18]

An interesting study was done by Afnan and Reid[19] of correlations between the triton binding energy, the doublet neutron-deuteron scattering length and the deuteron D-state probability. Such a comparison was done before for separable potentials and local potential models,[20,21] but they studied a set of realistic potentials in a uniform way, using the unitary pole approximation method. The interactions include Hamada-Johnston (HJ), Reid hard core (RHC), Reid soft core (RSC), alternate Reid soft core (RSCA), Sprung-de Tourreil-A, B and C (STA, STB, STC).

The result is a confirmation of the strong correlation between the triton binding energy and the n-d doublet scattering length for realistic potentials as shown in Fig. 2. This relation is called the Phillips line. The relation is however not necessarily a straight line. Other types of fits can be made which are of similar quality, the reason being the relatively small intervals for both variables involved in the plot.[22]

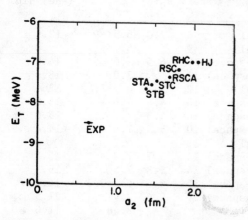

Fig. 2. Correlation between triton binding energy E_T and n-d doublet scattering length a_2 .

Also the observation, done earlier for model potentials, that a decrease in deuteron D-state probability corresponds to an increase in E_T was confirmed for this set of realistic potentials.

The correlations are obvious from Table II

Table II. Correlations between E_T, a_2 and P_D(deuteron)

potential	P_D	E_T	a_2
HJ	7.02	6.96	2.04
RHC	6.50	6.96	1.97
RSC	6.47	7.15	1.80
RSCA	6.22	7.32	1.68
STC	5.45	7.42	1.52
STA	4.43	7.52	1.46
STB	4.25	7.62	1.38

ELASTIC NEUTRON-DEUTERON SCATTERING

As an example the neutron-deuteron differential cross section is shown at E_n^{lab} = 14.1 MeV in Fig. 3. Note that the deuteron breakup channel already is open at this energy. From Fig. 3, one sees that a one term separable S-wave potential already reproduces the general features of the differential cross section (dashed line).[23]

An improvement in the forward direction is obtained by using a simple local S-wave potential. The Malfliet-Tjon I-III in particular which is a local Yukawa type S-wave potential with repulsion, compares rather well with the data (dotted line, continued into the solid line).[24]

The result for the Reid soft core interaction, including tensor force and P-waves ($^1S_0$, $^3S_1$-$^3D_1$, $^1P_1$, $^3P_0$, $^3P_1$, $^3P_2$) is hardly different (solid line)[25] from the former S-wave result. The same is true for Sprung-de Tourreil-C potential.[26] For the last case also the effect of the ^1D and ^3D components was calculated in purturbation and this slightly increases the forward peak in even better agreement with the data (dash-dotted line).

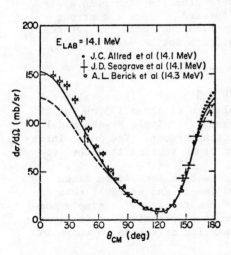

Fig. 3. Differential cross section for elastic n-d scattering at 14.1 MeV. Data are from Ref. 27.

At higher energy the more sophisticated details of the realistic potentials start paying off. At 46.3 MeV (Fig. 4) the simple S-wave local potential model (Malfliet-Tjon I-III) developes a minimum at 130 degrees (dashed line) which is substantially deeper than experiment.[24]

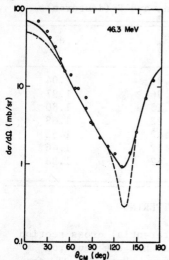

Fig. 4. Differential cross section for elastic n-d scattering at 46.3 MeV. Data are from Ref. 28.

In this case the Reid potential with tensor force and P-waves (solid line) substantially improves the fit to the data in the region of the minimum.[25] Also in the forward direction the cross section is increased and follows the data more closely.

The calculations discussed do not include the Coulomb interaction and are strictly valid only for n-d scattering. To compare with the more abundant data for p-d scattering the Coulomb force must be included because it will change the cross section, particularly in the forward direction.

Coulomb effects in elastic p-d scattering have been recently calculated by Alt at al.[29] at E_p^{lab} = 2 and 10 MeV. The results are very promising and mean a big step forward. Since the strong interaction they used, is not realistic, a final comparison with the data can not be made yet.

BREAKUP REACTION, nd → nnp AND pd → npp

The breakup channel opens at E_n^{lab} = 3.3 MeV. The total breakup cross section rises sharply to a plateau of about 175 mb, at 10 MeV, and after 20 MeV it slowly decreases.[30] Simple S-wave models reproduce this behaviour within about ten percent. The local interactions give a somewhat larger breakup cross section than the separable models.[30]

The interest of experimentalists has been focused on cases where two final particles are detected (angle and energy), thus determining the complete kinematics of the reaction. A simple model such as the Malfliet-Tjon I-III S-wave local potential again gives a good global fit to the data.[31] The question arises as to where in phase space such a model goes wrong.

One of the attempts to answer this question is a comparison between experiment and theory over the entire phase space for p-d breakup at 13.25 MeV by Wielinga et al.[32] and at 50 MeV by Blommestijn et al.[33] at I.K.O. Amsterdam. For both cases data were collected using a large set of simultaneous detectors located on a sphere around the target. This makes it possible to compare data and theory in a number of different plots as a function of various variables.

At 13.25 MeV a strong discrepancy between theory and experiment shows up if the neutron CM angle is used as a parameter.[32] In Fig. 5 the quantity (theory-experiment)/theory is plotted for the partially integrated cross section. Five dimensional phase space is projected onto a two-dimensional plot. Theory stands for a Monte Carlo simulation using as an input the theoretical amplitudes calculated from the local S-wave Malfliet-Tjon I-III potential.[31] The difference is as high as 100 percent at θ_n^{CM} = 140 degrees. If other variables are chosen for the axis to plot the results, no appreciable differences are observed. Another characteristic of this region is that the two final protons are symmetric relative to the neutron. This causes some of the spin amplitudes to vanish. The cross section has a deep minimum in this area. The precise physical interpretation of this kinematical configuration is not clear, but an analogy with the interference minimum in the elastic differential cross section at 110 degrees is possible.

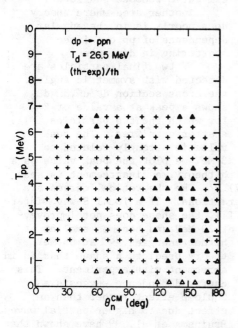

Fig. 5. Two dimensional comparison of experimental data and local potential model. The symbols denote (theory-experiment)/theory: □ (90±20)%, △ (50±20)%, + (0±30)%, ▲ (-50±20)%, ■ (-90±20)%.

The analysis of the 50 MeV data is still in progress. Preliminary results show the same trend as at lower energy. The largest discrepancy again occurs for a neutron CM angle of 140 degrees, but the discrepancy this time is even larger.

An experiment in the area of the minimum cross section has been done at the Manitoba Cyclotron at 23 and 39.5 MeV.[34] The data show also a minimum at the same position as theory, but not as deep (Fig. 6). It was expected that the minimum would be filled in by adding higher partial waves in the two-nucleon interaction. This was confirmed by calculations for the Reid soft core potential, including the tensor force and P-waves,[35] and also for separable potentials including the same components.[36] The results for the Reid potential

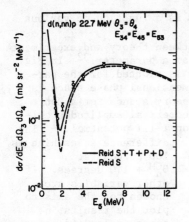

Fig. 6. Breakup cross section along the symmetric constant relative energy locus at 22.7 MeV. Data are from Ref. 34.

are shown in Fig. 6 at 22.7 MeV.

An aspect that should not be overlooked in comparing experiment and theory in areas of a deep and narrow minimum such as in Fig. 6, is the effect of the finite size of the experimental detectors. If the spread in detection angle is taken into account, the theoretical curve becomes shallower.

Another area where theory does poorly is in the angular dependence of pp quasifree scattering in d(p,pp)n.

If two final particles are detected with symmetric angles, the cross section $d\sigma/d\Omega_1 d\Omega_2 dE_2$ shows a peak at a value of E_2 for which the two particles scatter quasifree. The peak value is strongly related to free two-body scattering. If at 23 MeV we plot the peak value as a function of angle, and compare with theory, a disagreement is seen in the shape of the curve (Fig. 7).[37] The theoretical curves are for several S-wave separable interactions. If a local S-wave interaction is used (Malfliet-Tjon I-III), the shape does not improve.[31]

The disagreement is largest at 23 MeV and becomes smaller at

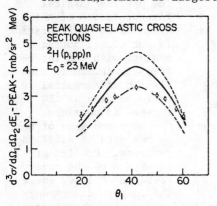

Fig. 7. Cross section at the QFS peak as a function of angle. For the data see Ref. 37.

higher energies.[38] At 65 MeV theory predicts a shape that is in agreement with experiment. This energy dependence suggests a Coulomb effect rather than an effect due to higher partial waves. Bruinsma et al.[36] have shown that higher partial waves indeed do not improve the shape very much. Preliminary results of Haftel et al.[39] indicate that more improvement can be obtained by including the Coulomb force.

A very interesting result extracted from the breakup reaction is the value of the neutron-neutron scattering length a_{nn}. The shape of the neutron-neutron final state interaction peak does not depend on the type of the interaction but only on the value of a_{nn}. Therefore the experimental shape can be fitted by a potential model using a_{nn} as a parameter. In Ref. 40 for the reaction d(n,2n)p at 18.4 MeV such a fit was made using for the neutron-neutron effective range the fixed charge-symmetric value

r_{nn} = 2.86 fm. The best fit was obtained for a_{nn} = -16.3 ± 1.0 fm. For other attempts to find a_{nn} see Ref. 41.

The analysis has been done using S-wave separable potential models. It is expected that more sophisticated interactions containing more partial waves, will not change these results appreciably because in this area of very low two-nucleon relative energy the higher partial waves do not contribute. Strictly speaking however this has not been tested.

ELASTIC POLARIZATION

For the spin observables in neutron deuteron scattering the measured quantities are the neutron polarization P_n, the deuteron vector polarization i T_{11} and the deuteron tensor polarizations T_{22}, T_{21}, T_{20}. Also a number of polarization transfer coefficients has been measured. For the definitions of the various observables see Ref. 42.

Again the first calculations for polarizations were for separable models containing S-waves, tensor force and P-waves.[43,44,45,36] The qualitative features of the available data were rather well reproduced, in particular the deuteron tensor polarizations.

Recently results for realistic potentials have been reported. In Ref. 26, for the interaction of Sprung-de Tourreil, type C, the $^1S_0$, $^3S_1 - ^3D_1$, $^1P_1$ and $^3P_{0,1,2}$ waves are treated exactly and the $^1D_2$ and $^3D_2$ waves are added as a perturbation. In Ref. 25,

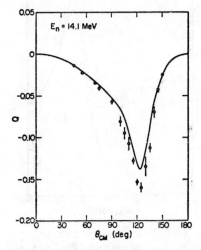

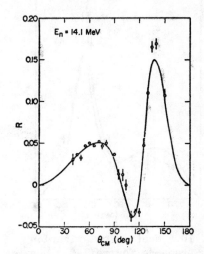

Fig. 8. Deuteron tensor polarization Q for Sprung-de Tourreil-C. Data are from Ref. 46.

Fig. 9. Deuteron tensor polarization R for Sprung-de Tourreil-C. Data are from Ref. 46.

400

for the Reid soft core potential, the tensor part and P-waves are treated in perturbation theory. The effect of using perturbation theory to determine spin observables may be that in some cases the exact solution is reproduced only qualitatively and not quantitatively.[35,44]

We mention here the results for the Sprung-de Tourreil C potential at 14.1 MeV. In Figs. 8 and 9 the deuteron polarizations $Q = (1/2\sqrt{2})\ (T_{20} + \sqrt{6}\ T_{22})$ and $R = (1/2\sqrt{2})\ (T_{20} - \sqrt{6}\ T_{22})$ are shown. The agreement with the data is remarkable. Also the neutron polarization P_n (Fig. 10) shows good agreement with the data. The solid curve does not include $^1D_2$ and $^3D_2$. The dashed curve is the result with the two D-waves included perturbatively. In this case their effect is important, while for the tensor polarizations their effect does not show on the scale of Figs. 8 and 9.

For the deuteron vector polarization (Fig. 11) the theory does not reproduce the minimum at 100 degrees. Also including $^1D_2$ and $^3D_2$ (dashed curve) gives no improvement in this region. Recent exact calculations for the Reid soft core potential indicate results for all polarizations which are not very different from the results for Sprung-de Tourreil-C.[48] Hence the deuteron vector polarization persists in showing a rather large disagreement between experiment and theory and is therefore the most interesting spin observable.

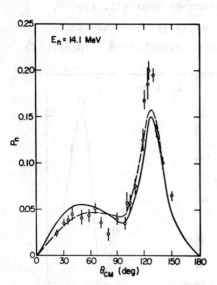

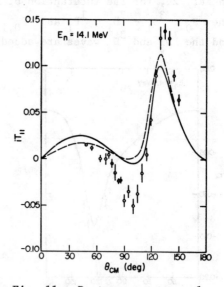

Fig. 10. Neutron polarization for Sprung-de Tourreil-C. Data are from Ref. 47.

Fig. 11. Deuteron vector polarization for Sprung-de Tourreil-C. Data are from Ref. 46.

The discrepancy for the deuteron vector polarization has been studied in separable potential models[45], and it was shown to be strongly correlated to the two-body tensor force parameters, $\delta(^3D_1)$ and ϵ_1. This has led to the suggestion[42] that here one may have a way to test the properties of the two-body tensor force in a three-nucleon reaction. Theoretical work is still in progress.

BREAKUP POLARIZATION

The proton polarization has been measured for d(p,p)d* at 22.7 MeV and the deuteron vector polarization for p(d,p)d* at the corresponding energy of 45.4 MeV by Rad et al.[49] Calculations of these quantities have been done for separable potentials[36] and for the Reid soft core potential.[35] In Fig. 12 the results are shown for the proton polarization A_y and the deuteron vector polarization iT_{11}. The curves are theoretical predictions for the Reid soft core potential.

The energy spectra of the detected protons show at the high energy end a peak due to neutron-proton final state interaction. The experimental data in Fig. 12 are obtained by integration of the peak area corresponding to a n-p pair (called d*) with relative energy $E_{np} \leq 1$ MeV. For the theoretical results the solid curve is the solution including S-waves, tensor force, P-waves, $^1D_2$ and $^3D_2$ where $E_{np} = 0.1$ eV. The dotted curve is the result for the same partial waves but $E_{np} = 1$ MeV and the relative momentum of the n-p pair parallel to the incoming beam. The curves show a significant dependence on the relative momentum of the n-p pair. This can be understood as follows. For $E_{np} = 0.1$ eV the cross section is mainly due to n-p final state scattering. Because of the large singlet scattering length the final state with the n-p pair in a spin singlet is dominant. At $E_{np} = 1$ MeV other final states such as the n-p pair in spin triplet are also important.

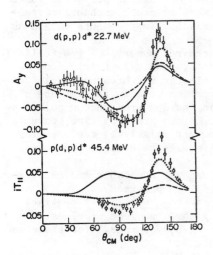

Fig. 12. Neutron·polarization and deuteron vector polarization for the breakup reaction with a final n-p pair in a low relative energy state. Results are for Reid soft core. Data are from Ref. 49.

To compare theory and experiment, the theoretical results for E_{np} values between 0 and 1 MeV should be integrated taking account of experiment dependent weight factors. This has not been done yet. It is interesting to see that the theoretical curve for $E_{np} = 1$ MeV is much closer to the data than the result for $E_{np} = 0.1$ eV.

402

This may indicate that the contribution of the pure singlet n-p pair
is less important than expected. It would be interesting if experimental data could be obtained including only n-p pairs with a relative energy substantially smaller than 1 MeV. One of the experimental observations of Ref. 49 is that the measured polarizations are
quite similar to the elastic analyzing powers, which may indicate
that the final n-p pair is mainly in a spin triplet state.

Finally, the dashed curve in Fig. 12 results from only S- and
P-waves and E_{np} = 0.1 eV. Comparing with the solid curve shows that
the tensor force and $^1D_2$ and $^3D_2$ waves are quite important here.
This is in contrast with the vector polarizations for elastic scattering where P-waves are relatively more important.

CONCLUDING REMARKS

We have seen that three-body calculations for realistic two-
nucleon interactions including all partial waves for $J \leqslant 2$ fail to
reproduce the three-nucleon binding energy by at least 1 MeV. The
charge form factor of ^{3}He is in good agreement with experiment at
low momentum transfer, but fails at higher momentum transfer in the
region of the dip and the second maximum. The potentials which produce the best binding energy, give the worst ^{3}He charge form factor.
An interesting experiment would be the measurement of the ^{3}H charge
form factor for momentum transfer of 10-20 fm^{-2}. Then this quantity
could also be included in the comparison of various potential models.
Moreover it would give an indication whether meson exchange corrections to the charge form factor give similar improvements for ^{3}H as
they seem to do for ^{3}He.

For neutron deuteron scattering the refinements of the realistic
interactions, compared to simpler models, result in fairly good
agreement with experimental data for elastic and breakup angular distributions and polarizations. An exception here is the failure of
the theory to reproduce the minimum in the deuteron vector polarization around 100 degrees. Since calculations are very time consuming,
we can only draw these conclusions for the relatively few energies
and kinematical conditions for which results have been reported at
present.

Some data for breakup polarizations are available and calculations for realistic potentials are still in progress.

This discussion has been restricted to conventional realistic
interactions. Among other interesting three nucleon problems are
two-body off-shell sensitivity and three-body forces. For these and
other aspects see Refs. 41, 33, 50, 51, 52.

ACKNOWLEDGEMENTS

I want to thank Drs. E. O. Alt, B. S. Bhakar, J. Bruinsma, H. E.
Conzett, B. F. Gibson, G. Gignoux, M. I. Haftel, D. R. Lehman, P. U.
Sauer, W. Sandhas, C. Stolk and J. A. Tjon for clarifying discussions,
suggestions and early communication of their work.

*Work performed under the auspices of the U. S. Energy Research and
Development Administration.

REFERENCES

1. T. Hamada and I. D. Johnston, Nucl. Phys. $\underline{34}$, 382 (1962).
2. R. V. Reid, Ann. Phys. (N. Y.) $\underline{50}$, 411 (1968).
3. R. de Tourreil and D. W. L. Sprung, Nucl. Phys. $\underline{A201}$, 193 (1973).
4. K. Erkelenz, K. Holinde, R. Machleidt, Phys. Lett. $\underline{49B}$, 209 (1974), K. Holinde and R. Machleidt, Nucl. Phys. $\underline{A247}$, 495 (1975).
5. C. N. Bressel, A. K. Kerman, B. Rouben, Nucl. Phys. $\underline{A124}$, 624 (1969).
6. B. Rouben, Ph.D. thesis, M. I. T. (1969).
7. Y. E. Kim and A. Tubis, Ann. Rev. Nucl. Sci. $\underline{24}$, 69 (1974).
8. L. M. Delves and M. A. Hennell, Nucl. Phys. $\underline{A168}$, 347 (1971).
9. C.-Y. Hu, Phys. Rev. C$\underline{3}$, 2151 (1971).
10. C. Gignoux and A. Laverne, Phys. Rev. Lett. $\underline{29}$, 436 (1972). A. Laverne and C. Gignoux, Nucl. Phys. $\underline{A203}$, 597 (1973).
11. R. A. Brandenburg, Y. E. Kim, A. Tubis, Phys. Lett. $\underline{49B}$, 205 (1974); Phys. Rev. C$\underline{12}$, 1368 (1975).
12. A. D. Jackson, A. Lande, P. U. Sauer, Phys. Lett. $\underline{35B}$, 365 (1971), M. A. Hennell and L. M. Delves, Phys. Lett. $\underline{40B}$, 20 (1972), V. F. Demin, Yu. E. Pokrovsky, V. D. Efros, Phys. Lett. $\underline{44B}$, 227 (1973), J. Bruinsma and R. van Wageningen, Phys. Lett. $\underline{44B}$, 221 (1973), M. R. Strayer and P. U. Sauer, Nucl. Phys. $\underline{A231}$, 1 (1974).
13. J. Bruinsma, R. van Wageningen, J. L. Visschers, Few Particle Problems in the Nuclear Interaction, ed. I. Slaus et al., Amsterdam, North Holland (1972) p. 368.
14. R. A. Brandenburg, P. U. Sauer, R. Machleidt, Hannover-Bonn preprint.
15. R. A. Malfliet and J. A. Tjon, Phys. Lett. $\underline{35B}$, 487 (1971).
16. R. A. Malfliet and J. A. Tjon, Ref. 13, p. 441.
17. J. S. McCarthy et al., Phys. Rev. Lett. 25, 884 (1970), M. Bernheim et al., Nuovo Cimento Lett. $\underline{5}$, 431 (1972).
18. W. M. Kloet and J. A. Tjon, Phys. Lett. $\underline{49B}$, 419 (1974), $\underline{61B}$, 356 (1976), M. I. Haftel and W. M. Kloet, Phys. Rev. C$\underline{15}$, 404 (1977).
19. I. R. Afnan and J. M. Read, Phys. Rev. C$\underline{12}$, 293 (1975).
20. A. C. Phillips, Nucl. Phys. $\underline{A107}$, 209 (1968).
21. R. A. Malfliet and J. A. Tjon, Ann. Phys. (N. Y.) $\underline{61}$, 425 (1970), B. L. G. Bakker, Ph.D. thesis, Free University, Amsterdam (1973).
22. D. R. Lehman and J. O'Connell, private communication.
23. R. Aaron, R. D. Amado and Y. Y. Yam, Phys. Rev. B$\underline{140}$, 1291 (1965).
24. W. M. Kloet and J. A. Tjon, Phys. Lett. $\underline{37B}$, 460 (1971), Ann. Phys. (N. Y.) $\underline{79}$, 407 (1973).
25. C. Stolk and J. A. Tjon, Phys. Rev. Lett. $\underline{35}$, 985 (1975).
26. J. J. Benayoun et al., Phys. Rev. Lett. $\underline{36}$, 1438 (1976).
27. J. C. Allred et al., Phys. Rev. $\underline{91}$, 90 (1953), J. D. Seagrave, Phys. Rev. $\underline{97}$, 757 (1955), A. C. Berick et al., Phys. Rev. $\underline{174}$, 1105 (1968).

404

28. J. L. Romero et al., Phys. Rev. C2, 2134 (1970).
29. E. O. Alt et al., Phys. Rev. Lett. 37, 1537 (1976).
30. G. Pauletta and F. D. Brooks, Nucl. Phys. A255, 267 (1975).
31. W. M. Kloet and J. A. Tjon, Nucl. Phys. A210, 380 (1973).
32. B. J. Wielinga et al., Nucl. Phys. A261, 13 (1976).
33. G. J. F. Blommestijn et al., Few Body Dynamics Conference
    Delhi, ed. A. N. Mitra et al., Amsterdam, North Holland (1976)
    p. 212.
34. A. M. McDonald et al., Phys. Rev. Lett. 34, 488 (1975).
35. C. Stolk and J. A. Tjon, Utrecht preprint.
36. J. Bruinsma and R. van Wageningen, Nucl. Phys. A282, 1 (1977).
37. E. L. Petersen et al., Phys. Rev. C9, 508 (1974).
38. M. I. Haftel et al., Nucl. Phys. A269, 359 (1976).
39. M. I. Haftel, private communication.
40. B. Zeitnitz et al., Nucl. Phys. A231, 13 (1974).
41. B. Kuhn, Few Body Problems in Nuclear and Particle Physics, ed.
    R. J. Slobodrian et al., Les Presses de l'Universite Laval,
    Quebec (1975) p. 122.
42. H. E. Conzett, Ref. 41 p. 566.
43. J. C. Aarons and I. H. Sloan, Phys. Rev. C5, 582 (1972); Nucl.
    Phys. A182, 369 (1972), A188, 193 (1972), A198, 321 (1972).
44. S. C. Pieper, Phys. Rev. Lett. 27, 1783 (1971); Nucl. Phys.
    A193, 529 (1972); Phys. Rev. C6, 1157 (1972), C8, 1702 (1973),
    C9, 883 (1974).
45. P. Doleschall, Phys. Lett. 38B, 298 (1972), 40B, 443 (1972);
    Nucl. Phys. A201, 264 (1973), A220, 491 (1974).
46. A. Fiore et al., Phys. Rev. C8, 2019 (1973).
47. J. C. Faivre et al., Nucl. Phys. A127, 169 (1969).
48. C. Gignoux, private communication.
49. F. N. Rad et al., Phys. Rev. Lett. 35, 1134 (1975).
50. D. D. Brayshaw, Phys. Rev. Lett. 32, 382 (1974), 34, 1478
    (1975).
51. M. I. Haftel and E. L. Petersen, Phys. Rev. Lett. 33, 1229
    (1974), 34, 1480 (1975).
52. H. P. Noyes, Phys. Rev. Lett. 23, 1201 (1969).

THREE-NUCLEON POTENTIALS DUE TO π, S AND ω AND NUCLEAR MATTER

T. Ueda, T. Sawada and S. Takagi
Osaka University, Toyonaka, Osaka, Japan

ABSTRACT

We derive the three-nucleon potentials due to π, S (isoscalar-scalar meson) and ω, which are exchanged between two nucleons via the third nucleon. We calculate the binding energy per nucleon in nuclear matter using the three-nucleon potentials with the two-nucleon OBEP of Ueda and Green (UGI), and find 0.4, 4.3 and -1.3 MeV at $k_F = 1.4$ fm^{-1} for the net contributions from the three-nucleon potentials due to π, S and ω, respectively. The sum of the two- and three-nucleon potentials gives approximately correct binding energy at the normal density. However, its density dependence is such that it produces overbinding at higher density. To remove the difficulty, we point out the possible importance of the Pauli principle effect in nuclear matter on the intermediate nucleons of an interacting two-pion pair, which is a part of the two-pion exchange potential.

INTRODUCTION, MODEL AND DEFINITION

The two-pion exchange (2πE) three-nucleon potential (3NP) due to Δ has been studied by many people[1]. However, it is obviously only a long range part of the total 3NP. Thus, we investigate (1) the role of heavy bosons as well, and (2) the effects other than Δ in the 2πE 3NP.

As is well known, the two-nucleon interactions are successfully described in terms of the one-boson-exchange potential (OBEP)[2,3]. Important ingredients in OBEP are one-π, one-S, (I=0, $J^P = 0^+$) one-ω and one-ρ exchange interactions. Three-nucleon interactions arise when those bosons, exchanged between two nucleons, interact with the third nucleon on their way. Interpreting S as the I = J = 0 part of the two pions and ω as the I = 0, J = 1 part of the π plus ρ system, we depict three-nucleon interactions due to S and ω as shown in figs. 1b and 1c. We may neglect the ρ contribution to the 3NP in nuclear matter (NM), since our previous estimate indicates it to be very small. Thus we take into account the three diagrams shown by figs. 1a, 1b and 1c as important mechanisms in NM and call the corresponding potentials as ππF, SSF and ωωF potentials.

We calculate the Feynman amplitude for each diagram and obtain the potential in the momentum space $V^{(3)}(\{\vec{p}_k'\}, \{\vec{p}_k\})$, where $\{\vec{p}_k'\}$ and $\{\vec{p}_k\}$ represent the momenta of the final and initial particles respectively. The 3NP in the $\vec{r}$-space, $V^{(3)}(\vec{r}_1, \vec{r}_2, \vec{r}_3)$, is obtained by the Fourier transformation. Integrating it over variables of the

ISSN: 0094-243X/78/405/$1.50 Copyright 1978 American Institute of Physics

third nucleon, we define the effective two-body potential as follows.

$$\bar{v}^{(3)}(\vec{r}_1 - \vec{r}_2) = \rho \int d\vec{r}_3 [\eta(\vec{r}_1 - \vec{r}_3) \; \eta(\vec{r}_2 - \vec{r}_3)]^2 \; v^{(3)}(\vec{r}_1, \vec{r}_2, \vec{r}_3), \quad (1)$$

where ρ is the density of NM, $\eta(r)$ is the correlation function, which is parametrized as $\eta(r) = 1 - \exp[-K^2 r^2]$ with K determined so as to reproduce the excitation parameter due to the 2NP of Ueda and Green[3], $v^{(2)}_{UG}$.

To calculate the $\pi\pi$F, SSF and $\omega\omega$F potentials we need the spin- and isospin-nonflip amplitude off the pion mass-shell. This amplitude can be described by four variables, ν, t, k^2 and k'^2, where k and k' are the four momenta of the incoming and outgoing pions, respectively, and ν and t are defined by $\nu = -(p' + p) \times (k' + k)/4m$, and $t = -(k' - k)^2$, where p and p' are the four momenta of the incoming and outgoing nucleons, respectively, and m is the nucleon mass.

We take into account the S and P wave scattering only, besides the nucleon-pole terms. Then we can write the amplitude as follows.

$$F^+(\nu, t, k'^2, k^2) = \tilde{K}(k'^2)\tilde{K}(k^2) \times [f_0 + f_1 \frac{1}{2}(k'^2 + k^2)$$

$$+ f_2 t + f_3 \nu^2 + \frac{g_\pi^2}{m} + \nu g_\pi^2 \{\frac{1}{(p' - k)^2 + m^2} - \frac{1}{(p + k)^2 + m^2}\}] \quad (2)$$

where g_π is the pion-nucleon coupling constant and $\tilde{K}(k^2)$ is the pionic form factor, $[\tilde{K}(k^2)]^2 = (-\mu^2 + \lambda_0^2)/(k^2 + \lambda_0^2)$; μ being the pion mass. We determine the coefficients f_i by requiring (i) that the amplitude on the pion mass-shell, $F^+(\nu, t, -\mu^2, -\mu^2)$, is the same as the Nielsen-Oades amplitude made on the basis of the scattering data and the dispersion relation, and (ii) that the amplitude at the soft-pion limit, $F^+(0, \mu^2, 0, -\mu^2)$, satisfies the partially conserved axial vector current (PCAC) hypotesis[4].

THE 3NP POTENTIALS

The $\pi\pi$F potential in the momentum space is given by

$$v^{(3)}_{\pi\pi F} = \frac{g_\pi^2}{(2\pi)^6} \frac{\vec{\tau}_1 \vec{\tau}_2}{4m^2} F(\vec{q}_1^2) F(\vec{q}_2^2) \times$$

$$(D_1 - D_2 \vec{q}_1 \vec{q}_2) \vec{\sigma}_1 \vec{q}_1 \vec{\sigma}_2 \vec{q}_2 / (\vec{q}_1^2 + \mu^2)(\vec{q}_2^2 + \mu^2), \quad (3)$$

where $F(\vec{q}_i^2)$ is the product of the pionic form factor $\tilde{K}(\vec{q}_i^2)$ in eq.(2) and the form factor $K(\vec{q}_i^2)$ at the vertex of the i th nucleon and pion, and is parametrized as $F(q^2) = (-\mu^2+\Omega^2)/(q^2+\Omega^2)$. Using eq. (2), we have $D_1 = f_0 = 0.91$ and $D_2 = f_1 + g_\pi^2/2m^3 = 0.81$ in units of μ. These are compared with the case where only Δ is considered for the πN scattering: $D_1 = 0$ and $D_2 = 1.04$.

Assuming the interaction Hamiltonian densities for the $S\pi\pi$ and NNS coupling as $H(S\pi\pi) = h_S\, S\partial_u\vec{\pi}\,\partial_u\vec{\pi}$, and $H(NNS) = g_S\,\bar{\psi}\psi S$, the SSF potential in the momentum space is given by

$$v^{(3)}_{SSF}= 0(\vec{p}',\vec{p},\vec{p}_3)\ \frac{3i}{(2\pi)^{10}}\, g_S^2\, h_S^2\, \frac{1}{(\vec{q}^2+m_S^2)^2}\, v^{(3)}(\vec{q}^2),\qquad (4)$$

where $\vec{q} = \vec{p}' - \vec{p}$, $\vec{p}' = \vec{p}_1' = -\vec{p}_2'$, $\vec{p} = \vec{p}_1 = -\vec{p}_1$ and

$$v^{(3)}(q^2) = \int d^4k\, F^+(\nu, 0, k^2, k^2)\, \frac{K(k^2, (k-q)^2)[k(k-q)]^2}{(k^2+\mu^2)^2\{(k-q)^2 + \mu^2\}},\qquad (5)$$

whith $q = (\vec{q}, 0)$ and

$$0(\vec{p}', \vec{p}, \vec{p}_3) = 1 - \frac{1}{2m^2}(\vec{p}'^2+\vec{p}^2+\vec{p}_3^2)+ \frac{\vec{q}^2}{4m^2} - \frac{i}{4m^2}(\vec{\sigma}+\vec{\sigma})\vec{p}' \times \vec{p}.\qquad (6)$$

$K(k^2, (k - q)^2)$ in eq. (5) is the product of the form factors in the $S\pi\pi$ vertices in fig. 1b, and parametrized as follows.

$$K(k_1^2, k_2^2) = (-\mu^2 + \lambda^2)^4/\{(k_1^2 + \lambda^2)(k_2^2 + \lambda^2)\}^2.\qquad (7)$$

To remove the uncertainty due to h_S, g_S and λ involved in eq. (4), we impose the condition that the 2NP represented by fig. 2 is equivalent to the one-σ exchange component in $v^{(2)}_{UG}$. The 2NP of fig. 2 is given by

$$v^{(2)}_{SS} = 0(\vec{p}',\vec{p},0)\ \frac{-g_S^2}{i(2\pi)^7}\cdot\frac{3h_S^2}{(\vec{q}^2 + m_S^2)^2}\ v^{(2)}(\vec{q}^2),\qquad (8)$$

where

$$v^{(2)}(q^2) = \int d^4k\ \frac{K(k^2, (k - q)^2)[k(k - q)]^2}{(k^2 + \mu^2)\{(k - q)^2 + \mu^2\}}\qquad (9)$$

The one σ-exchange component is given by

$$v^{(2)}_\sigma = 0(\vec{p}', \vec{p}, 0)\, v_\sigma(\vec{q}^2),\qquad (10)$$

where $v_\sigma(\vec{q}^2)$, is given in ref. 3.

Then the following condition is required.

$$V^{(2)}_{SS} = V^{(2)}_\sigma. \tag{11}$$

Thus we can write $V^{(3)}_{SSF}$ of eq. (4) as follows.

$$V^{(3)}_{SSF} = \frac{V^{(3)}_{SSF}}{V^{(2)}_{SS}} V^{(2)}_\sigma = 0(\vec{p}\,',\vec{p},\vec{p}_3) \frac{1}{(2\pi)^3} \frac{V^{(3)}(q^2)}{V^{(2)}(q^2)} v_\sigma(\vec{q}^2). \tag{12}$$

We note that, in eq. (12), g_S and h_S do not appear. Only one parameter λ is involved. However the condition of eq. (11) restricts severely the value of λ to be $3.8\mu < \lambda < 5\mu$.

The $\omega\omega F$ potential is constructed similarly to the SSF potential except that the interaction Hamiltonian densities for the $\omega\rho\pi$ and $NN\omega$ couplings are assumed as $H(\pi\rho\omega) = i\, h_\omega\, \varepsilon_{\alpha\beta\gamma\delta}(\mu_\alpha\, \omega_\beta)(\mu_\gamma\, \rho_\delta)\vec{\pi}$, and $H(NN\omega) = ig_\omega\, \bar{\phi}\, \gamma_\mu\, \omega_\mu\, \psi$.

DISCUSSIONS AND CONCLUSION

We calculate contributions due to the 3NP to the binding energy per nucleon, $\Delta(BE/A)$ in NM, using Wong and Sawada code[5]. We employ $V^{(2)}_{UG}$ for 2NP. Numerical results are shown in Table I and II with three choices of the form factors.

The total contribution due to the 3NP and 2NP yields the right value for $\Delta(BE/A)$ at $k_F = 1.4$ fm^{-1}. However we find overbinding at higher density with no trend of saturation. In this respect we note that the $2\pi E$ mechanism between two nucleons in NM is much different from that in free scattering. In NM, momenta k of the nucleons in the intermediate states when interacting two-pions are exchanged are to be restricted to $|k| > k_F$, whereas in the free scattering the restriction is not necessary. If the restriction is taken into account, the $2\pi E$ contribution in NM has to be considerably reduced. This implies that in the present calculation the scalar meson contributions in both 2NP and 3NP should be reduced by the amount due to the Pauli effect. This reduction increases rapidly with density. Very recently we have calculated the process in which two pions with $\pi\pi$ interaction are exchanged between two nucleons[6]. We assume at $k_F = 0$ that the one-σ exchange component in $V^{(2)}_{UG}$ is the same as the $I = J = 0$ part of the sum of the interacting $2\pi E$ exchange potential and the uncorrelated $2\pi E$ potential of Furuichi and Watari (with a cutoff). Then the effective coupling constant of σ to be used in NM with restriction $|k| > k_F$ is found. It is strongly dependent on k_F, decreasing remarkably as k_F increases beyond 1 fm^{-1}. Quantitatively the result depends strongly on the cutoff parameters. However, irrespective of the parameter values, we find that the Pauli principle effect of the

$I = J = 0$ part of the interacting $2\pi E$ potential acts to produce an additional repulsive force, increasing strongly with density.

In conclusion we remark that the SSF and $\omega\omega F$ potentials are the important ingredients in addition to the $\pi\pi F$ potential in NM at the normal density. At higher density, however, the SSF potential causes overbinding and the Pauli effect in the $2\pi E$ mechanism should be taken into account to remove the difficulty.

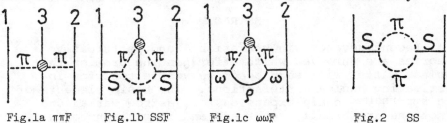

Fig.1a $\pi\pi F$ Fig.1b SSF   Fig.1c $\omega\omega F$ Fig.2   SS

Table I. $\Delta(BE/A)$ (MeV) at $k_F = 1.4$ fm^{-1} due to the 3NP.

| form factor | | $\pi\pi F$ | SSF | | $\omega\omega F$ |
Ω(MeV)	$\lambda_0(\mu)$	$\lambda = 3.8$	$\lambda = 5.0\mu$		
A 420	2.10	0.13	3.5	2.3	-0.92
B 600	2.94	0.43	5.1	3.4	-1.3
C 1800	13.0	1.2	9.5	7.4	-2.1

Table II.   The k_F-dependence of $\Delta(BE/A)$ (MeV) due to the $V_{\pi\pi F}$, V_{SSF} ($\lambda = 3.8\mu$) and $V_{\omega\omega F}$ with f.f. B, and BE/A due to $V^{(2)}_{UG}$.

k_F(fm^{-1})	0.7	1.0	1.3	1.4	1.5	1.7
$V^{(2)}_{UG}$	3.50	7.63	12.1	13.3	14.1	13.9
$V_{\pi\pi F}$	0.01	0.11		0.43		0.57
V_{SSF}	-0.03	0.35		5.1		18.1
$V_{\omega\omega F}$		-0.04		-1.3		-5.1

References

1. J. Fujita and H. Miyazawa, Prog. Theor. Phys. 17, 360 (1957).
2. S. Sawada, T. Ueda, W. Watari and M. Yonezawa, Prog. Theor. Phys. 28, 991 (1962). ; 32, 380 (1964). A.E.S. Green and T. Sawada, Rev. Mod. Phys. 39, 594 (1967).
3. T. Ueda and A.E.S. Green, Phys. Rev. 174, 1304 (1968).
4. T. Ueda, T. Sawada and S. Takagi, Nucl. Phys. A, (1977), to be published.
5. C.W. Wong and T. Sawada, Ann. Phys. 72, 107 (1972) and ref. 21-24 in this paper, K. Takada, S. Takagi and W. Watari, Prog. Theor. Phys., 38, 144 (1967).
6. T. Ueda and S. Takagi, Prog. Theor. Phys. (1977), to be pub.

NUCLEAR MATTER THEORY*

J. W. Negele[+]
Center for Theoretical Physics
Department of Physics and Laboratory for Nuclear Science
Massachusetts Institute of Technology
Cambridge, Massachusetts 02139

ABSTRACT

Recent advances in variational and perturbative
theories are surveyed which offer genuine promise that
nuclear matter will soon become a viable tool for investigating nuclear interactions. The basic elements of
the hypernetted chain expansion for Jastrow variational
functions are briefly reviewed, and comparisons of
variational and perturbative results for a series of
increasingly complicated systems are presented. Pro-
spects for investigating realistic forces are assessed
and the unresolved, open problems are summarized.

I. INTRODUCTION

The fictitious infinite translationally invariant
system of nucleons interacting without Coulomb forces
which has come to be known as nuclear matter is of
fundamental interest to theorists for a variety of
reasons. A definitive, converged microscopic calculation
of nuclear matter would be a major achievement in many-
body theory, a prelude to the systematic evaluation of
properties of finite nuclei, and a crucial first step in
exploring the equation of state of dense matter for
astrophysical purposes. Of particular relevance to this
conference is the fact that a reliable theory of nuclear
matter would provide an invaluable tool in exploring the
off-shell behavior of the nucleon-nucleon interaction.
Before focussing attention on the technical aspects
of nuclear matter theory and its use in investigating
nuclear interactions, it is perhaps worthwhile to
emphasize the significance of the many-body problem on
its own merits. For many years there has been a dog-
matic bias, especially on the part of solid state theorists, against any sort of microscopic calculation which
attempts to derive observables in terms of the underlying
two-body interaction. Now that we have at our disposal

----

* This work is supported in part through funds provided
  by ERDA under Contract Ey-76-C-02-3069.*000.

+ Alfred P. Sloan Foundation Fellow.

ISSN: 0094-243X/78/410/$1.50 Copyright 1978 American Institute of Physics

exceedingly powerful complementary techniques, it will
be possible to make a much stronger case for the convergence and the reliability of many-body calculations.
Thus the progress reviewed in this work is of direct
relevance to other areas of condensed matter physics,
and we shall freely address systems such as liquid He^3
and He^4 where appropriate.

It has long been taken for granted that nuclear
matter provides the best quantitative tool for studying
the off-shell behavior of the nuclear force. The
canonical argument is roughly the following. By virtue
of translation invariance, many-body theory is simpler
in nuclear matter than in finite nuclei both because
momentum conservation eliminates large numbers of
topologies of Goldstone diagrams and because there is
no need to determine single particle wave functions.
Since the semi-empirical mass formula reproduces the
systematics of the mass table, the reasonably unambiguous value for the volume binding energy per particle
succinctly quantifies the constraint on nuclear forces
provided by nuclear masses.

Although, as already stated, nuclear matter does
have significant value, it is nevertheless useful to
bear in mind its limitations and the advantages of
calculating finite nuclei directly. The semiempirical
mass formula tells us virtually nothing about the
saturation density of nuclear matter. Whereas the
Coulomb term yields some information about charge radii,
no direct information is provided concerning neutron
radii, so a model is required to learn about the total
density. Rather than invoke liquid drop or droplet
models, the most consistent means of inferring nuclear
matter density is to extrapolate using Hartree Fock
calculations with realistic or phenomenological forces
that reproduce energy and charge radius systematics.
Using the interactions summarized in Table II of Ref. 1
yields nuclear matter saturation at k_F = 1.30-1.34 fm^{-1}
corresponding to ρ = 0.148-0.163 fm^{-3} with E/A = 15.7-
16.6 MeV. Nevertheless, such extrapolation entails
unavoidable uncertainties to which the direct calculation
of finite nuclear energies and radii is not subject.
(In contrast, the liquid helium problem is much richer,
providing not only the equilibrium energy and density,
but also the compression modulus from the speed of sound,
the surface energy from the surface tension, and the
equilibrium two-body correlation function from neutron
or X-ray scattering.)

It is also not clear that nuclear matter is necessarily easier to calculate than optimally chosen finite
nuclei. The three-body problem can now be solved
extremely accurately by various techniques,[2,3] making it

clear, for example, that the Reid potential underbinds by the order of 1 MeV. Turning to heavier closed shell spherical nuclei, calculations using the e^S method[4] in ^{16}O and ^{40}Ca are technically comparable or superior to hole-line expansion calculations of nuclear matter. It is true that no comparable calculations exist in heavy nuclei, but nuclear matter, like ^{16}O and ^{40}Ca, is spin-isospin saturated and ^{40}Ca seems sufficiently large to provide a reasonable test of both the binding and saturation of nuclear forces.

Thus, although I shall concentrate my subsequent attention on the nuclear matter problem, in the final assessment of what the many-body problem teaches us about nuclear interactions, it will be useful to return to the complementary evidence provided by finite nuclei.

II. THEORETICAL APPROACHES

The three primary techniques for solving quantum many-body problems of the nature posed by nuclear matter or liquid helium are Monte Carlo solution of the Schrödinger equation, perturbation theory, and variational methods. These three approaches differ significantly in their goals, strengths, and limitations. The Monte Carlo method calculates the ground state wave function and energy for a finite number of particles in a box with periodic boundary conditions subject only to statistical sampling errors. To date, however, it has only been implemented for bosons.[5-7] Perturbation theory aspires to systematically calculate approximations of successively increasing accuracy to expectation values of few body operators. Its strength lies in the fact that it is systematic and describes the actual ground state energy. Its weakness, when applied to the nuclear matter problem, is that there are no rigorous convergence arguments formulated in terms of a suitable small parameter, and at best one must be satisfied by apparent numerical convergence. Finally, the variational method makes no pretense of systematically approximating the ground state energy, but rather provides a rigorous upper bound. By stationarity, physically sensible variational wave functions may be expected to yield bounds reasonably close to the true minimum, and the chief technical difficulty is the evaluation of the energy with a trial function.

A. Monte Carlo Solution of the Schrödinger Equation

The limitations of time and space preclude an adequate review of the Monte Carlo method, and interested readers are referred to the literature.[5,6] The basic idea, however, is the following. Letting R denote the set of coordinates $\vec{r}_1$, $\vec{r}_2$, $\cdots$ r_n, the Schrodinger equation, shifted by an arbitrary constant c

$$(T + V(R)+c)\psi(R) = (E+c)\psi(R) ,$$

may be rewritten

$$\psi(R) = (E+c)\int G(R,R')\psi(R')dR'$$

where $(T+V(R)+c)G(R,R_o) = \delta(R-R_o)$. If a sequence of functions is defined by

$$\psi^{(n+1)}(R) = (E+c)\int G(R,R')\psi^n(R')dR' ,$$

then the ground state is obtained in the limit of large n. In the Monte Carlo method, one samples sets of co-ordinates R' drawn at random from ψ^n. For any such R', a point R from the distribution $\psi^{(n+1)}$ is obtained by sampling $(E+c)G(R,R')$ considered as a density function for R conditional upon R'. The eigenvalue E is obtained from the condition that the normalization be asymptotic-ally stable. Sampling of the Greens function $G(R,R')$ is accomplished by writing an integral equation for G in terms of a known kernel which is solved by an iterative random walk procedure.

B. Perturbation Theory

Perturbation theory techniques developed over the last two decades are reviewed extensively in the litera-ture,[8-10] and hence only a few remarks are required here. Because of the repulsive interaction between nucleons at short range, the perturbation series is rewritten in terms of the reaction matrix, G, which sums the ladder diagrams representing all possible rescattering of two interacting nucleons into unoccupied intermediate states. Solving the integral equation

$$\psi = \phi - \frac{Q}{e} v \psi$$

where Q projects onto unoccupied states and e is the energy denominator, builds into ψ the short range cor-relations which are absent in the unperturbed wave function ϕ.

A crucial parameter in the theory is the "wound integral", κ, which specifies the probability that the correlated wave function contains components above the fermi sea. Defining $\zeta = \phi-\psi$,

$$\kappa = \frac{1}{A} \sum_{\substack{a,b>k_F \\ B,D<k_F}} |<ab|\psi_{BD}>|^2 = \rho<\psi|Q|\psi> = \rho<\phi-\zeta|Q|\phi-\zeta>=\rho<\zeta|\zeta>$$

414

and at nuclear matter density with the Reid potential, $\kappa \sim 15\%$. With this definition, a simple argument suggests that the contributions of Goldstone diagrams may be classified in powers of κ by the number independent hole lines. Consider

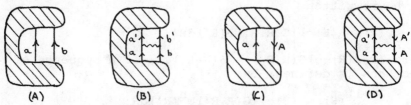

(A)   (B)   (C)   (D)

Figure 1.   Diagrams illustrating κ ordering.

the ratio of a diagram with an extra G interaction between two particle lines to the diagram without this interaction. Roughly estimating the ratio of diagram B to diagram A in Figure 1,

$$R \approx \sum_{a'b'>k_F} <a'b'|\frac{G}{e}|ab>\delta(\vec{a}'+\vec{b}'-\vec{a}-\vec{b}) = \sum_{a'b'} <a'b'|Q\frac{G}{e}|ab>\delta(\vec{a}+\vec{b}'-\vec{a}-\vec{b})$$

$$= \int d^3r \sum_k e^{ik \cdot r} \zeta_{ab}(r) = \zeta_{ab}(o) = 1.$$

Thus, creating an additional particle line by adding an interaction does not, in general, reduce the contribution of a diagram at all. In contrast, consider the effect of adding an independent hole line by calculating the ratio of diagram D to diagram C. Treating, for simplicity, only the repulsive core region and neglecting the effect of Q,

$$R \approx \sum_{\substack{A'<k_F \\ a>k_F}} <a'A|\frac{G}{e}|aA'>\delta(\vec{a}+\vec{A}'-\vec{a}'-\vec{A})$$

$$\approx \sum_{A'<k_F} \frac{1}{V} <\phi|\zeta> \approx \rho <\zeta|\zeta> = \kappa$$

Thus, the addition of an independent hole line diminishes the contribution by the order of κ, giving rise to the familiar hole line expansion in which diagrams are classified and calculated in groups having a specified number of hole lines. The next approximation beyond the 2-body reaction matrix is to calculate the three-hole line diagrams by solving the three-body Faddeev equation. Although most of the nuclear matter Faddeev results in the literature[8,9] involve inaccurate approximations and errors, recent results by Day[10,11] treat the tensor force quite accurately, and for most practical purposes, three hole-line terms are under control. Four hole-line contributions have only been estimated and remain a primary uncertainty in the perturbation approach.[12]

One flaw in the κ ordering argument, which is already apparent in our sketchy derivation above, concerns the fact that it does not really apply to the contributions of the long range tensor force. The contributions of long range forces to the RPA ring diagrams have always been suspect, and recent progress in understanding the spin-isospin instability known as "pion condensation" just beyond nuclear density suggests that these terms should be carefully investigated. A particularly appealing means of including ring graphs is the use of the e^S formalism,[4] in which the analog of the Faddeev equation automatically includes them to all orders, and work using this method is in progress at Argonne.[11]

C. Variational Bounds

Dramatic developments have occurred in recent years in the implementation of powerful techniques to evaluate variational bounds using Jastrow trial functions. The essential feature of the original Jastrow wave function[13]

$$\psi(r_1, r_2 \cdots r_n) = \prod_{i<j} f(|r_i - r_j|) \phi(r_1, r_2 \cdots r_n)$$

is the introduction of two-body correlations into an uncorrelated many body wave function ϕ of the proper Fermi or Bose statistics. For a central potential containing strong short range repulsion, f describes essentially the same depletion of probability at small interparticle separation that the defect function ζ describes in the reaction matrix approach. Whereas a partial wave decomposition of the reaction matrix renders it straightforward to describe the state dependence of ζ, it is rather difficult to generalize beyond a central, state independent f as we shall subsequently discuss. One could imagine a systematic hierarchy of successive approximations in which correlation functions involving increasing numbers of particles are introduced, yielding in next order,

$$\psi(r_1, r_2, \cdots r_n) = \prod g(r_\ell, r_m, r_n) \prod f(r_i, r_j) \phi(r_1, r_2 \cdots r_n)$$

This series, of course, approaches the true wave function only if no restrictions (like requiring $f = f|r_i - r_j|$) are imposed on the correlation functions. In practice, three body correlation functions are difficult to treat with sufficient accuracy to assure a valid bound and, by stationarity, one may aspire to extract useful physics using reasonable two body correlation functions.

Basically, two alternative techniques exist for evaluating the energy with Jastrow trial functions: Monte Carlo sampling, and summation of appropriate terms of a cluster expansion. The former is limited only by statistical sampling errors, and serves as an absolute check on the convergence of the latter.

The variety of alternative approaches in deriving and summing cluster expansions is so large[14-20] that I shall make no attempt at completeness or attribution of historical credit. Rather, I will concentrate exclusively on the hypernetted chain approach which has had the most direct impact on the nuclear matter problem.[19-22]

D. Irreducible Cluster Expansion

In evaluating

$$<\phi| \prod_{i<j} f_{ij} \; H \prod_{k<\ell} f_{k\ell} |\phi>/<\phi| \prod_{i<j} f_{ij} \prod_{k<\ell} f_{k\ell} |\phi>$$

one proceeds along the lines of the Mayer expansion in classical statistical mechanics.[23] Considering for the moment bosons, where $\phi=1$, and considering only the potential energy term in H, in both the numerator and denominator the product $\prod_{i<j} f_{ij}^2$ is rewritten as $\prod_{i<j}(1+h_{ij})$. Since f differs from unity only at short range, h contributes only over a restricted range and the product may be expanded in powers of h. range and the product may be expanded in powers of h. Following the argument of Ref. 20 gives rise to a very simple diagram expansion. Denoting h by a solid line, the denominator expansion is represented by the sum of terms obtained by connecting N numbered points with any number of lines in all possible ways as shown in low order in Figure 2.

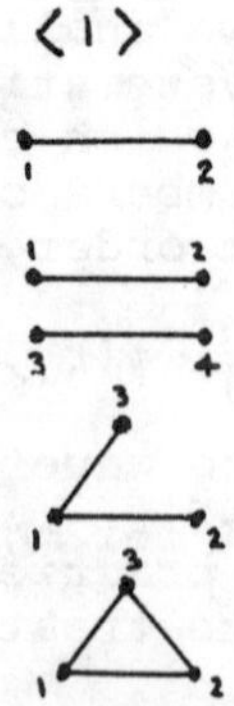

(A) $\langle 1 \rangle$ $= \Omega^N$

(B) $= \Omega^{N-1} \int h_{12} \, d^3r_{12}$

(C) $= \Omega^{N-2} \int h_{12} \, d^3r_{12} \int h_{34} \, d^3r_{34}$

(D) $= \Omega^{N-2} \int h_{12} d^3r_{12} \int h_{13} d^3r_{13}$

(E) $= \Omega^{N-2} \int h_{12} h_{13} h_{23} d^3r_{12} d^3r_{13}$

Figure 2. Diagrammatic expansion for $<\psi|\psi> = \int dr_1 \cdots dr_n \prod_{i<j}(1+h_{ij})$, where the solid line denotes h and the volume is Ω.

Note that an interesting feature arises in comparing
diagrams C and D. Although the topologies are different,
the integrals are identical and correspond to the simple
product of two independent factors. In general, any
diagram such as D which can be split into two disconnected subdiagrams by removing a single common point can
be factored into a product of two independent factors by
proper choice of relative variables, and is thus defined
to be reducible.

Turning now to the numerator, the expansion of
$\int dr_1 \cdots dr_n \prod_{i<j} f_{ij}^2 \sum_{m<n} v_{mn}$ is accomplished by grouping
f_{mn}^2 with v_{mn}, which is denoted by a double line, and
expanding all other terms $\prod_{i,j\neq m,n} f_{ij}^2 = \prod_{ij\neq m,n}(1+h_{ij})$.
Some leading order terms in this numerator expansion are
shown in Figure 3.

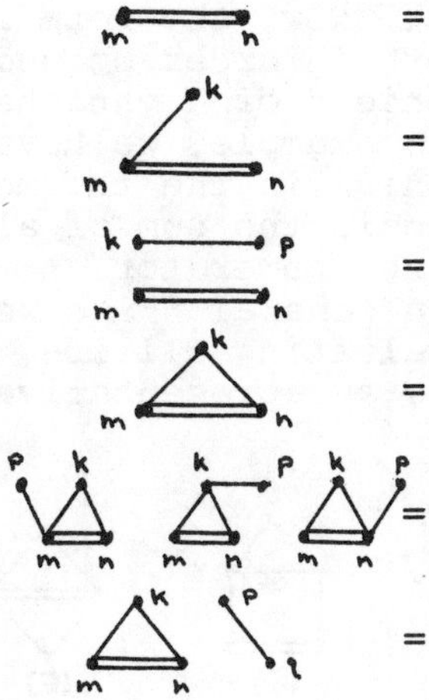

(A) $= \int dr_{mn}^3 \Omega^{N-1} f_{nm}^2 \, v_{nm}$

(B) $= \int dr_{mn}^3 \Omega^{N-2} f_{nm}^2 v_{nm} \int h_{km} d^3 r_{km}$

(C) $= \int dr_{mn}^3 \Omega^{N-2} f_{nm}^2 v_{nm} \int h_{kp} d^3 r_{kp}$

(D) $= \int dr_{mn}^3 \Omega^{N-2} f_{nm}^2 v_{nm} \int h_{mk} h_{nk} d^3 r_{km}$

(E) $= \int dr_{mn}^3 \Omega^{N-3} f_{nm}^2 v_{nm} \int h_{mk} h_{nk} d^3 r_{mk} \int h_{ip} d^3 r_{ip}$

(F) $= \int dr_{mn}^3 \Omega^{N-3} f_{nm}^2 v_{nm} \int h_{mk} h_{nk} d^3 r_{mk} \int h_{pq} d^3 r_{pq}$

Figure 3. Diagrammatic expansion for $\langle \psi | v | \psi \rangle =$
$\int dr_1 \cdots dr_n \prod_{i<j\neq mn}(1+h_{ij}) \sum_{m<n} v_{mn} f_{mn}^2$, where the
double solid line indicates $v_{mn} f_{mn}^2$.

In deriving a linked cluster expansion in many-body
perturbation theory, one usually cancels the unlinked
terms in the numerator against corresponding terms in
the denominator. In the present case, the situation is
evidently modified by the equality between certain
reducible diagrams and appropriate unlinked diagrams.
To leading order in h, the denominator is

$\Omega^N[1 + \frac{1}{\Omega} \frac{N(N-1)}{2} \int h_{12}d^3r_{12}]$ since there are $\frac{N(N-1)}{2}$
independent ways of picking points 1 and 2. Now,
consider the sum of diagrams D + F in Figure 2. Denoting
$\mathcal{E}_{IRRED.}$ as the contribution of the irreducible diagram D,
the sum of D+F yields D+F $= \mathcal{E}_{IRRED.}[1 + \frac{1}{\Omega} \frac{(N-3)(N-4)}{2}$ x
$\int h_{pq}d^3r_{pq}]$ since there are $\frac{(N-3)(N-4)}{2}$ ways of selecting
p<q distinct from m and n. Although the N^2 term in the
brackets cancels the N^2 term the denominator, the terms
of order N do not cancel and are finite in the limit
$N \rightarrow \infty, \Omega \rightarrow \infty, N/\Omega = \rho$. If in addition, the three diagrams
labelled E in Fig. 3 are included, one obtains

$$\mathcal{E}_{D+E+F} = \mathcal{E}_{IRRED.}[1 + \frac{1}{\Omega}(\frac{N^2}{2} - \frac{N}{2} - 3)\int h_{pq}d^3r_{pq}]$$

since there are (N-3) ways of selecting the additional
point in each of the 3 new diagrams. Now, the term in
brackets cancels the denominator both in order N^2 and in
order N, and the discrepancy independent of N vanishes
in the thermodynamic limit. In this example, we have
seen that aside from terms which vanish in the thermo-
dynamics limit, the denominator cancels the sum of all
reducible and unlinked diagrams in the numerator.
This cancellation is in fact true in general,[19] so we
are left with the task of simply evaluating all irreducible linked numerator diagrams, a representative
sample of which are shown in Figure 4.

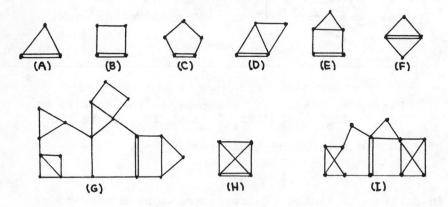

Figure 4. Irreducible linked diagrams.

E. Hypernetted Chain Expansion

The motivation for utilizing the hypernetted chain expansion arises from considering the connectivity of various diagrams. A particularly clear argument is given by Pandharipande and Bethe,[20] using the graph reproduced in Figure 5. The contribution of diagram A of Fig. 4, omitting the factor $\Omega^{N-1}f^2_{mn}v_{mn}$ and integral $\int dr_{mn}$ common to all diagrams, is

$$\mathcal{E}_A = \rho \int h_{mk}h_{kn}d^3r_k \equiv S_{mn},$$

where the factor ρ arises from the $1/\Omega$ and sum over labels for particle k. The functions $S_{mn} \equiv S(r_{mn})$ and $-h_{mn} \equiv -h(r_{mn})$ are drawn in Fig. 5 for liquid He^4 near equilibrium density. The contribution of diagram B, which contains one more link in a chain of h factors, is

$$\mathcal{E}_B = \rho^2 \int h_{mk}h_{k\ell}h_{\ell n}d^3r_k d^3r_\ell$$

$$= \rho \int S_{m\ell}h_{\ell n}d^3r_\ell$$

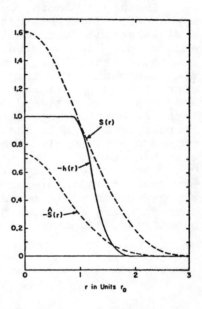

Figure 5. The functions h, S, and Ŝ defined in the text. (Ref. 20).

Since at high density S is everywhere larger than h, $\mathcal{E}_B > \mathcal{E}_A$ and the alternating sequence of chain diagrams A,B,C... must be summed by an integral equation.

With this sobering introduction, we ask if there are any other ways of adding a line to a diagram such that its contribution remains of the same order instead of decreasing. The contribution of diagram D is

$$\mathcal{E}_D = \rho^2 \int h_{nk} \ h_{mk}(h_{kp}h_{mp}d^3r_p)d^3r_k$$

$$= \rho \int h_{nk} \ h_{mk} \ S_{km} \ d^3r_k$$

Since $S \geq 1$ in the region where h is non-negligible, $\mathcal{E}_D$ is of the same order as $\mathcal{E}_A$. Similarly, diagrams E and F, in which additional chains are added between two particles in an existing chain, are also of the order of $\mathcal{E}_A$. Diagrams in which any two points of a chain may be connected by many other chains are called hypernetted chains, and a typical hypernetted chain is shown in Diagram G.

Instead of adding our additional line to D in such a way as to add or lengthen a chain, neither of which diminishes the contribution, suppose one generates diagram H. Its contribution is graphed as $\hat{S}$ in Fig. 5 and is smaller than A or D because all four particles must simultaneously be within the range of f. Thus, the connectivity of diagrams is crucial: the addition or lengthening of chains does not necessarily diminish the size of contributions since many particles are not forced thereby to be within the range of f, whereas adding more and more complicated cross links leads to significant reduction.

Such connectivity arguments lead naturally to the hypernetted chain hierarchy of approximations. The first approximation would be the chain approximation, which sums diagrams A,B,C,... of Fig. 4. The basic building block is G, denoted by a wavy line

Clearly, G satisfies the integral equation

$$G_{mn} = \rho \int h_{mk}[h_{kn} + G_{kn}] \, d^3r_k$$

and in terms of G, the potential energy is

$$\mathcal{E} = \int f_{mn}^2 \, v_{mn}(1 + G_{mn}) \, d^3r_{mn}$$

The next step is to sum hypernetted chains (HNC). Assume that we have a formula for a general chain G_{ab} such that each term involves a sum over at least one intermediate point. Then any number of G chains may be inserted between a and b by exponentiation:

Note that the n! in each term of the exponential simply cancels the n! different ways the wavy lines could be relabelled to yield the same irreducible graph. We now seek an integral equation for G analogous to the previous chain equation. Instead of using h as the basic source term, we would like to use a more general source term so that iteration grows diagrams like graph G of Fig. 4. The series $_e$G^{ab} is a good start, but has two shortcomings. By assumption there is at least one intermediate point in each chain comprising G, so there can be no h directly connecting a and b. Hence, from each diagram, we can generate an additional diagram by multiplication by h. Secondly, we must remove the 1 arising from the exponential. Hence, we replace the h in the chain equation by the source term

$$-1+(1+h)e^G = -1 + 1 + \text{〜〜} + \text{◯} + \text{◯}$$
$$+ \text{▬} + \text{◺} + \text{◲} + \text{◲} + \cdots$$

Thus, we obtain the von Leeuwen, Groeneveld and deBoer HNC integral equation[24]

$$G_{mn} = \rho \int [(1+h_{mk})e^{G_{mk}}-1][(1+h_{kn})e^{G_{kn}}-1 + G_{kn}]d^3r_k.$$

In terms of this G_{mn}, the potential energy is

$$\mathcal{E} = \int f^2_{mn} v_{mn} e^{G_{mn}} d^3r_{mn}$$

The next step in the hierarchy would be to evaluate the 4-point elementary diagram, graph H of Fig. 4 and to include it in the source term of the integral equation, thus generating terms like diagram I of Fig. 4. This is denoted the HNC/4 approximation and in principle one could go on to include n-point elementary diagrams up to any order.

F. Fermions

Having developed the basic ideas at some length, we will merely sketch the additional complications which arise in the case of fermions. The uncorrelated wave function in this case is a Slater determinant of plane waves. The product $\langle\phi|\phi\rangle$ occurring in the numerator and denominator is most conveniently written

$$\phi^*\phi = \begin{vmatrix} \psi_1(1)\cdots\psi_n(1) \\ \vdots \\ \psi_1(n)\cdots\psi_n(n) \end{vmatrix}^* \begin{vmatrix} \psi_1(1)\cdots\psi_1(n) \\ \vdots \\ \psi_n(1)\cdots\psi_n(n) \end{vmatrix} = \begin{vmatrix} \rho(1,1)\cdots\rho(1,n) \\ \vdots \\ \rho(n,1)\cdots\rho(n,n) \end{vmatrix}$$

$$= \sum_P (-1)^P \hat{\rho}(1,P1) \hat{\rho}(2,P2)\cdots\hat{\rho}(n,Pn) \quad \rho^N$$

where the normalized density matrix is defined

$$\hat{\rho}(r_1, r_2) \equiv \frac{1}{\rho} \sum_n \psi^*_n(r_1) \psi_n(r_2) = \frac{s}{(2\pi)^3 \rho} \int_{k<k_F} d^3k \; e^{ik\cdot r_{12}}$$

$$= \frac{3j_1(k_F r_{12})}{k_F r_{12}}$$

and s is the spin-isospin degeneracy (4 for nuclear matter). Thus, in addition to the dynamical correlation factors h treated previously, we must now keep track of the statistical correlation factors arising from all the possible permutations of the labels in the product of $\hat{\rho}$'s. The unit permutation yields one, since $\hat{\rho}(r_i, r_i) = 1$, and all other permutations yield products of the $\hat{\rho}$'s containing interchanged coordinates. Following Fantoni and Rosati,[19] the diagram notation is extended to account for a statistical correlation factor $-\frac{1}{s}\hat{\rho}_{ij}$ by a dashed line between points i and j. Since all permutations may be represented by interchanges, the dashed lines must form a series of closed polygons. In the special case of simple interchange of two particles (a 2-sided polygon) it is convenient to represent the product $-\frac{1}{s}\hat{\rho}^2_{ij}$ by a helical line. Extending the analysis of the boson case, Fantoni and Rosati have proved that the denominators again cancel against the unlinked and factorizable parts of the numerators to order 1/N. The potential is then given by the sum of all linked irreducible diagrams constructed according to the following rules:[19]

a) Each point is the extremity of at least one solid line.
b) The dashed and helical lines can be superimposed on the solid lines.
c) The dashed lines are arranged in closed polygons such that there are no common points between different polygons. Each polygon has a multiplicative factor -2s.
d) No helical line has a common point with another helical line or dashed line.

The formulation of fermi chain and fermi hypernetted chain (FHNC) equations is straightforward but somewhat tedious because of the topological restrictions contained in the preceding rules. The physical motivation is the same as before: since both h and $\hat{\rho}$ fall off at large distances, cross links of either kind between chains diminishes the contribution by the necessity of all the particles simultaneously being close together.

Whereas there was just one elementary 4-point diagram for bosons, our new set of solid, dashed, and helical tinker toys admits a large number of possibilities which can finally be reduced to 35 numerically different 4-point diagrams.[22] Given the fact that extension to FHNC/4 is so complicated, it is improbable that anyone will tackle the several thousand elementary diagrams required for FHNC/5.

G. Other Complications

The suspicious observer may perhaps have noticed that we have always considered the expansion of the potential energy instead of the total energy. Indeed, if we replaced $v(r_{ij})$ by the two body density operator, we have really only discussed the calculation of the two body radial distribtuion function, usually denoted $g(r)$. Letting F represent $\prod_{i<j} f_{ij}$ and applying the kinetic energy operator to the right, the kinetic energy may be written in the form used by Pandharipande and Bethe[20]

$$T_{PB}= -\frac{\hbar^2}{2mN\langle\psi|\psi\rangle} \sum_i \langle\phi|F[F\nabla_i^2 + (\nabla_i^2 F)+2(\nabla_i F)\cdot\nabla_i]|\phi\rangle.$$

The $\nabla_i^2 F$ term involves the three body operator $\nabla_i f_{ki}\cdot\nabla_i f_{\ell i}$ so its evaluation also requires the calculation of the 3-body radial distribution function. Alternative expressions for the kinetic energy are obtained by Jackson and Feenberg, T_{JF}, and by Clark and Westhaus, T_{CW}, by various integrations by parts. Although these formulations are equivalent if the irreducible cluster expansion is evaluated to all orders, truncation of the different expressions yields different results. Since no convincing argument exists as to why one form should be more accurate, it seems reasonable to use such discrepancies as an indication of the order of magnitude of truncation uncertainties.

Thus far, we have considered only spherically symmetric, state independent correlation functions $f(|r_i-r_j|)$. Clearly, if one allows f to have any momentum dependence or angular momentum dependence, the structure of the product $\langle\phi|\pi f_{ij} v\pi f_{mn}|\phi\rangle$ is immensely complicated by the fact that the differential operator for a given particle coordinate acts on very many other terms in the product. A more tractable generalization of the wave function is to write f_{ij} as the sum of different radial functions multiplying the operators $1, \sigma_i\cdot\sigma_j, \tau_i\cdot\tau_j, (\sigma_i\cdot\sigma_j)(\tau_i\cdot\tau_j), S_{ij}$, and $S_{ij}(\tau_i\cdot\tau_j)$, where S_{ij} is the usual tensor operator. If v is expressed in terms of these same operators, the algebra closes and a generalized

FHNC expansion may be derived.[21] Although this trial
function is not as general as one might like, it does
have considerable flexibility. The correlation
function can differ in singlet even, triplet even,
singlet odd and triplet odd partial waves, and the tensor
correlations appear reasonably well accommodated.

In general, it is difficult to assess the absolute
accuracy of various FHNC truncations. Hence it is
particularly refreshing to note the test provided by the
evaluation of the model kinetic energy.[22] Defining

$$T_{MOD} = \frac{1}{N} \frac{<\phi|T|\psi>}{<\phi|\psi>} = \frac{1}{N} \frac{<\phi|T_i \prod_{i<j} f_{ij}|\phi>}{<\phi|\prod_{i<j} f_{ij}|\phi>} ,$$

we note by integrating by parts that the exact result is
$3/5 \, k_F.^2$ However, writing $f_{ij} = 1+h_{ij}$ and expanding as
before, we can obtain the truncated result in the FHNC
and FHNC/4 approximations and compare with the exact result, thereby obtaining a quantitative measure of the
truncation error.

III. COMPUTATIONAL RESULTS

In order to display all the relevant features of the
three theoretical approaches discussed in the last
section, I will discuss four separate quantum many body
systems which progressively approach the nuclear matter
case.

A) Liquid He^4

When scaled in length by r_0, the radius of a sphere
containing one particle at equilibrium density, and in
energy by $\frac{\hbar^2}{mr_0^2}$, the nuclear S_0 potential and the He^4
potential are qualitatively similar. The core radius,
however, is twice as large for He^4, rendering the core
volume 8 times as large. Thus He^4 is a significantly
denser system than nuclear matter and hence a far more
stringent test of many body theory. Since κ in the hole
line expansion is of the order of 50%, we will not even
discuss perturbation theory results in this system.

In Fig. 6, I have collected for the first time on
one graph in consistent units the HNC and Monte Carlo
results for the Leonard Jones potential. Starting at
the bottom, the solid curve shows the experimental
saturation curve at and above equilibrium density.
(Envious theorists will probably also enjoy the experimental radial distribution function in Fig. 7). The
points with error bars denote the Greens function Monte
Carlo (GFMC) calculations. The three most accurate
points are recent unpublished results[7] and the other
three are from Ref. 6. The disagreement between GFMC and

Fig.6. Binding energy as a function of density for liquid He^4 where ρ_0 indicates the experimental equilibrium density $0.02185_A{-3}$.

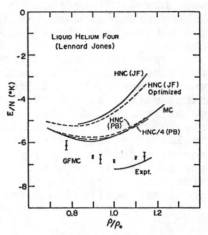

experiment clearly indicates that the Lennard Jones potential is slightly incorrect, so the goal of using condensed matter as a constraint on two body potentials has certainly been reached in this system. In fact, theoretical Born Oppenheimer calculations of the He potential and measurements of the second virial coefficient already indicate deficiencies in the simple Lennard Jones potential,[26] so we shall ignore the small discrepancy with experiment and simply regard this as a model problem to test other approximations.

There are two Jastrow calculations using the correlation function $f(r) = e^{-\left(\frac{b}{r}\right)^m}$. The curve labeled MC is a Monte Carlo calculation,[26,27] so the discrepancy with GFMC indicates the limitations of that form of trial function. Indeed, considering the high density and strong repulsion of the force, the accuracy of this Jastrow bound is quite impressive. The curve denoted HNC(JF) shows the hypernetted chain approximation to the energy using the Jackson Feenberg kinetic energy.[26] The discrepancy between this curve and MC shows the non-negligible error in the HNC approximation.

One might wonder how much the HNC energy may be lowered by solving an Euler-Lagrange equation for an optimal $f(r)$ instead of using the parameterization $e^{-\left(\frac{b}{r}\right)^m}$. The result obtained from a very elegant calculation[28] is labelled HNC(JF) OPTIMIZED. The extremely small increase in binding suggests that optimization of a central two body correlation function is quantitatively unimportant, and that the main deficiency of the Jastrow wave function is the omission of three body terms and state dependence.

Finally the HNC(PB) and HNC/4(PB) show the results using yet a different two body correlation function and the Pandharipande Bethe kinetic energy both omitting and including the 4-point elementary diagram.[35] Since HNC(PB) is far below the optimized HNC(JF) curve, it is clear that the difference arises from the PB kinetic energy prescription. In this case, the 4-point elementary diagram has very little effect, and both curves agree beautifully with the Monte Carlo curve. (We have already

426

seen how little effect the parameterization of f(r) had in the previous case, so it is reasonable that at least the HNC/4 curve should agree well with the Monte Carlo result.)

From this example, we conclude that a Jastrow wave function with a spherically symmetric f(r) yields very useful bounds, that different kinetic energy prescriptions matter at the HNC level and that HNC/4 appears to be extremely accurate. Just for fun, since nuclear physicists have yearned for a long time to integrate the inelastic electron scattering cross section over all energy transfer at fixed momentum transfer to extract the fourier transform of the two body correlation function, the result of the analogous sum rule with neutron and X-ray scattering from liquid He[4] is shown in Fig. 7.

Figure 7. The radial distribution function g(r) calculated in the HNC/4 approximation (dotted curve) and obtained from experiment (solid line). (Ref. 20).

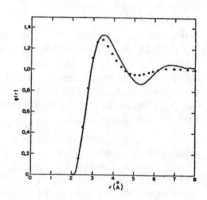

It is quite impressive how accurately at least the short range part is reproduced in a HNC/4 calculation.

B) Bose Gas Homework Problem

In order to compare various theories on a model problem which simulates certain features of the nuclear force, the $^3S_1$ component of the Reid potential treated as a central force has been widely used as a "homework problem". In Figure 8, we display HNC variational results[29] and hole-line expansion results[29,11] at and well beyond nuclear density for a Bose gas. Since the two hole line energy, E(2) has uncertainties in the potential energy on the order of κ, the resulting uncertainty in the energy is denoted by the error bars. Values of κ are also denoted at the positions of the arrows. Note that E(3) is, in fact, well within these error bars and furthermore it agrees quite well with HNC up to twice nuclear density. Thus, we observe satisfactory convergence in the hole line expansion and complete

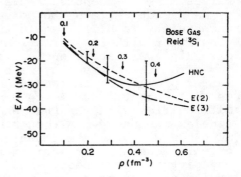

Figure 8. Binding energy as a function of density for the Bose gas homework problem.

consistency with the variational results.

C) Nuclear Matter Homework Problem

Turning now to fermions, we first treat the simple case of a pure central potential (again the $^3S_1$ Reid homework potential) so as to investigate the convergence of various fermion approximations without the added complications of realistic state dependence. In Fig. 9

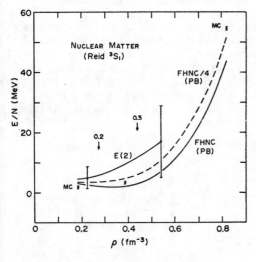

Figure 9. Binding energy as a function of density for the nuclear matter homework problem.

Monte Carlo variational calculations,[30] denoted by the error bars and labeled MC, and FHNC calculations with the same correlation function[22] are compared. Again, the FHNC convergence appears quite satisfactory, with little change between FHNC and FHNC/4 and with reasonable agreement between FHNC/4 and the Monte Carlo results.

For this system only two hole line perturbation theory results, E(2), are available.[29,11] As before,

values of κ are denoted above the arrows and error bars
of magnitude κ times the potential energy indicate the
uncertainty in E(2). Again the perturbation results are
consistent with the variational bounds for densities
extending well above nuclear density.

From these results, we conclude that the nuclear
matter problem is in fact under very good control for
central potentials. There don't seem to be any essential
difficulties with the FHNC expansion [31] or with the hole
line expansion .[32] Although variational calculations are
much harder for state dependent potentials, perturbation
calculations are not, and there is no apparent reason
why perturbation theory should suddenly go crazy with the
addition of state dependence. Hence, all evidence thus
far suggests that the hole line expansion should be
perfectly adequate for nuclear matter.

D) Nuclear Matter with Reid Potential

Nuclear matter results to date for the Reid potential are collected in Fig. 10. The two hole line
curve,[8] E(2), saturates at about 11 MeV. The error bars
on this curve are κ times the potential energy and values
of κ are indicated above the two arrows. The three hole
line curve,[11] E(3), is significantly different from
older results,[8] primarily due to more accurate treatment
of the tensor force. The error bar on this curve denotes
the uncertainty estimate given by κ^2 times the potential
energy. The four hole line values, E(4), are only
estimated,[12] and the band indicates Day's best guess as
to the probable uncertainties. Since all the hole line
curves are consistent within the uncertainties estimated
by κ arguments, there is every indication of satisfactory
convergence of perturbation theory.

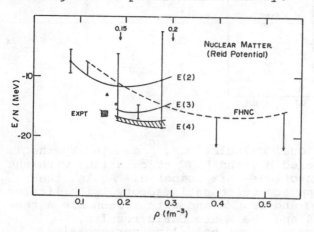

Figure 10. Binding energy as a function of density for the Reid potential.

At present, because of the complications of state dependence, variational results for nuclear matter are only preliminary. Earlier results suggesting a serious discrepancy with perturbation theory are now inoperative,[21] and the recent results of Pandharipande and Wiringa[33] with momentum independent correlation functions are labeled FHNC. These results should be regarded as a reasonably reliable but non-optimal bound, and it is certainly gratifying that nowhere do they conflict with perturbation theory. The downgoing arrows attached to this FHNC curve give a crude estimate of how much momentum dependence might be expected to lower the bound, based on the effect of momentum dependence on the two body contribution to the energy. Given the uncertainties in $E(4)$, even this estimate is not in serious conflict with perturbation theory.

Thus, although the variational results are not all in, present evidence is consistent with our expectation the perturbation theory should be adequate for nuclear matter.

IV. SUMMARY AND CONCLUSIONS

It is likely that this review has taught this audience more about many-body theory and less about nuclear forces than it might have wished. Hence, at least a few words about the Reid potential may be in order.

Based on the $1.2\pm.2$ MeV underbinding of the triton alone,[2,3] it is clear that the Reid potential (without relativistic or mesonic corrections) is incorrect. This underbinding is also consistent with the 3 MeV per particle underbinding of ^{16}O and the 2.5 MeV per particle underbinding of ^{40}Ca in the e^S calculations.[4] However, if one believes the four-hole line results in Fig. 10, there is a problem understanding why ^{16}O and ^{40}Ca underbind while nuclear matter overbinds. One means of attempting to resolve this inconsistency is to extrapolate the Reid ^{16}O and ^{40}Ca results to the nuclear matter limit in the spirit of the semi-empirical mass formula. This has been done[34] by adjusting the Skyrme II interaction such that the surface energy, symmetry energy, and effective mass were held fixed, while the bulk binding energy and saturation density of nuclear matter were varied to reproduce the calculated energy and radius of a given nucleus. The extrapolation for ^{16}O and ^{40}Ca are denoted by the triangle and circle respectively in Fig. 10. Unfortunately these results still remain contradictory, with finite nuclei suggesting saturation at about the right density with too little binding and nuclear matter saturating with too much binding at too high a density. Presumably, the extrapolation was too

crude, and light nuclei teach us something important
about the surface energy which could never be learned
from the study of nuclear matter alone.

In spite of the minimal impact nuclear matter
theory has made on our understanding of nuclear interactions to date, I believe that tremendous fundamental
progress in many-body theory has occurred. As demonstrated in Figs.6,8-10,the Greens function Monte Carlo,
variational, and perturbation theory approaches are
completely consistent in all cases in which they overlap.
At present, perturbation theory still appears to be the
most practical technique for finite nuclei and nuclear
matter with realistic forces, but its credibility is
inestimably enhanced by these thorough investigations of
complementary techniques.

Acknowledgements

Most of the recent calculations were reported at
the Illinois workshop on nuclear and dense matter in
May, and the author is grateful to both the organizer,
V. R. Pandharipande, and the participants for this
opportunity to synthesize the results of various
research groups. Special thanks are due to G. Chester,
B. Day and R. Wiringa for providing references and unpublished data used in preparing the figures.

Bibliography

1. J.L. Friar and J.W. Negele, Advances in Nuclear
   Physics, Vol. 8, Baranger and Vogt ed., 219 (1975).

2. M.R. Strayer and P.U. Sauer, Nucl. Phys. A231, 1
   (1974).

3. E.P. Harper, Y.E. Kim and A. Tubis, Phys. Rev. Lett.
   28, 1533 (1972).

4. H. Kummel, K.H. Luhrmann and J.G. Zabolitzky, Physics
   Reports, to be published; J.G. Zabolitzky, Phys. Rev.
   C14, 1207 (1976) and references therein.

5. M.H. Kalos, Phys. Rev. A2, 250(1970).

6. M.H. Kalos, D. Levesque, and L. Verlet, Phys. Rev.
   A9, 2178 (1974).

7. M.H. Kalos and P. Whitlock, private communication.

8. H.A. Bethe, Ann. Rev. of Nuc. Sci., Vol. 21, 93 (1971).

9. D.W.L. Sprung, Advances in Nuclear Physics, Vol. 5,
   Baranger and Vogt ed., 225 (1972).

10. B.D. Day, Rev. of Mod. Phys., to be published.

11. B.D. Day, private communication.

12. B.D. Day, Phys. Rev. 187, 1269 (1969).

13. R. Jastrow, Phys. Rev. 98, 1479 (1955).

14. F. Iwamoto and M. Yamada, Progr. Theor. Phys. 18, 345 (1957).

15. E. Feenberg, Theory of Quantum Fluids, Academic Press, N.Y. (1969).

16. E. Krotscheck, J. Low Temp. Phys. 27, 199 (1977).

17. M. Gandin, J. Gillespie, and G. Ripka, Nucl. Phys. A176, 237 (1971).

18. J.W. Clark and P. Westhaus, Phys. Rev. 141, 833 (1966); Phys. Rev. 149, 990 (1966).

19. S. Fantoni and S. Rosati, Lett. Nuovo Cimento 10, 545 (1974); Nuovo Cimento 20A, 179 (1974); Nuovo Cimento 25A, 593 (1975).

20. V.R. Pandharipande and H.A. Bethe, Phys. Rev. C7, 1312 (1973).

21. V.R. Pandharipande and R.B. Wiringa, Nucl. Phys. A266, 269 (1976), and to be published.

22. J.G. Zabolitzky, Phys. Lett. 64B, 233 (1976) and Argonne National Lab. preprint (1977).

23. J.E. Mayer and M.G. Mayer, Statistical Mechanics, John Wiley and Sons (1940).

24. J.M.J. van Leeuwen, J. Groeneveld and J. deBoer, Physica 25, 792 (1959).

25. H.W. Jackson and E. Feenberg, Ann. Phys. 15, 266 (1961).

26. R.D. Murphy and R.O. Watts, Jour. Low Temp. Phys. 2, 507 (1970).

27. D. Schiff and L. Verlet, Phys. Rev. 160, 280 (1967).

28. L.J. Lantto, A.D. Jackson and P.J. Siemens, Niels Bohr Institute preprint.

29. V.R. Pandharipande, R.B. Wiringa and B.D. Day, Phys. Lett. 57B, 205 (1975).

30. D. Ceperley, G.V. Chester, and M.H. Kalos, preprint.

31. B.A. Brandow, Phys. Lett. 61B, 117 (1976).

32. J.W. Clark, preprint.

33. R.B. Wiringa, private communication.

34. J.W. Negele, Comments in Nuclear and Particle Physics, VI, 149 (1976).

35. V.R. Pandharipande and K.E. Schmidt, Univ. Illinois preprint (1977).

THE PARITY VIOLATING NUCLEON-NUCLEON INTERACTION

BRUCE H.J. McKELLAR
School of Physics, University of Melbourne, Parkville, Vic
Australia, 3052

ABSTRACT

The difficulties in computing the parity violating nucleon-nucleon interaction from an underlying weak interaction theory are reviewed, and the extent to which the parameters of the interaction can be determined by the available data is discussed. A possible experiment in ^{21}Ne which can provide useful constraints on the parameters is outlined.

INTRODUCTION

Parity violating effects in nuclei were first unambiguously observed about 10 years ago [1]. They have yet to be unambiguously calculated from first principles. This has led to dispair of calculating the effects at all, and an attempt to parameterise the observed effects. In this brief talk I want to review three points. Firstly, I will explain why ab initio calculations are so difficult. Secondly, I want to describe the parameterisations which have been proposed. Finally, I will make some remarks about existing and possible future experiments.

Before embarking on the details, an overview is appropriate. The only parity violating interaction we know about at the moment is the weak interaction. It was suggested that the weak interaction would lead to parity violating interactions between nucleons long before these were observed, and we still try to explain the observations in terms of the weak interactions. About the time of the first observations it was realised that, if successful, these explanations would provide useful constraints on the weak interaction theory. This remains the basic reason for trying to understand the parity violating nucleon-nucleon interaction.

There has been considerable interest in our subject in recent years, with the great upsurge of activity in Weak Interactions since the development of gauge theories, and the discovery of neutral currents in the early seventies. One of the important open questions is the V, A structure of the neutral current. There is some hope of shedding light on this from the study of parity mixtures in nuclei. Analogous experiments in atoms may overtake us here, since in two important respects the atomic calculations are clearer, as we shall see.

CALCULATION FROM FIRST PRINCIPLES

We will assume that the weak interaction theory, whatever it is, has some gauge structure, and the weak interactions arise by exchange of massive bosons. At the low momentum transfers of nuclear physics we do not probe this exchange structure, and it suffices to consider a four fermion interaction. At the basic quark level this will be a four quark interaction, with terms like $(\bar{u}\,d)\times(\bar{d}\,u)$, $(\bar{u}\,s)\times(\bar{s}\,u)$, $(\bar{d}\,c)\times(\bar{c}\,d)$, etc., appearing (The space time operators γ_μ and $\gamma_5\gamma_\mu$ have been omitted here for simplicity).

Thus, as our first step we need to determine the strengths of these various terms in the effective Hamiltonian. A straight forward procedure, utilising operator product expansions and renormalisation group techniques is available to attack this part of the problem, and has been applied by Altarelli et al [2], Donoghue [3], and Galic et al [4]. This step has to be redone for each proposed underlying gauge theory - the effective Hamiltonian cannot be written down at sight. In particular the introduction of right handed currents can have dramatic effects on the effective Hamiltonian[5].

Even when we have H_{eff} as a product of currents we still must evaluate matrix elements like $\langle N_1\bar{N}_2|\bar{u}\,s\quad\bar{s}\,u|\pi\rangle$. I have chosen this as an example because it is just this matrix element which dominates the all important π exchange force.

If we are ever to constrain the weak interaction theory from the nucelar observations a calculation of the strength of the π force is crucial. This is because the $NN\pi$ parity violating amplitude derives from the isovector, G parity odd part of the weak Hamiltonian, [6,7] and is therefore very sensitive to the neutral current structure of the underlying theory.

The first attempt at this calculation was made using current algebra, relating the p.v. $NN\pi$ amplitude to the hyperon decay amplitudes.[7] This calculation has been questioned on two grounds.

(a) The vector meson pole terms illustrated in figure 1 extrapolate rapidly in going from pion mass shell (relevant to hyperon decays) to the soft pion point (relevant to pion exchange in nuclear physics). It is therefore argued that they should be subtracted before performing the current algebra calculation, thereby reducing the $NN\pi$ amplitude by a factor of at least 4.[3,8]

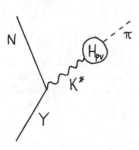

Fig. 1. K^* pole terms for hyperon decay.

(b) One can try to evaluate such matrix elements using quark model wavefunctions. This can be done because the renormalisation group calculation takes into account the important quark-gluon interactions. Because of the presence of $\bar{s}s$ terms in the Hamiltonian, this matrix element is proportional to the number of quarks in the sea, and is therefore

434

presumably small.[3]

The first argument is in my opinion on shaky ground, because <u>we don't know</u> the weak K*π coupling. In practice the approximation (**represented** in figure 2a)

$$<K^*|\bar{s}u\ \bar{u}d|\pi> \overset{\thicksim}{\thicksim} <K^*|\bar{s}u|u><o|\bar{u}d|\pi>$$

is used, which is without justification, (except that of simplicity).

Fig. 2a The factorisation diagram for $<\pi|H_w|K^*>$
2b t channel W exchange for $<\pi|H_w|K^*>$

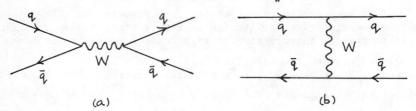

(a)   (b)

For example, the diagram of figure 2b will also contribute. At the very least a more serious calculation than has yet been performed is necessary to decide the issue.

The second objection is harder to overcome. It may be that the most reasonable way to reconcile the current algebra and quark model calculation is to accept that this basic matrix element $<N\bar{N}|\bar{u}s\ \bar{s}u|\pi>$ is indeed small, as proposed by Donoghue.[3] But, in my opinion, the magnitude of the NNπ coupling is still an open question.

I hope I have done enough to convince you that calculating hadronic weak interactions is fraught with difficulty, even for the relatively simple NNπ case where we have some hope of relating the NNπ amplitude to the YNπ amplitude. If more evidence of difficulty and confusion is needed, I refer you to the continuing saga of the $\Delta T=\frac{1}{2}$ rule for hyperon decays, for which there is still no satisfactory theory.[9]

The NNρ and NNω amplitudes are even more difficult to compute convincingly. So it is perhaps with relief that we turn to attempt a parameterisation of the parity violating nucleon-nucleon force.

One advantage of the atomic physics case is now obvious. Only semi leptonic weak interactions occur, and these can be calculated from first principles.

PARAMETERISATION OF THE PARITY VIOLATING NUCLEON-NUCLEON INTERACTION

The idea that the nucleon-nucleon weak interaction should be parameterised rather than calculated has its origins in the pioneering work of Danilov[10], was revived at the Mainz Conference[11], and has been developed in detail in two different ways by Box,

Gabric, Lassey and McKellar [12], and by Desplanques and Missimer [13].

Both BGLM and DM began with a potential of the form suggested by π and vector meson exchange, namely

$$V = V_V + V_\pi$$

with

$$V_\pi^{\Delta T=1} = (-\tfrac{1}{2}i\underset{\sim}{\tau}^{(1)} \times \underset{\sim}{\tau}^{(2)})_0 (\underset{\sim}{\sigma}^{(1)} + \underset{\sim}{\sigma}^{(2)}) \cdot [\underset{\sim}{p}, v_\pi(r)]_-,$$

where

$$v_\pi(r) = \frac{g_\pi f_\pi}{4\pi\sqrt{2} m_N} \frac{e^{-m_\pi r}}{r}$$

Here g_π is the strong $NN\pi$ coupling constant and f_π is the weak $NN\pi$ coupling constant. The vector meson exchange potential may be written as

$$V_V = V_V^{\Delta T=0,2} + V_V^{\Delta T=1}$$

Neglecting the ρ-ω mass difference we have

$$V_V^{\Delta T=1} =$$

$$\left\{ \frac{I}{6\sqrt{3}} (\tau_0^{(2)}\underset{\sim}{\sigma}^{(1)} - \tau_0^{(1)}\underset{\sim}{\sigma}^{(2)}) + \tfrac{1}{2}\sqrt{3} I'(\tau_1^{(1)}\underset{\sim}{\sigma}^{(1)} - \tau_0^{(2)}\underset{\sim}{\sigma}^{(2)}) \right\} \cdot [\underset{\sim}{p}, v_\rho(r)] +$$

$$+ \left\{ \frac{1+\mu_v}{12\sqrt{3}} I + \tfrac{1}{4}\sqrt{3}(1+\mu_s)I' \right\} (\tau_0^{(1)} + \tau_0^{(2)})(i\underset{\sim}{\sigma}^{(1)} \times \underset{\sim}{\sigma}^{(2)}) \cdot [\underset{\sim}{p}, v_\rho(r)]_-$$

$$+ \lambda(-\tfrac{1}{2}i\underset{\sim}{\tau}^{(1)} \times \underset{\sim}{\tau}^{(2)})_0 (\underset{\sim}{\sigma}^{(1)} + \underset{\sim}{\sigma}^{(2)}) \cdot [\underset{\sim}{p}, v_\rho(r)]_-,$$

$$V_V^{\Delta T=0,2} = \left\{ M(\tau) + (1/6)L \right\} (\underset{\sim}{\sigma}^{(1)} - \underset{\sim}{\sigma}^{(2)}) \cdot [\underset{\sim}{p}, v_\rho(r)]_+$$

$$+ \left\{ (1+\mu_v)M(\tau) + (1+\mu_s)(1/6)L \right\} (i\underset{\sim}{\sigma}^{(1)} \times \underset{\sim}{\sigma}^{(2)}) \cdot [\underset{\sim}{p}, v_\rho(r)]_-,$$

where

$$v_\rho(r) = - \frac{GG_A m_\rho^2}{4\pi\sqrt{2} m_N} \frac{e^{-m_\rho r}}{r}$$

and

$$M(\tau) = (1/6)(2H+K)\underset{\sim}{\tau}^{(1)} \cdot \underset{\sim}{\tau}^{(2)} + \sqrt{\frac{1}{6}}(K-H)[\underset{\sim}{\tau}^{(1)} \times \underset{\sim}{\tau}^{(2)}]_0^{(2)}.$$

The parameters f_π, H, I, I', K, L and λ can be related to the effective four fermion Hamiltonian if one is prepared to accept a model calculation. On the basis os such calculations one expects all parameters except possibly f_π to be of order 1.[5] Here we ask whether the parameters can be determined directly from experiments.

Immediately difficulties arise, partly because of the relatively small numbers of experiments and partly because of the uncertainties in the nuclear physics calculations leading from the potential to the parity violating observables. (The belief that we know more about atomic than nuclear wavefunctioning is another point in favour of the atomic case.)

BLGM and DM reacted to these difficulties in different ways. BLGM decided, arbitrarily, to reduce the number of parameters to 3, namely $\mathcal{A}$ = 2H+K, $\mathcal{B}$ = K-H and $\mathcal{C}$ = f_π / f_π^{CM} where f_π^{CM}(=4.6 x 10^{-8}) is the value of f_π in the standard charged current Cabibbo theory according to the calculation of McKellar,[7] which is used as a convenient normalisation. These parameters are fitted to the data as described in the next section. The difficulty with this approach is that the omitted parameters may in fact take significant values. The BLGM calculation was characterised by a direct use of nucleon-nucleon potentials, so that it was necessary for them to use nuclear physics calculations in which short range correlations were included. This meant that they were unable to include the parameter L because no calculation of the ^{16}O α decay using it and employing hard core correlations was available.

DM included more parameters in their fit. Moreover, they decided to avoid parameterising the N-N potential, and instead to parameterise the N-N scattering amplitudes. By an ingeneous nuclear matter calculation they are able to introduce an effective nucleon-nucleon interaction to replace the potential V_V in calculations involving large nuclei. This effective interaction is of the form

$$V_V^{eff} = \frac{4\pi}{M} \{\alpha(\sigma_1-\sigma_2)\cdot[p,\delta(r_1-r_2)]_+$$

$$+ \beta(\sigma_1+\sigma_2)\cdot[p,\delta(r_1-r_2)]_-$$

where

$$\alpha = \frac{1}{4}\bar{v}^{(0)}(3+\tau_1\cdot\tau_2) + \frac{1}{4}\bar{u}(1-\tau_1\cdot\tau_2) + \frac{1}{2}\bar{v}^{(2)}[\tau_1\times\tau_2]_0^{(2)}$$

$$+ \frac{1}{2}\bar{v}^{(1)}(\tau_1+\tau_2)_z$$

$$\beta = \frac{1}{2}\bar{w}^{(1)}(\tau_1-\tau_2)_z$$

This potential is designed for use between wavefunctions with no short range correlations. The relationship between the parameters $\bar{v}^{(i)}$, $\bar{u}$, $\bar{w}^{(1)}$ and the potential parameters given above depends on the strong interaction potential used. This enters in two places - in the nuclear matter calculation which relates the nuclear matter calculation to the nucleon-nucleon amplitude, and again when we relate the nucleon-nucleon amplitude to the nucleon-nucleon potential. Using the results of DM for the first step, and

Lassey and McKellar [14] and McKellar and Lassey [15] for the second we can express the DM parameters in terms of the potential parameters given above. The results are

$$m^{2-}_v{}^{(i)} = 4.8 \ m \ \rho^{(i)}$$

$$m^{2-}_u = 10.7 \ m \ \lambda_t = 10.7\{\mu(2H+K) + \nu \ L\}$$

$$m^{2-}_w{}^{(1)} = 10.7 \ m \ C^{(1)} = 10.7 \ C_2 \ (I' - \frac{1}{9}I) + 10.7 \ C_3\lambda$$

$$m\rho^{(0)} = -\frac{1}{3} \ \eta(2H+K) + \kappa L$$

$$m\rho^{(1)} = \rho(9I + I')$$

$$m\rho^{(2)} = -\frac{4}{\sqrt{6}} \ \eta(K-H)$$

The numerical parameters in the first 3 equations come from the nuclear matter calculation using soft core potentials of DM. The parameters $\eta,\kappa,\mu,\nu,\rho,C_2$ were computed by McKellar and Lassey and are given for a number of strong potentials in Table 1. C_3 is not given in that paper. It can be obtained by repeating the underlying calculation and is included in Table 1.

TABLE 1.

Expansion parameters for the p.v. amplitudes

Strong Potential

Amplitude	RPSC	GPD	GTG	EHM
$C_2 \times 10^8$	− 0.92	− 2.26	− 1.59	− 1.06
$C_3 \times 10^8$	− 3.55	− 3.01	− 2.32	− 3.32
$\mu \times 10^9$	− 8.60	−17.7	−13.3	− 9.45
$\nu \times 10^9$	− 1.90	− 1.02	− 2.19	− 2.08
$\eta \times 10^8$	+11.66	+23.3	+16.00	+17.05
$\kappa \times 10^8$	− 1.226	− 2.51	− 1.64	− 1.81
$\rho \times 10^8$	− 0.355	− 0.751	− 0.458	− 0.531

THE EXPERIMENTS AND THE PARAMETERS

A substantial number of experiments on parity mixing in nuclei are available [16]. Both BGLM and DM made use of a subset of these, because of the availability of reliable calculations. The same set of experiments were fitted by both groups. They were
 (i) The α **decay** of the 2^- state at 8.88 MeV in ^{16}O.

The α width Γ_α was measured in the classic experiment of the Mainz Group[17]. It is sensitive only to the $\Delta T=0$ potential and looked at naively suggests that the strength is approximately standard. As emphasised by DM there could however be cancellations between the ρ exchange and ω exchange parts of the potential i.e. between 2H+K and L, with each being individually large.

(ii) The circular polarisation of the photon emitted in the reaction $n+p \rightarrow d+\gamma$, which has been measured by Lobashov et al [18]. This is sensitive to both the $\Delta T=0$ and $\Delta T=2$ parts of the potential. A substantial enhancement of one or both parts of the potential is required to fit the value.

(iii) The analysing power of $p-p$ scattering for longitudinally polarised protons at 15 MeV, measured by Potter et al [19]. This experiment is sensitive to $\Delta T=0,1$ and 2 parts of the potential.

(iv) The asymmetry of the photon emitted in the decay of the 110 KeV $\frac{1}{2}^-$ state in ^{19}F, which was measured by Adelburger et al [20]. This is sensitive to $\Delta T=0$ and $\Delta T=1$ parts of the potential.

(v) The circular polarisation of the 1.2 MeV photon of ^{41}K, as measured by Lobashov et al [21].

(vi) The circular polarisation of the 396 KeV photon from ^{175}Lu, measured by Lobashov et al [22], and Vanderleeden and Boehm [23].

(vii) The circular polarisation of the 482 KeV photon from ^{181}Ta, again measured by Lobashov et al [22], and Vanderleeden and Boehm [23].

All of the last **three** cases depend on all three isospin components of the potential.

The techniques adopted by BGLM and DM were similar. The seven observables, if $\sqrt{\Gamma_\alpha}$ is used instead of Γ_α, can be expressed as linear combinations of the parameter involved. The coefficients in these expansions are given in Table II for the BGLM technique, and in Table III for the DM technique.

TABLE II

Expansion parameters $\alpha_i, \beta_i, \gamma_i$, experimental values y_i and results from the fitted parameters $\hat{y}_i$. The references are to the calculations from which the expansion parameters were determined.

Experiment and Scale Factor	Value	Ref.	α_i	β_i	γ_i	$\hat{y}_i^{(1)}$
^{16}O, $(\Gamma_\alpha \times 10^{10} eV^{-1})^{\frac{1}{2}}$	$\pm(1.01\pm0.13)$	24	0.52	0	0	-1.01
$n+p \rightarrow d+\gamma$, $10^6 P_\gamma$	$-(1.3\pm0.45)$	14	0.0	-0.036	0	-0.28
pp scattering, $10^7 A_\rho$	(1 ± 1.4)	15	0.19	-0.37	0	2.46
^{19}F, $10^5 A_\gamma$	$-(18\pm9)$	25	3.0	0	-2.0	-23.0
^{41}K, $10^5 P_\gamma$	(1.9 ± 0.3)	26	-0.27	0.039	0.14	1.97
^{175}Lu, $10^5 P_\gamma$	(5 ± 2)	26	-0.54	0.19	0.20	5.3
^{181}Ta, $10^6 P_\gamma$	$-(5\pm2)$	26	0.19	-0.19	-0.22	-5.0

TABLE III

10^6 times the coefficients of zero energy isospin parameters used in fits of the DM

Nucleus	f_π	$MC^{(1)}$	$M\rho^{(1)}$	$M\lambda_t$	$M\rho^{(0)}$	$\dfrac{M\rho^{(2)}}{\sqrt{6}}$
^{19}F	−430	−952	−427	−476	−641	
^{41}K	29.	52.4	23.0	26.2	34.8	−.48
^{175}Lu	52.	155	46.6	77.5	81.4	−23.
^{181}Ta	−6.0	−5.8	−1.0	−2.9	−2.3	1.6
^{16}O				30.0	40.3	
n+p→d+γ P_γ				.63	−.16	.32
A_{pp} (15 MeV)			−.48		−.48	−.48

The parameters of the potential are then determined by a least squares procedure. This step is in fact the one which is most open to criticism, in that no account is taken of the correlated errors in the coefficients. The DM calculation is a little better off here in that some of the correlation in the errors in the theoretical coefficients comes from uncertainties in the strong short range potential, and this enters the DM calculation after the fit is made, in going from their parameters to the potential parameters.

The parameters obtained by BGLM (updated to take account of the new calculation in ^{16}O of Apagyi et al[24] and the more stringent limit on the asymmetry in pp scattering) are

$$\alpha = -1.95 \qquad \beta = 7.67 \qquad C = 7.96$$

With a χ^2 of 7.3 for 3 degrees of freedom. The developments of the past year have significantly worstened the 3 parameter fit. DM have many more parameters. Since the expansion coefficients for ^{19}F and ^{41}K are almost proportional, one has in effect only 6 pieces of data and they have 6 parameters. Arbitrarily setting $C^{(1)}$, $\rho^{(1)}$ and $\rho^{(2)}$ to zero in turn they obtain the parameters given in Table IV, each set having a χ^2 of 2.3 for 1 degree of freedom.

TABLE IV

Five parameter fits to p.v. observables by DM. The parameters are in units of 10^{-6}.

f_π	$MC^{(1)}$	$M\lambda_t$	$M\rho^{(0)}$	$M\rho^{(1)}$	$\dfrac{M\rho^{(2)}}{\sqrt{6}}$	χ^2	P
I	.70		−1.11	1.08	−.24	−1.13	2.95	.23
II	.70	−.11	−1.04	1.02		−1.31	2.95	.23
III	.70	.67	−1.60	1.44	−1.73		2.97	.23

It is instructive to translate these parameters to the potential parameters using Table I. Choosing set I for the purpose of illustration we find, for the Reid Soft core potential

$$\mathcal{A} = 507, \quad \mathcal{B} = 15, \quad \mathcal{C} = 15.2 \quad L = -1711,$$

$$9 \ I + I' = 68.$$

With the additional parameters it was possible to reduce the χ^2 by using extremely large values of $\mathcal{A}$ and L with signs chosen to add in $n+p\rightarrow d+\gamma$, and subtract in ^{16}O. There is still a substantial contribution to p-p scattering and this is removed by a large value of the isovector potential $9 \ I+I'$. With more parameters the fit is of course better.

From the point of view of someone interested in weak interactions, neither the BGLM nor the DM parameters are expected. The strong enhancement of the $\Delta T=2$ part of the interaction obtained by BGLM conflicts with octet dominance. The very strong $\Delta T=0$ and $\Delta T=1$ parts of the DM potential are not in conflict qualitatively with octet dominance, but no one has previously suspected enhancements which are so large.

Thus a cautious approach suggests itself. These parameters certainly need confirmation in new experiments. And a more realistic attempt to include the nuclear theory uncertainties into the parameter fitting is required. I will conclude by briefly describing a possible new experiment in ^{21}Ne.

There are two $\frac{1}{2}$ states at 2.79 and 2.80 MeV - the separation is about 6 KeV, and they have opposite parity. Moreover the $\frac{1}{2}^- \rightarrow \frac{3+}{2}$ ground state E1 transition is so strongly hindered that the M2 $\frac{1}{2}^- \rightarrow \frac{5+}{2}$ transition is the large branch. So this looks an excellent prima facie case for a circular polarisation measurement. A quick calculation suggests that the mixing of the $(p_{\frac{1}{2}})^{-1}(sd_{\frac{1}{2}})^2$ component of the $\frac{1}{2}^-$ state with the $(sd_{\frac{1}{2}})^{-1}(sd_{\frac{1}{2}})^2$ component of the

$\frac{1}{2}^+$ state will dominate the mixing, but that there may also be a significant contribution from the $(f_{p_{\frac{1}{2}}})$ and $(sd_{\frac{1}{2}})$ components.

Taking the simplifying assumption that these fp components do not contribute significantly, we have the same mixing matrix element as in ^{19}F, but between neutron states. Thus putting the ^{21}Ne and ^{19}F results together we can find out the strengths of the $\Delta T=0$ and $\Delta T=1$ components.

Preliminary calculations indicate that the circular polarisation of the 2.79 MeV γ is in the range 10^{-2} - 10^{-3} - perhaps a larger P_γ, and certainly a more easily calculated P_γ than the celebrated ^{180}Hf case.

ACKNOWLEDGEMENT

This work was supported in part by the Australian Research Grants Committee.

REFERENCES

1. V.M. Lobashov, V.A. Nazarenko, L.F. Saenko, L.M. Smotritski, G.I. Kharevich, Sov. Phys.-J.E.T.P. Letters 3, 173(1966)
2. G. Altarelli, R.K. Ellis, L. Maiani and R. Petronzio, Nucl. Phys B88, 215(1975)
3. J. Donoghue, Phys. Rev D13 2064 (1976)
4. H. Galic and D. Tadic, Fizika 8, 99 (1976)
5. R.K. Ellis, Nucl. Phys B108, 239 (1976)
   H. Fritzsch, P. Minkowski, Phys. Lett. 61B, 275 (1976)
6. G. Barton, Nuovo Cim 19, 512 (1961)
7. B.H.J. McKellar, Phys. Lett 26B, 107 (1967)
8. M. Gronau, Phys. Rev. Lett 28, 188 (1972), Phys Rev D5, 118 (1972)
   A. Andrasi, B. Eman, J. Missimer and D. Tadic, Phys. Rev. D11, 2484 (1975)
9. J.F. Donoghue, E. Golowich and B.R. Holstein. Phys. Rev D15, 1341 (1977)
10. G.S. Danilov, Phys. Lett. 18, 40 (1965)
11. M. Simonius, Interaction Studies in Nuclei
    (ed. H. Jochim, B. Ziegler, North Holland, Amsterdam(1975) p 3.
    B.H.J. McKellar, M.A. Box, A.J. Gabric and K.R. Lassey, ibid p 61. see also B.H.J. McKellar, Nucl. Phys. A254, 349(1975)
    B.H.J. McKellar, K.R. Lassey, M.A. Box & A.J. Gabric
    Few Body Dynamics (ed. A.N. Mitra, I. Slaus, V.S. Bhasin and V.K. Gnpta. North Holland. Amsterdam(1976) p 368
    J. Missimer, Phys. Rev C14, 347 (1976)
12. M.A. Box, A.J. Gabric, K.R. Lassey and B.H.J. McKellar, J. Phys. G. 2, L107 (1976)
13. B. Desplanques and J. Missimer, Carnegie Melton preprint COO-3066-82(1976)
14. K.R. Lassey and B.H.J. McKellar, Nucl. Phys. A260, 413 (1976)
15. B.H.J. McKellar and K.R. Lassey, University of Melbourne preprint UM-P-76/12 (1976).

16. E. Fiorini, Rev. Nuovo Cim 4, (1974)
17. N. Neubeck, H. Schober and H. Wäffler, Phys. Rev. $\underline{C10}$, 320(1974)
18. V.M. Lobashov et al., Sov. J. Nucl. Phys. $\underline{15}$, 632 (1972); Nucl. Phys. $\underline{A197}$, 241 (1972)
19. J.M. Potter et al Phys. Lett. $\underline{33}$ 1307 (1974)
20. E.G. Adelburger, H.E. Swanson, M.D. Cooper, J.M. Tape & T.A. Trainor, Phys. Rev. Lett. $\underline{34}$ 402, (1975)
21. See reference 18.
22. See reference 1., and V.M. Lobashov et al, Sov. Phys - JETP Letters $\underline{5}$, 59 (1967)
23. J.C. Vanderleedeon & F. Boehm, Phys. Lett. $\underline{30B}$, 467(1969), Phys. **Rev.** $\underline{C2}$, 748 (1970)
24. B. Apagyi, G. Fai & J. Nemeth, Nucl. Phys $\underline{A272}$ 317 (1976)
25. M.A. Box, A.J. Gabric and B.H.J. McKellar, Nucl Phys $\underline{A271}$, 412 (1976)
26. B. Desplanques, These. Universite de Paris Sud. (1975).

STUDY OF PARITY VIOLATION BY NEUTRON CAPTURE GAMMA RAYS

J.F. Cavaignac, P. Liaud,[*] R. Steinberg,[**] B. Vignon
Institut des Sciences Nucléaires, Grenoble, France

E. Jeenicke,[†] R. Wilson
Harvard University, Cambridge, MA 02138

Presented by Richard Wilson

A program of studying parity violation by neutron capture is in progress at the Institute Laue-Langevin at Grenoble. A schematic of the experimental apparatus is presented in figure 1.

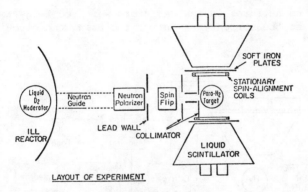

Figure 1

Neutrons from the 56 MW reactor are moderated in a liquid deuterium moderator in the reactor vessel and guided into the experimental hall over a distance of 80 metres by a curved guide tube which cuts off all neutrons of wavelength less than 2.8 Å. This removes all γ rays and fast neutrons.

The neutrons impinges on a guide tube of width 5 mm and height 5 cm with magnetized Fe-Cu mirrors[1] which has a transmission of 40% for one neturon polarization at $\lambda = 5$ Å. The average polarization is (70 ± 7)% with a total flux 3.5 x 10^8 sec^{-1}. The first part of the beam line and the polarizer were originally set up for a time reversal experiment in neutron decay.[2]

*Present address: Universite de Technologie de Compiègne, 25, rue Eugène-Jacquet, B.P. 233, 60206 Compiègne, France.
**Present address: Dept. of Physics, University of Maryland, College Park, MD 20742.
†Present address: Bahnhofstrasse 4, P.F. 129, 7141 Schwieberdingen, Stuttgart, Germany.

ISSN: 0094-243X/78/443/$1.50

The neutron spin is adiababically rotated into the beam direction and reversed every 0.8 second by a magnetic spin flip system originally proposed by Dropkin.[3]

For np capture, we used a target of liquid parahydrogen, which has a molecular angular momentum of zero; neutrons with energies less than the energy of the first rotational state will not change their spin direction.

Two liquid scintillation detectors of large solid angle located on either side of the beam were used. We estimate that 2.2 mev gamma ray produces pulses with an amplitude within about 30%; this pulse height spread does not appreciably affect the statistical accuracy of the experiment.

The count rates in the detectors were too large to measure directly, so the current was integrated using the circuitry of figure 2.[4]

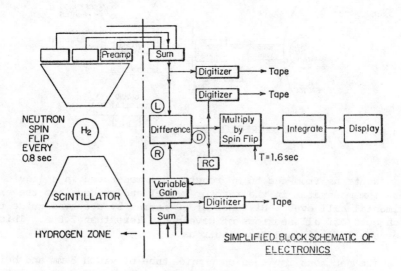

Figure 2

After correction for low neutron polarization and for angular resolution, the final asymmetry in np cature was found to be[5]

$$A_\gamma = (0.6 \pm 2.1) \times 10^{-7}$$

Spurious asymmetries due to misalignments, Stern-Gerlach effect due to inhomogeneous fields, are calculated to be less than 10^{-8}.

This number is to be compared with predictions of theory of $A_\gamma = 10^{-8}$ with no parity violating neutral currents and $A_\gamma \approx 10^{-7}$ with neutral currents.

We hope to increase the aperture of the polarizer a factor of 6, increase the running time a factor of 2, and the polarization to 95% and hence be in a position to measure the calculated neutral currents.

Meanwhile we are attempting to measure the asymmetry in neutron capture γ rays from a variety of nuclei: Cd^{114}, Sn^{116}, Te^{124}, Cs^{134} using Ge(Li) detectors and a counting technique[7] and hope to resolve outstanding discrepancies.[8]

We are grateful for the help and encouragement of the staffs of ILL, ISN and Harvard University and in particular, Mr. Barnoux (ISN).

REFERENCES

1. K. Berndorfer, Z. Phys. 243, 188 (1971).

2. R.I. Steinberg, P. Liaud, B. Vignon, and V. Hughes, Phys. Rev. Letts. 33, 41 (1974).

3. P. Liaud, R.I. Steinberg, and B. Vignon, Nucl. Inst. and Methods, 125 (1975).

4. For details, see Ph.D. Thesis, Harvard University, 1976 of E. Jeenicke who designed and constructed this apparatus.

5. J.F. Cavaignac, B. Vignon, and Richard Wilson, Phys. Letts. 67B, 148 (1977).

6. B. Guberina and D. Tadić, Report IRB-TP-2-77, Institute "Rudjer Bošković" and University of Zagreb, Zagreb, Croatia, Yugoslavia.

   M. Gari and J. Schlitter, Phys. Letts. 59B, 118 (1975).

   B. Desplanques and E. Hadjimichael, Nucl. Phys. B107, 125 (1976).

7. In this we are joined by Mr. H. Ben Koula and Dr. Kwong.

8. Y.G. Abov, P.A. Krupkhitsky, M.I. Bulganov, O.I. Ermakov, and I.L. Karpikhin, Sov. J. Nucl. Phys. 16, 670 (1973).

PROTON-PROTON BREMSSTRAHLUNG AT SMALL ANGLES*

J.L. Beveridge, D.P. Gurd,† J.G. Rogers, and H.W. Fearing
TRIUMF, Vancouver, B.C., Canada V6T 1W5

A.N. Anderson, J.M. Cameron and L.G. Greeniaus
University of Alberta, Edmonton, Alta., Canada T6G 2N5

C.A. Goulding,‡ J.V. Jovanovich and C.A. Smith
University of Manitoba, Winnipeg, Man., Canada R3T 2N2

A.W. Stetz
Oregon State University, Corvallis, OR, U.S.A. 97331

J.R. Richardson
University of California, Los Angeles, CA, U.S.A. 90024

R. Frascaria
Institut de Physique Nucléaire, Orsay, France

ABSTRACT

In principle nucleon-nucleon bremsstrahlung remains one of the
most direct and least ambiguous ways of investigating the off-energy-
shell behaviour of the NN interaction. Model calculations, such as
that by Heller,[1] indicate that in the symmetric Harvard geometry[2]
the off-shell effects are largest for high bombarding energies and
small proton opening angles. Heller's calculation also shows that
for the particular model and geometry he used, off-shell effects are
fairly small for the conditions of previous experiments. Thus motivated we have measured ppγ cross sections at a somewhat higher bombarding energy (200 MeV) and over proton angles significantly smaller
than in previous experiments. Two sets of data were taken covering
intervals centered at 13° and 16.3°, respectively. Preliminary results from the 16.3° data are reported here.

EXPERIMENTAL TECHNIQUE

The measurement utilized an ultrapure natural hydrogen gas target at 0.95 atm pressure. The proton beam from the TRIUMF cyclotron
entered the gas through a 20 μm thick titanium window 2.3 m upstream
of the scattering chamber and exited through a 25 μm thick steel window 3.6 m downstream. A carefully designed system of five collima-
tors in the beam pipe upstream of the scattering chamber was used to
eliminate beam halo, especially that caused by scattering in the
entrance window.

The two protons from the p + p → p + p + γ reaction exited from
the target gas through 76 μm thick Kapton windows and were detected

*Work supported in part by the National Research Council of Canada.
†University of Alberta, Edmonton, Canada
‡Present address: LAMPF, Los Alamos, New Mexico, U.S.A.

ISSN: 0094-243X/78/446/$1.50 Copyright 1978 American Institute of Physics

in coincidence in two identical detector telescopes. A permanent magnet was placed inside the target chamber to deflect delta rays. Each telescope provided two independent measures of the proton's energy: the times of flight over a 2.5 m flight path and the energies deposited in a thick plastic scintillator. A thin (0.793 mm) plastic scintillator at the entrance to the telescope provided timing resolution without destroying trajectory information. The thick scintillator was sufficient to stop the ppγ protons but not the elastic protons. This feature allowed a veto detector to be used behind the total energy detector to eliminate most coincidences involving elastic protons. The proton trajectories were determined using two x-y multiwire counters in each telescope. Helium bags minimized the multiple scattering in the flight path between the multiwire counters. The detector telescopes limited the effective length of the target to 25 cm along the beam direction and the total angular acceptance to ±0.2° in θ and ±4° in φ. Figure 1 shows the layout of the experiment.

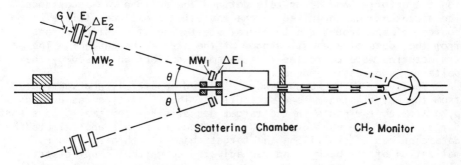

Fig. 1. Layout of ppγ experiment showing collimator system and makeup of proton telescopes.

An on-line computer recorded 60 words of information on magnetic tape for each detected event. In addition to the ppγ events the computer simultaneously recorded elastic scattering events defined by small scintillators located beyond the veto counters. These were acquired independently in the left and right counter telescopes in order to measure indirectly the combined target thickness, beam intensity and electronics dead time. An independent measure of the normalization was also made using a direct beam current monitor.[3]

The experiment utilized beam currents of 10 to 20 nA which gave a ppγ counting rate of approximately 10 per hour. About 300 h of cyclotron beam time was required for both data-taking and calibration.

PRELIMINARY DATA ANALYSIS

We have made the conventional "θγ spectrum" choice, expressing $d^5\sigma/d\Omega_{p1}d\Omega_{p2}d\theta_\gamma$ as a continuous function of the photon polar angle θ_γ for fixed proton angles (Ω_{p1} and Ω_{p2}). ppγ events were identified by requiring simultaneous detection of two protons with measured energies approximately satisfying the ppγ kinematics, i.e. that the missing mass[4] M_γ be approximately zero. The missing-mass distribution consisted of a flat background underlying a ppγ peak centered at $M_\gamma^2=0$, whose width reflected primarily the energy resolution of the detectors. In this preliminary analysis the criterion for acceptance was that the missing mass lie within the observed ppγ peak, while θ_γ was taken to be simply the polar angle of the missing momentum. Errors introduced by this procedure are estimated to be small even near $\theta_\gamma=90°$, where the uncertainty in the energy calibration is greatest.

The conversion factor from scintillator pulse height to proton energy was determined as a function of position across the large scintillators by utilizing pp elastic scattering from a CH_2 target. Corrections for energy loss effects were included in both pulse height and time-of-flight calculations. The conversion to energy was verified both by comparison with time-of-flight and by noting that the centroid of the missing-mass spectrum was near zero mass over the entire allowed ppγ kinematic region.

The coordinates from the multiwire detectors determined the proton trajectories and accurately defined the solid angle acceptance. The acceptance was adjusted in the analysis to exclude visible sources of background due to plural scattering of elastic protons from the edges of the exit window of the scattering chamber. The trajectories were corrected for the small deflection caused by the delta-ray suppression magnet.

Two types of background were measured separately and subtracted from the ppγ missing-mass distribution. The first type was due to p,2p from contaminants in the target gas arising from small leaks and outgassing of the chamber walls. This background was approximated by data acquired after filling the target chamber with air. The normalization of the background was adjusted to match the ppγ distribution in a region far from the ppγ peak. The second type of background consisted of accidental coincidence events, the contribution of which was determined from data in which the protons originated during adjacent cyclotron rf cycles.

The most significant source of inefficiency was dead time in the multiwire and scintillation counters resulting from the high singles rates. These dead times were monitored during data acquisition by using a pulser to randomly trigger each detector in the entire system. Contributions to the dead time could then be measured by counting the number of pulser signals surviving various cuts as a fraction of the number presented. A small correction was made for reactions in the scintillators using the calculation of Measday and Richard-Serre.[5]

Two independent techniques to measure the absolute (beam-target) normalization were employed. In the first technique, integrated beam current was measured directly by a CH_2 scattering monitor which had been previously calibrated with a Faraday cup.[6] The temperature and pressure of the target gas were measured by standard transducers. As a second technique, prescaled elastic scattering events from the hydrogen gas target were acquired along with the ppγ data. This allowed computation of the ppγ cross section relative to the known hydrogen elastic cross section. The two methods of measuring the beam-target normalization agree within 5%, which we take to be the estimated overall normalization error.

RESULTS

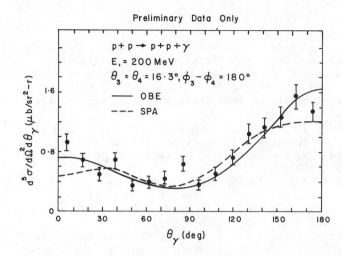

Preliminary Data Only

$p + p \rightarrow p + p + \gamma$

$E_{,} = 200 \, MeV$

$\theta_3 = \theta_4 = 16 \cdot 3°, \phi_3 - \phi_4 = 180°$

—— OBE

-- - SPA

Fig. 2. Preliminary ppγ cross sections for 16.3° compared with soft photon approximation (SPA) of Fearing and One-Boson-Exchange model (OBE) of Kamal and Szyjewicz.[7]

Figure 2 shows the θ_γ spectrum of the data analyzed to date, which are about 2/3 of the total data acquired at 16.3°. The curve labeled OBE is the one-boson exchange model calculation of Kamal and Szyjewicz,[7] which is an improved version of the earlier calculation of Baier, Kuhnelt, and Urban.[8] The dashed curve is the soft-photon approximation to the ppγ cross section, computed at TRIUMF by Fearing.[9]

The statistical uncertainties of the data are indicated by the error bars. In addition there is a ±5% normalization uncertainty. The statistical uncertainties of the analyzed data will become smaller as more of the data recorded on tape are analyzed. System-atic errors in the preliminary analysis are estimated to be small and will be investigated fully in later analysis. Corrections which have not yet been made involve primarily improvements in energy cal-ibration and the method of determining θ_γ. It is not expected, how-ever, that the 16.3° data will ultimately change by more than one standard deviation.

Because of the finite angular acceptance of the counter system, the measured cross sections are actually the integral of the differ-ential cross section over a small region of phase space. Of the cal-culated cross sections presented here, only the soft-photon calcula-tion of Fearing has been integrated to simulate this averaging effect, which was found to be negligible.[10] We would therefore expect the effect to be negligible for other theoretical calculations as well.

From the analysis performed so far we conclude that the data are in qualitative agreement with the theoretical calculations. The slight tendency to favour the OBE calculation over SPA near 0° and 180°, where off-shell effects should be largest, is encouraging as it indicates that our data may be precise enough to actually distinguish the presence of non-soft-photon terms. Data at smaller angles (13°)

remain to be analyzed. Off-shell effects should be larger there and so may be more easily distinguishable.

REFERENCES AND NOTES

1. These (unpublished) calculations performed by Leon Heller and M. Rich have been discussed briefly at several conferences. They are mentioned most recently by Heller in: Few Body Problems in Nuclear and Particle Physics, edited by R.J. Slobodrian, B. Cujec, and K. Ramavataram (Les Presses de L'Université Laval, Québec, 1975) p. 206.

2. "Symmetric Harvard Geometry" means that the protons are detected in counters placed symmetrically on either side of the beam and in a plane containing the beam. The γ-ray is not detected. See Ref. 1, p. 195.

3. Details on beam current normalization are contained in A.W. Stetz, J.M. Cameron, D.A. Hutcheon, R.H. McCamis, C.A. Miller, G.A. Moss, G. Roy, J.G. Rogers, C.A. Goulding, W.H. van Oers, Nucl. Phys. A290, 285 (1977).

4. The missing mass M_γ is calculated by $M_\gamma^2 = E_M^2 - P_M^2$ where E_M and $\vec{P}_M$ are the difference between the initial and final energy and momentum, respectively, of the two scattering protons. A non-zero value of M_γ implies a corresponding error in θ_γ.

5. D.F. Measday and C. Richard-Serre, CERN report 69-17 (1969).

6. A Faraday cup similar to the one we used was found to have an accuracy of $\pm 1\%$. See R.J. Barrett, B.D. Anderson, H.B. Willard, A.N. Anderson, and Nelson Jarmie, Nucl. Instr. & Meth. 129, 441 (1975).

7. A.N. Kamal and Adam Szyjewicz, contribution to these proceedings and private communication; Kamal and Szyjewicz, Nucl. Phys. A (in press).

8. R. Baier, H. Kuhnelt, and P. Urban, Nucl. Phys. B11, 675 (1969).

9. H.W. Fearing, contribution to these proceedings.

10. The soft-photon approximation results were numerically integrated using a coarse mesh over the angular acceptance and the resulting cross section was found to differ negligibly from the differential cross section calculated at the centre of the detector system.

NUCLEON-NUCLEON BREMSSTRAHLUNG EXPERIMENTS

J.V. Jovanovich
Cyclotron Laboratory, Department of Physics,
University of Manitoba, Winnipeg, R3T 2N2

ABSTRACT

This paper reviews the present status of recent bremsstrahlung experiments and their comparison with theoretical calculations. A significant disagreement with predictions based on the Hamada-Johnston potential is observed. It is also pointed out that experimental techniques exist which would make it possible to perform very accurate and comprehensive bremsstrahlung experiments.

A. INTRODUCTION

There have been several reviews[1-7] written about the nucleon-nucleon bremsstrahlung (NNB) process but only one of these was devoted chiefly to experiments. It was given by M.L. Halbert, a bremsstrahlung veteran, at the Gull Lake Symposium on the Two-Body Force in Nuclei[4]. As Halbert has done an excellent job of reviewing the experimental situation up to that time, I shall review here only those experiments which have been performed since 1971. However, before doing this, I shall make some general comments about the experimental techniques used in bremsstrahlung experiments.

In a typical p-p (or n-p) bremsstrahlung experiment, one detects two protons (or one proton and one neutron) in coincidence and measures their angles and energies. The photon direction and energy are then inferred using momentum conservation. The additional constraint due to energy conservation is usually used to reduce the random and/or prompt background. This is done either by explicitly computing a goodness-of-fit parameter, χ^2, or, simply, by making a scatter plot of the energies of the detected particles and comparing the position of representative points with the kinematically predicted loci.

The chief difficulty in performing bremsstrahlung experiments stems from the smallness of cross section ($d^4\sigma/d\Omega_1 d\Omega_2 \approx 1 - 10 \ \mu b/st^2$ for ppB and about 10 times larger for npB) and the fact that the final state nucleons have smaller energies than in the case of elastic scattering and, consequently, can be easily confused with background due to protons or neutrons whose energies have been degraded. These very small bremsstrahlung cross sections require that high beam intensities and/or thick targets be used in order to get appreciable counting rates. Unfortunately, the bremsstrahlung rate depends linearly on the product of the beam intensity and target thickness, while the random rate increases quadratically. Therefore, the ppB rate cannot be increased by simply using more beam or thicker targets. The experimentalist has to be very clever and introduce additional pieces of equipment that suppress random coincidences by

ISSN: 0094-243X/78/451/$1.50 Copyright 1978 American Institute of Physics

a very large factor (10^5 or greater) while the true ppB event rate
stays unaffected. Various schemes have been used in the past to
achieve this goal[4].

In NNB experiments performed so far, the angular measurements
of the detected protons was done in two different ways, which naturally divide the experiments into two distinct categories. In all of
the early experiments[4] (except the one done at Rochester[8] at 204
MeV) the proton angles were determined by collimation. This was
generally done either by using slits which physically removed all
those protons which were not within the desired solid angle, or by
using the "inverse of slits", that is scintillation counters through
which protons had to pass in order to be detected. In these experiments the angular resolution was determined by the size of solid
angles used. This experimental arrangement was relatively simple,
but a compromise always had to be made between good angular resolution and large solid angles. As a result the counting rates were
very small, generally only a few events per hour. Because of these
low rates, the experiments were done only at a few selected angle
pairs. Thus only a small part of the available phase space was
studied in these experiments. I shall call all these valuable pioneering experiments "first generation" or "selective" experiments.

In the Rochester experiment (which was an exception among the
early experiments) and in the latest three ppB experiments (see
below), some kind of position sensitive proton detectors (spark,
wire spark, or multi-wire-proportional chambers) were used. This
additional measurement of the proton coordinates added an entirely
new dimension to bremsstrahlung experiments. Besides measuring the
proton energies, trajectories for each of the protons were fully determined, thus allowing the solid angle defining slits (or counters)
to be completely eliminated and the background contribution from
the target walls (if such target was used) to be clearly identified.
The use of these detectors allowed large solid angles to be used
while, at the same time, achieving very good angular resolution.
The use of large solid angles meant that a large fraction of the
phase space available to the reaction was observed simultaneously
thus increasing the data taking rates by up to two orders of magnitude. Although, strictly speaking, the early Rochester experiment
falls into this category, the Manitoba 42 MeV experiment[9-14] is a
true representative of these, which I term, "second generation", or
"comprehensive" bremsstrahlung experiments.

Just as experiments naturally divide into first and second generation (selective and comprehensive) experiments, the theoretical-
numerical calculations also divide into two categories.

The first generation numerical calculations were done to suit
the first generation experiments. Since, in these experiments, the
data were collected at a few selected angles, the theoretical calculations were done in a similar manner. As a result the experiment
to theory comparison was easy and straight forward. At the Gull
Lake Symposium, Halbert was able to summarize almost all experimental and almost all theoretical results in a table two pages long.

The comprehensive, second generation experiments, which yield much more detailed data, naturally require extensive theoretical calculations to be performed in order to be able to make a meaning-ful comparison between experiment and theory. A good example of this is the experiment we performed at Manitoba. In order to com-pare our results with a particular theoretical calculation, we computed a set of approximately 6000 $d^5\sigma/d\Omega_1 d\Omega_2 d\psi_\gamma$ cross sections covering Θ_1 and Θ_2 ranges from 14^0 to 42^0 and the full ranges of non-coplanarities and photon angles, ψ_γ. Since this mass of numbers is not very practical to work with, we developed a Monte Carlo sim-ulation procedure[13,14] (the "global" method) which we used for a detailed comparison of the experimental results with the theoretical calculations. Some unpublished results of this analyses are pres-ented later in this paper.

B. REVIEW OF EXPERIMENTS

B.1 Status of NNB at the Gull Lake Symposium

In order to put the new experiments into their proper perspec-tive, I shall first review the status of NNB at the time of the Gull Lake Symposium.

In Tables I and II of Halbert's paper[4], most of the results of all ppB and npB experiments performed by that date are listed and in Figures 3 - 8 the comparison is made with the Hamada-Johnston and one boson exchange calculations. The agreement between exper-iment and theory was generally good except in the case of the McGill experiment[15]. (This disagreement was considered not to be serious because the actual differences were thought to be within the errors of the theoretical calculations.) However, these Figures do not, themselves, tell the whole story. First, the total number of data points measured was quite small and most of the experiments quoted were performed only in the Harvard geometry[16] and at rather large proton opening angles, that is, not very far from the elastic limit. (It should also be pointed out that these large opening angles correspond only to center of mass polar angles between 70^0 and 90^0, thus the range of CM angles explored was also very small.) Second, in most of the experiments, only the $d^4\sigma(\Delta\phi=o)/d\Omega_1 d\Omega_2$ cross sections were measured, thus giving us no information about the dependence of the cross section on photon energy and direction (i.e., event non-coplanarity). Finally, the statistical accuracy of most of the ex-periments was quite low, on the average only of about 20%.

Keeping in mind that bremsstrahlung process depends on five in-dependent variables (four in the case of an unpolarized target and beam), that before the Gull Lake Symposium only a very limited (in fact, as it now appears, the least interesting) ranges of proton polar angles were explored, that essentially no non-coplanar cross sections were measured, that no polarized beams or targets were used and that the accuracy of all experiments (except McGill) was very low, it is fair to say that at that time only a qualitative and limited comparison between experiment and theory existed. For critical tests of the validity of potential models OFES much more accurate and much more comprehensive experiments were (and still are) needed, or to paraphrase Halbert, "departures from well-trodden paths of the late 60's are needed". As we shall see later in this paper, some experiments have departed from well-trodden paths, but the distance yet to travel on these new paths is very long indeed.

B.2 Summary of NNB Experiments Since the Gull Lake Symposium

At the time of the Gull Lake Symposium the Orsay[17] experiment and the second part of the Manitoba [11-14] experiment were under way. After the completion of these experiments, the flurry of bremsstrahlung activity that existed in the late 60's waned considerably. In Table I, all ppB, npB and πpB experiments performed since the Gull Lake Symposium are summarized and, in Figure 1, the nominal ranges of the polar angles of the massive particles, which were detected in these experiments, are depicted. Most of the columns in Table I are self-explanatory. In column 6, the ratio (or the range of the ratios) of the sum of the polar angles of the outgoing massive particles divided by the opening angle for the case of elastic scattering is shown. This ratio can be taken as a measure, in a geometrical sense, of how far an experiment is from the elastic limit. This ratio should not be confused (or identified) with the notion of "far off-energy-shell" (far OFES) as an experiment can go OFES in ways other than decreasing the sum of the polar angles of the outgoing particles. The range of percentage errors shown in the last column was determined by using the errors from the most accurate and the least accurate data points and no attempt was made to find an average error. A single number preceded with "greater than" (>) sign means that some of the quoted errors are larger than 100%.

The two first generation experiments[18,19] performed at Van de Graaf energies and the one[20] at 52 MeV are in good agreement with theoretical calculations. However, these experiments were done at rather large angles (see Table I), thus, are not very far from the elastic scattering limit.

Since the Gull Lake Symposium, only one npB experiment[23] has been done. This experiment was plagued by a large number of random coincidences and a low intensity neutron beam so the measured cross sections have sizable errors.

So far there has been only one πpB experiment done[24]. It was a predecessor to the UCLA ppB experiment performed at 730 MeV[22]. As the πpB process is outside the range of topics implied by the title of this paper, I shall not make any more comments about this fine experiment.

The TRIUMF experiment[21] was performed at 200 MeV with the use of MWPC chambers but the proton detectors subtended relatively small solid angles. As this experiment was already described at this conference, I shall not discuss it further.

The Orsay and Manitoba experiments strongly disagree with theoretical calculations and the Manitoba and UCLA experiments are of the

Institution	Experiment	Beam Energy (MeV)	Θ_1 (Proton)	Θ_2 (Other)	$\frac{\Theta_1+\Theta_2}{(\Theta_1+\Theta_2)}$	Photon Detected?	Coordinates Measured?	Type Of Cross Sections Measured	Number Of Data Points Measured	Range Of Percentage Errors	References
Laval	ppB	6.92	38°	38°	0.85	no	no	$d^4\sigma(\phi_r=0)/d\Omega_1 d\Omega_2$	1	41%	18
			37°	37°	0.82	no	no	-"-	1	42%	18
ETH, Zurich	ppB	11	30°	30°	0.67	no	no	-"-	1	7%	19
		13	30°	30°	0.67	no	no	-"-	1	11%	19
Manitoba	ppB	42	16°-40°	16°-36°	0.36-0.85	no	yes	$d^5\sigma(\psi_\gamma)/d\Omega_1 d\Omega_2 d\psi_\gamma$	250	>14%	11
								$d^4\sigma(\phi_r)/d\Omega_1 d\Omega_2$	145	> 9%	11
								$d^2\sigma(\Theta_S,\Theta_D)/d\Theta_1 d\Theta_2$	17	4%-27%	11
								Also "global" analyses (see text)			12
Tokyo	ppB	51.8	33°	33°	0.74	no	no	$d^5\sigma(\phi_r=0)/d\Omega_1 d\Omega_2$	1	16%	20
			8.5°	8.5°	0.19	no	no	Incomplete - upper limit only			20
Orsay	ppB	156	15°	$18^\circ,21^\circ$ $24^\circ,27^\circ$	0.37-0.48	yes	no	$d^5\sigma(\psi_\gamma)/d\Omega_1 d\Omega_2 d\psi_\gamma$	12	13%-45%	17
TRIUMF	ppB	200	14°-18°	14°-18°	0.32-0.41	no	yes	$d^5\sigma(\phi_r=0)/d\Omega_1 d\Omega_2 d\psi_\gamma$	16	~15%	21
			11°-14°	11°-14°	0.25-0.32	no	yes	not yet analysed			
UCLA	ppB	730	10°-60°	43°-57°	0.68-0.99	yes	yes	$d^5\sigma(\phi_r=0)/d\Omega_p d\Omega_\gamma dk$	60	16%-44%	22
Harwell	npB	130	$20^\circ,32^\circ$	$23^\circ,26^\circ$ $29^\circ,38^\circ$	0.5-0.8	no	no	$d^4\sigma(\phi_r=0)/d\Omega_p d\Omega_n$	8	>18%	23
UCLA	πpB	298	10°-60°	43°-57°	-	yes	yes	$d^5\sigma(k)/d\Omega_p d\Omega_\gamma dk$	268	>13%	24

TABLE 1

A Summary of Bremsstrahlung Measurements
Performed Since the Gull Lake Symposium

456

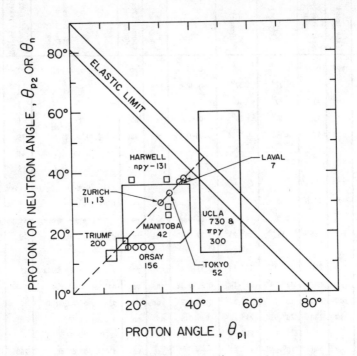

Fig.1   Polar angles of the detected massive particles are shown
for all ppB, npB and πpB experiments performed since the Gull Lake
Symposium. The experiments which were done without measuring the
coordinates of the charged particles are shown as circles (ppB) or
small squares (npB). The nominal angular ranges detected in the
second generation experiments are shown as large squares or rectangles. The numbers underneath the institution name give the incident beam energy in MeV. In the case of the UCLA experiment, the
same equipment was used to perform a ppB experiment at 730 MeV and a
πpB experiment around 300 MeV. In the case of Tokyo and Harwell
experiments, only those polar angle pairs yielding $d^4\sigma/d\Omega_1 d\Omega_2$ cross
sections with errors smaller than 50% are shown. References to all
experiments are given in Table I. The two heavy lines labeled "elastic limit" define the magnitude of the opening angle for N-N elastic scattering in the nonrelativistic limit and at 730 MeV incident energy.

second generation type. Consequently, I shall comment on each of
them separately and in some detail.

B.3 Orsay Experiment at 156 MeV

The Orsay experiment[17] was a typical experiment of the first
generation. In many respects it resembled the original Harvard

experiment[4] except that proton detectors were positioned at much smaller angles and a very large Cerenkov counter was used always in coincidence. A total of 4x3 $d^5\sigma/d\Omega_1 d\Omega_2 d\psi_\gamma$ cross sections were measured with an accuracy ranging from 12% to 28%.

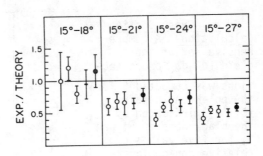

The results of this experiment were compared with three different calculations[25,26,27], all of them using the Hamada-Johnston potential. The calculation done by Brown[25] included the double scattering term and was done for the coplanar case only. (Brown also used the Bryan-Scott potential). Celenza et al[26] included relativistic corrections and non-coplanar effects but did

Fig.2   A summary of the comparison of Orsay results[17] with calculations by Brown[25], (◐), Celenza et al[26] (∓) and Heller and Rich[27] (◒) done with the use of the Hamada-Johnston potential. In the comparison done by Heller and Rich, the experimental data have been split into three bins according to the range of photon angles detected. Comparisons done by Brown and Celenza et al are for the averages of all three bins taken together.

not include the double scattering term or Coulomb corrections. Heller and Rich[27], on the other hand, included Coulomb corrections, the double scattering term and non-coplanar effects, but did not include relativistic corrections. In Figure 2, I have summarized the experiment to theory ratios for all three Hamada-Johnston calculations. The discrepancy at the largest angles is about a factor of two with a statistical significance of 4-5 standard deviations. Although none of the three calculations included all corrections, it seems unlikely that doing so will remove this discrepancy. At present, this discrepancy is not understood, but since it is based on a small number of points from one experiment, and on calculations which are known to be deficient it would be premature to conclude on the basis of this data alone that the Hamada-Johnston potential is not able to fit the data.

B.4. Manitoba Experiment at 42 MeV

This was a comprehensive, second generation experiment. A wire chamber spectrometer, especially constructed for this purpose[9], was used and event rates of over 100 ppB events/hour were achieved. A total of about 13000 events were collected in the two experiments performed[10,11]. The data were analysed and compared extensively with a theoretical calculation in two different ways[11-14] which are termed "conventional" and "global".

458

a. Cross Section Calculation. A comprehensive set of the $d^5\sigma/d\Omega_1 d\Omega_2 d\psi_\gamma$ cross sections was computed[28] for the polar angle pairs (Θ_1, Θ_2) indicated by the black dots in Figure 3, for five values of relative non-coplanarities and in 10^0 increments of photon angle ψ_γ.

These calculations were performed[29] in the laboratory system using the Hamada-Johnston (H-J) potential; they included internal rescattering terms and relativistic spin corrections but did not include Coulomb effects and exchange terms.

It is difficult to observe general trends in any set of data, if that data set is very large. So, to help in the interpretation of our data, we have computed, with the use of the "global" procedure, the following cross sections.

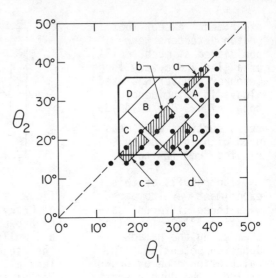

Fig.3 Full dots represent the Θ_1-Θ_2 proton angle pairs at which the numerical computations of the $d^5\sigma/d\Omega_1 d\Omega_2 d\psi_\gamma$ cross sections were made. The shaded rectangles labeled a, b, c and d are the areas of the Θ_1-Θ_2 plane over which are averaged the cross sections and the squares of matrix elements presented in Figures 4 and 5, respectively. The outside rectangle outlines the ranges of polar angles detected in the experiment and the areas A, B, C and D give the ranges of integration when making the experiment to theory comparison as outlined in Figure 9.

$$\frac{d\sigma(\psi_\gamma)}{d\psi_\gamma} = \int_{\Theta_S'}^{\Theta_S''} d\Theta_S \int_{\Theta_D'}^{\Theta_D''} d\Theta_D \int_{\phi_r'}^{\phi_r''} d\phi_r \int_{\phi_1'}^{\phi_1''} d\phi_1 \ \times \ \frac{d^5\sigma}{d\Theta_S d\Theta_D d\phi_r d\phi_1 d\psi_\gamma} \tag{1}$$

where $\Theta_S = \Theta_1 + \Theta_2$, $\Theta_D = |\Theta_1 - \Theta_2|$, ϕ_r is the relative non-coplanarity (see ref. 10 for the definition) and ψ_γ is the photon angle as defined in ref. 30. In Figure 4, a set of these cross sections is

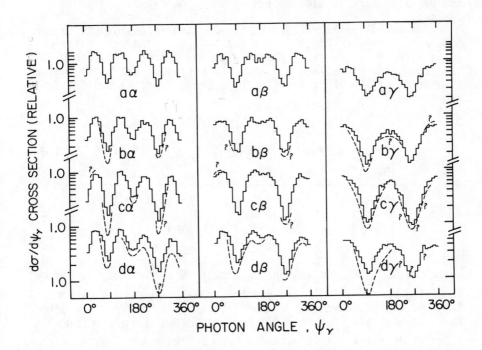

Fig.4   The $d\sigma/d\psi_\gamma$ cross sections as defined by eq. (1) are plotted for four shaded regions, $a(66^0 < \Theta_S < 78^0$ and $\Theta_D < 2^0)$, $b(46^0 < \Theta_S < 58^0$, and $\Theta_D < 4^0)$, $c(32^0 < \Theta_S < 44^0$ and $\Theta_D < 4^0)$ and $d(46^0 < \Theta_S < 58^0$ and $9^0 < \Theta_D < 13^0)$ indicated in Figure 3 and for three different ranges of the non-coplanarity $\alpha(\phi_r < 0.3)$, $\beta(0.35 < \phi_r < 0.65)$ and γ $(0.7 < \phi_r < 1.0)$ as a function of the photon angle ψ_γ.   The dashed curves below or above the histograms indicate the regions and the approximate size of the disagreements between experiment and theory. The question marks placed near certain portions of the curves indicate that disagreement is only suggestive.

plotted in relative units.   It is interesting to note that, in the first approximation, the overall shape of the $d\sigma/d\psi_\gamma$ cross sections seem to be independent of Θ_S **and** Θ_D but is strongly dependent on the relative non-coplanarity.   The four quadrupole bumps that are so distinct in the case of the coplanar cross sections degenerate into two bumps for the case of maximally non-coplanar events.   In detail, one can notice that the peak to valley ratio increases as Θ_S decreases, and that, in case of large Θ_D, the bumps at large ψ_γ are depressed relative to the bumps at small ψ_γ.   It is also evident that the overall cross section decreases as Θ_S decreases and Θ_D increases.

The global method allows us to investigate easily some other aspects of the theoretical cross sections and kinematic relationships between various variables. With the use of this method, we have computed the average value of the square of matrix element as a function of the laboratory photon energy, k, using the following expression.

$$|M(k)|^2 = \frac{\int_{\Theta_S'}^{\Theta_S''} d\Theta_S \int_{\Theta_D'}^{\Theta_D''} d\Theta_D \int_{\phi_r'}^{\phi_r''} d\phi_r \int_{\phi_1'}^{\phi_1''} d\phi_1 \ |M(k,\Theta_S,\Theta_D,\phi_r)|^2 \ F(k,\Theta_S,\Theta_D,\phi_r)}{\int_{\Theta_S'}^{\Theta_S''} d\Theta_S \int_{\Theta_D'}^{\Theta_D''} d\Theta_D \int_{\phi_r'}^{\phi_r''} d\phi_\theta \int_{\phi_1'}^{\phi_1''} d\phi_1 \ F(k,\Theta_S,\Theta_D,\phi_r)} \qquad (2)$$

where $F(k,\Theta_S,\Theta_D,\phi_r)$ is the phase space density factor. In Figure 5, histograms of $|M(k)|^2$ are plotted on a log-log scale as a function of the photon energy in the laboratory system. It is evident that, in the case of large Θ_S and at photon energies between 7 and 11 MeV, the matrix elements have a $1/k$ dependence; at energies smaller than 7 MeV the results are inconclusive but consistent with a $1/k^2$ dependence setting in. At energies above about 11 MeV, the shape of the curves becomes anything but proportional to $1/k$ or $1/k^2$. A word of caution is needed here. What is plotted in Figure 5 is an <u>average</u> of the square of the matrix element. This is a good approximation to the matrix element at a given point only if the matrix element itself is fairly constant over the range of integration. How much this mathematical approximation distorts the shapes of curves in Figure 5 is hard to say without more detailed calculations, but the approximate independence of $d\sigma/d\psi_\gamma$ cross sections as a function of Θ_S and Θ_D suggests that these distortions should not be serious.

It might be difficult to say what the true meaning of these curves is but they suggest that at photon energies above about 10 MeV, the low energy theorem is not applicable. As a corollary, it would follow that the off-energy-shell contributions to the cross sections (or matrix elements) may not be negligible in this range.

b. Conventional Analysis. In Figure 6, a set of measured and calculated $d^5\sigma/d\Omega_1 d\Omega_2 d\psi_\gamma$ and $d^4\sigma/d\Omega_1 d\Omega_2$ cross sections is presented. Comparison of the experimental data with the theoretical calculation (heavy line histogram) reveals disagreements in several distinct places, but due to the relatively poor statistics on the individual points, it is difficult to get a general pattern of disagreement.

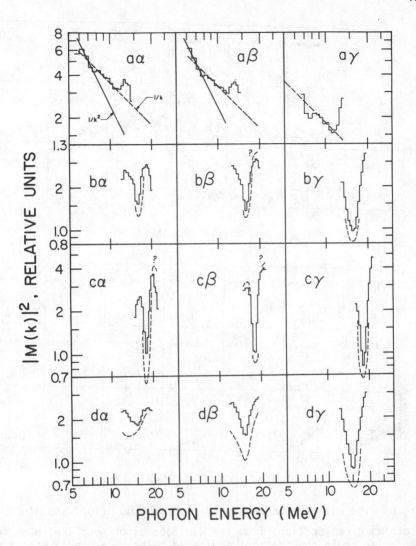

Fig.5   The squares of the matrix elements as defined by eq. (2) are plotted as a function of photon energy, k, for the same phase space regions as in Fig. 4. The meaning of the dotted curves and adjacent question marks is the same as in Figure 4. The heavy and dashed lines in the top three panels represent the $1/k^2$ $1/k$ dependence, respectively, as is expected from the first two terms of the soft photon approximation.

In Figure 7, the ratios of the experimentally measured and theoretically simulated $d^2\sigma/d\Theta_1 d\Theta_2$ cross sections are presented[11]. This figure exhibits a significant discrepancy between the experimental results and the theoretical calculations. In general, the

462

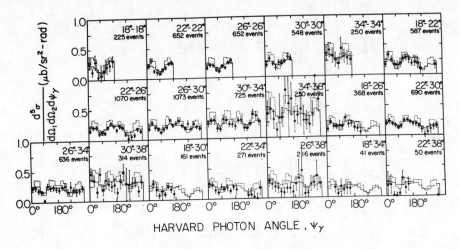

HARVARD PHOTON ANGLE, ψ_γ

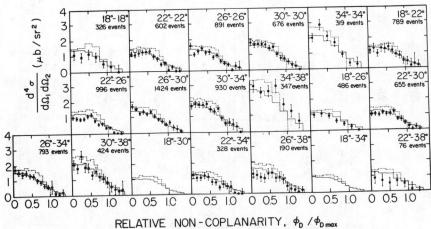

RELATIVE NON-COPLANARITY, $\phi_D / \phi_{D\,max}$

Fig.6   The measured and computed $d^5\sigma/d\Theta_1 d\Theta_2 d\psi_\gamma$ (top) and $d^4\sigma/d\Omega_1 d\Omega_2$ (bottom) cross sections from the Manitoba experiment are shown for the polar angle combinations indicated. The polar angle bins are 4^0 wide in both Θ_1 and Θ_2. The $d^5\sigma/d\Omega_1 d\Omega_2 d\psi_\gamma$ cross sections were integrated over ϕ_r from 0 to 0.7. The experimental data (points with error bars) were not corrected for the effects of angular and energy resolutions of the spectrometer but were corrected for the geometrical efficiency of the spectrometer and the detection efficiency of the counters. The error flags shown include statistical and estimated systematic errors. There is an uncertainty of ±6% in the vertical scale due to normalization errors. The dotted histograms represent the theoretical cross sections averaged over the same angular ranges as the experimental data, while the solid histograms represent the same theoretical cross sections but with all angular and energy resolutions folded in.

agreement is good for large values of Θ_S and small values of Θ_D, but, as Θ_S decreases and/or Θ_D increases, the experimental results become significantly lower than theory. It is of interest to note that the Orsay data (see also Figure 2) show the same trend of discrepancy increasing with Θ_D.

c. Global Analysis. In brief, the procedure used in the global analysis is the following: ppB events are produced by the computer, randomly distributed in phase space. Then, to each event is attached a weight which is proportional to the square of the matrix element computed at that point in phase space where that particular event is situated. After that, the events are traced through the experimental equipment, energy losses and multiple scattering being taken into account, and events are tested as to whether or not they pass through the detecting equipment. Finally, these events are reconstructed in the same way as events detected in the real experiment. Events simulated in such a manner correspond very closely to the experimentally detected events. The comparison between experiment and theory is then made by comparing equivalent distributions made from the simulated and real events.

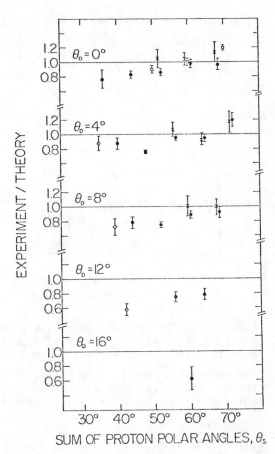

Fig.7 Comparison of Manitoba results with Hamada-Johnston calculation. The ratio of the experimental to theoretical $d^2\sigma/d\Theta_1 d\Theta_2$ cross sections as a function of $\Theta_S = \Theta_1 + \Theta_2$ for different values of $\Theta_D = |\Theta_1 - \Theta_2|$ is presented. The points x and ● are due to the first and second experiments, respectively. The results of the McGill[20] and Orsay[22] experiments are also shown (denoted by ▢ and ◇, respectively) in a form suitable for comparison to data from Manitoba experiment.

Some results of the global analysis have been reported previously[12]. There, it was shown that the largest discrepancy between experimental and simulated data existed for the most asymmetric events

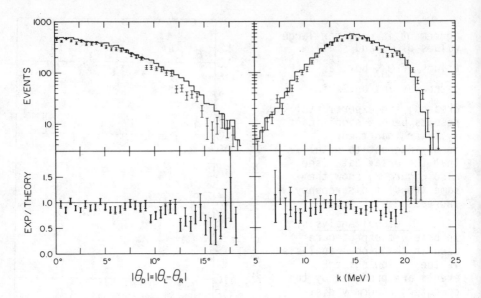

Fig.8   In the top two panels, all experimental (points with error bars) and simulated (histograms) data as a function of the difference in proton polar angles, Θ_D, and photon energy, k, are shown. In the bottom two panels, the ratios of the above two distributions are shown. The heavy error bars are due to statistical errors only, while the light bars are due to estimated systematic errors.

and events with photon angle ψ_γ around 80° (see Fig.2 of ref.12).

Also, analysis in the CM system revealed that under certain conditions, the discrepancy was increasing with decreasing photon energy (see Fig.3 in ref.12).
In Figure 8, the distributions and ratios for all real and simulated events as a function of the difference in proton polar angles, Θ_D, and photon energy, k, are shown. The lack of events at the larger values of Θ_D is evident, as well as the statistically significant difference in the shapes of the photon energy distributions. In order to understand the origin of this disagreement, the data has been split into four regions, A, B, C, and D as indicated in Figure 3, and then each of these into three regions depending on the non-coplanarity. In some of these regions, the agreement between the simulated and experimental data is excellent, in others, it is very poor. In Figure 9, are shown the ψ_γ and k distributions and their ratios for region Cα which exibits quite an interesting discrepancy.

Although the experiment to theory comparison is made only in terms of event distributions, it is possible to find an approximate correspondence between those distributions and the cross sections (or matrix elements) as given in Figures 4 and 5. The argument is as follows: the shaded area, c, in Figure 3, for which the cross sections and matrix elements were computed is essentially contained within the area C for which the experimental and simulated data were analysed. Figure 4 shows that the $d^5\sigma/d\Omega_1 d\Omega_2 d\psi_\gamma$ cross sections do not change very fast as a function of Θ_S and Θ_D, and, therefore, it is reasonable to assume that these cross sections will have about the same shape whether they are averaged over the small area, c, or over the larger area, C. In Figure 9, the resolution in ψ_γ is roughly equivalent to the bin size so there exists a good correlation between the ψ_γ's in Figures 9 and 4. As a result, it is reasonable to assume that if a disagreement exists in Figure 9, the same kind of disagreement will also exist in Figure 4. This disagreement is indicated in the $C\alpha$ panel of Figure 4 by a dashed line.

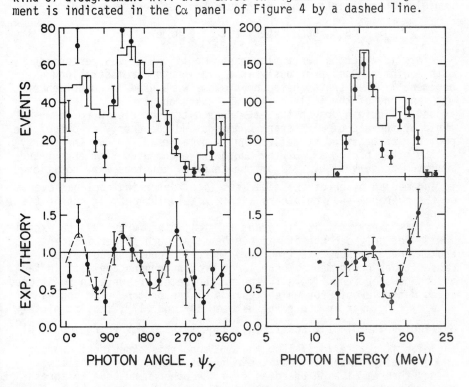

Fig.9   In the top two panels, the experimental and simulated data as a function of the photon angle, ψ_γ, and photon energy, k, are shown for the region C of Figure 3 and for the non-coplanarity range $\alpha(\phi_r < 0.3)$. In the bottom two panels, the ratios of the above two distributions are shown.

Because of the approximations involved, this should be considered only as having semi-quantitative significance in that it indicates the general regions where the disagreement is observed and its approximate size. Similar arguments can be made for all of the other subsets of data in Figure 4.

The reasoning presented above for the ψ_γ distributions is essentially valid for the $|M(k)|^2$ distributions shown in Figure 5. The difference is that the $d\sigma/dk$ cross sections depend more strongly on Θ_S and Θ_D, than the $d\sigma/d\psi_\gamma$ do, thus the averaging procedures are less reliable. The results of this kind of analysis are also shown in Figure 5 by the dashed lines.

e. Phenomenological Off-Shell t-Matrix. In a short paper, Celenza et al[31] have obtained, for the first time, off-shell matrix elements implied by experimental data, without the use of a potential model. In that work, they have used 80 data points from the first Manitoba experiment[10]. By varying 8 parameters they have reduced the overall χ^2 value from 284 to 135. Although more experimental data now exists, this procedure has not been repeated.

B.5 UCLA Experiment at 730 MeV

This is a second generation experiment[22] which was performed with equipment used previously in a pion-proton bremsstrahlung experiment[24]. Two proton detector systems were used in coincidence with any one of the 16 lead glass Cerenkov counters positioned around a liquid hydrogen target. Proton trajectories were determined by the use of wire spark chambers and the energy of one of the protons was measured by deflection in a magnetic field. The energy of the other proton and of the photon were computed from energy and momentum conservation. Six of the Cerenkov counters were positioned in the median plane of the apparatus (i.e. detecting coplanar events on the average) and the others approximately 19° and 38° out of the plane.

This experiment is different from all other previous ppB experiments not only because it was performed at much higher incident proton energy, thus detecting very energetic photons, but also because it was performed at very large proton opening angles. In fact, as can be seen from Figure 1, the proton detectors were so positioned as to detect events for which the sum of the proton polar angles $\Theta_S = \Theta_1 + \Theta_2$, was in the range between 54° and 117°, thus completely overlaping the elastic scattering range (at 730 MeV energy, $\Theta_S = 81°$ for elastic scattering). For that reason, this experiment appears to be ideally suited to test the range of validity of Low's low energy theorem at high energies. On the other hand, due to threshold effects in the Cerenkov counters, only events with photons more energetic than 20 MeV were detected, thus the region of small photon energy E_γ was not explored. In order to illustrate how close to the elastic limit that experiment is we have computed at Manitoba the distribution of potentially detectable events as a function of Θ_S. The results are presented in Figure 10 assuming zero energy threshold (dots) and a 20 MeV threshold (crosses) in the photon detection. This calculation was done using the MANSA program[32] and involved the following major approximations: (1) phase space distribution of events, (2) point hydrogen target, (3) solid angles of proton detectors as given in ref.22 with proton detection efficiency equal to 100%, and (4) no selection of events on the basis of photon direction.

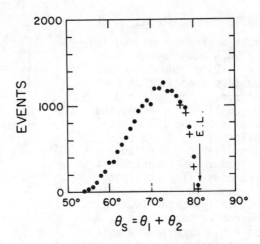

Fig.10 The distribution of simulated events as a function of the sum of the proton polar angles, $\Theta_S = \Theta_1 + \Theta_2$ is shown for the geometry defined by the proton detectors in the UCLA experiment[22] (see text for details).

The data from this experiment were presented in the form of $d^5\sigma/d\Omega_p d\Omega_\gamma dk$ differential cross sections averaged over the system acceptance for each proton counter and 20 MeV bins in photon energy. There are 66 data points, each measured with about 20% accuracy. The general trend of the cross sections is the same for all photon counters, a decrease following roughly 1/k dependence for photon energies smaller than about 80 MeV and then a significant rise. To illustrate this 1/k behaviour of the cross sections, the data for G10 Cerenkov counter from ref.22 have been replotted on a log-log scale in Figure 11. Nefkens and Sober[33] have tried to explain the 1/k dependence by postulating the existence of external emission dominance (EED) which allowed them to neglect all the terms in Low's soft photon expansion except the first (1/k) term. The full and dashed lines in Figure 11 are replotted from Fig. 3 of ref.22 and represent the EED cross sections averaged over the experimental acceptance as described in ref.22.

On the other hand, a different explanation to this result can be offered. If it is assumed, as Signel has pointed out in ref.2, p. 281, that "the Low expansion is effectively one in powers of the photon energy divided by the average NN kinetic energy" then for all photon energies smaller than about 80 MeV or so, this dimensionless expansion parameter is on the order of 0.1 or smaller. Thus, it should not be too surprising to find that the result is so well fitted by the first term in the Low's expansion.

Fig.11   The $d^5\sigma/d\Omega_p d\Omega_\gamma dk$ cross sections measured in the UCLA experiment[22] for one (G10) of the Cerenkov counters used in that experiment. The full and dashed lines represent the results of the EED calculation of the cross sections averaged over the acceptance of experimental apparatus. The straight line labeled 1/k is plotted just to indicate visually the energy dependence of the first term in the soft photon approximation.

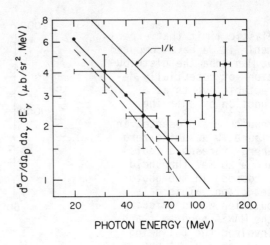

C.   CONCLUSIONS AND COMMENTS

C.1   Is the Hamada-Johnston Potential Good Off-Shell?

At present, the UCLA experiment has little, if any, relevance to the question of the validity of nucleon-nucleon potentials off-energy-shell (OFES). In fact, there exists no theoretical calculation of the ppγ cross sections based on a realistic potential model which the UCLA results could be compared with. Also, it is doubtful that any calculation using the concept of non-relativistic potentials could be meaningful at 730 MeV. (A field theoretic one boson exchange calculation has been performed and reported at this conference[34]). As far as the lower energy experiments are concerned, most of them (except Orsay and Manitoba) were performed at rather large proton scattering angles (see Secs. B.1 and B.2) and are in good agreement with H-J calculations, so I will not discuss them in this section.

The results of the Orsay[17] and Manitoba[11-14] experiments, when compared with calculations of the ppB cross sections based on the H-J potential, exhibit a significant discrepancy, up to a factor of two, with a statistical significance of four to five standard deviations. Clearly, if the experiments and calculations are taken at their face value, then the only possible conclusion is that H-J potential is unable to describe OFES effects.

But should the calculations and the experiments be taken at their face value?

I do not wish to comment in detail on how good the theoretical calculations are, except to say that it is believed to be unlikely that the cross sections will decrease by a factor of two once all the corrections are included in the calculations. As far as the present experiments are concerned, I can say the following. I have discussed at great length the details of the Orsay experiment with

Professor Marty and her collaborators and have no reason to suspect that their results are wrong. At Manitoba, Smith and myself have performed many tests in order to convince ourselves that the discrepancy observed at 42 MeV is not due to systematic errors in our experiment. In the end, we have convinced ourselves. Therefore, I have no reason to suspect that either of the two experiments are in fault by a factor of two.

What are the salient points of the observed disagreements?

First, the Orsay and Manitoba experiments, although performed at different energies and under different experimental conditions, qualitatively agree among themselves in their common disagreement with the H-J calculations. As it can be seen from Figures 2 and 7, the discrepancy increases as the polar angle asymmetry, $\Theta_D = |\Theta_1 - \Theta_2|$, increases. At present, it is not clear whether this correlation in the disagreements is just fortuitous, or has some common, physically significant, origin.

Results presented in this paper and in ref.12 indicate that the disagreements at 42 MeV can be correlated with particular ranges of certain variables. Figure 2 in ref.12 together with figures 4 and 9 in this paper point out to large disagreements around $\psi_\gamma \approx 80^{\circ}$ and $\psi_\gamma \approx 280^{\circ}$. It is interesting to note that this is the region of photon angles where the first two terms of soft photon expansion are equal to zero[35]. Other regions of discrepancy can be broadly correlated with asymmetric events ($\Theta_D \gtrsim 10^{\circ}$), relatively small sums of proton polar angles ($\Theta_S \lesssim 55^{\circ}$) and non-coplanar events ($\phi_r \gtrsim 0.7$). When all of these conditions are combined, one notices that discrepancy is the most pronounced where the cross sections are relatively small, not where the photon energy is the largest. This is also illustrated in Figure 5 where the largest discrepancies are observed to be in the valleys.

As a conclusion, I feel it is reasonable to say that the existing results throw serious doubt on the validity of the H-J potential OFES. It is my personal opinion that more complete calculations, and possibly more accurate experiments, are needed before it can be considered as definite that the H-J potential is not good OFES. For instance, it is possible to imagine that a more complete theoretical calculation will bring the cross sections down for about 30%, and that the new experimental results will be up by roughly one to two standard deviations in which case the theory and experiment would again be in agreement.

C.2 Some Characteristics of the Second Generation NNB Experiments

I feel this review would not be complete if I did not make some comments of a general nature about the comprehensive, second generation experiments.

From the point of view of experimental techniques, one can identify three ranges of incident proton energies. These ranges correspond to three different technologies that have to be used in order to perform comprehensive and accurate ppB experiments.

In the energy range between approximately 35 MeV and 350 Mev wire chambers with plastic scintillation counters (as used in the Manitoba[9] and TRIUMF[21] experiments) are very suitable. Due to various reasons (multiple scattering, energy losses, size of elastic and reaction cross sections, etc.) the easiest range to perform ppB experiments with this technology is roughly between 100 MeV and 200 MeV.

At energies below 30 - 40 MeV the use of wire chambers is not suitable as the energy losses are too large. At these energies it would be necessary to use a setup with some kind of drift chambers, to a large extent similar to the one tested at Manitoba[36]. With such equipment accurate ppB experiments could be done even at Van de Graaf energies.

At energies above 300 - 400 MeV, the technology must be different again. Due to the high reaction cross sections for protons, it is necessary to measure the energies of the protons by bending them in a magnetic field, rather than by stopping them in a scintillation counter. As discussed in Sec. B.5, this technology was used in the UCLA experiment[22].

Whichever of these three technologies is used, one has to notice that the second generation experiments have all the characteristics of a typical high energy experiment. Superficially, this might look rather unimportant, but the practical consequences are quite far reaching. The experiments are much more elaborate and costly and require much more manpower to set up, execute and analyze. In the case of low energy accelerators ($\lesssim$ 100 MeV), where wire and drift chambers are not in everyday use, a complete retooling is necessary to do these experiments. This is rarely a very popular undertaking, and definitely quite unpopular in times of financial restraints. This is all rather obvious. However, what is much less obvious, and too often underestimated by the non-high-energy physicists (I emphasize "non-high-energy") is the software support, computer time and overall effort needed to analyze the data from these experiments and compare them with theoretical calculations. In this respect, the second generation bremsstrahlung experiments are more similar to a typical bubble chamber experiment than to a typical first generation ppB experiment. To illustrate this point, which is so relevant to the future bremsstrahlung experiments, I shall say that at Manitoba, we have spent well over three quarters of our overall effort on software development and data analysis (which, of course, included the global analysis). In comparison, building and debugging the wire chamber spectrometer was a relatively easy task to perform . This may be an exceptionally large fraction of effort as we had no software at all when we started the experiment, but even if all software existed, I still estimate that the data analysis effort will always constitute more than half of the overall effort.

A further comment is needed with respect to what I have called the "second generation" theoretical calculations. As we have seen, a comprehensive second generation experiment requires a very comprehensive numerical calculation. Who should do these calculations, theorists or experimentalists? I have talked with a number of my colleagues, theorists, and I did not find anyone very anxious to grind 6000 cross sections through a computer. At 42 MeV we did our own numerical calculation, thanks to both M.K. Liou who generously left his fine program with us and to the University of Manitoba who did not charge us for the computer time. Clearly, though, our experiment should be compared with other calculations as well. How is that going to be done? In fact, this question is much more general. The real question is: How are the comprehensive, second generation experiments and theory going to be compared in the future? Who is going to do this work, theorists, experimentalists or "ppB phenomenologists", a breed of physicists that does not exist yet.

C.3 Future Outlook

Fourteen years have elapsed since the first calculation of the ppB cross sections was done by Sobel and Cromer[37] and twelve years since the first experiment was performed at Harvard[38]. A number of experiments and calculations have been done since, but much has not been learned about the basic nucleon-nucleon interaction from these experiments and calculations. The reasons are the following: Early calculations indicated that the ppB cross sections depended very strongly on the OFES behaviour of the phenomenological N-N potentials, thus suggesting that even very crude experiments, experiments in which only a few hundred events would be collected, would be able to distinguish between different potentials. As a result all "first generation" experiments were "selective" and of low accuracy. Unfortunately, more sophisticated theoretical analyses of the NNB process later showed that this was not so, thus, only much more accurate and comprehensive experiments could be informative. However, until the late sixties, the technology (wire chambers) for performing such experiments did not exist and since then only a few comprehensive "second generation" experiments have been performed. One of these (Manitoba experiment) shows definite and detailed disagreements with a theoretical calculation, thus demonstrating that the accurate and comprehensive second generation experiments can provide useful information about the OFES effects of the N-N interaction, the information that the first generation experiments were not able to give.

For these reasons, I think that the era of the first generation experiments is over. To be more specific, I feel that experiments in which 10 points will be measured, each with 10% statistics, are of little use, as they will provide only very limited and insufficiently complete information about the N-N forces OFES. Therefore, I sincerely doubt that they are worth doing.

What kind of second generation experiments could be done, at least as far as the experimental technologies of today are concerned?

I would like to answer this question starting with our own experience. At 42 MeV, we collected data at a rate in excess of 100 good ppB events per hour. That was with equipment designed in 1966. I see no reason (other than monetary) why an experiment could not be done today with data taking rates of over 500 ppB events/hour. Therefore, it should not be difficult to collect 250,000 events at 42 MeV

in a new experiment, thus giving, for instance, 250 data points with
3% accuracy. As the incident proton energy increases above 42 MeV,
but stays lower than about 200 MeV, the ppB experiments become easier
to perform because the ppB cross section increases and the p-p elastic cross section decreases. Experiments at 100 MeV or 200 MeV in
which 10^6 events are to be collected (say 400 data points with 2%
accuracy) are definitely possible to do. At Van de Graaf energies
ppB experiments would be more difficult to do, but I feel that, by
using drift chambers, one should be able to do at least as comprehensive an experiment as was done at Manitoba at 42 MeV.

npB experiments are also more difficult to do. I have not done
any, so I am not as confident as in the case of ppB experiments. On
the other hand, I am optimistic. For instance, with such fine equipment as the BASQUE group has assembled at TRIUMF, with a neutron beam
such as that produced at TRIUMF by a 10 µA proton beam, and with
Cerenkov counters similar to those used in the Orsay and UCLA experiments, I believe that one could collect 10^5 events within a reasonable amount of the data taking time, thus measuring hundred data
points with, say, 3% accuracy. On the other hand, the low energy npB
experiments are almost impossible to do within the framework of existing technologies.

To summarize my personal view on this subject: the technologies
exist which would enable experimentalists to collect a few thousand
data points with 2 - 3% accuracy, thus giving us at least as much experimental information about the NNB process as we have today about
the N-N elastic scattering process.

But how meaningful and useful to theorists would this wealth of
experimental information be? Would it not be preferable to perform
just a few selective experiments, done at a certain, carefully chosen,
set of kinematic conditions, rather than a set of comprehensive,
"shotgun" experiments?

Theorists should, of course, answer these questions in detail.
I shall make only some general comments from an experimentalist's
point of view.

1) So far, the theorists have not been able to give us, the experimentalists, the precise recommendations under what experimental
conditions, namely, at what incident energies, what outgoing angles
and with what accuracy should NNB experiments be done[39]. The only
advice that we have received from them is of a very qualitative
nature, namely that the OFES effects are, in general, expected to be
larger at higher incident energies and at smaller exit proton
angles[39,40]. This, of course, should not be interpreted as a proof
that no useful information can be obtained at lower energies and
other than the smallest scattering angles. There are several reasons
for this: (a) There is more than one way to go OFES and several
different variables can be used to study these effects. The photon
energy is clearly one of these, but others, like the event non-coplanarity, should not be ignored either. The NNB matrix elements are
composed of several distinct terms and the resulting destructive or
constructive interference between them could enhance or depress the
OFES effects. (b) In Sec. 4.4 of ref.2, Signell has summarized some

results based on exact potential model calculations, on-shell approximations and soft photon approximations. He has found sizable differences between some of these calculations even at incident energies as low as 10 MeV. These calculations are not very extensive nor are they complete, but from this basis it is not obvious that a given potential model which describes well elastic scattering can also accurately describe the ppB process at low energies. Therefore, it should be interesting to explore, both theoretically and experimentally, this low energy region as well. (c) Often, for practical reasons, it is much easier to perform a more accurate experiment for one set of experimental conditions, than a less accurate experiment for another set. For instance, low energy accelerators are more accessible than the higher energy ones, thus it might be easier, from a practical point of view, to test the validity of the H-J potential at 20 MeV proton energy, than at 200 MeV. As a result, I feel that, in the absence of "hard" information which theorists do not seem to be able to give us at present, we experimentalists should do a few good, accurate and comprehensive NNB experiments, at several, high and low, incident nucleon energies.

2) Celenza et al[31] have shown that phenomenological off-shell matrix elements can be obtained from the experimental data without the use of a potential model. With better data, clearly better matrix elements could be obtained and the existence of phenomenological ppB matrix elements would be, presumably, as useful for the understanding of the N-N interaction OFES, as the elastic scattering phase shifts are for the understanding of the N-N interaction ONES. Thus, again, new and comprehensive results would be the most useful.

3) There exists a point of view among some theorists and experimentalists, that comprehensive NNB experiments should not be done until theorists find out exactly what kind of information about the N-N force could be extracted from such experiments. I do not share this point of view. I feel that the bremsstrahlung process is sufficiently important and that experiments should proceed even without the "hard" guidance from the theorists. Theorists are an imaginative group of people, once they have good experimental data, they will find good ways to use it.

I shall conclude this paper by repeating my view that it would be interesting and informative to perform few as good and precise bremsstrahlung experiments as the presently available technology would allow us to do.

D. ACKNOWLEDGEMENTS

It is my pleasure to acknowledge contributions and full participation of Dr. C.A. Smith in the analysis of data from the Manitoba 42 MeV experiment; without his diligent effort this analysis would have never been done. I am also very thankful to Dr. Smith for carefully reading this manuscript and making many useful comments.

Mr. R. Penner has been most helpful in assisting with the computer analysis done in order to produce Figures 4, 5 and 9.

Mr. T.W. Millar has produced the data for Figure 10 by using the

474

first version of the program MANSA.

I acknowledge the support of the University of Manitoba through a computer grant which has enabled us to perform all necessary calculations on the university computer free of charge.

I am thankful to the Cyclotron director, Dr. J.S.C. McKee, for providing secreterial and drafting services so much needed in the preparation of this paper.

REFERENCES AND FOOTNOTES

1. P. Signell, Proc. Int. Conf. on Light Nuclei, Few Body Problems and Nuclear Forces, Brela, Yugoslavia, June 26-July 5, 1967, Gordon and Breach, New York 1968.

2. P. Signell, Adv. Nucl. Phys. $\underline{2}$, 223 (1969).

3. E.M. Nyman, Physics Reports, $\underline{9}$, 179 (1974).

4. M.L. Halbert, Proc. of the Gull Lake Symposium on the Two-Body Force in Nuclei, Sept. 7-10, 1971, (S.M. Austin and G.M. Crawley, eds.) Plenum, New York (1972).

5. M.J. Moravcsik, Rep. Prog. Phys. $\underline{35}$, 587 (1972).

6. M.K. Liou, Proceedings of the Int. Conf. on Few Body Problems in Nuclear and Particle Physics, at Laval University, Quebec City, Canada, 1974, edited by R.J. Slobodrian, B. Cujec, and K. Ramavataram (Les Presses de l'Université Laval, Quebec City, Canada, 1974), p.193.

7. M.K. Srivastava and D.W.L. Sprung, Adv. Nucl. Phys. $\underline{8}$, 121 (1975).

8. K.W. Rothe, P.F.M. Koehler, and E.H. Thorndike, Phys. Rev. $\underline{157}$, 1247 (1967).

9. J. McKeown, L.G. Greeniaus, J.V. Jovanovich, W.F. Prickett, K.F. Suen and J.C. Thompson, Nucl. Inst. Meth. $\underline{104}$, 413 (1972) and four papers following this one.

10. J.V. Jovanovich, L.G. Greeniaus, J. McKeown, T.W. Millar, D.G. Peterson, W.F. Prickett, K.F. Suen, and J.C. Thompson, Phys. Rev. Lett. $\underline{26}$, 277 (1971)

11. L.G. Greeniaus, J.V. Jovanovich, R. Kerchner, T.W. Millar, C.A. Smith, and K.F. Suen, Phys. Rev. Lett. $\underline{35}$, 696 (1975).

12. J.V. Jovanovich, C.A. Smith, and L.G. Greenaius, Phys. Rev. Lett. $\underline{37}$, 631 (1976).

13. J.V. Jovanovich, in Proceedings of the International Conference on Few Body Problems in Nuclear and Particle Physics, Laval University, Quebec City, Canada, 1974, edited by R.J. Slobodrian, B. Cujec, and K. Ramavataram (Les Presses de l'Université Laval, Quebec City, Canada, 1974), p. 218-227; see also Cyclotron Laboratory, University of Manitoba, Report No. 74-755A (unpublished).

14. C.A. Smith, J.V. Jovanovich, L.G. Greeniaus, "Measurement of Proton-Proton Bremsstrahlung Cross Sections at 42 MeV and Its Comparison with a Theoretical Calculation" (to be published).

15. F. Sannes, J. Trischuk, and D.G. Stairs, Nucl. Phys. $\underline{A146}$, 438 (1970).

16. There exists some confusion in literature as to the exact def-

inition of "Harvard geometry". In this paper, I shall assume that Harvard geometry means that both proton polar angles in the lab system are equal to each other ($\Theta_1 = \Theta_2$) and that the events are coplanar ($\phi_2 - \phi_1 - \pi = 0$) irrespective of whether gamma ray is detected or not. Another term "Rochester geometry" has been used also to designate those experiments where the photon, as well as the protons, are detected. Since, in all such experiments, the photons were detected mainly in order to reduce the background, there exists no substantial difference between experiments in which the photons are detected and those where they are not. Therefore, I shall not use the term "Rochester geometry" at all.

17. A. Willis, V. Comparat, R. Frascaria, N. Marty, M. Morlet, and N. Willis, Phys. Rev. Lett. <u>28</u>, 1063 (1972).

18. B. Frois, M. Irshad, C.R. Lamontagne, U. Von Moellendorff, R. Roy and R.J. Slobodrian, Phys. Lett. <u>53B</u>, 341 (1974).

19. M. Suter, W. Wolfli, G. Bonani, Ch. Stoller and R. Miller, Phys. Lett. <u>58B</u>, 36 (1975); Helevtica Physica Acta, <u>49</u>, 863 (1976).

20. J. Sanado, K. Kondo and S. Seki, Nucl. Phys. <u>A203</u>, 388 (1973).

21. J.L. Beveridge, D.P. Gurd, J.G. Rogers, H.W. Fearing, A.N. Anderson, J.M. Cameron, L.G. Greeniaus, C.A. Goulding, J.V. Jovanovich, C.A. Smith, A.W. Stetz, J.R. Richardson and R. Frascaria, Contribution to these proceedings.

22. B.M.K. Nefkens, O.R. Sander and D.I. Sober, Phys. Rev. Lett. <u>38</u>, 876 (1977).

23. J.A. Edgington, V.J. Howard, I.M. Blair, B.E. Bonner, F.P. Brady, and M.W. McNaughton, Nucl. Phys. <u>A218</u>, 151 (1974).

24. D.I. Sober, M. Arman, H.C. Ballagh, Jr., P.F. Glodis, R.P. Haddock, B.M.K. Nefkens, and D.I. Sober, Phys. Rev. <u>D14</u>, 698 (1976).

25. V.R. Brown, Phys. Rev. <u>C6</u>, 1110 (1972).

26. L.S. Celenza, M.K. Liou, and M.I. Sobel, Phys. Rev. <u>C8</u>, 838 (1973).

27. L. Heller and M. Rich, Phys. Rev. <u>C10</u>, 479 (1974).

28. L.G. Greeniaus, J.V. Jovanovich, R. Kerchner, M.K. Liou, T.W. Millar, P. O'Connor, C.A. Smith, and K.F. Suen. Manitoba Cyclotron Report 75-11 (unpublished).

29. For details of these calculations see M.K. Liou and K.S. Cho, Nucl. Phys. <u>A160</u>, 417 (1971). We are very grateful to M.K. Liou for providing us with the computer program we used to calculate these theoretical cross sections.

30. M.K. Liou and M.I. Sobel, Ann. Phys. <u>72</u>, 323 (1972).

31. L.S. Celenza, B.F. Gibson, M.K. Liou and M.I. Sobel, Phys. Lett. <u>42B</u>, 331 (1972).

32. T.W. Millar and J.V. Jovanovich, Physics in Canada, <u>33</u>, No. 3, 12 (1977).

33. B.M.K. Nefkens and D.I. Sober, Phys. Rev. <u>D14</u>, 2434 (1976).

34. Adam Szyjewicz and A.N. Kamal, ppB Workshop, Nucleon-Nucleon Interaction Conference, Vancouver, June 27th to July 1st, 1977.

35. H. Fearing, ppB Workshop, Nucleon-Nucleon Interaction Confer-

476

ence, Vancouver, June 27th to July 1st, 1977; also private communication.

36. R. Pogson and J.V. Jovanovich, Bul. Am. Phys. Soc. II, $\underline{20}$, 601 (1975).
37. M.I. Sobel and A.H. Cromer, Phys. Rev. $\underline{132}$, 2698 (1963).
38. B. Gottschalk, W.J. Shlaer and K.H. Wang, Phys. Lett. $\underline{16}$, 294 (1965).
39. For instance, Srivastava and Sprang on p. 191 of reference 7, write "It is thus clear that bremsstrahlung will be able to provide useful information about the nuclear interaction only when more precise experiments are performed . . . What is needed are experiments in the 200-300 MeV range and sufficiently far off-shell, i.e., at small proton exit angles so that most of the energy goes into the photon". A statement like this one is not very useful to the experimentalist since it does not say how precise the experiments need to be, how small the angles have to be, or how "sufficiently far off-shell" is defined. The kinematics of NNB are such that the photon energy increases only a few percent when the proton angles decrease from 15^0 to 0^0, but experimental difficulties (accidental background) increase inversely proportional to about the 10th power of proton angles! Clearly, the experimentalist would like to know just <u>how small</u> the angles <u>really</u> need to be as it is much easier to perform a 1% experiment at 20^0 - 20^0, then a 10% experiment at 5^0 - 5^0.
40. L. Heller, Bull. Am. Phys. Soc. $\underline{17}$, 480 (1972). See also P. Signell, in Proceedings of the International Conference on Few Particle Problems in the Nuclear Interaction, Los Angeles, California, 1972, edited by I. Slaus et al. (North-Holland, Amsterdam, 1973), p. 1 - 25.

PROTON-PROTON BREMSSTRAHLUNG AT 730 MeV AND THE LOW THEOREM

B. M. K. Nefkens
University of California, Los Angeles, California 90024

Nearly 30 years ago Ashkin and Marshak[1] suggested that proton-proton bremsstrahlung (PPB) is a useful reaction for probing the nucleon-nucleon interaction. Now, after many PPB experiments have been carried out and nearly 30 theoretical analyses have appeared, it is not untimely to assess the progress that has been made in the field. Yesterday, we were treated to a multicolor display of PPB results presented using manifestly noncovariant and not terribly relevant variables that seem to have been selected for optimum concealment of the possible physics. Many participants of this conference are wondering: "What is the physics that we have learned from PPB?"

Today I would like to discuss a recent PPB experiment and present the results in a more transparent way than was done yesterday. After all, there is a solid, model-independent standard to which experimental results as well as theoretical calculations can be compared. That standard is, of course, the soft photon approximation (SPA) based on the Low theorem,[2] as streamlined by Burnett and Kroll[3] and put in manageable form for PPB by Fearing.[4]

The bremsstrahlung amplitude M can be expanded in a power series in the photon energy k as follows:

$$M = \frac{a}{k} + b + ck + \ldots \tag{1}$$

the cross section $d\sigma$ is then

$$d\sigma = |M|^2 \, \phi \sim \frac{a^2}{k} + 2ab + (b^2 + 2ac)k + \ldots \tag{2}$$

ϕ is a phase space factor that is proportional to k; thus, $\phi = k\beta$, where β is a known kinematical factor.

The Low theorem states that in the limit $k \to 0$, the coefficients a and b can be calculated from the (known) elastic scattering amplitudes. SPA is the calculable part of Eq. (2). Thus, there is only new physics to be learned from PPB when an experiment or a model calculation shows a substantial deviation from SPA. As befits a low-energy theorem, the range in photon energy over which the Low theorem is applicable cannot be clearly defined by mere theoretical consideration, but needs experimental input. I report here on an experiment that was done in part to explore the usefulness of the Low theorem for PPB for finite photon energies, and to test the range of validity of external emission dominance[5] (EED) which is a speedy calculation of the first term of Eq. (2). In the notation of Ref. 5, EED is the relation

$$|M_{pp\gamma}|^2 = -e^2 \, A^\mu A_\mu \, |M_{pp}(\bar{s},\bar{t})|^2 ,$$

$$A^\mu = -\frac{P_1^{\ \mu}}{P_1 \cdot K} - \frac{P_2^{\ \mu}}{P_2 \cdot K} + \frac{P_3^{\ \mu}}{P_3 \cdot K} + \frac{P_4^{\ \mu}}{P_4 \cdot K}$$

$$\bar{s} = \frac{1}{2}\,[(P_1 + P_2)^2 + (P_3 + P_4)^2]$$

$$\bar{t} = \frac{1}{2}\,[(P_1 - P_3)^2 + (P_2 - P_4)^2] \quad .$$

The experiment was conducted by the UCLA group consisting of O. Sander, D. Sober, and B. Nefkens. Some of the considerations that led to our choice of the incident beam energy and the detector configuration are:

a) All previous PPB experiments have been made at low incident beam energy where the proton-proton total cross section, $\sigma_t(pp)$, is entirely due to elastic scattering. Figure 1 shows $\sigma_t(pp)$ and $\sigma(pp{\to}pp)$ as function of the incident proton energy; the vertical arrows on the abscissa mark the beam energies of the highest energy PPB experiments. At 730 MeV, which is our choice for the incident beam, $\sigma_t(pp)$ is nearly half inelastic.

b) In the previous experiments, the elastic scattering cross section $d\sigma/dt$ is nearly constant. At 730 MeV there is a rapid variation in $d\sigma/dt$.

c) To enhance bremsstrahlung by direct emission, the momentum

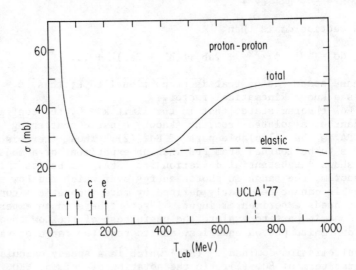

Fig. 1. Total cross section for proton-proton interactions and for elastic pp → pp scattering versus the incident proton energy. The vertical arrows on the abscissa indicate the energy of the highest energy PPB experiments.

transfer t should be large, favoring direct collisions over peripheral ones. Our choice gives $t \simeq - 20 \, m_\pi^2$.

d) To investigate systematically possible deviations from the Low theorem when the photon energy is finite, the differential PPB cross section must be measured as a function of increasing photon energy, preferably at various well-defined photon angles.

e) The detection system must have an excellent background rejection, in particular against the large π^0 background.

In our experiment we detect all three outgoing particles. We used a modified version of the original UCLA pion-proton bremsstrahlung setup.[6] It is shown in Fig. 2 and includes 16 lead glass Cherenkov counters, a large magnetic spectrometer for the scattered protons and a recoil proton detector made of large spark chambers.

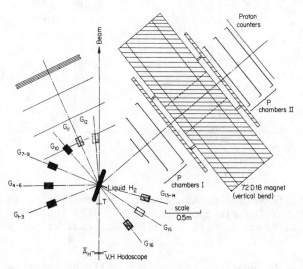

Fig. 2. Experimental setup used by the UCLA group for PPB at 730 MeV.

Shown in Fig. 3 are some results of our PPB experiment[7] for an incident beam of 730 MeV and a proton scattering angle in the lab of 50°. We have plotted the center-of-mass differential cross section $d^5\sigma/d\Omega_p \, d\Omega_\gamma \, dk$ for selected photon counters as a function of the photon energy. The dashed curves are the predictions of external emission dominance.

The differential cross sections up to $E_\gamma \sim 80$ MeV for 14 photon counters are in quantitative agreement with EED. The results for the two counters G11,12 are somewhat higher than EED; however, the EED prediction is varying rapidly over the solid angle subtended by these two counters. For $E_\gamma \gtrsim 80$ MeV, several date points are well above the EED predictions. The b-term of Eq. (1) is being evaluated for

480

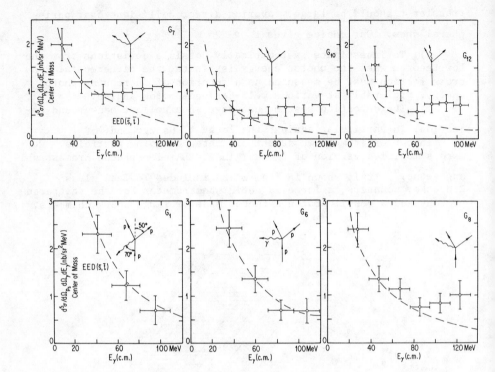

Fig. 3. Differential cross sections in center-of-mass for
PPB at 730 MeV for six photon counters. The location of the
detectors is given in Ref. 7. EED = external emission
dominance, Ref. 5.

our experiment by H. Fearing, and, as you will see tomorrow at the
PPB workshop, his tentative calculations for the three photon counters
G7,8,10 indicate reasonable agreement between the full SPA calculations of Eq. (2) and our experimental results up to the highest photon energy reported, $E_\gamma \sim 140$ MeV in c.m., or about 180 MeV in the lab.

We can summarize the status of bremsstrahlung physics as follows. In all direct confrontations with PPB experiments, the soft
photon approximation based on the Low theorm has been capable of describing the data within the margin of the errors. This includes the
early PPB work discussed by Nyman,[8] the new TRIUMF experiment at 200
MeV reported at this conference, and our results at 730 MeV up to
$E_\gamma \sim 80$ MeV. Furthermore, $\pi^{\pm}p$ bremsstrahlung at 300 MeV covering
the pronounced $P_{33}(1236)$ resonance, can be described satisfactory by
EED,[5] up to the highest energy measured $E_\gamma \sim 120$ MeV (lab) or
E_γ (c.m.) ~ 160 MeV, See Fig. 4.

The physics we have learned from PPB is different than was
originally anticipated. Rather than measuring large off-mass-shell

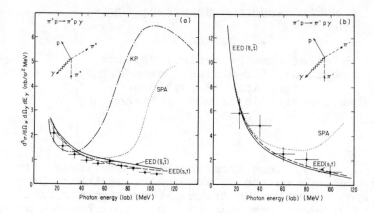

Fig. 4. Differential cross sections in lab. frame for
$\pi^+ p \rightarrow \pi^\pm p\gamma$ at 290 MeV from Refs. 5 and 6.

effects in PPB, we have found that the Low theorem is useful over a
large range in photon energy. Thus, it is now possible to test one's
snappy new potential or fancy off-mass-shell amplitude by checking if
they give the same PPB result as SPA. It is often possible--when the
photon energy is not too high, say up to $E_\gamma \sim 80$ MeV for 730 MeV incident protons--to reduce the complicated full SPA to the simple EED
calculation. PPB is perhaps not glamorous but it certainly has justified a place in the study of the strong interaction.

REFERENCES

1.   J. Ashkin and R. E. Marshak, Phys. Rev. 76, 58 (1949); 76, 989
     (E)(1949).
2.   F. E. Low, Phys. Rev. 110, 974 (1958).
3.   T. Burnett and N. Kroll, Phys. Rev. Lett. 20, 87 (1968).
4.   H. W. Fearing, Phys. Rev. C 6, 1136 (1972).
5.   B. Nefkens and D. Sober, Phys. Rev. D 14, 2434 (1976).
6.   D. Sober, M. Arman, D. Blasberg, R. Haddock, K. Leung, B.
     Nefkens, B. Schrock, and J. Sperinde, Phys. Rev. D 11, 1017
     (1975).
7.   B. Nefkens, O. Sander, and D. Sober, Phys. Rev. Lett. 38, 876
     (1977).
8.   E. M. Nyman, Phys. Rev. 170, 1628 (1968).

NUCLEON–NUCLEON BREMSSTRAHLUNG*

G.E. Bohannon

Center for Theoretical Physics
Laboratory for Nuclear Science and
Department of Physics
Massachusetts Institute of Technology
Cambridge, Massachusetts   02139

ABSTRACT

The nucleon-nucleon bremsstrahlung amplitude through order k depends on the NN wavefunction within the strong interaction volume and on the two-body magnetic dipole moment, both of which are unknown. Meson exchange currents have been included by several authors to account for parts of the magnetic moment. The wavefunction is of major interest. We show that the cross section at E_L= 200 MeV is very sensitive to unitary transformations of the Reid soft core wavefunction within 1.5 fm. Results are also shown at 42 MeV.

Most effort on nucleon-nucleon bremsstrahlung has been within the framework of the potential model, i.e. the Schrödinger equation.[1] My discussion will also be based upon this model. The justification for placing emphasis on this particular approach is found in the importance that potentials have in our attempts to understand the physics of nuclei. One primarily wishes to obtain information from bremsstrahlung about the intermediate range NN wavefunction, although the magnetic dipole moment operator of the NN system is another, perhaps significant, unknown quantity.

I will first summerize the various parts of the bremsstrahlung amplitude which may yield information on the strong interaction. An important consideration will be the use of current conservation to limit the number of unknown quantities to those mentioned above. We will then see some calculational results which suggest the extent to which the amplitude depends on the intermediate range wavefunction.

One usually expresses the amplitude as a sum of external and internal emission terms. The external emission amplitude is just

$$\vec{M}_E = <\phi_f | \vec{J}_1 (\vec{k}) | \chi_i^{(+)} > + < \chi_f^{(-)} | \vec{J}_1 (\vec{k}) | \phi_i > , \tag{1}$$

*This work is supported in part through funds provided by ERDA under Contract EY-76-C-02-3069.*000.

where $\vec{J}_1(\vec{k})$ is the sum of the electromagnetic current densities of two free nucleons. (The subscript 1 means that it is a one body operator.) The wavefunctions which we wish to study are $\chi_i^{(+)}(\vec{r})$ and $\chi_f^{(-)}(\vec{r})$, which are the scattered parts of the Schrödinger wavefunctions
$$\psi^{(\pm)} = \phi + \chi^{(\pm)}.$$

The internal emission amplitude is
$$\vec{M}_I = <\chi_f^{(-)} | \vec{J}_1(\vec{k}) | \chi_i^{(+)}> + <\psi_f^{(-)} | \vec{J}_2(\vec{k}) | \psi_i^{(+)}>, \tag{2}$$

where $\vec{J}_2$ is a two body operator defined as the difference between the total electromagnetic current density and $\vec{J}_1$. The total amplitude is just the distorted wave Born approximation (the term $<\phi_f | \vec{J}_1 | \phi_i>$ does not contribute).

$$\vec{M} = \vec{M}_E + \vec{M}_I = <\psi_f^{(-)} | \vec{J}_1(\vec{k}) + \vec{J}_2(\vec{k}) | \psi_i^{(+)}>. \tag{3}$$

The long wavelength electric dipole and quadrapole parts of the $\vec{J}_2$ matrix element can be expressed uniquely in terms of the potential if the charge density is known. This is usually referred to as Siegert's theorem.[2,3] The matrix elements are

$$\vec{M}_2^{E1} = \frac{1}{2} ie <\psi_f^{(-)} | [V, (Q_1 - Q_2)\vec{r}] | \psi_i^{(+)}>, \tag{4}$$

$$\vec{M}_2^{E2} = \frac{1}{8} e <\psi_f^{(-)} | [V, (Q_1 + Q_2)\vec{r}\vec{r}\cdot\vec{k}] | \psi_i^{(+)}>, \tag{5}$$

where Q_j is the charge operator of nucleon j. These differ from the E1 and E2 matrix elements as usually defined in terms of vector spherical harmonics[4] by terms of order k^2 and higher, which we will not consider.

The dipole term, $\vec{M}_2^{E1}$ vanishes for ppγ but is very important for npγ, as V. Brown and J. Franklin[5] have shown. Brown and Franklin have also shown that for the potentials they used, which were the Hamada-Johnston and Brian-Scott ones, it is the exchange character of the interaction rather than the nonlocality which causes this term to be significant. That is, using only the first term of

$$[V, (Q_1 - Q_2)\vec{r}] = [V, (Q_1 - Q_2)]\vec{r} + (Q_1 - Q_2)[V, \vec{r}] \tag{6}$$

gives almost the complete effect. Of course, one automatically includes this contribution along with part of the rescattering amplitude if one adds to the external emission amplitude a term which makes the complete amplitude gauge invariant (the additional term is unique in the soft photon limit).[6,7,8] The amplitude M_2^{E2} contributes to ppγ and npγ if the potential is nonlocal, and must be included to maintain gauge invariance. To derive these results one assumes the charge distribution to be that of spherically symmetric protons (not necessarily point charges).

By adding the contributions from $\vec{J}_2$ to the rescattering amplitude we obtain the internal emission amplitude. For ppγ it may be written in the c.m. frame as

$$\vec{M}_I = \vec{M}_I^{E1} + \vec{M}_I^{E2} + \vec{M}_I^{M1} + O(k^2) , \tag{7}$$

with

$$\vec{M}_I^{E1} = 0 , \tag{8}$$

$$\vec{M}_I^{E2} = \frac{e}{2} \vec{\nabla}_{p_f} \vec{k} \cdot \vec{\nabla}_{p_f} \langle \vec{p}_f s_f \mu_f | T_1(\varepsilon_i) | \vec{p}_i s_i \mu_i \rangle$$

$$-\frac{e}{2} \vec{\nabla}_{p_i} \vec{k} \cdot \vec{\nabla}_{p_i} \langle \vec{p}_f s_f \mu_f | T_1(\varepsilon_f) | \vec{p}_i s_i \mu_i \rangle$$

$$-\frac{1}{4} ek \langle \chi_f^{(-)} | \vec{r}\vec{r} \cdot \vec{k} | \chi_i^{(+)} \rangle , \tag{9}$$

$$\vec{M}_I^{M1} = i\vec{k} \times \{ \langle \psi_f^{(-)} | \vec{m}_2 | \psi_i^{(+)} \rangle$$

$$+ \frac{e}{2m} \langle \chi_f^{(-)} | [\vec{j} + (2\mu_p - 1)\vec{s}] | \chi_i^{(+)} \rangle \} , \tag{10}$$

where T_1 is the isotriplet T-matrix, $\vec{j}$ and $\vec{s}$ are the total angular momentum and spin ($\mu_p = 2.79$) and $\vec{m}_2$ is the two-body magnetic dipole moment operator,

$$\vec{m}_2 = \frac{1}{2} i [\vec{\nabla}_k \times \vec{J}_2(\vec{k})]_{k=0} . \tag{11}$$

The corresponding expressions for npγ have been given elsewhere.[9]

Most bremsstrahlung calculations have excluded the rescattering amplitude because of the difficulty involved in computing it. It has been included by V. Brown,[10] Heller and Rich,[11] and Brown and Franklin[5] (the last was for npγ). The expressions above show, however, that one can include the electric part of the ppγ rescattering amplitude through order k easily.

The dipole moment $\vec{m}_2$ is a source of uncertainty in bremsstrahlung calculations. There is no way to obtain $\vec{m}_2$ uniquely from a phenomenological potential if gauge invariance is one's only guide since the solenoidal part of the current remains arbitrary.[12] This was discussed

recently by Heller.[13] The contributions to $\vec{m}_2$ shown in
Fig. 1 have been calculated from particle exchange models.
These include the ρπγ current by Ueda[14] and the ωπγ and
Δ-excitation currents by Kamal and Szyjewicz[15] for ppγ,
evaluated in plane wave Born approximation, and the one-
pion-exchange[16,17] and Δ-excitation[17] currents by Heller,
Thompson and Bohannon for npγ. In the case of the npγ
one-pion-exchange current one must subtract off the part
which remains at k=0 to avoid double counting it, since
the k=0 value is included in $\vec{M}_2^{E1}$. The ppγ one-pion-exchange
current, which vanishes in the nonrelativistic limit, was
included in the one-boson-exchange calculation of Ref. 15.

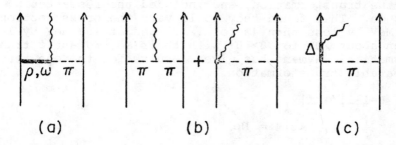

Fig. 1.   Meson exchange currents: (a) ρπγ and ωπγ,
          (b) one-pion-exchange, and (c) Δ-excitation.

The calculations indicate that at 200 MeV these exchange
currents are almost undetectable with present ppγ experimental accuracy. (However, the size of the ppγ one-pion-
exchange current by itself has not been established.)
     For some kinematical situations the final nucleons
have a relative energy near threshold. Heller[18] has
pointed out that in these situations one should ensure
that the interaction currents included in npγ calculations
are consistent with the deuteron photo and electrodisintegration data, for which the theory has had some sucess.
Indeed, the isovector Ml pion exchange current is seen
clearly in the Mainz electrodisintegration data near
threshold.[19]
     The sensitivity of the bremsstrahlung observables to
the short and intermediate distance wavefunction can be
estimated by making unitary changes of the wavefunction.
Actually it is possible to bound the magnitudes of $\vec{M}_E$, $\vec{M}_I^{E2}$,
and the rescattering part of $\vec{M}_M^{M1}$ in terms of the elastic
scattering observables when the potential is independent
of energy. This gives bounds on the cross section which
must be satisfied by all unitary transformations. I hope
to discuss this at the workshop. However, unitary transformations give some control over the wavefunctions to be

investigated. Since our knowledge of the interaction decreases with radius, it would, nevertheless, make little sense to study sensitivity to the wavefunction if the cross section were unbounded.

Unitary transformations were first considered for bremsstrahlung by Signell and coworkers.[20] They found very little effect on the symmetric 30° cross section at 158 MeV when they changed the potential inside about one fermi with a Baker transformation. More recently Heller and Rich[21] have investigated the effect of Baker transformations. They enforced gauge invariance by introducing a minimal substitution into the p^2 dependence generated by the transformation, and included the rescattering amplitude. They found they could vary the cross section at 42 MeV by less than 4% for $\theta > 10°$ and at 158 MeV by less than about 20% for $\theta > 10°$ with their most drastic transformation. Nyman[1] has given results for the one term separable transformation

$$U = 1 - 2 |\Phi\rangle\langle\Phi| , \tag{12}$$

$$\langle\Phi|\Phi\rangle = 1, \quad \langle\vec{r}|\Phi\rangle = Ne^{-\beta r}$$

applied to the Yamaguchi separable potential. This calculation was for spinless particles and was made gauge invariant by a minimal substitution. The results showed a large dependence on the range β^{-1} of the transformation.

We have studied the effect produced by a separable transformation applied to the Reid soft core potential.[22] The transformation was applied only to the $^1S_0$ state. The wavefunction becomes

$$u_0(r) \rightarrow \tilde{u}_0(r) = u_0(r) - 2\phi(r) \int_0^\infty dr' u_0(r')\phi(r') \tag{13}$$

and we have chosen

$$\phi(r) = \tilde{N}r^4 e^{-\alpha\mu r} \qquad (\mu = \text{pion mass}). \tag{14}$$

The amplitude was made gauge invariant through order k by including the order k parts of $\vec{M}_I^{E2}$, but no attempt was made to include $\vec{M}_I^{M1}$. The transformation is quite easy to use since one can express the spin-singlet T-matrix element as

$$t_\ell(q,p,p^2) = \left(\frac{q}{p}\right)^{\ell+1}\sin\delta_\ell(p)$$
$$+ p^{-1}(q^2 - p^2)\int_0^\infty dr F_\ell(qr)[u_\ell(p,r) - \omega_\ell(p,r)] \tag{15}$$

where F_ℓ is the regular Ricatti spherical Bessel function and u_ℓ is the radial wavefunction, which becomes ω_ℓ outside the range of the potential. All spin and angular momentum dependence of the potential is retained. Fig. 2 shows the wavefunctions generated by this transformation when applied to the Reid wavefunction at 42 MeV lab energy.

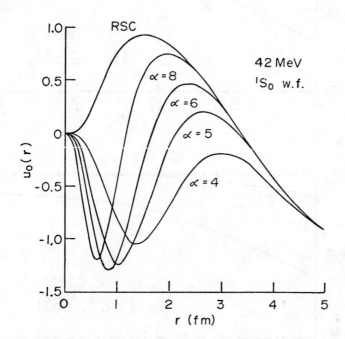

Fig. 2. $^1S_0$ wavefunctions at E_L=42 MeV obtained by applying the unitary transformation with α=4,5,6, and 8 to the Reid soft core wavefunction. The Reid wavefunction is labelled RSC.

The case α=4 is extreme in that it changes the wavefunction even at 3 fm. The α=8 case is perhaps more realistic since it changes the wavefunction primarily inside 1.5 fm. The dramatic modification of the wavefunction which this type of transformation produces should avoid our underestimating the sensitivity and thereby unduly discouraging possibly interesting experiments.

The Reid cross section at 42 MeV is reduced by the α=4 transformation but changed negligably by the α=8 one. The integrated cross sections divided by that obtained from the Reid soft core potential are shown in Fig. 3.

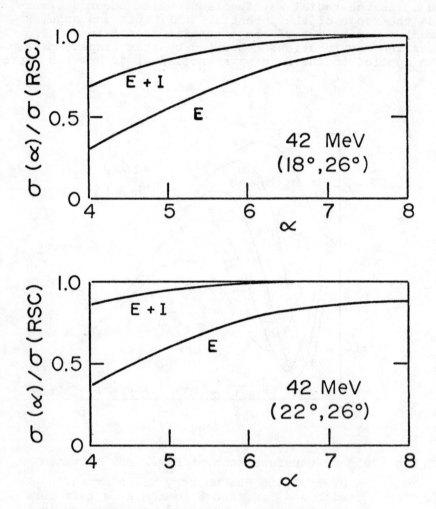

Fig. 3.   Ratio of the cross section $d\sigma/d\Omega_1 d\Omega_2$
          for the transformed potential to that for
          the Reid potential as a function of α.
          The curve E includes only external emis-
          sion whereas the curve E+I also includes
          the order k part of M_I^{E2}.

The upper curves include the order k part of the electric quadrapole internal emission amplitude, while the
lower curves are the external emission alone. All $\ell > 0$,
$J \leqslant 2$ partial waves were computed with the Reid potential.
There is a large reduction in the cross section if the
longer ranged $\alpha = 4$ transformation is used but the reduction

is much less if the $\alpha=8$ one is used. One sees that the effect of the transformations is reduced when the internal emission is included. A similar effect was found by Heller and Rich[21] and by Nyman[1] when they introduced the minimal substitution into their nonlocal potentials. The two sets of (θ_1, θ_2) shown in Fig. 3 are cases where a significant disagreement exists between theory and experiment.[23]

The transformations have a larger effect on the cross section at 200 MeV. Fig.4 shows the $\alpha=6$, $\alpha=8$, and Reid soft core $^1S_0$ wavefunctions at 200 MeV.

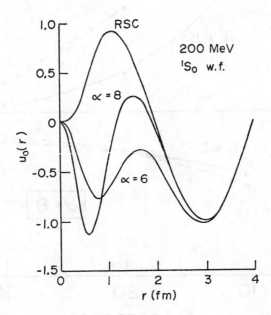

Fig. 4. $^1S_0$ wavefunctions at $E_L=200$ MeV for $\alpha=6$, $\alpha=8$, and the Reid potential.

One should keep in mind that the final state nucleons have lower energy. For example, for $\theta_1=\theta_2=10°$ and $\theta_\gamma=0$ the final nucleons together have 2 MeV in their c.m. frame. Fig. 5 shows the ratio of the $\alpha=6$ and $\alpha=8$ cross sections to the Reid cross section at 200 MeV as a function of the exit angle of one proton. The other proton is at $10°$. An interesting point is that at 200 MeV the $\alpha=6$ transformation increases the cross section while the $\alpha=8$ one decreases it. For a symmetric geometry at $10°$ one finds a 40% reduction from the $\alpha=8$ transformation. As at 42 MeV, the shorter range transformations produce smaller internal emission amplitudes (recall that we have not included the internal magnetic dipole radiation). For 200 MeV the

2<J≤4 partial waves were included as computed from the Hamada-Johnston potential.

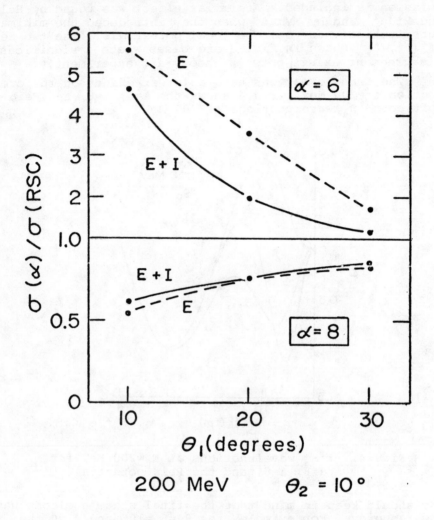

Fig. 5.   Ratio of the α=6 and α=8 cross sections
           dσ/dΩ₁dΩ₂ to the Reid one as a function
           of θ₁ when θ₂=10°. The labels E and
           E+I are as in Fig. 3.

The photon angular distribution is shown in Fig. 6 for the symmetric 10° case.

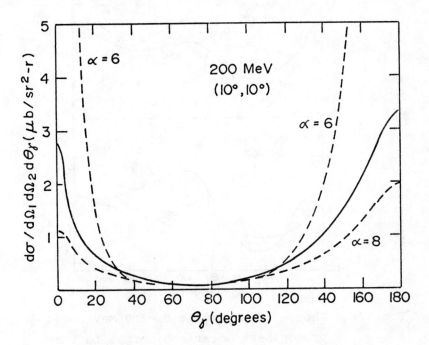

Fig. 6.   Coplanar $d\sigma/d\Omega_1 d\Omega_2 d\theta_\gamma$ at E_L=200 MeV,
$\theta_1=\theta_2=10°$. The dashed curves are for
α=6 and α=8. The solid curve is for
the Reid potential.

One can show that because of the Pauli principle the $^1S_0$ state does not contribute to the cross section at θ_γ=0 and 180° if only the convection current is used for J_1. Thus, the spin current (proton magnetic moments) is very important here.

The asymmetry plotted in Fig. 7 also shows a strong effect from the transformation. This asymmetry is defined as the difference between the number of photons detected at an angle θ_γ to the right of the beam direction and the number detected at θ_γ to the left, divided by the sum, when the beam is polarized normal to the plane.

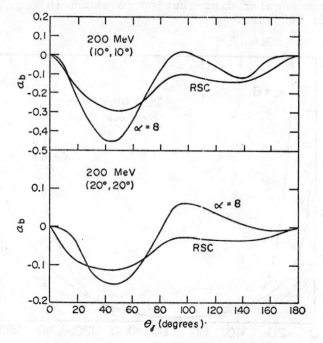

Fig. 7.   Photon asymmetry at E_L=200 MeV for a
proton beam polarized normal to the
scattering plane. The α=8 and Reid
values are shown.

A few concluding remarks may be made about the
influence the wavefunction has on the ppγ cross section.
At 42 MeV we have found no indication that the wavefunction inside 1.5 fm. has a measurable influence on the
cross section at the kinematics of the Manitoba experiment.[2,3]
At larger distances the specific form of the wavefunction
becomes more important, but the theoretical uncertainties
in the wavefunction are smaller.

At 200 MeV, where the TRIUMF experiment is being
run, there is substantial dependence on the wavefunction
at less than 1.5 fm., especially at the smaller angles,
near (10°, 10°). Even at (30°, 10°) the dependence is
significant. We have also performed calculations for
(20°, 20°) and found the ratio plotted in Fig. 5 to be
very close to the values for (30°, 10°).

The interpretation of experimental results is complicated by another unknown quantity, the solenoidal part of
the current. Meson exchange calculations suggest that this
may not be an unmanageable problem. The most satisfying
theoretical approach would be to compute that part of the
bremsstrahlung amplitude which corresponds to the interme-

diate and long range strong interaction consistently from a field theoretic model. The one-boson-exchange calculations[15,24] are steps in this direction.

Relativistic effects also complicate the interpretation of data. The one-boson-exchange calculations and the soft photon calculations of Nyman[25] and Fearing[26] have the advantage of being completely relativistic. Celenza, Liou, Sobel, and Gibson[27] have shown that relativistic effects are already very important at 156 MeV. A more complete treatment of relativistic effects in bremsstrahlung is certainly needed.

REFERENCES

1. E. M. Nyman, Physics Reports $\underline{9}$, 179 (1974).
2. A. F. Siegert, Phys. Rev. $\underline{52}$, 787 (1937).
3. R. G. Sachs and N. Austern, Phys. Rev. $\underline{81}$, 705 (1951).
4. For example, T. deForest and J. D. Walecka, Adv. in Phys. $\underline{15}$, 1 (1966).
5. V. R. Brown and J. Franklin, Phys. Rev. C$\underline{8}$, 1706 (1973).
6. F. E. Low, Phys. Rev. $\underline{110}$, 974 (1958).
7. S. L. Adler and Y. Dothan, Phys. Rev. $\underline{151}$, 1267 (1966).
8. M. K. Liou and M. I. Sobel, Phys. Rev. $\overline{C4}$, 1597 (1971).
9. G. E. Bohannon, to be published.
10. V. R. Brown, Phys. Rev. $\underline{177}$, 1498 (1968).
11. L. Heller and M. Rich, Phys. Rev. C$\underline{10}$, 479 (1974).
12. L. L. Foldy, Phys. Rev. $\underline{92}$, 178 (1953).
13. L. Heller in The Two-Body Force in Nucleii, edited by S. M. Austin and G. M. Crawley (Plenum, New York 1972), p. 79.
14. Y. Ueda, Phys. Rev. $\underline{145}$, 1214 (1966).
15. A. N. Kamal and Adam Szyjewicz, preprint, University of Alberta, Edmonton, Alberta, Canada; these proceedings.
16. R. H. Thompson and L. Heller, Phys. Rev. C$\underline{7}$, 2355 (1973).
17. G. E. Bohannon, L. Heller, and R. H. Thompson, Phys. Rev. C (to be published).
18. L. Heller, private communication.
19. G. G. Simon et al., Phys. Rev. Lett. $\underline{37}$, 739 (1976).
20. M. D. Miller, M. S. Sher, P. Signell, N. R. Yoder, and D. Marker, Phys. Lett. 30B, 157 (1969).
21. Comment by L. Heller in Intl. Conf. on Few Body Problems in Nuclear and Particle Physics, Univ, Laval, Quebec, Canada (1974).
22. R. V. Reid, Jr., Ann. Phys. $\underline{50}$, 411 (1968).
23. L. G. Greeniaus et al., Cyclotron report 75-11, University of Manitoba, Winnipeg, Canada; J.V. Jovanovich et al., Phys. Rev. Lett. $\underline{37}$, 631 (1976).
24. R. Baier, H. Kuhnelt, and P. Urban, Nucl. Phys. B$\underline{11}$, 675 (1969).
25. E. Nyman, Phys. Rev. $\underline{170}$, 1628 (1968).
26. H. Fearing, NNγ workshop, in these proceedings.
27. L. S. Celenza, M. K. Liou, M. I. Sobel and B. F. Gibson, Phys. Rev. C$\underline{8}$, 838 (1973).

BOUNDS ON THE MODEL DEPENDENCE OF THE ppγ AMPLITUDE

G. E. Bohannon
Center for Theoretical Physics,
Laboratory for Nuclear Science,
Massachusetts Institute of Technology
Cambridge, Massachusetts 02139

ABSTRACT

Bounds on the magnitude of the external emission pp bremsstrahlung amplitude are derived, assuming the nucleon-nucleon potential to be independent of energy. Similar bounds on the order k electric internal emission amplitude are briefly discussed. Numerical results are shown for the upper and lower bounds on the cross section computed from the convection current part of the external emission amplitude at 42 MeV and 200 MeV.

The nonrelativistic external emission pp bremsstrahlung amplitude for an energy independent potential is bounded by elastic scattering observable quantities. The internal emission amplitude is similarly bounded through first order in k, the photon energy, except for the magnetic dipole contribution. We do not include relativistic corrections in this discussion. Relativistic effects are important already at 160 MeV.[1] However, our discussion of the external emission amplitude can be easily extended to include the relativistic corrections of Liou et al.[2,1] These bounds can be made tighter by assuming the potential outside some distance R, but within the interaction volume, to be known. We shall first prove these statements and then examine the convection current part of the external emission amplitude numerically.

The external emission amplitude can be written as a sum of post-emission and pre-emission terms as follows.

$$\vec{M}_E = \vec{M}_E^{(post)} + \vec{M}_E^{(pre)} \tag{1}$$

$$\vec{M}_E^{(post)} = \sum_\mu \int \frac{d^3q}{(2\pi)^3} \; \langle \vec{P}_f \; s_f \; \mu_f | \vec{J}_1(\vec{k}) | \vec{q} \; s_i \; \mu_i \rangle$$

$$\times \frac{m}{(p_i^2 - q^2)} \; \langle \vec{q} \; s_i \; \mu | T(\varepsilon_i) | \vec{P}_i \; s_i \; \mu_i \rangle, \tag{2a}$$

$$\vec{M}_E^{(pre)} = \sum_\mu \int \frac{d^3q}{(2\pi)^3} \, \langle \vec{p}_f \, s_f \, \mu_f | T(\epsilon_f) | \vec{q} \, s_f \, \mu \rangle$$

$$\times \frac{m}{(p_f^2 - q^2)} \, \langle \vec{q} \, s_f \, \mu | \vec{J}_1(\vec{k}) | \vec{p}_i \, s_i \, \mu_i \rangle, \tag{2b}$$

where $\vec{J}$ is the one-body current density and T is the pp T-matrix. We wish to relate T to the NN wavefunction. The following technique for doing this was used by V. Brown.[3] The interacting wavefunction $\psi^{(+)}$ and the non-interacting one satisfy

$$(\nabla^2 + p^2) \psi_{p s \mu}^{(+)}(\vec{r}, s, \mu') - \langle \vec{r} \, s \, \mu' | mV | \psi_{p s \mu}^{(+)} \rangle = 0, \tag{3a}$$

$$(\nabla^2 + q^2) \exp(-i\vec{q} \cdot \vec{r}) = 0. \tag{3b}$$

Multiply Eq. (3a) by $\exp(i\vec{q} \cdot \vec{r})$ and Eq. (3b) by $\psi^{(+)}$ and subtract. Then integrate over the interaction volume of radius R (outside of which V is negligable).

$$\int^R d^3r \, e^{-i\vec{q} \cdot \vec{r}} \, \langle \vec{r} \, s \, \mu' | V | \psi_{p s \mu}^{(+)} \rangle$$

$$= \frac{1}{m} (p^2 - q^2) \int^R d^3r \, e^{-i\vec{q} \cdot \vec{r}} \, \psi_{p s \mu}^{(+)}(\vec{r}, s, \mu')$$

$$+ \frac{1}{m} \int^R d^3r \, \vec{\nabla} \cdot (e^{-i\vec{q} \cdot \vec{r}} \, \vec{\nabla} \psi_{p s \mu}^{(+)}(\vec{r}, s, \mu')$$

$$- \psi_{p s \mu}^{(+)}(\vec{r}, s, \mu') \, \vec{\nabla} e^{-i\vec{q} \cdot \vec{r}}). \tag{4}$$

The left hand side of Eq. (4) is the half-shell T-matrix element. Applying the divergence theorem gives

$$\langle \vec{q} \, s \, \mu' | T(\epsilon_p) | \vec{p} \, s \, \mu \rangle$$

$$= \frac{1}{m} (p^2 - q^2) \int^R d^3r \, e^{-i\vec{q}\cdot\vec{r}} \, \psi_{ps\mu}^{(+)}(\vec{r},s,\mu')$$

$$+ \frac{R^2}{m} \int d\Omega_R \left(e^{-i\vec{q}\cdot\vec{R}} \, \frac{\partial}{\partial R} \, \psi_{ps\mu}^{(+)}(\vec{R},s,\mu') \right.$$

$$\left. - \psi_{ps\mu}^{(+)}(\vec{R},s,\mu') \, \frac{\partial}{\partial R} \, e^{-i\vec{q}\cdot\vec{R}} \right). \qquad (5a)$$

This separates the dependence of T on the wavefunction inside the interaction region and the wavefunction outside it where ψ is known from elastic scattering. One may derive in a similar way

$$\langle \vec{p} \, s \, \mu | T(\epsilon_p) | \vec{q} \, s \, \mu' \rangle$$

$$= \frac{1}{m} (p^2 - q^2) \int^R d^3r \, \psi_{ps\mu}^{(-)*}(\vec{r},s,\mu') \, e^{i\vec{q}\cdot\vec{r}}$$

$$+ \frac{R^2}{m} \int d\Omega_R \left(e^{i\vec{q}\cdot\vec{R}} \, \frac{\partial}{\partial R} \, \psi_{ps\mu}^{(-)*}(\vec{R},s,\mu') \right.$$

$$\left. - \psi_{ps\mu}^{(-)*}(\vec{R},s,\mu') \, \frac{\partial}{\partial R} \, e^{i\vec{q}\cdot\vec{R}} \right). \qquad (5b)$$

The two terms on the right hand sides of each of these equations suggest a natural way to define, for our purposes, a model dependent and a model independent part of the bremsstrahlung amplitude. We now introduce the nonrelativistic expression for the one-body current density.

$$\vec{J}_1(\vec{k}) = \frac{-2ie}{m} \sin(\vec{k}\cdot\vec{r}/2) \, \vec{p}$$

$$+ i\vec{k} \times \left(e^{-i\vec{k}\cdot\vec{r}/2} \vec{\mu}_1 + e^{i\vec{k}\cdot\vec{r}/2} \vec{\mu}_2 \right), \qquad (6)$$

where $\qquad \vec{\mu}_j = \frac{e\mu}{2m} \vec{p} \, \vec{\sigma}_j \, , \quad \mu_p = 2.79.$

The model dependent parts of the amplitudes are then

$$\vec{M}_{E,MD}^{(post)} = -\frac{2ie}{m}\delta_{s_f s_i} \ \vec{p}_f \ S_i^{(+)}(\vec{k},R,\mu_f)$$

$$+\frac{ie\mu}{2m}P \ \sum_\mu (<s_f \ \mu_f|\vec{k}\times(\vec{\sigma}_1+\vec{\sigma}_2)|s_i \ \mu> \ C_i^{(+)}(\vec{k},R,\mu)$$

$$-i<s_f \ \mu_f|\vec{k}\times(\vec{\sigma}_1-\vec{\sigma}_2)|s_i \ \mu> \ S_i^{(+)}(\vec{k},R,\mu)),$$

$$(7a)$$

$$\vec{M}_{E,MD}^{(pre)} = -\frac{2ie}{m}\delta_{s_f s_i} \ \vec{p}_i \ S_f^{(-)*}(\vec{k},R,\mu_i)$$

$$+\frac{ie\mu}{2m}P \ \sum_\mu (s_f \ \mu|\vec{k}\times(\vec{\sigma}_1+\vec{\sigma}_2)|s_i \ \mu_i> \ C_f^{(-)*}(\vec{k},R,\mu)$$

$$-i<s_f \ \mu|\vec{k}\times(\vec{\sigma}_1-\vec{\sigma}_2)|s_i \ \mu_i> \ S_f^{(-)*}(\vec{k},R,\mu)),$$

$$(7b)$$

where, for $\alpha=i,f$,

$$S_\alpha^{(\pm)}(\vec{k},R,\mu)=\int^R d^3r \ \sin(\vec{k}\cdot\vec{r}/2) \ \exp(-i\vec{p}_\alpha\cdot\vec{r})$$

$$\times\psi_{p_\alpha s_\alpha \mu_\alpha}^{(\pm)}(\vec{r},s_\alpha,\mu), \qquad (8a)$$

$$C_\alpha^{(\pm)}(\vec{k},R,\mu)=\int^R d^3r \ \cos(\vec{k}\cdot\vec{r}/2) \ \exp(-i\vec{p}_\alpha\cdot\vec{r})$$

$$\times\psi_{p_\alpha s_\alpha \mu_\alpha}^{(\pm)}(\vec{r},s_\alpha,\mu). \qquad (8b)$$

An upper bound to the magnitude of each term can be obtained by applying the Schwarz inequality.[4]

$$|S_\alpha^{(\pm)}(\vec{k},R,\mu)|^2 \leq f_s(k,R) \ N_\alpha(R), \qquad (9a)$$

$$|C_\alpha^{(\pm)}(\vec{k},R,\mu)|^2 \leq f_c(k,R) \ N_\alpha(R), \qquad (9b)$$

where

$$f_s(k,R) = \int^R d^3r \, \sin^2(\vec{k}\cdot\vec{r}/2),$$

$$= \frac{2}{3}\pi R^3 \left(1+3\,\cos(kR)/(kR)^2 - 3\,\sin(kR)/(kR)^3\right), \qquad (10a)$$

$$f_c(k,R) = \int^R d^3r \, \cos^2(\vec{k}\cdot\vec{r}/2),$$

$$= \frac{2}{3}\pi R^3 \left(1-3\,\cos(kR)/(kR)^2 + 3\,\sin(kR)/(kR)^3\right), \qquad (10b)$$

$$N_\alpha(R) = \int^R d^3r \sum_\mu |\psi^{(\pm)}_{p_\alpha s_\alpha \mu_\alpha}(\vec{r},s_\alpha,\mu)|^2. \qquad (11)$$

The sum on μ was introduced into N_α because with it $N_\alpha(R)$ may be written in terms of the wavefunction at R. The derivation of this relation parallels that for the T-matrix element given above, but one uses $\psi_{qs\mu}$ rather than $\exp(i\vec{q}\cdot\vec{r})$. Letting $q \to p$ one obtains

$$N_\alpha(R) = \frac{R^2}{2p_\alpha} \sum_\mu \int d\Omega_R \left(\left[\frac{\partial}{\partial R} \psi^*_{p_\alpha s_\alpha \mu_\alpha}(\vec{R},s_\alpha,\mu) \right] \right.$$

$$\times \left[\frac{\partial}{\partial p_\alpha} \psi_{p_\alpha s_\alpha \mu_\alpha}(\vec{R},s_\alpha,\mu) \right]$$

$$\left. - \psi^*_{p_\alpha s_\alpha \mu_\alpha}(\vec{R},s_\alpha,\mu) \frac{\partial^2}{\partial R \partial p_\alpha} \psi_{p_\alpha s_\alpha \mu_\alpha}(\vec{R},s_\alpha,\mu) \right). \qquad (12)$$

R may be taken to be within the interaction volume by adding to the model independent part of the amplitude a contribution from $V\psi$ for $r < R$.

Equations (9,10,12) provide bounds on the external emission amplitude. The internal emission amplitude to order k may be separated into E2 and M1 parts. These have been given previously.[5] The E2 part can be written as gradient operators acting on the half-shell T-matrix elements. Since the gradients are with respect to the off-shell momenta one can bound the E2 internal emission amplitude just as we did the external emission amplitude.

The M1 part of the rescattering amplitude is

$$\vec{M}_R^{M1} = \frac{ie}{2m} \vec{k} \times < \chi_f^{(-)} | [\vec{j} + (2\mu_p - 1)\vec{s}] | \chi_i^{(+)} >, \tag{13}$$

where $\vec{j}$ is the total angular momentum and $\vec{s}$ is the total spin. Since $\vec{j}$ commutes with the potential one can show that

$$\int^R d^3r \sum_{\mu'} | < \vec{r} \; s \; \mu' | \hat{\epsilon} \cdot \vec{j} | \psi_{ps\mu}^{(\pm)} > |^2$$

depends only on the wavefunction at the distance R. We also have

$$| < \vec{r} \; 1 \; \mu' | \hat{\epsilon} \cdot \vec{s} | \psi_{p1\mu}^{(\pm)} > |^2$$

$$\leq \sum_{\mu''} | < 1 \; \mu' | \hat{\epsilon} \cdot \vec{s} | 1 \; \mu'' > |^2 \sum_{\mu'''} | \psi_{p1\mu}^{(\pm)} (\vec{r}, 1, \mu''') |^2. \tag{14}$$

Bounds on $\vec{M}_R^{M1}$ follow upon separating it into r<R and r>R contributions.

Figures (1) and (2) show numerical results for these bounds when only the convection current external emission amplitude is included. The Reid soft core potential was used to generate the wavefunctions outside the distance R referred to in the figures. At 200 MeV the Hamada-Johnston potential was used for partial wavefunctions having $2<j\leq 4$ at all distances. The shaded regions are the limiting values for the cross section if the wavefunction is unknown within 0.5 fm. At 42 MeV this is a very narrow band showing that, at least for this part of the amplitude, all energy independent potentials which differ only inside 0.5 fm. will give essentially the same bremsstrahlung cross section as long as the elastic amplitude is correct. At 200 MeV the band is still not large. If the wavefunction is unknown inside 0.75 fm. then the bounds open to the upper and lower solid curves. These are still moderately restrictive at 42 MeV but not so at 200 MeV. At 42 MeV we show by dashed lines the results assuming the wavefunction to be known only outside one fm.

The cross section bounds must be extended numerically to include the spin current and, to the extent possible, relativistic corrections and internal emission before comparisons can be made with experimental results. A measurement which lay outside the bounds for a particular value of R would suggest several interesting possibilities, such as changing the potential at distances greater than

R, introducing interaction currents, or giving up the concept of an energy independent potential.

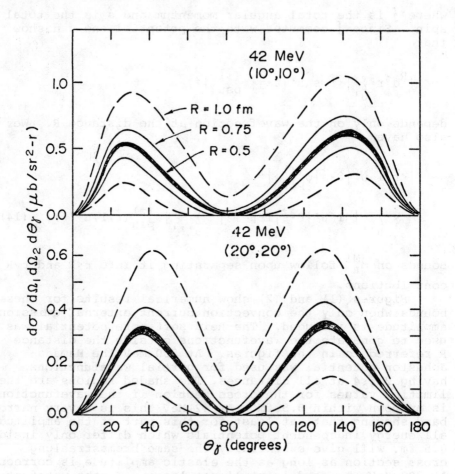

Fig. 1.   Upper and lower bounds on the
symmetric coplanar cross section
at 42 MeV when the wavefunction
is equal to the Reid soft core
one outside R but unknown otherwise. Only the convection current external emission amplitude
is included. Labels for R on
the upper graph also apply to
the lower one.

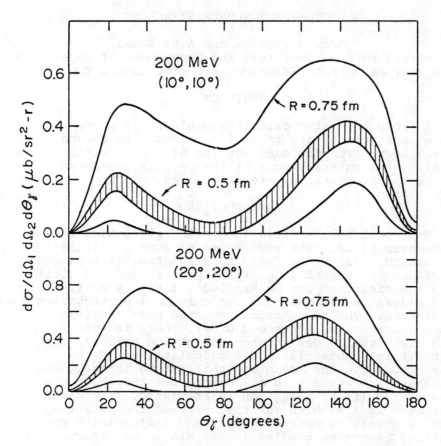

Fig. 2.   Same caption as for Fig. 1 but
at 200 MeV.

REFERENCES

1. L.S. Celenza et al., Phys. Rev. C 8, 838 (1973).
2. M.K. Liou and M.I. Sobel, Phys. Rev. C 4, 1597 (1971).
3. V.R. Brown, Phys. Rev. 177, 1498 (1969).
4. See for example J. von Neumann, Mathematical Foun-
    dations of Quantum Mechanics (Princeton, 1955), p. 40.
5. G.E. Bohannon, Second International Conference on the
    Nucleon-Nucleon Interaction, Vancouver, British
    Columbia (June, 1977).

502

PROTON-PROTON BREMSSTRAHLUNG

Adam Szyjewicz and A.N. Kamal
Theoretical Physics Institute and Department of Physics
University of Alberta, Edmonton, Alberta, Canada T6G 2J1

ABSTRACT

A field theoretic calculation of ppγ is presented. Strong interactions are taken into account to second order. Apart from the contribution of OBE external emission, the contributions of the ω-radiative-decay process and Δ excitation are calculated.

INTRODUCTION

We report here a field-theoretic calculation for pp bremsstrahlung. The advantages of such a calculation are that (i) it is fully relativistic, (ii) gauge invariance is imposed from 'first principles' and (iii) N and Δ excitations can be handled relatively easily. The disadvantage is that one can only do a perturbation calculation. The calculation reported here handles strong interactions to second order. Form factors, which are higher order strong-interaction effects, are not built in. Potential model calculations, on the other hand, have the advantage that strong interactions are taken into account to all orders but they have the disadvantage (i) of being non-relativistic, (ii) gauge invariance is a problem for that part of the potential which is purely phenomenological and (iii) N and Δ excitations cannot be handled in any simple way short of doing a multi-channel NN scattering problem.

MODEL AND RESULTS

All calculations reported here were done in the laboratory system with symmetric and coplanar geometry. The calculations were done in the following chronological order.

1. OBE external emission [1] (Fig. 1).

In the soft photon limit these processes are of $O(\frac{1}{K})$. The interaction Lagrangian density is given by,

$$\mathcal{L}_{int} = \sqrt{4\pi}\left\{g_\pi\bar{\psi}\gamma_5\vec{\tau}\cdot\vec{\phi}_\pi\psi + g_\eta\bar{\psi}\gamma_5\psi\phi_\eta + g_\sigma\bar{\psi}\psi\phi_\sigma + g_\omega\bar{\psi}\gamma^\mu\psi\omega_\mu\right.$$

$$\left. + \bar{\psi}[g_\rho\gamma^\mu\vec{\tau}\cdot\vec{\rho}_\mu + \frac{f_\rho}{4M}\sigma^{\mu\nu}\vec{\tau}\cdot\vec{\rho}_{\mu\nu}]\psi \right\} \tag{1}$$

SU(3) with Zweig's rule $(g_\phi = 0)$ implies $g_\omega^2 = 9 g_\rho^2$ with
ideal nonet mixing angle for 1^- mesons. With pseudo-
scalar mixing angle $\theta_p = 0$ one also has

$$g_\eta^2 = \frac{(3-4\alpha)^2}{3} g_\pi^2 ,$$

where $d/f = \alpha/(1-\alpha)$. Using[2] $d/f = 1.8$, one finds $g_\eta^2 \simeq \frac{1}{16} g_\pi^2$.
The η-coupling can therefore be ignored. The strong
coupling parameters we have[3] used are those given by
Arndt, Bryan and MacGregor[3] which were chosen to fit
the peripheral partial waves $(L \geq 1)$ in the pp-elastic
scattering. The electromagnetic vertex used was
$e(\gamma_\mu - i\sigma_{\mu\nu}k^\nu \frac{\kappa}{2M})$ with $\kappa \simeq 1.79$. No form factors were
used.

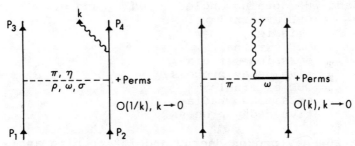

Fig. 1. OBE external   Fig. 2. ω radiative
emission graph.   decay graph.

2. ω-radiative-decay contribution (Fig. 2)

For soft photons this process is of $O(k)$. As $g_{\omega PP}^2 = 9 g_{\rho PP}^2$ and also[4] $\Gamma_{expt}(\omega \to \pi\gamma) \simeq 25 \Gamma_{expt}(\rho \to \pi\gamma)$ one expects[5]
the ω-radiative-decay process to be more important than
the ρ-radiative-decay process. The $\omega \to \pi\gamma$ vertex is given
by $g_{\omega\pi\gamma}\varepsilon_{\mu\nu\rho\sigma}\varepsilon^\mu(\omega)\varepsilon^\nu(\gamma)p^\nu k^\sigma$ where p and k are the pion and
photon momenta respectively. We use $g_{\omega\pi\gamma}^2/4\pi = 0.864 \times 10^{-3}$
m_π^{-2} which implies $\Gamma(\omega \to \pi\gamma) = 880$ KeV.

Figs. 3 and 4 show the result of the first two
stages of calculation. As one does not a priori know the
relative sign of the OBE and ω-radiative-decay amplitudes
we have shown the results with both relative signs. The
effect of ω-radiative decay amplitude is clearly more
important at higher proton energies.

3. Δ excitation (Fig. 5)

This process is of $O(k)$ for soft photons. So far
we have calculated the effect of the π-exchange graph
only. To get a measure of the importance of Δ we evaluated $(p_i-k)^2$, $i = 1,2$ and $(p_j+k)^2$, $j = 3,4$ at different

504

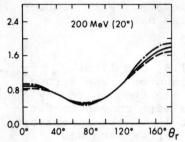

Fig. 3. $d\sigma/d\Omega \, d\Omega_4 d\theta_\gamma$ ($\mu b/sr^2 - \gamma$) vs θ_γ at 200 MeV and 20° opening angle.

——— = OBE external emission,

- - - = OBE external emission – ω radiative decay process,

–·–·–· = OBE external emission + ω radiative decay process.

Fig. 4. Same as Fig. 3 at 700 MeV and 20° opening angle.

θ_γ. At 200 MeV proton energy and 12° opening angle the smallest and the largest values of these invariants were 0.64 GeV^2 and 1.06 GeV^2. Note that the on-mass-shell value is 0.88 GeV^2. Thus the off-mass-shell excursions are not large and certainly not close to $m_\Delta^2 =$ 1.52 GeV^2 where a large Δ contribution is expected. On the other hand at 700 MeV and 12° opening angle the smallest and the largest values of these invariants are -0.033 GeV^2 and 1.55 GeV^2. Thus one can 'sit' on the Δ peak and pick up a large Δ contribution.

The interaction Lagrangian we use is [6]

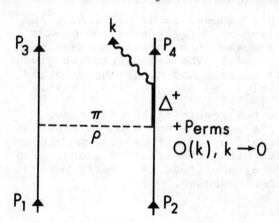

Fig. 5. Δ excitation graph.

$$\mathcal{L}_{\Delta N\pi} + \mathcal{L}_{\Delta N\gamma} = G^* \sum_i \bar{\psi}_\mu^i \psi \partial^\mu \phi_\pi^i - eC\bar{\psi}_\mu^3 \gamma_\nu \gamma_5 \psi F^{\mu\nu} \qquad (2)$$

with $\qquad G^{*2}/4\pi = 18.4 \times 10^{-6} \text{ MeV}^{-2}$ $\qquad\qquad$ (3)

and $Cm_\pi = 0.315$.   (4)

Figs. 6 and 7 show the effect of the Δ excitation due to π exchange relative to that of π-exchange external emission. Notice that the π-exchange external emission cross-section is about a factor of 2 larger than the OBE external emission cross-section and that the Δ excitation plays an important role at 700 MeV and 20° opening angle. As the momentum transfers involved are large at 700 MeV one might expect a suppression of cross-sections due to the form factors. When Δ excitation by ρ exchange is included we expect a considerable cancellation of the Δ excitation by π exchange. This will be the next stage of our program.

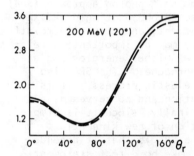
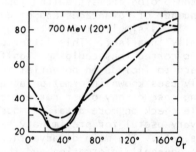

Fig. 6. $d\sigma/d\Omega_3 d\Omega_4 d\theta_\gamma$ ($\mu b/sr^2 - \gamma$) vs θ_γ at 200 MeV and 20° opening angle.
———— = π exchange exter-
            nal emission.
- - - = π exchange exter-
            nal emission +
            Δ excitation.

Fig. 7. $d\sigma/d\Omega_3 d\Omega_4 d\theta_\gamma$ ($\mu b/sr^2 - \gamma$) vs θ_γ at 700 MeV and 20° opening angle.
———— = Δ excitation.
- - - = π exchange exter-
            nal emission - Δ
            excitation.
-·-·-· = π exchange exter-
            nal emission + Δ
            excitation.

We wish to thank Joel Rogers and Harold Fearing for several discussions.

REFERENCES

1. R. Baier, H. Kühnelt and P. Urban, Nucl. Phys. **B11**, 675 (1969).
2. M. Gronau, Phys. Rev. **D5**, 118 (1972).
3. R.A. Arndt, R.A. Bryan and M.H. MacGregor, Phys. Letters **21**, 314 (1966).
4. B. Gobbi et al., Phys. Rev. Letters **33**, 1450 (1974).
5. A.N. Kamal and A. Szyjewicz, Nucl. Phys. (in press).
6. M.G. Olsson and E.T. Osypowski, Nucl. Phys. **B87**, 399 (1975).

SOME COMMENTS ON THE SOFT-PHOTON APPROACH TO
PROTON-PROTON BREMSSTRAHLUNG*

Harold W. Fearing
TRIUMF, Vancouver, B.C., Canada V6T 1W5

ABSTRACT

General features of proton-proton bremsstrahlung as obtained from soft-photon calculations are discussed, and numerical results of such calculations are compared with data from the TRIUMF, Manitoba, and UCLA experiments.

The soft-photon, or Low, theorem is well known and has been used for a variety of radiative processes. Here we discuss its application to proton-proton bremsstrahlung ($pp\gamma$). The soft-photon approach (SPA) has a number of advantages and of course some compensating disadvantages. It is simple, trivially gauge invariant, and trivially relativistically invariant. This latter property may be important as relativistic corrections could be significant at higher energies and are very hard to include in potential model approaches. The SPA also directly uses known information about the elastic process. It is a limiting case of any other correct calculation and so serves as a valuable check on more detailed models. Finally, since it is simple it is very useful for preliminary exploration of new kinematic regions, etc. The main disadvantage of the SPA is that it depends only on on-shell information and thus can make no direct statement about the off-shell interaction, which is the primary motivation for studying $pp\gamma$.

One's philosophy toward the SPA is thus ambivalent. It contains a core of what must be present and correct and yet one hopes it fails, in fact fails miserably, thus increasing the possibility of learning interesting off-shell information.

In any case, one can use the SPA to learn about a number of general features and results in $pp\gamma$. Here we will discuss these general results and then see how numerical results in this approach compare with potential models and with data.

Recall first how the Low theorem is derived.[1-4] One begins with the amplitudes for the external emission graphs, which contain all of the pieces of $O(1/k)$ in the photon momentum k, and expands the off-shell nucleon-nucleon (NN) amplitude and the electromagnetic vertex functions about an on-shell point. Some $O(k^\circ)$ terms are then added to make the result gauge invariant. The choice of such terms is, as usual, not unique but is made in the simplest way and has the effect of including some contributions from the off-shell interaction, from internal radiation, etc., but contributions which are fixed by gauge invariance in terms of known on-shell quantities.

The resulting amplitude can be written schematically as $M_{NN\gamma}$ = $A/k + B + Ck$. Here, A and B contain the elastic NN amplitudes and, in the case of B, also derivatives of these amplitudes, evaluated at an

---

*Work supported by the National Research Council of Canada.

ISSN: 0094-243X/78/506/$1.50 Copyright 1978 American Institute of Physics

on-shell point. They also contain known, simple functions of the radiative variables arising from the kinematics and thus implicitly depend on k. A contains only charge terms whereas B contains anomalous magnetic moment terms as well. A and B depend on the choice of the on-shell point about which the expansion is made (taken as an average energy and momentum transfer in the Low prescription), i.e., on the choice of variables for the elastic amplitude. They may individually vary significantly for different choices, but it is a general result that the sum $A/k + B$ changes for different choices of the on-shell point only by amounts of $O(k)$, i.e., by terms which contribute only to C.[5-7] For larger k, when initial and final proton energies may differ significantly, or particularly when the final nucleons are left with low relative momentum, it may be preferable to use the Feshbach-Yennie[8,9] prescription, which essentially avoids the expansion about an average energy.

The Ck term (we have lumped all higher-order terms into C so that C is an explicit function of k), contains a number of contributions, including in particular the unknown off-shell terms coming from the NN and γN interactions. However, it also contains higher-order on-shell terms, contributions from internal radiation, the ambiguity in the choice of on-shell point, etc. It therefore should be very clear that a measurement of C is not a direct measurement of the off-shell parts of the NN interaction. Thus C large allows, but does not require, sizable off-shell contributions. Conversely $C \equiv 0$ does not, strictly speaking, mean off-shell effects are small, since in principle there could be cancellations among various large contributions. Such a situation seems unlikely over any finite region, however, as the kinematic dependences of the various terms are different. Thus one presumedly should look for regions where C is large and then resort to more detailed models to decide whether off-shell effects are large in those regions or whether the large terms are coming from some other piece of C.

The cross section can be obtained by squaring $M_{NN\gamma}$ and is of the form $d\sigma \sim k|M_{NN\gamma}|^2 \sim A^2/k + 2ReAB* + (B^2+2ReAC*)k + 2ReBC*k^2 + C^2k^3$ where the A^2 and AB terms and the B^2 part of the $O(k)$ term are given in terms of on-shell information by the SP theorem.

The first two terms in the cross section, $A^2/k + 2ReAB*$, can be expressed in a rather neat way via the Burnett-Kroll theorem[5] as an operator function of the kinematics acting on the unpolarized elastic cross section. Thus a knowledge of the elastic NN cross section is sufficient to determine the first two terms of the NNγ cross section. More detailed knowledge, i.e. phase shifts, is needed first for the B^2 part. This operator is independent of the anomalous magnetic moments so we have the general result that anomalous magnetic moment terms do not contribute through the leading two orders in k, but appear first in the B^2 piece.

So far we have discussed unpolarized cross sections. It is an important question whether or not measuring spins gives additional information. Observe first that the Low theorem is an amplitude relation. Hence knowledge of the elastic NN amplitudes, i.e., of the phase shifts, allows calculation of A/k and B for any spin combination and thus calculation of the leading cross section terms $A^2/k + 2ReAB* + B^2k$ for any spins.

508

One can also generalize the Burnett-Kroll theorem to obtain the leading two terms in the polarized NNγ cross section as an operator acting on the polarized NN cross section.[7] Again, it is thus sufficient to know the elastic cross section, albeit for all polarization states, to obtain the polarized bremsstrahlung cross section through the first two terms. As before, to calculate B^2 one needs the phase shifts.

Thus one cannot escape the SP theorem by measuring spins, i.e., one still wants regions where C is large. However, one does investigate different combinations of amplitudes when spins are measured. For example, now radiation from anomalous magnetic moments enters in $0(k°)$, i.e., in the ReAB* term. Thus the sensitivity may be different. Furthermore, there is the intriguing possibility of finding specific spin correlations for which the leading terms vanish, thus allowing a direct measurement of the higher-order pieces.[10]

Next we investigate the relative sizes of the various terms contributing to the SP cross section. Such a separation into individual terms proves rather instructive and, although there have been several previous SP calculations,[2,3,11] the separation seems not to have been made before. In Fig. 1(a) we show the various contributions for a particular symmetric coplanar case, one of those investigated in the

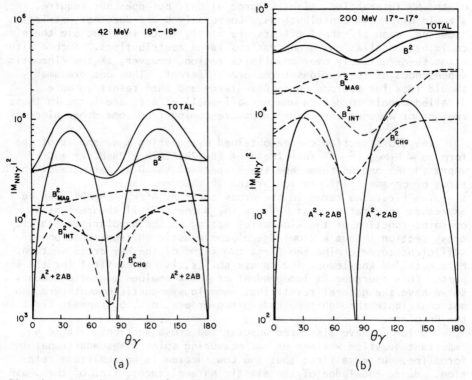

Fig. 1. Contributions of individual terms in the SP expansion to the square of invariant ppγ amplitude.

42 MeV Manitoba experiment.[12] At this energy the A^2 + 2ReAB* and the B^2 terms are comparable over much of the region except at 0°, 180°, and ~80° where the A^2 + 2ReAB* term vanishes. The B^2 term is relative smooth and contains a sizable magnetic contribution. For the TRIUMF experiment at 200 MeV [see Fig. 1(b)] results are qualitatively similar with the very important exception that now the A^2 + 2ReAB* term is quite small and the cross section is given almost entirely by the B^2 contribution.

The vanishing of the A^2 + 2ReAB* term at certain points is an interesting general feature and has important consequences. The combination vanishes not because of cancellations but because the amplitude A vanishes. The proof of this for 0° and 180° is simple. Note that A is proportional to the kinematic quantity $\vec{\epsilon} \cdot (\vec{p}_1/k \cdot p_1 + \vec{p}_2/k \cdot p_2 - \vec{p}_3/k \cdot p_3 - \vec{p}_4/k \cdot p_4)$, with $\vec{\epsilon}$ the photon polarization vector and the $\vec{p}$'s nucleon momenta. In the lab system for $\theta_\gamma = 0°$, 180° we have $\vec{p}_2 = 0$ and $\vec{p}_1 \sim \hat{k}$. Here also for symmetric geometry the four-vector combinations $k \cdot p_3$ and $k \cdot p_4$ are equal and $\vec{p}_3 + \vec{p}_4 \sim \hat{k}$. Thus at these points the entire term is proportional to $\vec{\epsilon} \cdot \hat{k}$ which is zero so that $A \equiv 0$.

This result has the important implication that at these points the remaining $O(k)$ term, i.e., that proportional to ReAC*, vanishes as well and thus here the entire $O(k)$ contribution is given by the known B^2 term. Thus for this particular geometrical situation the SP theorem is good to one higher order in k than usual. By continuity arguments there will be regions in the vicinity of these points, and also a region involving asymmetrical angles where the result is approximately true and where one must look to the fourth term, which is $O(k^2)$, to find a deviation from the soft-photon theorem.

For the UCLA geometry,[13] where one proton and the gamma angle and energy are measured, one gets a sampling of both previous energy regions (cf. Fig. 2). At low k the A^2 + 2ReAB* term dominates but as k increases this term falls rapidly until eventually the relatively constant B^2 term becomes most important. Again B^2 contains a large component of magnetic radiation.

Another interesting insight can be obtained by looking at the A^2 and 2ReAB* terms individually. Recall that we have mentioned two ways of calculating these terms. One can construct the amplitudes A and B separately using on-shell variables for the invariant functions in A and B and radiative variables for the explicit kinematic factors and spinor wave functions, or one can use the Burnett-Kroll theorem which requires the cross section evaluated at on-shell variables. These two approaches amount to slightly different choices of the on-shell variables and hence may give different A^2 and ReAB* terms, but the combination must be the same up to terms which contribute to C.[5-7] In Fig. 3 we show the A^2, 2ReAB*, and A^2 + 2ReAB* terms for both choices for the UCLA geometry.[13] The individual terms do vary somewhat but the sum is essentially identical. Note also that the ReAB* term is very small. Thus for this case keeping just the A^2 term in the Burnett-Kroll approach—which is just the external emission dominance (EED) model of the UCLA group[13]—gives a good approximation to the cross section which will fail only when the B^2 term becomes important. Note also that in the Burnett-Kroll approach a portion of the "ReAB*" term is put into the "A^2" term, as compared with the alternative

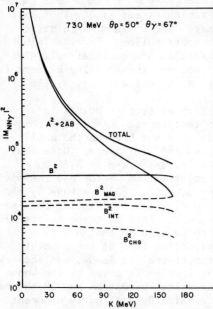

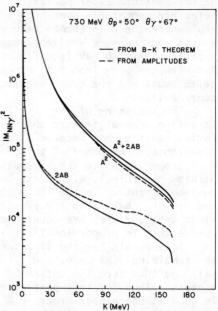

Fig. 2. Contributions of individual SP terms as in Fig. 1,
plotted versus laboratory photon
energy.

Fig. 3. A^2 and AB contributions
for the two different methods of
calculation described in the text
plotted versus photon energy.

approach, so that the A^2 term is a better approximation to the sum.
These effects are more pronounced for the TRIUMF experiment but unimportant there since the B^2 term dominates.

It is now of interest to see how the predictions of the SPA compare with model calculations and with the data. We should re-empha-
size that the differences between data or model calculations and the
SP prediction reflect the size of C (and perhaps, when comparing with
potential models, relativistic corrections to the leading terms) but
are not a direct measure of off-shell effects since C contains contributions from a number of sources. Thus a region of large C may or may
not indicate large off-shell effects. However, once found, such
regions should be explored using models which allow one to vary off-
shell input independently. On the other hand regions of small C are
probably less interesting, as then either off-shell effects are small
or there has been some elaborate cancellation which, while possible,
seems unlikely.

Figure 4 shows a comparison between the SP calculation and a
Hamada-Johnston (HJ) potential calculation[14] at 42 MeV. At 0°, 180°,
and ~80° where $A \equiv 0$ the results differ very little thus indicating
that at these points the higher-order $2ReBC*k^2 + C^2k^3$ piece of the
cross section must be small. This, together with our earlier observation that $A \gtrsim B$, probably means that the difference at the peaks is
due mostly to the $2ReAC*$ term.

In comparing with data, some caution must be exercised as the
data are averaged over sizable non-coplanarity while the theory is not.

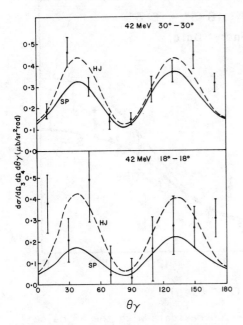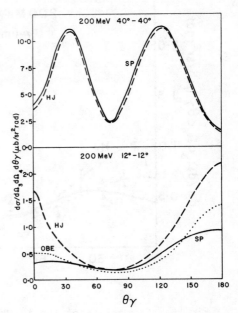

Fig. 4. Comparison of SP calculation with the data and a HJ potential calculation of the Manitoba group at 42 MeV.

Fig. 5. Comparison of SP calculation with HJ calculation of Bohannon and a OBE calculation of Kamal and Szyjewicz at 200 MeV.

When this is taken into account the integrated cross section tends to fall below the HJ prediction particularly at small and/or asymmetric angles.[12] One can conclude from the graph, however, that the difference between SP and HJ predictions is of the same order of magnitude as the size of the error bars.

At 200 MeV the situation is completely different (see Fig. 5). Here differences between HJ,[15] a one-boson exchange (OBE) calculation,[16] and SP are most pronounced at 0° and 180°. Since at these points $A \equiv 0$ (recall as well that $A << B$ over the whole region) the difference must be due mainly to the $2ReBC*k^2$ and C^2k^3 terms. Thus at this energy one is measuring a different set of unknown terms than at 42 MeV. Compared with the preliminary data of the TRIUMF group[17] as shown in Fig. 6 the SP calculation seems a bit low and the HJ calculation a bit high, though again the differences are not too much greater than the size of the error bars.

Finally, consider the UCLA experiment[13] (Fig. 7). Here the geometry is entirely different and the energy is much higher, 730 MeV. Since for these kinematics generally $A >> B$ and since the ReAB* term

512

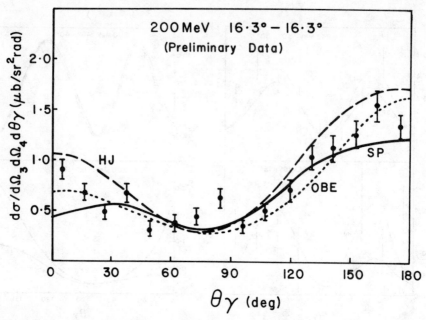

Fig. 6. Comparison of the TRIUMF 200 MeV data with the SP calculation, HJ calculation of Bohannon (Ref. 15) and OBE calculation
of Kamal and Szyjewicz (Ref. 16).

is small (cf. Figs. 2 and 3), it is perhaps not a surprise that the
EED model[13] describes the cross section well over a sizable region. At
larger k, however, the B^2 term dominates and it, together with a rising phase space, produces an increase in the cross section. Thus the
SP result, although generally a bit low, differs from the data, at
least for the cases plotted, by amounts again of the order of the size
of the error bars. Again one must be a bit cautious in interpreting
this as the theory has not been integrated over the rather large
acceptance of counters, and in addition the phase shifts used in constructing the elastic amplitudes are not as well known here as at the
lower energies.

   Thus to summarize, we have examined the soft-photon approach to
ppγ and some of the general results one can obtain from it, looking in
particular at the contributions of individual terms. We find that the
particular geometrical situation and energy determine to some extent
which of the unknown terms are likely to be most important. Thus for
the symmetric coplanar geometry at low energies one apparently probes
the ReAC* term and should find effects in the middle regions of θ_γ,
whereas by 200 MeV the unknown contributions come predominately from
the ReBC* and C^2 terms and are most pronounced at $\theta_\gamma = 0°$ and 180°.
In the UCLA geometry one is looking primarily at the leading terms,
with the B^2 term becoming important only at larger k. Finally, in all
cases considered here the SPA reproduces the qualitative features of
the data and differs quantitatively by amounts not too much greater
than the sizes of the error bars.

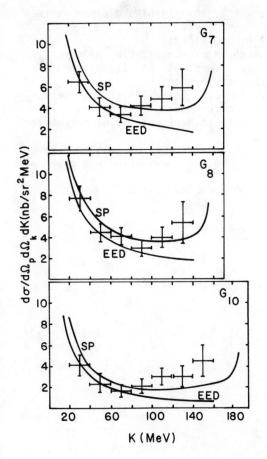

Fig. 7. Comparison of the SP calculation with the UCLA data at 730 MeV.

ACKNOWLEDGEMENTS

The author would like to thank G. Bohannon, A.N. Kamal, and A. Szyjewicz for making the results of their calculations available to him. He has also benefited greatly from conversations with many colleagues including in particular those above, A. Anderson, L.G. Greeniaus, J.V. Jovanovich, B.M.K. Nefkens, and J.G. Rogers.

REFERENCES

1. F.E. Low, Phys. Rev. <u>110</u>, 974 (1958).
2. E.M. Nyman, Phys. Rev. <u>170</u>, 1628 (1968).
3. H.W. Fearing, Phys. Rev. <u>C6</u>, 1136 (1972).
4. H.W. Fearing, E. Fischbach, and J. Smith, Phys. Rev. D2, 542 (1970).
5. T.H. Burnett and N.M. Kroll, Phys. Rev. Lett. <u>20</u>, 86 (1968).
6. J.S. Bell and R. Van Royen, Nuovo Cimento <u>60A</u>, 62 (1969).
7. H.W. Fearing, Phys. Rev. <u>D7</u>, 243 (1973).
8. H. Feshbach and D.R. Yennie, Nucl. Phys. <u>37</u>, 150 (1962).
9. E.M. Nyman, Phys. Letters <u>40B</u>, 323 (1972).

10. M.J. Moravcsik, Phys. Letters <u>65B</u>, 409 (1976) and paper at this conference.
11. G. Felsner, Phys. Letters <u>25B</u>, 290 (1967).
12. L.G. Greeniaus, J.V. Jovanovich, R. Kerchner, T.W. Millar, C.A. Smith, and K.F. Suen, Phys. Rev. Lett. <u>35</u>, 696 (1975), and J.V. Jovanovich, paper at this conference.
13. B.M.K. Nefkens, O.R. Sander, and D.I. Sober, Phys. Rev. Lett. <u>38</u>, 876 (1977), and paper at this conference.
14. L.G. Greeniaus *et al.*, University of Manitoba Cyclotron Laboratory report #75-11.
15. G. Bohannon, private communication and paper at this conference.
16. A.N. Kamal and A. Szyjewicz, private communication and paper at this conference.
17. J.L. Beveridge *et al.*, paper at this conference.

POSSIBLE USE OF POLARIZATION IN OFF-SHELL STUDIES*

Michael J. Moravcsik
Institute of Theoretical Science
University of Oregon
Eugene, Oregon 97403

SUMMARY

This work was published in Physics Letters 65B, 409 (1976), and hence will be reproduced here only in the form of a summary, stressing those aspects which are relevant to this conference.

As it is explained in Harold Fearing's talk, the separation of on-shell and off-shell effects is a subtle process which in general cannot be carried out unambiguously and model-independently. It would therefore be of great value to find a circumstance in which the on-shell contribution is definitely known to be zero and yet the off-shell contribution is in general non-zero, because in such a case the latter could be determined unambiguously and model-independently.

Such a circumstance is created by conservation laws and exists if we look at certain types of polarization quantities instead of the differential cross section. For example, in pion-nucleon bremsstrahlung, the on-shell contribution (from pion-nucleon scattering) is zero for any experiment in which the polarization of the initial or of the final nucleon is in the scattering plane and the other nucleon is unpolarized. In the corresponding experiment for pion-nucleon bremsstrahlung, however, (that is, in the same experiment with an additional unpolarized photon emerging) the polarization need not be zero, and thus all of it will come from off-shell contributions. The reason why the four- and five-particles reactions behave so differently as far as the vanishing and non-vanishing of these polarization quantities are concerned is that in a reaction containing more than four particles the rank-zero tensors on which the amplitudes can depend will be partly scalars and partly pseudoscalars, while in a four-particle reaction they are guaranteed to be scalars. In order to take advantage of this effect, the bremsstrahlung process must be measured in a non-coplanar geometry. Similar considerations apply also to nucleon-nucleon bremsstrahlung. There are a large number of different polarization quantities of varying complexity, that can be used for the purpose described here.

* Research supported in part by the U.S. Energy Research and Development Administration

A NEW PROGRAM FOR BREMSSTRAHLUNG STUDIES *

M.K. Liou

Department of Physics and Institute of Nuclear Theory
Brooklyn College of the City University of New York
B'klyn, N.Y. 11210

KINEMATICS AND CROSS SECTION

We consider photon emission accompanying the scattering of two particles A and B:

$$A(q_i^\mu) + B(P_i^\mu) \longrightarrow A(q_f^\mu) + B(P_f^\mu) + \gamma(K^\mu) \quad (1)$$

Here, q_i^μ (q_f^μ) and P_i^μ (P_f^μ) are the initial (final) four-momenta of particles A and B, respectively, and K^μ is the four momentum of the emitted photon. In the laboratory frame, these five four-momenta are defined, in the spherical system, as

$$q_i^\mu = (E_i, 0, 0, q_i),$$

$$P_i^\mu = (M, 0, 0, 0),$$

$$q_f^\mu = (E_q, q_f \sin\theta_q \cos\phi_q, q_f \sin\theta_q \sin\phi_q, q_f \cos\theta_q)$$

$$P_f^\mu = (E_p, P_f \sin\theta_p \cos\phi_p, P_f \sin\theta_p \sin\phi_p, P_f \cos\theta_p)$$

$$K^\mu = (K, K\sin\theta_\gamma \cos\phi_\gamma, K\sin\theta_\gamma \sin\phi_\gamma, K\cos\theta_\gamma)$$

where m and M are the masses of particles A and B, respectively, $E_i = (m^2 + \vec{q}_i^2)^{1/2}$, $E_q = (m^2 + \vec{q}_f^2)^{1/2}$ and $E_p = (M^2 + \vec{P}_f^2)^{1/2}$. They satisfy equations of the energy momentum conservation:

$$q_i^\mu + P_i^\mu = q_f^\mu + P_f^\mu + K^\mu \quad (3)$$

For a given incident energy E_i, there are nine variables in these equations. If five of them are chosen to be independent variables, then the other four can be determined in terms of these five independent variables by solving Eqs.(3). The bremsstrahlung cross section can be expressed as a function of the five independent variables. The choice of these five kinematical variables is strongly influenced by experimental considerations. Two most typical geometries used in bremsstrahlung experiments are the Harvard geometry and the Rochester geometry.[1,2] Harvard geometry refers to any experimental arrangement in which two final-state particles are detected and the photon momentum is calculated from Eq.(3). Rochester geometry refers to an experimental arrangement in which all three final-state particles are detected.(photon is detected in this geometry). The

*Supported in part by PSC-BHE Research Award Program of the
 City University of New York.

expression for the bremsstrahlung cross section is then dependent upon which set of five variables is chosen to be the independent variables. Since many different sets of independent variables can be chosen from these geometries, many different kinds of cross sections have been obtained. These cross sections can be classified into two classes: Harvard cross sections and Rochester cross sections. They are defined as follows:

(i) Harvard cross sections: $\sigma_H = d^5\sigma / d\Omega_q \, d\Omega_P \, dX$
Here, $d\Omega_q = \sin\theta_q \, d\theta_q \, d\phi_q$, $d\Omega_P = \sin\theta_P \, d\theta_P \, d\phi_P$
and X is chosen from one of the following variables:
q_f , E_q , P_f , E_P , K , θ_γ , ψ_γ , ϕ_γ .
The independent variables for Harvard cross section are, therefore, θ_q , ϕ_q , θ_P , ϕ_P and X.

(ii) Rochester cross sections: $\sigma_R = d^5\sigma / d\Omega_\gamma \, dK \, dY \, dZ$
Here, $d\Omega_\gamma = \sin\theta_\gamma \, d\theta_\gamma \, d\phi_\gamma$, Y can be either ϕ_q or ϕ_P , and Z is chosen from one of the following variables:
q_f , E_q , P_f , E_P , θ_q , θ_P and either ϕ_P (if $Y = \phi_q$) or ϕ_q (if $Y = \phi_P$). The independent variables are θ_γ , ϕ_γ , k, Y and Z.

One should not confuse Rochester (Harvard) <u>cross sections</u> with Rochester (Harvard) <u>geometry</u> since Rochester (Harvard) cross section can also be measured from Harvard (Rochester) geometry. There are other types of cross section which can not be classified into these two classes. These types of cross section will not be discussed here mainly because they have never been studied experimentally or theoretically.

SOFT-PHOTON EXPANSION AND THE SOFT-PHOTON THEOREM

We all know that there exists a powerful basic principle for bremsstrahlung studies. This principle, known as the soft-photon theorem, Low's Theorem or the low-energy theorem, was first introduced by Low[3] and was generalized and extended later by many other authors[4]. Unfortunately, due to some misunderstandings in the past, this beautiful theorem was not found to be very useful in either predicting theoretical cross sections or analyzing and interpreting experimental results. Most of these misunderstandings have been clarified recently.[5,6] Our new program, which is designed for both theoretical and experimental studies of the bremsstrahlung problems, is mainly based upon a new understanding of the soft-photon theorem. The soft-photon theorem is based upon the existence of the expansion of the bremsstrahlung cross section in powers of photon energy k:

$$\sigma = \frac{\sigma_{-1}}{K} + \sigma_0 + \sigma_1 K + \cdots \qquad (4)$$

where $\sigma_{-1} = \lim_{K \to 0} (K\sigma)$

$\sigma_0 = \lim_{K \to 0} \frac{\partial}{\partial K} (K\sigma)_{x_i}$
. . . .

Here, the χ_i refer to the set of observables which are held constant in carrying out the partial differentiation. This expansion is the soft-photon expansion. We should emphasize that all coefficients in Eq.(4) are independent of k. This expansion is possible only if the kinematics and dynamics of the bremsstrahlung process are expanded consistently and completely. In the past, various kinematical terms were not expanded consistently. As a result, the expansion obtained previously was not appropriate to the soft-photon theorem. The incomplete expansion can be written as

$$\sigma = \frac{\overline{\sigma_{-1}}(k)}{k} + \overline{\sigma_0}(k) + \overline{\sigma_1}(k)k + \cdots \quad (5)$$

This kind of expansion should not be considered as a "soft-photon expansion" since all coefficients in Eq.(5) are still functions of k.

The expansion given by Eq.(4) cannot be used directly for all cross sections since some cross sections are either zero or undefined in the $k \rightarrow 0$ limit. For example, it was shown recently that the soft-photon expansion given by Eq.(4) can be derived for σ_R , but it cannot be obtained physically for σ_H.[6] Note that while σ_H is either zero or undefined around $k = 0$, it can be obtained for finite k. Therefore, an expansion which is similar to Eq.(5) could be obtained for σ_H .

In terms of the soft-photon expansion, Eq.(4), the soft-photon theorem states that the coefficients σ_{-1} and σ_0 are independent of the off-mass-shell (or off-energy-shell) effects and they can be evaluated from the amplitude of the corresponding nonradiative process (elastic process) and its derivatives. Thus the theorem provides us with an approximate method for calculating the bremsstrahlung cross section:

$$\sigma_R^{SPA} \equiv \frac{\sigma_{-1}^R}{k} + \sigma_0^R \quad . \quad (6)$$

If the incomplete expansion given by Eq.(5) is used then the cross section calculated from the first two terms of Eq.(5) will be

$$\overline{\sigma_R} = \frac{\overline{\sigma_{-1}^R}(k)}{k} + \overline{\sigma_0^R}(k) \quad . \quad (7)$$

Similarly, we have the following approximation for σ_H :

$$\overline{\sigma_H} = \frac{\overline{\sigma_{-1}^H}(k)}{k} + \overline{\sigma_0^H}(k) \quad . \quad (8)$$

COMPARISON BETWEEN HARVARD CROSS SECTION AND ROCHESTER CROSS SECTION

We have now specified the Harvard cross section σ_H and Rochester cross section σ_R . We must find out which one is better for our study of bremsstrahlung problems. Our choice is the Rochester cross section. In order to understand the reasons behind this choice, we have to show that σ_R has many advantages over σ_H. We will show this from the experimental point of view (1a - 3b) and from the theoretical point of view (4a - 6b):

(1a): The soft-photon expansion exists for σ_R. This means that σ_R has the expansion given by Eq.(4) with coefficients which are independent of k. If σ_R is plotted against 1/k or if $k\sigma_R$ is plotted as a function of k, then we expect that the experimental data in the soft-photon region form a straight line which is characterized by two constants, σ_{-1}^R and σ_0^R. Thus, σ_{-1}^R and σ_0^R can be determined from the measurement of σ_R in the low photon energy region. The details of our method are as follows: (A): If data for σ_R are plotted as a function of 1/k and if a χ^2 fit is used to determine the best straight line in the soft-photon region, we obtain a straight line which represents Eq.(6). σ_{-1}^R can be obtained from the slope of this line and σ_0^R can be evaluated from the expression,

$$\sigma_0^R = \sigma_R^{SPA} - \sigma_{-1}^R / K .$$ Fig.(1) represents an hypothetical case with $\sigma_0^R > 0$.

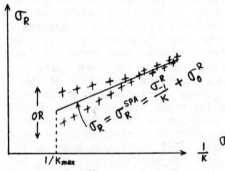

Fig. (1 a): σ_R as a function of 1/k

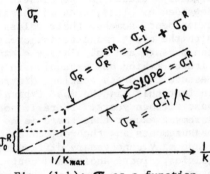

Fig. (1 b): σ_R as a function of 1/k

(B): If data for $K\sigma_R$ are plotted as a function of k, and they are fitted to a straight line in the soft photon region, then we obtain a straight line which represents Eq.(6) written in different form: $K\sigma_R^{SPA} = \sigma_{-1}^R + K\sigma_0^R$. The slope of this line gives σ_0^R and the value of $K\sigma_R$ on this line at $K = 0$ will give σ_{-1}^R. See Fig.(2)

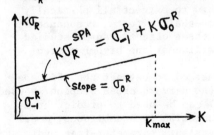

Fig.(2): $K\sigma_R$ as a function of k.

(C): If σ_R is plotted as a function of k, we would obtain a hyperbola in the soft-photon region(See Ref.5). This hyperbola can also be used to determine σ_{-1}^R and σ_0^R. This method is more complicated than those methods described in (A) and (B). We, therefore, do not recommend this scheme.

520

(1b): The soft-photon expansion does not physically exist for σ_H in the sense discussed above. The only expansion σ_H can have is the expansion similar to Eq.(5). All coefficients of the expansion will be functions of k. Thus, if σ_H is plotted against 1/k or if $k\sigma_H$ is plotted against k, we do not expect to obtain a straight line in the soft-photon region. Therefore, $\overline{\sigma}_{-1}^{H}(K)$ and $\overline{\sigma}_{0}^{H}(K)$ cannot be measured directly and easily from the experimental data.

(2a): Once the values of σ_{-1}^{R} and σ_{0}^{R} are determined from experiment by the method discussed in (1a), these values can be checked by comparison with the elastic scattering cross sections. Note that σ_{-1}^{R} is directly related to the elastic scattering cross section, and σ_{0}^{R} can be evaluated from the amplitude of the elastic process. Furthermore, for unpolarized particles, σ_{0}^{R} will depend only on the unpolarized nonradiative (elastic) cross section (Burnett-Kroll theorem)[7]. In this way elastic cross section can be used to check the bremsstrahlung cross section, and vice versa.

(2b): Besides the difficulties in determining $\overline{\sigma}_{-1}^{H}(k)$ and $\overline{\sigma}_{0}^{H}(K)$ from experiment as discussed in (1b), $\overline{\sigma}_{-1}^{H}(K)$ and $\overline{\sigma}_{0}^{H}(k)$ are not directly related to the elastic cross section because of their k-dependence. Thus, even if these quantities can be determined from experiment, somehow their values cannot be checked by comparing with the elastic scattering cross section. One may argue that we can measure and calculate the exact bremsstrahlung cross section without making any expansion in powers of k. For example, we can measure the cross section, $d^5\sigma/d\Omega_e\,d\Omega_p\,d\theta_\gamma$ as a function of θ_γ. In fact this was the most typical cross section measured during the past fifteen years. There is nothing wrong with this type of measurement except that there is no easy way to check the experimental measurements and theoretical calculations. Lack of such checks can cause many unnecessary difficulties. We believe that many ambiguities, discrepancies and contradictions which have occurred in the past were due to the fact that we did not have a easy and reliable scheme to check our experimental and theoretical results. This is the main reason why we choose σ_R rather than σ_H for our study of bremsstrahlung problems.

(3a): The straight line determined in (1a) is characterized by σ_{-1}^{R} and σ_{0}^{R} : $\sigma_{R}^{SPA} = \sigma_{-1}^{R}/k + \sigma_{0}^{R}$. Since σ_{-1}^{R} and σ_{0}^{R} are independent of the off-mass-shell effects, this line will provide us with the "on-shell, model independent cross section." This line has a very important application. It can be used to extract all physically interesting effects, such as off-mass-shell effects, resonance effects etc. The contribution from these effects can be studied from the deviation of data from this line at the higher photon energies.

(3b): There is no reliable and easy method to extract the off-mass-shell effects and other effects if the Harvard cross section is used.

(4a): On the theoretical side, Eq.(6) can be used to predict bremsstrahlung cross sections. The calculation is unambiguous, model independent, relativistic, gauge invariant, parameter-free and correct to order K^0. Eq.(6) was used to predict $\pi^{\pm}p\gamma$ at 298 MeV and was found in excellent agreement with UCLA data.[5] On the other hand, when Eq.(7), which is based upon an incomplete and inconsistent expansion, was used, the result was found to be in total disagreement with the data at higher photon energies.

(4b): When σ_H is expanded around a finite value of k, we obtain an approximation given by Eq.(8). Since $\overline{\sigma_{-1}}^H(K)$ and $\overline{\sigma_0}^H(K)$ in Eq.(8) are still functions of k, they cannot be evaluated without any ambiguities from the amplitude of the elastic process. The calculation based upon Eq.(8) may sometimes give reasonable results for certain energies and angles, but these calculations may also give very poor results for some cases. Because of the ambiguities involved with this calculation, the results cannot be used to extract off-mass-shell effects and other effects.

(5a): One may perform an exact, model dependent calculation without any expansion. This type of calculation is correct to all order in k and it includes, at least in principle, all possible effects. The soft-photon approximation, on the other hand, is valid only to order K^0 and it gives no useful information about the off-mass-shell effects and other effects. Thus, the exact, model dependent calculation has many advantages over the soft-photon calculation. Unfortunately, the model dependent calculation also has its own difficulties. The internal scattering terms in this calculation are very difficult to evaluate and for some complicated terms, they are almost impossible to calculate in practice. Furthermore, the exact exchange current is very difficult to obtain and Heller has shown that the expression for it is not unique. Many different approximations have been developed to obtain an approximate exchange current for practical calculations. These approximations should be tested. If these approximations are used to predict σ_R in the soft-photon region, then σ_{-1}^R and σ_0^R can be obtained from these calculations. The results can be checked by comparing with the experimental elastic cross section (or elastic scattering amplitude). Due to the contribution from the off-mass-shell effects and other effects, this kind of check cannot be done at higher photon energies.

(5b): There is less possibility to make checks if we use exact, model dependent calculations to predict σ_H. A typical cross section calculated is $d^5\sigma / d\Omega_q \, d\Omega_p \, d\theta_\gamma$. The shape of this cross section when it is plotted as a function of θ_γ is . There is practically no way to check this type of calculation. As a result, the errors introduced from the approximation are mixed with the contributions from all physically interesting effects. Related to this are unsolved ambiguities, discrepancies and contradictions.

(6a): For σ_R, the non-coplanar calculation is about the same as the coplanar calculation.

(6b): For σ_H, the non-coplanar calculation is much more complicated than the coplanar calculation.

NEW PROGRAM

Based upon the arguments given above, we now discuss our program in the following:

(A) The bremsstrahlung process $A + B \longrightarrow A + B + \gamma$ and its corresponding nonradiative process (elastic process) must be studied together.

(B) The bremsstrahlung cross sections to be measured or calculated should be Rochester cross sections. These cross sections can be measured by using either Rochester <u>geometry</u> or Harvard <u>geometry</u>.

(C) Bremsstrahlung data can be presented by either plotting σ_R against $1/k$ or plotting $k\sigma_R$ as a function of k.

(D) The bremsstrahlung spectrum should be measured for the entire photon energy range. The data in the soft-photon region should be analyzed to determine a straight line and two constants, σ_{-1}^R and σ_0^R . These two constants should be checked by comparing them with the values for the quantities obtained from the elastic scattering data. Any measurement which fails to obtain consistent values of σ_{-1}^R and σ_0^R (assuming the elastic cross section is well-known) is not acceptable. The straight line should be presented along with the data so that the deviation of the data from this line at higher photon energy can be clearly seen.

(E) Eq.(6) can be used to predict the model independent cross section. The result should be compared with experiment as described in (D).

(F) All model dependent calculations should have their validity checked. This can be done by comparing their predictions in the very soft-photon region with the result obtained in (D) or (E). Any theoretical calculation which fails to produce correct σ_{-1}^R and σ_0^R is not acceptable.

(G) If two models are to be distinguished by bremsstrahlung experiments, the on-shell differences between these two models must be subtracted out. This can be done by performing the following calculation:

$$\sigma_R = \sigma_R^{SPA} - \left(\sigma_R^{SPA}\right)_{model} + \left(\sigma_R\right)_{model}$$

Here $\left(\sigma_R\right)_{model}$ is the model dependent calculation, σ_R^{SPA} is the model independent calculation using Eq.(6) as described in (E), and $\left(\sigma_R^{SPA}\right)_{model}$ is the calculation using Eq.(6) but with σ_{-1}^R and σ_0^R calculated from the model.

(H) Partial wave analysis is mainly based on the elastic scattering data (cross sections and polarizations). The phase shifts obtained from this analysis are not unique and involve ambiguities. These ambiguities and nonuniqueness cannot be resolved from elastic scattering cross section and polarizations.[8] Since the bremsstrahlung measurements will provide an additional on-shell information σ_0^R which is related to the derivatives of the on-shell amplitude, the bremsstrahlung with polarized beam or target may be able to resolve the ambiguities. Furthermore, the off-mass-shell effects may be able to distinguish between two different sets of phase shifts.

OTHER USEFUL PROJECTS

Resonance effects were not seen in the $\pi^{\pm} P \gamma$ experiments, but they were observed in the $P-C^{12}$ bremsstrahlung.[9] More experimental and theoretical studies are needed in order to understand these effects. Finally, can this program be applied to study other processes, such as inelastic bremsstrahlung, radiative absorption, radiative decays and other production processes?

REFERENCES

1. M.L. Halbert, Proceedings of the Gull Lake Symposium on the Two-Body Force in Nuclei, (Plenum Press, New York, 1972).
2. M.K. Liou, Proceedings of the International Conference on Few Body Problems in Nuclear and Particle Physics, (Les Presses De L'Universite Laval, Quebec, 1975).
3. F.E. Low, Phys. Rev. 110, 974 (1958).
4. See references quoted in Ref. 2.
5. M.K. Liou and W.T. Nutt, preprint (to be published in Phys. Rev. D), and papers in preparation.
6. M.K. Liou, papers in preparation.
7. T.H. Burnett and N.M. Kroll, Phys. Rev. Letts. 20, 86 (1968).
8. F.A. Berends and J.C.J.M. Van Reisen, Phys. Letts. 64B, 183(1976).
9. C. Maroni, I. Massa and G. Vannini, Phys. Letts. 60B, 344 (1976).

BREMSSTRAHLUNG IN THE p-d SYSTEM AT LOW ENERGY[*]

R. Roy, C. Rioux and R.J. Slobodrian
Université Laval, Département de Physique,
Laboratoire de Physique Nucléaire, Québec, G1K 7P4, Canada

B. Frois
Service de Physique Nucléaire à Hautes Energies,
CEN Saclay, France

Bremsstrahlung cross sections have been measured for the p-d system at three deuteron bombarding energies. The energies were 6.3, 6.6 and 6.9 MeV at the center of the target. The two particles were detected in coincidence for $\theta_d = \theta_p = 20°$ with fast timing provided by two Ortec 130 systems[1] and analysed with a slow coincidence from the linear energy signals[2]. The three signals (E_1, E_2, time) were recorded on-line with a PDP-9 computer and the events analyzed off-line to give the final two-dimensional energy spectra as shown on Figure 1. The analysis is performed using a window on timing peaks of time spectra previously corrected for time-of-flight differences of the particles.

Below 10 MeV, only three bremsstrahlung cross sections have been previously reported[3] at 7.0, 8.0 and 9.0 MeV for angles of $20° - 35°$, above the break-up threshold. The present measurements go through the threshold region and the data show interesting variations. Experimental cross sections seen on Figure 2 are corrected for noncoplanarity. They are much larger than the measurements[3] above the three body threshold for asymmetrical angles.

Theoretical calculations using the Signell's formula[4] based on the first term of the Feshbach and Yennie expansion[5] are in progress. It should be interesting to see if experimental cross section values and variations with energy can be reproduced.

REFERENCES

1) Ortec Inc., Oak Ridge, Tennessee.
2) B. Frois et al., Phys. Rev. C8, 2132 (1973) and references therein.
3) J. Hall, W. Wölfli and R. Müller, Phys. Lett. 37B, 53 (1971).
4) P. Signell, Advances in Nuclear Physics, Vol.2 (Plenum Press, New York, 1969).
5) H. Feshbach and D.R. Yennie, Nucl. Phys. 37, 150 (1962).

*Supported in part by the NRC of Canada and The Ministery of Education of Québec.

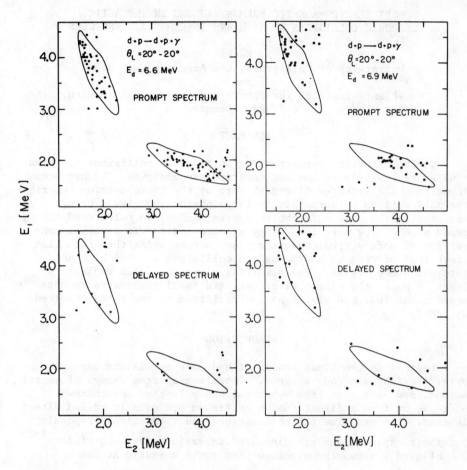

Figure 1. $E_1 - E_2$ spectra of prompt and random coincidence events
at 6.6 and 6.9 MeV, with kinematics boundaries.

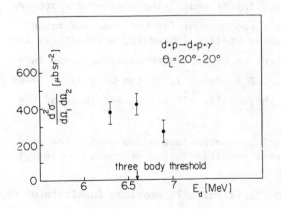

Figure 2.

Experimental coplanar cross
sections for the reaction
p+d → p+d+γ

526

SOFT ELECTROMAGNETIC BREMSSTRAHLUNG IN INELASTIC HADRONIC COLLISIONS AT HIGH AND INTERMEDIATE ENERGIES

R. Rückl
University of California, Los Angeles, Ca. 90024
and
Max-Planck-Institut für Physik und Astrophysik, Munich,
Fed. Rep. Germany

ABSTRACT

Electromagnetic bremsstrahlung in hadronic collisions has been studied extensively at low and intermediate energies. It has been found that the infrared divergent term of the cross section describes the data well up to surprisingly large photon energies. Using essentially the same soft photon approximation, we calculated the production of low mass-low energy electron pairs via internal conversion of soft virtual bremsstrahlung accompanying the production of charged hadrons in hadron-hadron collisions at very high and intermediate energies. The resulting electron yields explain, at least in part, the direct electrons with small transverse momenta seen at the ISR, and are in no contradiction to the rates observed at LAMPF.

INTRODUCTION

Leptons in the final states of hadronic collisions may originate from numerous sources. Those coming from decays of vector mesons, new particles like W-bosons, heavy leptons and charmed hadrons, or from a direct photon continuum are usually called direct leptons. The experimental observations and theoretical proposals for their explanations are discussed in various review articles.[1-3]

Figure 1 shows the electron-pion ratio measured at ISR energies.[3-5] At large transverse momenta ($p_T \geq 1$ GeV/c) vector mesons,[6] charmed mesons[6] and hard photons (produced via Drell-Yan mechanism[7] and bremsstrahlung[8] in the underlying quark-quark process or via virtual vector mesons[9]) can account for the observed rates. The rise of the spectrum towards small p_T, however, necessitates low-mass parents[1,8,9] or a frequent decay of a more massive parent into a three-body final state (e.g. $D \to K^* e \nu$ [10]). A two-body decay leads to a single lepton spectrum which falls off for $p_T \lesssim \frac{1}{2}$ (mass of the parent).

We propose[11] soft virtual bremsstrahlung accompanying the production of charged hadrons as a possible source of these low energy electrons.

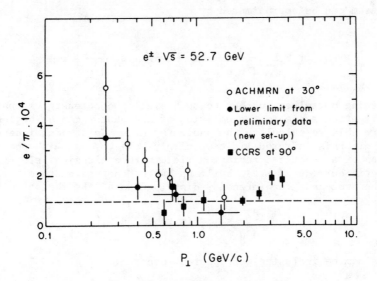

Fig. 1. Direct electrons observed in pp collisions at the ISR, normalized to the corresponding pion rates.

DESCRIPTION OF THE MODEL

The basic process in our picture is pp → "γ" + hadrons with internal conversion of the virtual photon into an electron pair

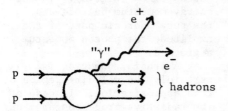

Fig. 2. Direct electron production via conversion of virtual photons.

(Fig. 2). Accordingly the electron pair spectrum can be written in the form

$$E_+ E_- \frac{d\sigma^{e^+ e^-}}{d^3 p_+ d^3 p_-} = \left(\frac{\alpha}{4\pi^2} \right)^2 \frac{W_{\mu\nu} m^{\mu\nu}}{q^4}$$

where $W_{\mu\nu}(m_{\mu\nu})$ is the squared matrix element of the hadronic (leptonic) current and the photon momentum $q = p_+ + p_-$.

For sufficiently small photon masses the hadronic tensor $W_{\mu\nu}$ is determined by the inclusive real photon spectrum $q_0 d\sigma^\gamma/d^3 q$:

$$\lim_{q^2 \to 0} \sum_{pol} \frac{\alpha}{4\pi^2} W_{\mu\nu} \varepsilon^\mu \varepsilon^\nu = q_0 \frac{d\sigma^\gamma}{d^3 q}$$

(ε^μ is the polarization vector of the photon) and the electron pair spectrum can be approximated by

$$E_+E_- \frac{d\sigma^{e^+e^-}}{d^3p_+ d^3p_-} \simeq \frac{\alpha}{2\pi^2} \frac{1}{q^2} \left(q_o \frac{d\sigma^\gamma}{d^3q} \right) .$$

Confining ourselves further to such small photon energies that all recoil effects due to the emission of the photon can be neglected, we are able to calculate the photon spectrum in a model independent way. In this infrared limit we simply have to deal with the inter-action of a quantized radiation field with a classical electro-magnetic current $J(x,t)$ built up by the charged hadrons involved in the process. The intensity distribution of the radiation field

$$\frac{dI}{d^3q} = \left| \int J(x,t) \, e^{iqx} d^4x \right|^2$$

gives us the inclusive soft photon spectrum:

$$q_o \frac{d\sigma^\gamma}{d^3q} = \sigma^{pp}_{inel} \cdot \frac{q_o dN^\gamma}{d^3q} = \sigma^{pp}_{inel} \cdot \frac{dI}{d^3q} .$$

In the more formal language of Feynman graphs this approximation is equivalent to the radiation of the photon from the outer charged hadron lines (Fig. 3). In the energy momentum balance and in the pure hadronic part of the amplitude the photon momentum is neglected.

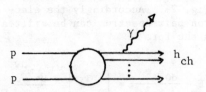

A crucial question is certainly the validity region of this approximation. In case of the radiative processes $pp \rightarrow pp\gamma$ and $\pi^\pm p \rightarrow \pi^\pm p\gamma$ at low and inter-mediate energies, one finds that the "external-emission dominance model"[12,13] which is based on essentially the same approxima-tion, agrees with the data up to

Fig. 3. External emission amplitude.

relatively large photon energies (Fig. 4). We do not see any obvious reason why this infrared approach should fail in multiparticle pro-duction processes.* Therefore, we expect the soft photon approxima-tion described above to allow a reliable calculation of the direct electron rates for $q^2 \lesssim m_\pi^2$ and $q_o \lesssim$ few hundred MeV.

*As far as the kinematics is concerned, the photon certainly plays a much less important role in such a highly inelastic process.

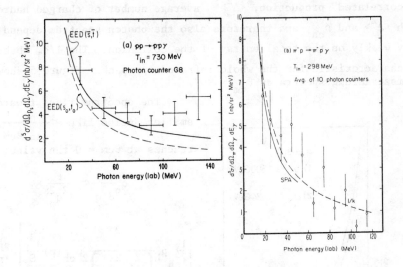

Fig. 4. Comparison of the "external emission dominance model" (EED) with the data.[12,13]

SOFT BREMSSTRAHLUNG SPECTRUM

Due to the factorization between the photon emission and the pure hadronic process, the photon spectrum can be expressed in terms of simple inclusive hadron spectra[14] (the total inelastic cross section σ^{pp}_{inel}, the density distribution of a particular charged hadron and the 2-particle correlation functions). Using this sum rule one can show[15] that the bremsstrahlung at large angles in the center of mass system is controlled by the charge correlations among the final hadrons and by the charged hadron density:

$$q_o \frac{d\sigma^\gamma}{d^3q} \simeq \frac{\alpha}{\pi^2 q_T^2} (<\Delta^2(y_\gamma)> + 0.35 \; \rho_{ch}(y_\gamma)) \; \sigma^{pp}_{inel} \; .$$

$y_\gamma = \ell n \cot \frac{\theta_{cm}}{2}$ is the photon rapidity, θ_{cm} the production angle of the photon and ρ_{ch} is the density distribution of the charged hadrons. $<\Delta^2(y_\gamma)>$ is the mean square of the charge (Q) transfer across the photon rapidity:

$$<\Delta^2(y_\gamma)> = <[Q_{final}(y_{hadr} < y_\gamma) - Q_{initial}(y_{hadr} < y_\gamma)]^2> \; .$$

(For example, $<\Delta^2> = 0$ for local charge conservation; in the case of "uncorrelated" production, $\langle\Delta^2\rangle$ ~ average number of charged hadrons.) Both $\langle\Delta^2\rangle$ and ρ_{ch}, and therefore also the photon spectrum depend only very weakly on the total energy of the collision. The $1/q_T^2$-behavior is characteristic for the whole infrared approach. Figure 5 shows bremsstrahlung spectra at θ_{cm} = 90°.

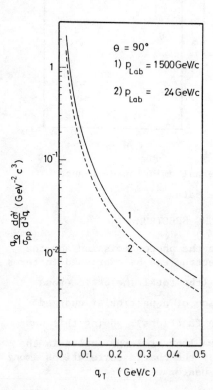

Fig. 5. Soft bremsstrahlung spectrum in pp-collisions at high energies.

The spectrum rises towards smaller angles like ~ $\frac{1}{\sin^2\theta cm}$ and reaches at θcm = 0 the value:

$$q_o\frac{d\sigma^\gamma}{d^3q} \approx \frac{\alpha}{\pi^2 q_o^2} \times$$

$$\left[\frac{1}{2}\sum_{ch} P_L^2 \frac{p_T^2}{\left(p_T^2+M_{hadr}^2\right)^2}\rho_{ch}(p)\frac{d^3p}{E}\right]\sigma_{inel}^{pp} .$$

Since at high energies most of the final hadrons are produced with small angles to the beam axis this angular dependence simply reflects the peaking of the soft bremsstrahlung near the direction of the radiating particles. We obtain in this way a soft electron pair spectrum without any free parameter.

DIRECT ELECTRON SPECTRA

Finally, to find the single electron spectrum we have to integrate the pair spectrum over one electron momentum or the photon momentum q:

$$E\frac{d\sigma^e}{d^3p} = \frac{\alpha}{2\pi^2}\int\frac{dq^2}{q^2}\int\left(q_o\frac{d\sigma^\gamma}{d^3q}\right)\delta(q^2-2pq)\,\theta(q_o-E)\frac{d^3q}{q_o} .$$

The resulting electron-pion ratio at $\sqrt{s}$ = 53 GeV and θ_{cm} = 30° is shown in Figure 6 together with the observed ratio[4] and the expected contribution from known vector mesons. We see that our picture leads to a direct electron yield of the right shape and order or magnitude.

As mentioned above, the bremsstrahlung spectrum decreases only very slightly with decreasing total energy. On the other hand, the

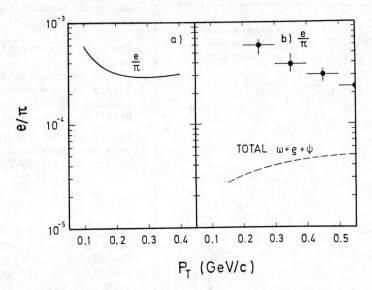

Fig. 6. Electron-pion ratio at $\sqrt{s}$ = 52.7 GeV and θ_{cm} = 30°.

    a)   contribution from soft virtual bremsstrahlung

    b)   observed rates[4] and contribution from vector mesons.

positron-pion ratio at LAMPF energy ($\sqrt{s} \approx$ 2.2 GeV) has been reported[16] to be of the order of 10^{-6}. This rate, however, does not include the contribution from the conversion of bremsstrahlung. All positrons originating from any photon source with internal conversion coefficient similar to that of the π^{o} are subtracted out. We can now calculate what we expect from soft virtual bremsstrahlung (Fig. 7). The electron rate depends somewhat on the cutoff in the photon mass, but is in any case on the 10% level compared with the observed raw rate. Therefore, our picture is not inconsistent with this experiment.

SUMMARY

Our conclusions are:
1. Soft virtual bremsstrahlung seems to be an important source of low energy direct electrons.
2. Our infrared approach based on "external emission dominance" is consistent with the existing data at very high and intermediate energies.

Further tests of the model could be provided by
1. a measurement of the low mass-low energy electron pair spectrum,
2. a measurement of soft direct muons (due to the relatively large muon mass, the soft muon yield should be more than an order of

532

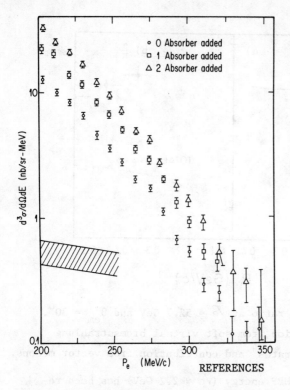

magnitude smaller than the electron yield).

The soft bremsstrahlung contribution to the real photon spectrum in inelastic hadron-hadron collisions can only compete with the huge contribution from the π^o-decay at very low transverse momenta ($q_T \lesssim 10$ MeV/c).

Fig. 7. Raw electron data at $\sqrt{s} \approx 2.2$ GeV. The shaded area is the contribution expected from soft virtual bremsstrahlung (experimental cuts not taken into account).

REFERENCES

1. L. M. Lederman, Proc. Int. Symp. on Lepton and Photon Interactions, Stanford 1975, p. 265.
2. J. D. Sullivan, AIP Conference Proceedings No. 30, 1976, p. 142.
3. P. A. Piroué, Review Talk at the 1976 Particles and Fields Conference at BNL, Princeton University preprint.
4. L. Baum et al., Phys. Lett. 60B (1976) 485.
5. F. W. Büsser et al., Nucl. Phys. B113 (1976) 189.
6. M. Bourquin and J. M. Gaillard, Phys. Lett. 59B (1975) 191.
7. S. D. Drell and T. M. Yan, Phys. Rev. Lett. 25 (1970) 316.
8. G. R. Farrar and S. C. Frautschi, Phys. Rev. Lett. 36 (1976) 1017.
9. N. C. Craigie and D. Schildknecht, Nucl. Phys. B118 (1977) 311.
10. H. Hinchliffe and C. H. Llewellyn Smith, Phys. Lett. 61B (1976) 472.
11. R. Rückl, Phys. Lett. 64B (1976) 39.
12. D. I. Sober et al., Phys. Rev. D11 (1975) 1017.
13. B. M. K. Nefkens et al., Phys. Rev. Lett. 38 (1977) 876.
14. P. Rotelli, Phys. Rev. D7 (1973) 230.
15. R. Rückl, Ph.D. Thesis, University of Munich, 1976.
16. A. Browman et al., Phys. Rev. Lett. 37 (1976) 246.
17. C. M. Hoffman, private communication.

NN̄ RESONANCES AND p̄D DYNAMICS

G. Alberi
Istituto di Fisica Teorica della Universita degli Studi
di Trieste, Trieste, Italy

L.P. Rosa
International Centre for Theoretical Physics, Trieste, Italy
and
Universidade Federal do Rio de Janeiro, Rio de Janeiro, Brazil

Z.D. Thomé
Universidade Federal do Rio de Janeiro, Rio de Janeiro, Brazil

There is definite experimental evidence for a narrow resonance of mass 1932 MeV, in p̄p total cross section measurement.[1,2] This resonance is also seen in p̄n annihilation,[3] where the target neutron is bound in the deuteron. This second evidence is extremely important for the determination of isospin, being p̄n a pure isospin 1 system. However, the price we have to pay to observe a pure isospin state is to introduce the complications of the deuteron dynamics.

Using the formalism, previously derived,[4] we have calculated the rescattering effect for the distribution in the mass M_X of the annihilation product. The expression is:

$$\frac{d\sigma}{dM_X} \sim M_X \frac{\lambda^{1/2}(M_X^2,m^2,m^2)}{\lambda(s,M^2,m^2)} \int_{p_s^-}^{p_s^+} |T|^2 \, p_s \, dp_s$$

(M, m are the deuteron and nucleon mass)

where

$$|T|^2 = \sigma_2(M_X^2) \left\{ \sum_{ij} A_i A_j \exp\left[-(\alpha_i+\alpha_j)p_s^2\right] C_{ij}(p_s^\perp) + \psi_2^2(p_s) \right\}$$

where

$$C_{ij}(p_s^\perp) = 1 - \frac{\sigma}{2\pi}\frac{1}{\gamma_i}\exp(y_i p_s^{\perp 2}) + \frac{\sigma^2}{32\pi^2}\frac{\varepsilon_i\varepsilon_j \exp(z_{ij}p_s^{\perp 2})}{(b+\alpha_i)\varepsilon_i + (b+\alpha_j)\varepsilon_j}$$

(see Ref. 4 for more details)

The expression of C_{ij}, valid for a multi-Gaussian representation of the deuteron wave function, shows that since $\sigma(\bar{p}p)$ is very large >100 mb, the effect can be quite large. As seen from Fig. 1, where we show our preliminary results, the effect is overall suppression of the mass distribution: the input in the calculation is the behaviour of $\sigma_2(M_X^2)$ which we assume to be $1/q$, where q is the p̄n c.m. momentum, as it would be for a purely absorptive case in S-wave. The comparison with unpublished data[5] of the Roma-Trieste collaboration for the pure spectator model (Fig. 1) and the corrected version with rescattering (Fig. 2) shows that the data are compatible with a non-resonant behaviour, as recently suggested.[6]

ISSN: 0094-243X/78/533/$1.50 Copyright 1978 American Institute of Physics

534

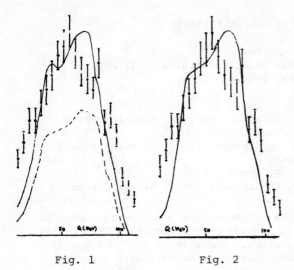

Fig. 1   Fig. 2

1. A.S. Carroll et al.,
   Phys. Rev. Lett. 32,
   247 (1974).
2. A.S. Carroll et al.,
   Phys. Lett. 61B, 303
   (1976).
3. T.E. Kalogeropoulos
   and G.S. Tzanakos,
   Phys. Rev. Lett. 34,
   1047 (1975).
4. G. Alberi et al., Phys.
   Rev. Lett. 34, 503
   (1975).
5. E. Castelli, private
   communication.
6. F. Myhrer and A. Gersten,
   CERN preprint TH 2170.

Hyperspherical method applied to the quark model of nucleons[*]

K. K. Fang
Saskatchewan Accelerator Laboratory,
University of Saskatchewan, Saskatoon, Canada   S7N OWO

The technique, hyperspherical harmonics (h.h.) method[1,2,3], used a great deal in the calculation of nuclear three-body problems is introduced to the nonrelativistic quark model of hadrons[4]. In particular, we will use h.h. to find the ground state wave functions of Nucleon systems using the harmonic model potential and following the symmetry, O(3) x SU(6) of the systems. The charge form factor is discussed and numerical results will be given for both Fermi and parastatistics. Furthermore, in parallel approach of Brayshaw's[5] work, by keeping the lowest term (the grand orbital zero) in the expansion of the ground state wave function, we find an analytical form of the proton wave function using the well-known dipole fit of the charge form factor. Hyperpotential of the system is then found and quark-quark two-body interaction is derived from the hyperspherical method.

1. M. Fabre de la Ripelle and J. S. Levinger, Nuov. Cim. 25A (1975) 555.
2. M. Vallieres et al., Nucl. Phys. A271 (1976) 95.
3. K. Myers et al., Phys. Rev. C, to be published.
   K. K. Fang, Phys. Rev. C, to be published.
4. R. Horgan and R. H. Dalitz, Nucl. Phys. B66 (1973) 135.
5. D. D. Brayshaw, in Los Angeles Conference, "Few Particle Problems"
   ed. I. Slaus et al., Amer. Elsevier, 1972.

* Work supported by the National Research Council of Canada.

THE n-n SCATTERING LENGTH FROM THE γ SPECTRUM OF THE REACTION π⁻ + d → 2n + γ

J.C. Alder, W. Dahme, B. Gabioud, W. Hering, C. Joseph, J.F. Loude,
N. Morel, H. Panke, A. Perrenoud, J.P. Perroud, D. Renker,
G. Strassner, M.T. Tran, P. Truöl, E. Winkelmann and C. Zupancic
Lausanne-München-Zürich Universities collaboration

Experimental set-up and measurement. With the SIN γ pair spectrometer[1] and a target which could be filled with liquid H_2 or D_2, we have measured ∿120,000 events from the 129.4 MeV γ line ($π^- + p → γ + n$) and in between ∿430,000 deuterium events (Fig. 1).

Method of analysis. The measured hydrogen line (0.9 FWHM resolution) is used to fold the theoretical deuterium spectrum on the spectrometer energy scale. This folding involves only the theoretical energy difference between deuterium and hydrogen photons. It takes fully account for the response of the spectrometer and fixes for the theoretical spectrum exactly the same energy scale as for the measured one if there is no energy shift during the measurement. Comparison of both spectra (Fig. 1) is performed by $χ^2$ test with two possible parameters for the theoretical spectrum: the a_{nn} value and an energy shift. The sensitivity of the method and the errors are determined from the variation of $χ^2$ in function of the parameters. A 10 keV energy shift induces a change of 0.7 fm. Monitoring of the spectrometer magnetic field and of the MWPC gas pressure and temperature which affect the pair energy loss have set a dispersion on a_{nn} of ±0.2 fm. Subtraction of the contribution from the gas hydrogen impurity leads to an uncertainty of at most ±0.1 fm.

Results. The results of a preliminary analysis, using Bender's theory[2] and r_{nn} = 2.8 fm, are the following

$E_γ$ (MeV)	q_r^* (MeV/c)	events	a_{nn} (fm)	$σ_{stat}$ (fm)	$σ_{a_{nn}}$ (fm)
130.5 - 132	< 31	50,968	-17.45	0.41	0.46
130.0 - 132	< 38	74,603	-17.72	0.34	0.39
129.5 - 132	< 44	96,179	-17.52	0.30	0.37
129.0 - 132	< 50	116,301	-17.14	0.27	0.35
128.5 - 132	< 54	135,731	-16.54	0.25	0.34

*) q_r: neutrons relative momentum

In the energy range 129.5 - 132 MeV, the variation of $χ^2$ in function of an energy shift gives -0.5 ± 3.5 keV, confirming the expected stability of the spectrometer. One should notice that the theoretical spectrum differs strongly from the measured one below 129.5 MeV. This leads to a lower extracted a_{nn} value. This preliminary analysis suggests a value for a_{nn} of -17.5 ± 0.4 fm close to a_{pp} = 17.1 ± 0.3 fm. Bander's theory[2] used here allows an uncertainty on the a_{nn} extraction lower than 1 fm for q_r < 50 MeV/c. Analysis using the

ISSN: 0094-243X/78/535/$1.50 Copyright 1978 American Institute of Physics

spectrum from Gibbs et al.[3] (0.3 fm for q_r < 35 MeV/c) will also be made.

1. J.C. Alder et al., contributed paper p.204, International Topical Conference on Meson-Nuclear Physics, Carnegie-Mellon University, 1976.
2. M. Bander, Phys. Rev. 134B, 1052 (1964).
3. W.R. Gibbs et al., Phys. Rev. C11, 90 (1975).

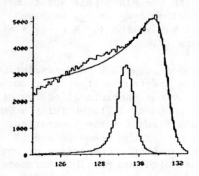

Fig. 1: Experimental H and D spectra with Bander's folded spectrum (-17.5 fm)

A STATISTICAL MODEL OF HADRONS

A. Chatterjee
Physics Department, Calcutta University

S. Ray
Physics Department, Kalyani University

S.K. Ghosh
Saha Institute of Nuclear Physics, Calcutta

The statistical model of hadron rest-mass level density is discussed. It is shown that the experimentally observed spectrum is roughly reproduced in magnitude and trend using a simple model of non-interacting pions as building blocks with all statistical parameters predetermined. The simple model thus approximately fulfills the desired saturation effect in the observed spectrum. A method of evaluating the level density of the building blocks (constituent hadrons) is discussed. Possible improvements of the simple model, introducing interactions amongst the constituent blocks, are suggested and partially analyzed. Results have been compared with the bootstrap model spectrum in the energy range between 100 and 2000 MeV.

ISSN: 0094-243X/78/536/$1.50

MEASUREMENT OF THE TOTAL CROSS-SECTION DIFFERENCE
IN pp SCATTERING IN LONGITUDINAL SPIN STATES AT
1.2 to 6.0 GeV/c*

I.P. Auer, A. Beretvas, E. Colton, D. Hill, K. Nield, B. Sandler,
H. Spinka, D. Underwood, Y. Watanabe, and A. Yokosawa
Argonne National Laboratory, Argonne, IL 60439

We have measured the total cross-section difference for pp scattering in initial spin states parallel to the beam direction at beam momenta of 1.2, 1.5, 1.7, 2.0, 2.5, 3.0 and 6.0 GeV/c. This measurement was done in a standard transmission experiment. A striking energy dependence is observed with a maximum difference of -16.9 mb at P_{lab} = 1.47 GeV/c.

ISSN: 0094-243X/78/537/$1.50 Copyright 1978 American Institute of Physics

MEASUREMENTS OF THE SPIN-SPIN CORRELATION PARAMETERS C_{SS},
C_{SL} AND C_{LL} IN pp ELASTIC SCATTERING AND SCATTERING
AMPLITUDES AT 6 GeV/c*

I.P. Auer, A. Beretvas, E. Colton, D. Hill, K. Nield, B. Sandler,
H. Spinka, D. Underwood, Y. Watanabe, and A. Yokosawa
Argonne National Laboratory, Argonne, IL 60439

The elastic spin-spin correlation parameters for pp scattering C_{SS}, C_{SL} and C_{LL} have been measured at 6 GeV/c. Polarized protons from the ZGS with spins in the $\vec{N}$ direction were rotated by a solenoid into the $\vec{S}$ direction. A bending magnet further precessed the spins from $\vec{S}$ to $\vec{L}$. Target spins in the $\vec{S}$ and $\vec{L}$ directions were provided by a new polarized target magnet. The C_{SS} measurement determines the imaginary part of the U_2 amplitude (corresponding to π exchange). The data points are all negative and the absolute values increase with $|t|$. The C_{LL} measurement determines the imaginary part of the U_0 amplitude (corresponding to A_1 exchange). The results are compared with some existing attempts to describe the pp scattering process. The experimental program to determine the scattering amplitudes is discussed.

---

*Work supported by the U.S. Energy Research and Development Administration

ISSN: 0094-243X/78/537/$1.50 Copyright 1978 American Institute of Physics

np CHARGE EXCHANGE SCATTERING FOR NEUTRON ENERGIES 300-800 MeV*

B.E. Bonner and J.E. Simmons
Los Alamos Scientific Laboratory, University of California
Los Alamos, New Mexico 87545

G. Glass and M. Jain
Texas A & M University, College Station, Texas 77843

C.L. Hollas, C.R. Newsom, and P.J. Riley
University of Texas, Austin, Texas 78712

In the 1960's it was noted[1] that the shape of the np angular distribution for backward neutron c.m. angles could be fit very well by a sum of two exponentials in the square of the invariant momentum transferred between the incoming neutron and recoil proton: $d\sigma/dt = \alpha_1 e^{\beta_1 t} + \alpha_2 e^{\beta_2 t}$. It was somewhat surprising when the data from the PPA[2] in 1969 showed a very striking peak around 800 MeV/c in the logarithmic slope of this cross section as well as a sharp decrease in the cross section at t = 0 beyond about 1 GeV/c. That this behavior was already evident in the phase shifts of MacGregor et al.[3] was claimed by Londergan and Thaler[4] in 1970. More recent data from Saclay[5] indicated that the t = 0 cross section did not decrease so sharply, but their measurement of the logarithmic slope was consistent with a peak as reported by the PPA.

At LAMPF we have measured the np charge exchange cross section from 300-800 MeV using a continuum neutron beam in conjunction with a liquid hydrogen radiator and multiwire proportional chamber spectrometer. Time-of-flight measurement on the incident neutron in conjunction with momentum determination in the spectrometer sufficed to separate elastically scattered protons from other processes. Good resolution on the proton momentum determination ($\sim$1%) then allowed the events to be binned according to the incident neutron momentum calculated from two-body kinematics. Absolute normalization in the range 300-800 MeV was obtained from the deuterons detected simultaneously from the reaction np $\rightarrow$ dπ^0. Data have been analyzed for angles 0°-16° and incident energies 300-800 MeV at the present time.

Our results indicate that over the entire range the logarithmic slope $\beta = (\alpha_1\beta_1 + \alpha_2\beta_2)/(\alpha_1 + \alpha_2)$ is constant and equal to $\sim$80 (GeV/c)$^{-2}$. In addition the quantity P_L^2 $d\sigma/dt(t=0)$ is constant ($\sim$160 mb), the ratio α_2/α_1 falls from about 1.4 at 300 MeV to 1 at 800 MeV, β_1 is constant ($\sim$160 (GeV/c)$^{-2}$), and β_2 falls smoothly from 8 to 6 over the range 300 to 800 MeV. The present results are in good agreement with our previously reported results[6] at 650 MeV where a single energy measurement was made. These results will be compared to those mentioned previously as well as calculations using one boson exchange potentials.[7]

1. J.L. Friedes et al., Phys. Rev. Letters 15, 38 (1965); R. Wilson, Ann. Phys. 32, 193 (1965).
2. R.E. Mischke et al., Phys. Rev. Letters 23, 542 (1969).
3. M.H. MacGregor et al., Phys. Rev. 182, 1714 (1969) and earlier references therein.
4. J.T. Londergan and R.M. Thaler, Phys. Rev. Letters 25, 1065 (1970).
5. G. Bizard et al., Nucl. Phys. B85, 14 (1975).
6. M.L. Evans et al., Phys. Rev. Letters 36, 497 (1976).
7. B.S. Bhakar, private communication.

*Work supported by U.S. ERDA.

MEASUREMENT OF NEUTRON-PROTON ANALYSING POWER AT 14 MeV

J.E. Brock, A. Chisholm, J.C. Duder and R. Garrett
Physics Department, University of Auckland, Auckland, New Zealand

Polarized neutrons of energy 14.1 MeV are generated by the $T(\vec{d},\vec{n})^4He$ reaction, using vector polarized deuterons of energy about 150 keV produced by an atomic beam polarized ion source. To estimate the neutron polarization, P_n, the source is operated periodically in modes to produce tensor polarized deuterons and the quantity P_{zz} measured; we take $P_n = \frac{2}{3}P_{zz}$. Also, P_n is monitored continuously throughout the n-p measurements by observing the asymmetry in the scattering from carbon. P_n is reversed at regular intervals by changing the R-F transitions induced in the neutral atomic beam section of the ion source.

A cone of neutrons is defined by detecting the alpha particles from the $T(d,n)^4He$ reaction. The target protons are those in NE213 scintillator. Neutrons scattered to left and right are detected in four pairs of plastic scintillators. For each event satisfying a triple-coincidence requirement, six parameters are recorded on magnetic tape, and the multiparameter spectra are later treated off-line. A Monte Carlo simulation has demonstrated that the only significant background can be attributed to multiple scattering in the proton target, and that this does not give a significant spurious asymmetry.

Results are given in the table, and on the graph are compared with the Yale-IV and Livermore-X phase-shift predictions. The values given have been corrected for small effects due to finite geometry. The errors shown are statistical; the overall scale error due to uncertainties in P_n is probably less than 5%.

Further data taking is proceeding.

θ_{c-m}	$A(\theta_{c-m})$
61.8	0.0221 ± 0.0032
68.6	0.0233 ± 0.0032
80.8	0.0206 ± 0.0037
102.6	0.0154 ± 0.0029
121.4	0.0094 ± 0.0026
141.0	0.0029 ± 0.0044

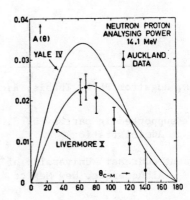

ISSN: 0094-243X/78/539/$1.50 Copyright 1978 American Institute of Physics

THE ROLE OF THE NN-NΔ AMPLITUDE IN PION PRODUCTION FROM DEUTERIUM*

Ian Duck, K. R. Hogstrom,[†]and G. S. Mutchler
T. W. Bonner Nuclear Laboratories
Rice University, Houston, Texas, U.S.A.

A Glauber theory of pion production including single and double Glauber scattering has been used to analyze the experimental results for pd→dπ⁺n at 585 MeV.[1] The kinematical conditions of the experiment (Θ_d=35°, Θ_π=35°, 50°, and 90°) were selected such that the neutron recoil momentum was 400 MeV/c or greater. In this region of phase space the reaction is dominated by Glauber double scattering amplitudes which involve the excitation of the Δ(1236) resonance.

The three-body cross sections exhibit a definite peak in the deuteron momentum spectrum that decreases in magnitude and shifts to greater momentum with increasing pion angle. The general magnitude of the cross section is reproduced by Glauber double scattering amplitudes with Δ production at the first or second hard scattering (NN-NΔ vertex). These amplitudes interfere coherently and require knowledge of the NN-NN as well as NN-NΔ amplitudes for a rigorous treatment of the problem. The peak in the deuteron spectrum is due to the Glauber single and double scattering amplitudes with the pp→dπ⁺ reaction at the last vertex. The two contributions are combined incoherently to provide reasonable good global fits to the data. The pion production experiments on deuterium provide an instance where the full NN scattering matrix, including the inelastic channels, must be included in a Glauber description of the scattering amplitude.

---

[1] K. R. Hogstrom Ph.D. Thesis, Rice University (1976) unpublished.

* Work supported in part by U. S. Energy Research & Development Administration

[†] Present address: University of New Mexico School of Medicine, Albuquerque, New Mexico

PARITY VIOLATIONS IN INELASTIC PROTON-PROTON SCATTERING

G.N. Epstein
Department of Physics, University of Pittsburgh,
Pittsburgh, PA 15260

One of the most promising ways to detect parity violations in the pp system is to measure the asymmetry

$$A_L = \frac{\sigma_+ - \sigma_-}{\sigma_+ + \sigma_-},$$

where $\sigma_+(\sigma_-)$ is the total cross section for beam protons with +ve(-ve) helicity. The Los Alamos experiment[1] at 15 MeV lab energy yields $A_L = (-1.7 \pm 3.0) \times 10^{-7}$. We draw attention here to the possibility that $\Delta\sigma_{p.v.} \equiv \sigma_+ - \sigma_-$ may be large and rapidly varying in the $\Delta(1232)$ production region. The figure shows the results of a simple model calculation of the contribution of pp $\rightarrow$ NΔ, to $\Delta\sigma_{p.v.}$ Our model is an extension of the Δ production model of Riska and Epstein,[2] to include parity violations. It takes π exchange to be the p.v. mechanism. The ΔNπ p.v. coupling is neglected and the NNπ p.v. coupling is fixed at the Weinberg-Salam model value: $f_\pi = 0.95 \times 10^{-6}$. Distortions are included for the pp state by using simple unitarity arguments. The NΔ final state interaction is neglected. We are extending the model to include p.v. ρ exchange. Certainly it appears promising to do p.v. experiments in the $\Delta(1232)$ production region.

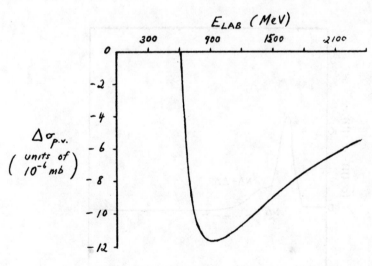

1. D.E. Nagle, in H. E. Physics and Nuclear Structure, Conference 1975 Santa Fe.
2. G.N. Epstein and D.O. Riska, contribution to this Conference.

ISSN: 0094-243X/78/541/$1.50

542

Δ(1232) PRODUCTION AND THE SPIN DEPENDENCE OF THE pp TOTAL CROSS SECTION

G.N. Epstein
Department of Physics, University of Pittsburgh, Pittsburgh, PA 15260

D.O. Riska
Department of Physics, Michigan State University,
East Lansing, MI 48824

The recent discovery of the strong energy dependence of the difference, Δσ, between the total pp cross sections for transversely polarized protons with spins anti-parallel and parallel to the target protons at lab. momentum > 1 GeV/c is of great interest.[1] Thomas and Kane have shown that the effect is consistent with existing phase shift solutions and reflects inelasticity in the pp $^1D_2$ wave.[2] This suggests that single Δ production is probably the underlying mechanism. We calculate both pp → NΔ and pp → ΔΔ contributions to Δσ in a simple relativistic model which includes both π and ρ exchange and extends in a natural way a very successful recent model of $\pi^+d \to pp$.[3] We use hadronic vertex form factors and coupling constants consistent with analyses of independent processes such as nucleon-nucleon elastic scattering. Our results are shown in the figure and are in good agreement with the data. We find that π-exchange pp → NΔ is the dominant mechanism. We also find that a non-relativistic version of our model is remarkably accurate.

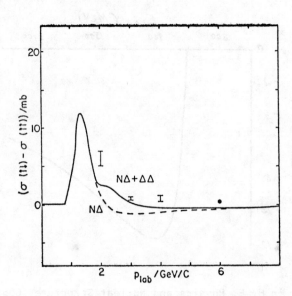

1. W. de Boer et al., Phys. Rev. Lett. 34, 558 (1975).
2. G.H. Thomas and G.L. Kane, Phys. Rev. D13, 2944 (1976).
3. D.O. Riska et al., Phys. Lett. 61B, 41 (1976).

ISSN: 0094-243X/78/542/$1.50 Copyright 1978 American Institute of Physics

NEUTRON EXPERIMENTS AT SIN*

Th. Fischer, W. Hagedorn, G. Hammel, W. Hürster
K. Kern, P.R. Kettle, M. Kleinschmidt, L. Lehmann, E. Rössle,
H. Schmitt, and D.M. Sheppard†
Fakultät für Physik, Universität Freiburg, Germany

A neutron beam facility together with a flight path of 60 m has been installed at SIN. It permits a wide range of experiments with unpolarized neutrons within an energy range of $150 \lesssim E_n < 600$ MeV covered simultaneously with an energy resolution of better than 1.5%.

The experimental facility includes a magnet spectrometer of large momentum and angular acceptance which is equipped with a set of drift chambers. Particle identification is achieved by the measurement of time-of-flight and dE/dx. Total cross sections on protons and deuterons have been measured with the aim of extracting the neutron-neutron cross section in order to provide a complete test of isospin invariance. The magnet spectrometer has been used in these measurements as a neutron detector via the charge exchange np-scattering. In addition first measurements of momentum spectra of charged particles (π^+, π^-, p, d) emitted in forward direction have been performed. Of particular concern are here: the cross section of the neutron-proton charge exchange scattering and its angular dependence at small momentum transfer, proton spectra from inelastic scattering, production of positive and negative pions on protons and deuterons, deuteron spectra arising from backward elastic scattering and from the pion producing reaction $n + d \rightarrow d + N + \pi$.

*Work supported by Bundesministerium für Forschung und Technologie
†University of Alberta, Alta., Canada

THE INCLUSIVE REACTION, np → pX at 800 MeV*

G. Glass and M. Jain†
Texas A&M University, College Station, Texas 77843

B.E. Bonner
Los Alamos Scientific Laboratory, University of California,
Los Alamos, New Mexico 87545

C.L. Hollas, C.R. Newsom and P.J. Riley
University of Texas, Austin, Texas 78712

The data presented here are the preliminary results of recent experiments performed to study various final states associated with 800 MeV (1457 MeV/c) neutron bombardment on a proton target. Results are presented for the reaction np → pX, where X is all possible combinations of pions and a nucleon at this incident energy. The proton is the detected particle at angles between 0° and 19°. The reactions for protons at 0° serve to complement measurements reported in an earlier paper[1] on pp → npπ+ and enhance the coverage of the Saclay measurements[2] at incident momenta of 1390 and 1560 MeV/c.

One pion exchange is thought to be the principal mechanism for these reactions. The possibility of a nucleon-nucleon final state interaction (FSI) also exists, and introduces many unknowns, since the two nucleons can be off-shell and the presence of a pion creates further complications. However, a calculation (SGG of Ref. 1) does seem to fit the pp data[1] quite well and will be generalized and tested for the data presented here.

In Fig. 1 we show six proton spectra for central angles from 0° to 17.19°. The principal feature of these data is clearly the peak which moves from near 1020 MeV/c down to about 900 MeV/c as the angle increases to 17°. The peak is undoubtedly a manifestation of the 1232 MeV/c^2 πN resonance. More subtle features are indicated in the 0° spectrum by the shoulder centered near 800 MeV/c and the slight enhancement near 1200 MeV/c. Both of these features persist for all angles, moving to lower momenta as the angle increases. The shoulder on the low momentum side is probably due to the zero energy nucleon-nucleon FSI. It has been pointed out[3] that, since the enhancement on the high momentum side corresponds to a nucleon-nucleon invariant mass between 2000 and 2100 MeV/c^2, it possibly could be the result of a di-nucleon $^3F_4$ resonance near 2085 MeV/c^2 recently predicted by Lomon using the Feshbach-Lomon potential. However, his model is not refined and the predicted mass is still uncertain.

In Fig. 2 we have added our measurement to the averaged cross sections as defined in Fig. 7 of Ref. 1. The present result, although consistent with the Saclay results, shows that more nonresonance production is involved in pion production at 800 MeV than was estimated in Ref. 1.

*Work supported by the U.S. Energy Research and Development Administration.
†Currently at Los Alamos Scientific Laboratory, Los Alamos, New Mexico.

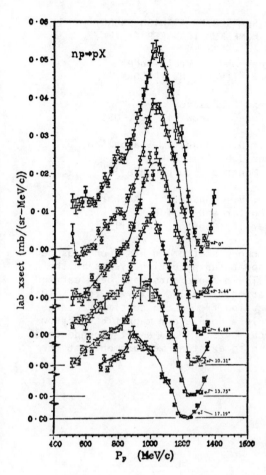

Fig. 1 (Left). Proton spectra from np→pX at central angles (from top to bottom) 0°, 3.4°, 6.9°, 10.3°, 13.7° and 17.2°, respectively. Lines are drawn through points for clarity only. Vertical scales are same for all angles but displaced as indicated.

Fig. 2 (Below). Cross sections averaged over the inelastic peak at 0° for various incident energies.

REFERENCES

1. G. Glass et al., Phys. Rev. D15, 36 (1977).
2. G. Bizard et al., Nucl. Phys. B108, 189 (1976).
3. E. Lomon, private communication.

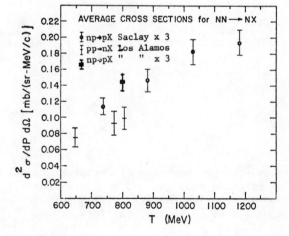

ELASTIC PROTON-PROTON SCATTERING AT HIGH ENERGY AND VERY LARGE MOMENTUM TRANSFER

J. Hartmann, J. Orear and J. Vreislander
Cornell University, Ithaca, NY 14850

P. Baranov and S. Rusakov
Lebedev Physical Institute, Moscow 117 924, USSR

S. Connetti, C. Hojvat,† D.G. Ryan, K. Shahbazian
D.G. Stairs and J.M. Trischuk
McGill University,* Montréal, Québec H3A 1A4

W. Faissler, M. Gettner, J. Johnson,‡ T. Kephart,
E. Pothier, D. Potter and M. Tautz
Northeastern University, Boston, MA 02115

The cross section for elastic p-p scattering has been measured at 200, 300 and 400 GeV with an apparatus accepting events in the region $4 \leq |t| \leq 17.5$ (GeV/c)^2. The measurements have been carried out using an extracted beam operating at an intensity up to 8×10^{11} protons per pulse. Particles emerging from a ten centimetre target of liquid hydrogen were analyzed in a double-arm magnetic spectrometer equipped with proportional chambers. The s and t dependence of the data are discussed in relation to existing measurements at the ISR and Fermilab.

*Supported by the National Research Council of Canada, The Institute of Particle Physics, and Le Ministère de l'Éducation du Québec.
†Now at Fermilab, Batavia, Illinois.
‡Now at SLAC, Stanford, California.

THE HIGH RESOLUTION SPECTROMETER FACILITY AT LAMPF*

G.W. Hoffmann, D. Madland, C.L. Morris, J. Pratt, J.E. Spencer,
N. Tanaka, and H.A. Thiessen
Los Alamos Scientific Laboratory, Los Alamos, NM 87545

B. Zeidman
Argonne National Laboratory, Argonne, IL 60439

T. Kozlowski, H. Palevsky, and R. Sutter
Brookhaven National Laboratory, Upton, NY 11973

T.H. Bauer, J. Fong, G.J. Igo, R. Ridge, R. Rolfe, C.A. Whitten, Jr.
University of California, Los Angeles, CA 90024

N. Hintz, G. Kyle, and M. Oothoudt
University of Minnesota, Minneapolis, MN 55455

P. Lang and K.K. Seth
Northwestern University, Evanston, IL 60201

D. McDaniels and P. Varghese
University of Oregon, Eugene, OR 97403

G. Blanpied and R. Liljestrand
University of Texas, Austin, TX 78712

The HRS facility has been operational since late 1976. The system consists of a beam line (Line C) optically matched to a QDD spectrometer[1] in such a way that, for a given-type nuclear reaction or scattering process, the outgoing particles signifying the population of particular final nuclear states appear on the focal plane of the spectrometer in the dispersed direction at discrete positions independent of the incident particle momenta or the scattered particle momenta. Focal plane position in the non-dispersed direction is directly proportional to the scattering angle. Studies of the operating characteristics of the system have been interwoven with elastic and inelastic scattering experiments on a variety of nuclei. Experimental studies of Line C demonstrated that the beam line had a resolving power greater than 15000 (80 keV, FWHM, for 800 MeV protons). Experience has shown that the entire beam line optics can be tuned to this resolving power in a straightforward manner by starting with the theoretical solution and slightly adjusting certain magnets in conjunction with visual observation of phosphors located along the beam line. Simple linear combinations of various Line C quadrupoles have been found which only affect the dispersion, focus, and twist, respectively, at the target to enable experimental matching of the beam line to the HRS for operation in the energy-loss mode. The HRS

---

*Work supported by the United States Energy Research and Development
Administration

focal plane position and angle information is provided by four 60-cm
two-dimensional delay-line chambers, two of which are used only for
calibration purposes. The event trigger and time-of-flight system
presently consists of six scintillation counters. Two of these are
VETO's which define a sharp focal plane (dp/p = ±1.4%) while the rest
provide pulse height information for dE/dx cuts and flight time
information over a distance of 3 m for particle identification. Preliminary angular distributions and extended energy spectra
(E_x < 25 MeV) have been obtained for proton elastic and inelastic
scattering at 800 MeV on a variety of nuclei. Energy resolution as
good as 80 keV FWHM has been observed with full phase space from the
accelerator and the full solid angle of the spectrometer.

1. We are indebted to K. Brown, H. Enge, K. Halbach, and S. Kowalski
    for their design work and consultations provided for the HRS
    project over the past years.

ISOSPIN NONCONSERVATION IN THE N-N INTERACTION

A. Gersten
Division de Physique Théorique,* Institut de Physique
Nucléaire, and Laboratoire de Physique Théorique des
Particules Elémentaires
Université Pierre et Marie Curie, 75230 Paris, France

A relativistic formalism of the N-N scattering including the
electromagnetic contributions is presented. Relativistic electromagnetic phase shifts are defined and evaluated. The difference between
the nuclear phase shifts of the p-p and n-p interaction is evaluated
taking into account the interference between strong and Coulomb
forces, electromagnetic n-p contributions and the mass differences
between the charged and neutral pions.

The n-p interaction is described with the aid of six amplitudes.
Neutron-proton observables are calculated with these six amplitudes.
The mixing angle of the singlet-triplet transitions is evaluated.
Experiments which should show these transitions are discussed.

*Laboratoire associé au C.N.R.S.

INFLUENCE OF THE πNN FORM FACTOR ON TWO-NUCLEON DATA

K. Holinde

Institut für Theoretische Kernphysik, Bonn, West Germany

We start from a q-space OBEP $V(\vec{q}',q) \sim \sum_{\alpha} g_{\alpha}^2 F_{\alpha}^2 (\Delta^2 - m_{\alpha}^2)^{-1}$, $\alpha = \pi, \eta, \sigma, \delta, \rho, \omega, \phi$. The form factor F_{α} is chosen as $F_{\alpha} = [(\Lambda_{\alpha}^2 - m_{\alpha}^2)/(\Lambda_{\alpha}^2 - \Delta^2)]^n$ with $n = 1$ for $\alpha = \pi, \eta, \sigma, \delta$ and $n = 3/2$ for $\alpha = \rho, \omega, \phi$. (We refer to Ref. 1 for explicit details.) A good fit to the NN-data is obtained ($\chi^2 = 2.87$) with a common value of 1530 MeV for Λ_{α}; thus F_{π} has much shorter range compared with direct determinations,[2,3] which predict $F_{\pi} \sim 700$-1000 MeV. Using $\Lambda_{\pi} = 1000$ MeV ($\Lambda_{\alpha \neq \pi} = 1530$ MeV as before) the resulting fit is not satisfactory (especially for ε_1, $^3P_1$). This is traced back to the strong suppression of the OPE-tensor force in the inner region. Using $\Lambda_{\pi} = 1265$ MeV, the results are acceptable. The situation is the same for the deuteron [$Q = 0.281$ (0.263) fm^2, $P_D = 5.18$ (3.63)% for $\Lambda_{\pi} = 1530$ (1000) MeV]. Meson exchange current corrections to Q (~ 0.009 fm^2, see Ref. 4) can only partially remove the discrepancy occurring for $\Lambda_{\pi} = 1000$ MeV.

The results show a strong sensitivity of Q, P_D to details of the πNN form factor. These deuteron properties play a large role in fixing properties of nuclei and nuclear matter. Thus, it is absolutely necessary to determine the πNN form factor as accurately as possible. Furthermore, it appears that the results of Refs. 2,3 for F_{π} (which are derived only for the outer-range part) cannot simply be extrapolated into the inner region, at least in the OBE-frame. Nevertheless, it seems that meson theory favors rather small values for P_D ($\lesssim 5\%$)

REFERENCES

1. K. Holinde and R. Machleidt, Nucl. Phys. A247, 495 (1975).
2. K. Bongardt et al., Phys. Lett. 52B, 271 (1974).
3. W. Nutt and B. Loiseau, Nucl. Phys. B104, 8 (1976).
4. A.D. Jackson et al., Phys. Lett. 55B, 23 (1975).

A MEASUREMENT OF THE DEUTERON SPECTRUM IN np → d(ππ)° AT 800 MeV*

C.L. Hollas, C.R. Newsom, and P.J. Riley
University of Texas, Austin, Texas 78712

B.E. Bonner
Los Alamos Scientific Laboratory, University of California
Los Alamos, New Mexico 87545

G. Glass
Texas A & M University, College Station, Texas 77843

The "ABC" effect in double pion production reactions was first observed[1] as an enhancement in the missing mass spectrum from the reaction pD → ^{3}He(ππ)°. Since then various models have evolved to explain this effect in pD reactions and also in the simpler reaction np → d(ππ)°. The two most extensively developed theories are the two nucleon exchange model of Anjos, Levy, and Santoro[2] and the one-pion exchange model with double delta (ΔΔ) excitation of Bar-Nir, Risser, and Shuster.[3] An interesting extension[4] of the latter model to allow for deep binding of the ΔΔ system appears to account for the recent observation[5] of a peak in the polarization of the proton from deuteron photodisintegration as well as the measurements so far reported in double pion production.

These models yield predictions of the energy dependence and the detailed shape of the enhancements in the missing mass spectra. Data to test these models for np → d(ππ)° presently consist of the bubble chamber total cross section measurements[6] at eight momenta from 1.75 to 3.5 GeV/c and one deuteron momentum spectrum[7] at 4.5° and 1.88 GeV/c.

At LAMPF we have performed an experiment designed to measure the asymmetry in the reaction np → dπ° at 800 MeV. The unexpectedly large cross section we observe for the np → d(ππ)° process means that we also obtained the momentum spectrum of the deuterons from this reaction over the entire angular range. At this time we have analyzed only a small fraction of this data however.

Although detailed comparison of our results with the published calculations has not been made yet, the magnitude of the cross sections appear to be about a factor of four larger than that given in Ref. 3 and in rough agreement with the calculation of Ref. 2.

1. A. Abashian et al., Phys. Rev. Lett. 5, 258 (1960).
2. J.C. Anjos, D. Levy, and A. Santoro, Nuovo Cimonto 33A, 471 (1977).
3. I. Bar-Nir, T. Risser, and M.D. Shuster, Nucl. Phys. B87, 109 (1975).
4. T. Kamae and T. Fujita, Phys. Rev. Lett. 38, 471 (1977).
5. T. Kamae et al., Phys. Rev. Lett. 38, 468 (1977).
6. I. Bar-Nir et al., Nucl. Phys. B54, 17 (1973).
7. G. Bizard et al., Proc. 5th Int. Conf. on High-Energy Physics and Nuclear Structure, Uppsala, Sweden, 1973.

*Work supported by U.S. ERDA.

PRODUCTION OF THE Δ(3/2,3/2) in the REACTION pp→pπ⁺n at 800 Mev*

J. Hudomalj-Gabitzsch, Ian Duck, G. S. Mutchler, J. M. Clement,
R. D. Felder, T. M. Williams, W. H. Dragoset, & G. C. Phillips
T. W. Bonner Nuclear Laboratories
Rice University, Houston, Texas
and
E. V. Hungerford, B. W. Mayes, L. Pinsky, M. Warneke, J.C.Allred
University of Houston, Houston, Texas

Preliminary results of production of the $\Delta(3/2,3/2)$ resonance in proton-proton interaction in the reaction pp→pπ⁺n at 800 Mev are reported. At this incident energy the Δ^{++} formation is the dominant feature of the reaction,[1,2] although reaction mechanisms involving the neutron-proton final-state interaction and the Δ^+ resonance are also possible. A kinematically-complete experiment was performed in order to study the Δ^{++} production in kinematical conditions which provide its optimal separation from the other mechanisms. The 800 Mev external proton beam of LAMPF impinged on a liquid hydrogen target. The emerging pions and protons were detected in coincidence with a system of multiwire proportional counters and scintillators interfaced to a PDP-11/45 computer. A magnetic spectrometer was used to detect one of the charged particles while time-of-flight and pulse-height analysis was applied to both particles to enable their identification. The momentum resolution of the system was 1.5%, the angular resolution 0.25° and the angular opening 5.5° for the magnet and 5° for the other arm. The overall systematic error due to beam normalization, misalignment and detector efficiency was estimated to be ∿7%.

The differential cross section $d^5\sigma/dpd\Omega_1d\Omega_2$ was measured as a function of the pion angle and momentum and also the proton angle and momentum over an angular range from 15° to 50° in the lab. The three body cross section was reduced to a pp→Δ^{++}n two body cross section $d\sigma/d\Omega$ by assuming the dominance of the isobar model.[3] Angular distributions ($d\sigma/d\Omega_n$) for pp→Δ^{++}n are examined for evidence of neutron-proton final state interactions as well as effects of other amplitudes not included in the isobar analysis, such as pp→Δ^+p. A comparison to the one pion exchange model (OPE)[4] is included.

---

*Work supported in part by the U. S. Energy Research & Development Administration
[1]J. Hudomalj-Gabitzsch, et al., Phys.Lett. 60B (1976) 215.
[2]G. Glass, et al., Phys.Rev.D 13 (1977) 136, and R. D. Felder, Rice University Ph.D. Thesis (1977) unpublished.
[3]S. Mandelstam, Proc.Roy.Soc.(London) A244 (1958) 491, and S. J. Lindenbaum and R. M. Sternheimer, Phys.Rev. 105 (1959) 1974.
[4]E. Ferrari and F. Selleri, Nuovo Cimento 27 (1963) 1450, and D. Drechsel and H. J. Weber, Nucl.Phys. B25 (1970) 159.

552

THE ω-RADIATIVE DECAY CONTRIBUTION TO PROTON-PROTON BREMSSTRAHLUNG

A.N. Kamal and Adam Szyjewicz
Department of Physics, Theoretical Physics Institute
University of Alberta, Edmonton, Alberta, Canada

The contribution to the proton-proton bremsstrahlung due to the ω-radiative decay is calculated. Numerical calculations at T_{lab} = 158 MeV, 200 MeV and 400 MeV and various proton opening angles show that this contribution is small compared to the external radiation in single pion exchange. The calculation is fully relativistic. A non-relativistic reduction is made and relativistic corrections are found to be very important even at T_{lab} = 158 MeV. Finally, the space structure of the amplitude is extracted.

ON THE PARIS NUCLEON-NUCLEON POTENTIAL

M. Lacombe, B. Loiseau, J.M. Richard, R. Vinh Mau
Division de Physique Théorique, Institut de Physique Nucléaire
and Laboratoire de Physique Théorique des Particules Elémentaires
Université Pierre et Marie Curie, 75230 Paris, France

P. Pirès and R. de Tourreil
Division de Physique Théorique, Institut de Physique Nucléaire
91406 Orsay Cedex, France

In view of practical calculations on nuclear structure, the semi-phenomenological N-N potential found recently,[1] is parametrized in a simple analytical form. This parametrization consists of a regularized discrete superposition of Yukawa type terms. Results on nuclear matter parameters obtained with this potential in a G matrix calculation are presented.

1. New semi-phenomenological soft core and velocity-dependent N-N potential by M. Lacombe, B. Loiseau, J.M. Richard, R. Vinh Mau, P. Pirès and R. de Tourreil, Phys. Rev. D12, 1495 (1975).

NUCLEAR FRAGMENT PRODUCTION AT HIGH ENERGY*

J. Lee-Franzini, R.L. McCarthy, R.D. Schamberger Jr.,
and P.M. Tuts
SUNY at Stony Brook, Stony Brook, NY 11790

S. Childress and P. Franzini
Columbia University, New York, NY 10027

We have observed copious production of nuclear fragments from bombarding 300 GeV protons on carbon (~100 μb/MeV/sr). The production ratio of p:d:t at 20 MeV is about 3.5:2:1. The angular distribution from 45° to 90° in the laboratory system for these fragments is approximately isotropic, $f(\theta) \approx 1 + \cos\theta$. We observe equal abundance of ^{3}He and ^{4}He, a result not previously seen at lower energies. We compare our results with those predicted by the evaporation model and find agreement wanting.

*Research supported in part by the National Science Foundation.

AN EXACT ANALYTICAL SOLUTION TO THE TWO-BODY PROBLEM WITH
A SEPARABLE CENTRAL AND TENSOR INTERACTION

Clarence E. Lee
Los Alamos Scientific Laboratory
Los Alamos, New Mexico 87545

Franz Mohling
Department of Physics and Astrophysics
University of Colorado, Boulder, Colorado 80309

A two-nucleon, non-relativistic, spin-dependent, non-local and separable, central and tensor interaction, including a repulsive hard core, is investigated. Exact analytical solutions to the Schrödinger wave equation are obtained for all angular momentum states. For states coupled by the tensor interaction two sets of boundary conditions are considered, those due to Blatt and Biedenharn and those of Stapp, Ypsilantis and Metropolis. All phase-shift parameters for those two cases are exhibited explicitly in terms of known functions. A general class of separable interaction components is treated, but only the spatially exponential interaction is examined in detail.

IS THE D STATE OF THE DEUTERON ABOUT 4%?*

E. Lomon
Massachusetts Institute of Technology, Cambridge, MA 02139

Recent accurate measurements of the $\pi^+ + D \to p + p^1$ angular distribution for 50 MeV pions is not well fit by known models with D state > 4.5%, and the 40 MeV data may require a smaller D state. An analysis of sub-Coulomb (d,p) reactions[2] shows that the inclusion of the deuteron D state is required to approximate the cross section; but the Reid wave functions used overestimate the cross section and a realistic potential with a smaller per cent D state seems to be needed. Calculations will be done with the Feshbach-Lomon potentials of 4.6% - 7.5% D state to test the sensitivity. The cross sections for the forward photodisintegration of the deuteron at low energy has recently been shown[3] to be much smaller than predicted by most models. Again, a 4.6% D state improves the fit substantially especially at the upper part of the energy range.[4] Elastic e,d scattering and tensor polarization at $5f^{-2} < q^2 < 30f^{-2}$ can provide further information on the deuteron structure including the D state. Using $q^2 = 5 - 7f^{-2}$ will minimize the ambiguities due to relativistic and meson current corrections, and data at higher q^2 will further discriminate between models. Graphs will illustrate the variability.

1. B.M. Preedom et al., Phys. Lett. 65B, 31 (1976).
2. L.D. Knutson and W. Haeberli, Phys. Rev. Lett. 35, 558 (1975).
3. R.J. Hughes et al., Nucl. Phys. A267, 329 (1976).
4. Phys. Lett. B, to be published.

*This work is supported in part through funds provided by ERDA under Contract EY-76-C-02-3069.*000

RELATIONS BETWEEN THE DEUTERON FORM FACTORS AND THE WAVE FUNCTIONS

L. Mathelitsch and H.F.K. Zingl
Institut für Theoretische Physik,
Universität Graz, A-8010 Graz, Austria

We have calculated the deuteron electromagnetic form factors directly in momentum space with some local and separable potentials. We classify these potentials as Hulthen-like, soft- and hardcore potentials. Important differences came from the zeros of the momentum space wave functions, especially from the location of the first one. With this we define the high energy component of the deuteron. We derive a relation between the structure of the charge form factor and of the S-wave function. If the form factor can be measured, the high energy component can be estimated within small errors. Furthermore we find that the height of the quadrupole form factor is related to the D-state probability. Therefore this important quantity can be measured in the near future in a polarization experiment in electron-deuteron scattering.

ISSN: 0094-243X/78/555/$1.50 Copyright 1978 American Institute of Physics

JASTROW TYPE CALCULATIONS IN NUCLEAR MATTER BY MEANS OF THE NORMALIZATION CONDITION

E. Mavrommatis
Nuclear Research Center "Demokritos",
Aghia Paraskevi Attiki, Greece

Results are given from constrained Jastrow calculations for the energy per particle and the density of nuclear matter using mainly the test potentials IY and OMY and several correlation functions. The constraint chiefly employed is the normalization condition in first order, a condition entering naturally in Jastrow formalism. The energy in the minimization is approximated by its first two and in some cases by its first three terms of the FIY cluster expansion. In addition to the three body terms in FIY expansion, the corresponding ones of AHT expansion are also calculated for comparison. From the above it seems that at least in some cases the normalization condition constitutes a quite satisfactory constraint leading when used in low order calculations to reliable results. Besides conclusions are drawn for the sensitivity of the results on the used constraint, correlation function and minimizing energy expression.

ISSN: 0094-243X/78/555/$1.50 Copyright 1978 American Institute of Physics

556

N-P FINAL STATE INTERACTIONS IN P-P PION PRODUCTION at 800 MeV.*

G. S. Mutchler, R. D. Felder, J. Hudomalj-Gabitzsch, I.Duck,J.Clement,
T. M. Williams, K. R. Hogstrom, W. H. Dragoset, G. C. Phillips (Rice
University, Houston, Texas); E. V. Hungerford, M. Warneke, B. W. Mayes,
L. Y. Lee, and J. C. Allred (University of Houston, Houston, Texas).

A kinematically-complete measurement of $pp \rightarrow d\pi^+$ and $pp \rightarrow p\pi^+n$ has
been performed, using the LAMPF 800 MeV proton beam. Kinematic con-
ditions were chosen to study the role of np final state interactions
(FSI) in the three-body process $pp \rightarrow d*\pi^+ \rightarrow pn\pi^+$ where d* refers to the
np system at small relative energy. Fig.1 gives the measured c.m.
cross section for $pp \rightarrow d\pi^+$ along with previous data at 810 MeV.[1] The
solid curve is a fit to the present data with the parameterization
given in Fig.1. From this fit, the total $pp \rightarrow d\pi^+$ cross section at
800 MeV was found to be 1.25 ± .09 mb.

Three-body $pp \rightarrow p\pi^+n$ cross sections which are proportional to the
square of the matrix element were calculated by projecting the data
on the np relative momentum axis (k_{np}) and dividing by phase space.
Using the Goldberger-Watson singlet and triplet enhancement factors
for np FSI, these cross sections were fit to an incoherent sum of the
contributions from $^1S_0$ and $^3S_1$ scattering. By summing over k_{np},
singly differential $pp \rightarrow d\pi^+$ cross sections were obtained for singlet,
triplet, and total d* production (see Fig.2).The k_{np},Mev/c, integration
limits are indicated in Fig.2; the $pp \rightarrow d\pi^+$ cross section is included
for comparison. The shapes of the triplet and total d* angular distributions closely approximate that for $pp \rightarrow d\pi^+$ --evidence for a dominant $^1D_2 \rightarrow {}^3S_1P_2$ transition.

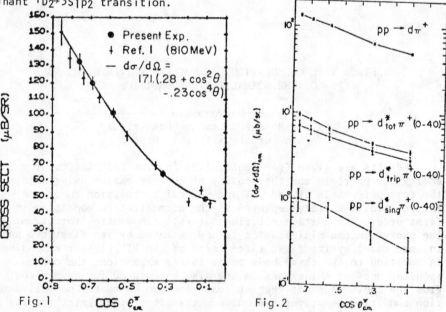

Fig.1

Fig.2

* Work supported by U. S. E. R. D. A.
[1] C. Richard-Serre et al., Nucl.Phys. B20 (1970) 413.

NUCLEON-NUCLEON POTENTIALS FROM REGGE POLES

M.M. Nagels, T.A. Rijken and J.J. de Swart
Institute for Theoretical Physics, University of Nijmegen,
Nijmegen, The Netherlands

We use nucleon-nucleon potentials derived from the Regge-pole model.[1] The contributions of the exchanged Regge-trajectories are for low energy baryon-baryon scattering strongly dominated by the lowest J values. This explains the success of the OBE-models. The linearity of the trajectories leads to the usual OBE propagators. The shapes of the diffraction peaks at high energies strongly suggest exponential form factors for the meson vertices. In addition to the usual pseudoscalar, scalar and vector meson contributions one gets now also contributions from the pomeron and the $J = 0$ contributions of the tensor meson trajectories (f, f', A_2). Because of the needed ghost-killing factors in the residue functions the propagators are absent and the signs of the potentials are opposite to the ones of the scalar mesons. Therefore one obtains Gaussian repulsive potentials $\sim \exp (-m^2 r^2)$, where the "Pomeron mass" m is essentially determined by the steepness of the elastic diffraction peak at high energies.

Fitting these soft core, partially nonlocal, potentials to the NN data using only 14 free parameters (12 coupling constants, 2 masses) we obtain χ^2/data < 2, compared to the Livermore χ^2 surface, the lowest χ^2 for meson theoretic potentials we know of. The values for the couplings and mass of the Pomeron ($\sim$ 300 MeV) as well as the cutoff mass for the mesons agree well with estimates from high energy fits.

1. T.A. Rijken, Ph.D. Thesis, University of Nijmegen, 1975;
   T.A. Rijken, OPE-models and Regge Poles, Québec, 1974.

IS THERE A NUCLEON-NUCLEON POTENTIAL?*

H. Pierre Noyes
Stanford Linear Accelerator Center, Stanford, California 94305, USA

The nucleon-nucleon interaction exhibits the maximum complexity allowed by the conservation laws — differing as it does in each spin-parity state, and exhibiting significant non-locality or velocity dependence as well as tensor and spin-orbit couplings. The maximal "locality" which can be assumed is to assign a different radial function for the "potential" in each separate state. There is no guarantee that the "potential" so defined is appropriate to use for systems with more than two nucleons, or even for two nucleon electromagnetic effects.

The two main groups that have persisted in their efforts to compute the "potential" from a combination of field theory and dispersion-theoretic techniques (i.e. Paris and Stonybrook-Michigan State) have come to a reasonable consensus, although there are differences in detail; for example the Paris group introduces a velocity-dependent model. Both, however, assume that the problem is to calculate that function which, when used in some non-relativistic Schroedinger-like equation will reproduce the two-nucleon on-shell scattering observables. Again there is no theoretical guarantee that this is the correct object to use for other calculations. For example, a much simpler a priori calculation using a covariant boundary condition model for the $NN\pi$ system below pion threshold gives the correct binding to the two S-states within a percent with no adjustment of parameters; it also gives accurate results for πd scattering, etc. Yet this model does not even define a Lippmann-Schwinger equation, except in the unrealistic limit where the pion mass becomes infinite compared to the scattering energy.

An alternative to conventional approaches to the problem is to start with a covariant n-hadron theory and discover in detail for what problems and to what accuracy the interaction can be represented by 2-body (and/or n-body) "potentials" once the particulate degrees of freedom which cannot be excited as free particles have been integrated out. The Brayshaw boundary condition approach is not convenient for this purpose, as it does not allow two-body "left-hand cuts", but we have recently shown that the zero-range limit of the Karlsson-Zeiger equations does not suffer from this defect. This provides a non-relativistic "correspondence limit" for covariant n-hadron systems. Thus we can define the "potential" in a new way, but so far have no guarantee of the range of validity of the result.

* Work supported by the Energy Research and Development Administration.

An Effective Hamiltonian for the ABC Effect

A. Reitan

Fysisk institutt, Universitetet i Trondheim,
N-7000 Trondheim, Norway

The ABC effect [1] is an enhancement in the missing-mass spectrum of the two pions created in reactions like $pn \to d(\pi\pi)^0$ and $pd \to {}^3He(\pi\pi)^0$. This enhancement appears at missing-mass values near 300 MeV in the isospin 0 channel and was originally regarded as evidence for a $J^P=0^+$, $T=0$ $\pi\pi$ resonance. It has later become clear that this interpretation is incorrect, and that the ABC effect is simply another manifestation of the Δ resonance in the πN system, the cross section for $\pi\pi$ production then being largest when each of the pions is close to the resonant energy with respect to one of the nucleons [2].

We focus here our attention on the reaction $pn \to d(\pi\pi)^0$, the aim being to describe the process in terms of an effective two-body interaction Hamiltonian in the nuclear coordinates. In particular, we are interested in studying the cross section as a function of energy down to threshold, as part of a feasability study regarding future experiments at the meson factories. As in the Risser-Shuster model [3] we assume that the reaction takes place by a one-pion exchange between the two nucleons, but rather than treating the intermediate πN system as a particle (the Δ) we describe the pion rescattering on a nucleon in terms of the off-shell pion-nucleon scattering amplitude; this also makes it possible to include other partial waves in a systematic manner.

1. A. Abashian, N. E. Booth and K. M. Crowe, Phys. Rev. Lett. 5 (1960) 258.
2. G. W. Barry, Nucl. Phys. B85 (1975) 239.
3. I. Bar-Nir, T. Risser and M. D. Shuster, Nucl. Phys. B87 (1975) 109.

ISSN: 0094-243X/78/559/$1.50

QUASI-ELASTIC CHARGE EXCHANGE IN nd → pnn AT 800 MeV*

P.J. Riley and C.W. Bjork
University of Texas, Austin, Texas 78712

B.E. Bonner, J.E. Simmons, and J. Wallace
Los Alamos Scientific Laboratory, University of California
Los Alamos, New Mexico 87545

M.L. Evans, G. Glass, J.C. Hiebert, M. Jain, and L.C. Northcliffe
Texas A & M University, College Station, Texas 77843

C.G. Cassapakis
University of New Mexico, Albuquerque, New Mexico 87131

We have measured the proton spectrum at lab angles from 0° to 18° resulting from 800 MeV neutron bombardment of deuterium. The LAMPF 800-MeV neutron beam and the multiwire proportional counter spectrometer system in the Nucleon Physics Laboratory at LAMPF were used.[1] Particle separation was unambiguous with contamination less than 1%, and the momentum resolution $\Delta P/P$ was 1% FWHM. Absolute normalization of the cross sections was obtained from a separate measurement of the np CEX cross section at 800 MeV and has an overall uncertainty of ±7%. Measurements were made with the spectrometer at angles of 0°, 4°, 8°, and 16° to the incident neutron beam. The angular resolution of the spectrometer was about ±2 mrad. The sharp peak in the proton spectrum observed at 0° gradually broadens with increasing angles. At 16° the width of the peak is consistent with that calculated from the Fermi momentum distribution of the struck nucleon in the deuteron folded with kinematic broadening due to the finite acceptance of the spectrometer. At smaller angles $^1S_0$ final state interaction between the two unobserved neutrons sharpens the peak dramatically. For free np CEX the t variation of the cross section is described over a wide range of incident energies by a sum of two exponentials: $d\sigma/dt = \alpha_1 e^{\beta_1 t} + \alpha_2 e^{\beta_2 t}$.[2,3] The present quasi-elastic np CEX data are also fit very well by this function. We obtained the following values for the four parameters: $\alpha_1 = 12.58$ mb/(GeV/c)^2, $\beta_1 = 121.7$ (GeV/c)$^{-2}$, $\alpha_2 = 24.52$ mb/(GeV/c)^2, and $\beta_2 = 5.37$ (GeV/c)$^{-2}$. The ratio of the quasi-elastic CEX cross section at t = 0 to the corresponding elastic cross section is 0.56 ± 0.04 at 800 MeV. A plain-wave-impulse-approximation analysis of the quasi-elastic CEX cross section provides a good fit to the data.

1. C.W. Bjork et al., Phys. Lett. 63B, 31 (1976).
2. M.L. Evans et al., Phys. Rev. Lett. 36, 497 (1976).
3. G. Bizard et al., Nucl. Phys. B85, 14 (1975).

---

*Work supported in part by U.S. ERDA.

CHEW-LOW MODEL FOR P-WAVE PION ABSORPTION ON TWO NUCLEONS AND LOW-ENERGY PION-NUCLEUS SCATTERING

R. Rockmore, P. Goode, and E. Kanter
Rutgers University, New Brunswick, N.J. 08903

The Chew-Low model for P-wave pion absorption on two nucleons has been used to construct the P-wave optical potential $\sim \rho^2(r)$ usually introduced phenomenologically in studies of low-energy pion-nucleus scattering. We find

$$V^{(2)}(r) = -i \; \frac{64\pi^2}{3} \; \frac{M(ME_\pi)^{1/2}}{E_\pi} \left(\frac{\mu^2}{ME_\pi + \mu^2}\right)^2 \left(\frac{f}{\mu}\right)^2 \frac{\sin^2\delta_{33}(E_\pi)}{p_\pi^6}$$

$$\times \; (\hbar c)^6 A^2 \left\{ p_\pi^2 \rho^2 + (\hbar c)^2 \left((\rho')^2 + \rho\rho'' + \frac{2}{r}\rho\rho' \right) \right\},$$

in the local Laplacian form, with the density $\rho(r)$ normalized to 1; as a result of our approximations, only the graph with a doubly rescattered pion makes a nonvanishing contribution to $V^{(2)}$. (The small, model-dependent, real part of $V^{(2)}$ has been neglected.) Results of an optical-model study of $\pi^- - {}^{12}C$ scattering in the energy interval 60 MeV $\leq T_\pi \leq$ 120 MeV in the local Laplacian model (with Fermi-averaged b_0 and b_1) with and without the optical potential $V^{(2)}$ are shown in Table I. It is found that including effects of retardation at $T_\pi = 60$ MeV reduces $V^{(2)}$ there by 28%; however, 1/A corrections to $V^{(2)}$ scale up $V^{(2)}$ by the factor $(1+2/(A-1))$.

Table I $\quad \pi^- - {}^{12}C$ Reaction Cross Sections

T_π (MeV)	σ(mb) (without $V^{(2)}$)	σ(mb) (with $V^{(2)}$)
60	177	196
80	264	279
100	350	359
120	408	411

ISSN: 0094-243X/78/561/$1.50 Copyright 1978 American Institute of Physics

ON SOLITARY WAVE EXCHANGE N-N POTENTIALS

Mesgun Sebhatu
Department of Physical Sciences
Pensacola Jr. College, Pensacola, FL 32504

Solitary-wave exchange N-N potentials (SWEPs) are obtained by employing solitary-wave propagators (propagators constructed from exact particular solutions of nonlinear meson field equations[1,2] in the usual second order perturbation theory derivation of N-N potentials. Examples of such potentials are the $\lambda\phi^4$ SWEP[3] and the sine-Gordon SWEP.[4] As is shown by the plots of the $^1S_0$ state $\lambda\phi^4$ SWEP reproduced below, the SWEPs approach the OPEP tail at large ranges (mr $\gtrsim$ 1.5), are strongly attractive at intermediate ranges (0.5 $\lesssim$ mr $\lesssim$ 1.5), and they smoothly become repulsive at short ranges (0 $\lesssim$ mr $\lesssim$ 0.5). These are characteristic features of successful OBEP and popular phenomenological potential models (e.g. Reid soft core). The SWEPs achieve these without undue number of parameters. In fact, the $\lambda\phi^4$ and S.G. SWEP involve only one parameter (the self-interaction coupling constant, λ) in addition to those associated with OPEP (g and m). Encouraged by the S.G. and $\lambda\phi^4$ SWEP results, we are now deriving potentials based on more general solitary-wave propagators.[5] The more general SWEP will have three parameters (two self-interaction coupling constants and a factor in the exponent of the nonlinear terms introduced by the field equation $\partial_\mu\partial^\mu\phi + m^2\phi + \lambda_1\phi^{2q+1} + \lambda_2\phi^{4q+1} = 0$). The new SWEP is more flexible and may be suitable for comparison with N-N data. The expressions for the SWEP along with plots for some of its special cases will be shown and their implications discussed.

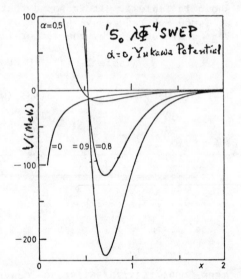

$^1S_0$ $\lambda\phi^4$ SWEP
$\alpha=0$, Yukawa Potential

1. P.B. Burt, Acta Phys. Pol. B7, 617 (1976).
2. P.B. Burt and M. Sebhatu, Acta Phys. Pol. B7, 729 (1976).
3. M. Sebhatu, Nuovo Cim. 33A, 538 (1976).
4. M. Sebhatu, Lett. Nuovo. Cim. 16, 463 (1976).
5. M. Sebhatu and P.B. Burt, in preparation.

Δ RESONANCE ADMIXTURES IN THE NN PROBLEM AND THE ELECTRODISINTEGRATION OF THE DEUTERON

Bernd Sommer

Institut für Theoretische Physik, Ruhr-Universität Bochum
Universitätsstrasse 150, Postfach 2148, D-4630 Bochum 1,
West Germany

The treatment of Δ-admixtures in the Schrödinger equation is strongly dependent on the basic assumptions of the choice of transition potentials (N → Δ) and of the interaction between Δ's. The Feynman diagrams for the OBE contributions require vast assumptions concerning energy distributions which may lead to potentials of infinite range in the (NΔ → ΔN) channel.

Using unitary transformations to decouple the pure baryon (N,Δ) states and the baryon-meson states we derive a prescription for Δ potentials in the OBE-model which is consistent with recent two-pion-exchange calculations.[1] These refined potentials decrease the Δ-admixtures in the wave functions. In contradiction to other authors we find that the Δ's lead to a slightly lower cross section at momentum transfers < 16 fm^{-2} which is mainly due to the renormalization of the pure NN partial waves because of additional Δ channels.

1. J.W. Durso et al., Isobars, Transition Potentials and Short Range Repulsion in the NN interaction, preprint, 1976, Stony Brook.

THE PHOTON CIRCULAR POLARIZATION IN THE np → dγ PROCESS AND THE DEUTERON P STATE ARISING FROM THE SMALL COMPONENT OF THE DIRAC SPINOR

Shin-ichi Morioka and Tamotsu Ueda
Faculty of Engineering Science, Osaka University
Toyonaka, Osaka, Japan

The photon circular polarization P_γ in the np → dγ process is studied in the relativistic one-nucleon-exchange model in which the deuteron has the P state component arising from the small components of the Dirac spinor. We obtain the remarkable result that a considerable amount of P_γ appears from the existence of this P state component without the parity violation due to the weak interaction. At very low energies the value of P_γ depends mainly on the ratio of the P-state to the $^3S_1$-state wavefunction in the low frequency region of their Fourier components and also on the phase difference between these two states. Assuming reasonable values for these quantities, we find the value of $10^{-4} \sim 10^{-2}$ for P_γ. This is quite large, compared with the result obtained in the weak interaction model which predicts the value of $10^{-7} \sim 10^{-9}$. Our result is only preliminary. Adopting more realistic ones for both the P-state wavefunction and the phase difference, we expect a result consistent with the experimental value for P_γ of the order 10^{-6}.

COULOMB CORRECTIONS IN PROTON-PROTON SCATTERING

L. Streit
Fakultät für Physik, Universität Bielefeld,
D-48 Bielefeld, Germany

J. Fröhlich, H. Zankel, H. Zingl
Institut für Theoretische Physik,
Universität Graz, A-8010 Graz, Austria

In the scattering of two protons, electromagnetic effects in addition to the hadronic i.e. "strong" interaction are present. To express these contributions to the total interaction in terms of the phase shifts, one uses the decomposition:

$$\delta_\ell = \delta_{C,\ell} + \delta_{S,\ell} + \delta_{R,\ell}$$

with $\delta_{C,\ell}$ the pure Coulomb phase shift, $\delta_{S,\ell}$ the pure hadronic phase and $\delta_{R,\ell}$ the so-called "rest phase", which originates from the interference of the strong interaction with the Coulomb force. Since in the framework of nonrelativistic scattering no satisfactory theory of the Coulomb corrections, as they appear in the rest phase, exists, we derived a simple description, independently from any potential model. First of all we were able to show that these Coulomb corrections are not sensitive to a screening of the Coulomb potential. Consequently an explicit expression to first order in e^2 for $\delta_{R,\ell}$ was derived, which leads to essential improvements over the results obtained by means of a WKB approximation in an earlier paper for all partial waves.

Pion Production in Neutron-Proton Collisions at 790 MeV

W. Thomas, C. Cassapakis, B. Dieterle, C. Leavitt, D. Wolfe,
University of New Mexico; M. Evans, G. Glass, Mahavir Jain,
L. Northcliffe, Texas A&M University; B. Bonner, J. Simmons, LASL.

An experiment has been performed at LAMPF to look at pion
spectra from the reactions $np \rightarrow pp\pi^-$ and $np \rightarrow nn\pi^+$ for 790 MeV
neutrons. Approximately 10^5 pions of both signs were collected
from center-of-mass angles ranging from 4 to 113 degrees. We have
searched for a T=0 contribution to these reactions by three differ-
ent methods. Following Rosenfeld $\sigma_{01} = 2\sigma(np \rightarrow \pi^{\pm}) - \sigma(pp \rightarrow \pi^\circ)$ we
have also created, via Monte Carlo, a model where the reaction
$np \rightarrow N\Delta \rightarrow \pi^{\pm}NN$ is simulated. Finally we have looked for asymmetries
in the π^+ and π^- double differential cross sections. All three of
these methods indicate the existence of a significant departure
from a pure T=1 matrix element, although the first method leads to
inconclusive results because of uncertainties in the $pp \rightarrow pp\pi^\circ$
total cross section at 790 MeV.

The angular distribution of pions in the center of mass is
given by

$$123\pm3[1 + (.72\pm.04) \cos{}^2\theta]\mu \text{ b/str.}$$

1) A.H. Rosenfeld, Phys Rev 96, 139 (1954)

ISSN: 0094-243X/565/$1.50 Copyright 1978 American Institute of Physics

Precision Measurements of The Analyzing Power for n-p Scattering in the 14-17 MeV Region

W. TORNOW,[+] P.W. LISOWSKI, R.C. BYRD and R.L. WALTER

Department of Physics, Duke University, Durham, North Carolina 27706
and
Triangle Universities Nuclear Laboratory,[*] Durham, North Carolina 27706

Precise measurements of the analyzing power $A_y(\theta)$ for n-p scattering were obtained between 14 and 17 MeV (lab energy) by scattering polarized neutrons from ^{1}H contained in an organic scintillator. The method for producing the polarized neutron beam employed the ^{2}H$(\vec{d},\vec{n})^3$He polarization transfer reaction. At an angle of 0°, this reaction gives a clean source of neutrons with a polarization of 62% for our (typical) deuteron beam polarization of 73%. Furthermore, with this polarized neutron source, the hydrogenous scatterer could be located in the intense forward peak of the neutron flux, i.e., at 0°. Four measurements for $A_y(\theta)$ were obtained at 90° c.m. between 14 and 16.9 MeV and an angular distribution was obtained at 16.9 MeV. The final $A_y(\theta)$ values were determined to about ±0.002 or better. To our knowledge, this is the highest accuracy available to date for a neutron scattering experiment. The data are reasonably consistent with other data in this energy range, but are finally accurate enough to show that the $A_y(\theta)$ calculations from the available n-p phase shift sets are too high in magnitude and possess the wrong shape for the angular distribution at 16.9 MeV. These findings necessitate a new parameterization for n-p scattering at least in the region between 10 and 20 MeV.

[+] Supported in part by the Deutsche Forschungsgemeinschaft
[*] Supported by the U. S. Energy Research and Development Administration

Similarity between $A_y(\theta)$ for $^2H(p,p)^2H$ and $^2H(n,n)^2H$ at 12 MeV

W. TORNOW,[+] P.W. LISOWSKI, R.C. BYRD and R.L. WALTER

Department of Physics, Duke University, Durham, North Carolina 27706
and
Triangle Universities Nuclear Laboratory,[*] Durham, North Carolina 27706

Using the high polarization transfer capability of the $^2H(\vec{d},n)^3He$ reaction to provide a means of producing an intense flux of highly polarized neutrons, the analyzing power $A_y(\theta)$ for n-d elastic scattering was measured at 12 MeV. The ^2H was contained in a deuterated organic scintillator.

The uncertainty in the measured analyzing power values ranged from ±0.006 to ±0.013. These data form the most accurate measurement for an angular distribution of the analyzing power for n-d scattering. To within the accuracy of the two sets of data, the $A_y(\theta)$ distribution measured by Clegg et al.[1] at 12 MeV for p-d scattering agrees with the results obtained in the present work. This agreement suggests that if any differences exist in $A_y(\theta)$ for these two charge symmetric scattering systems, they must be of the order of ±0.01 to ±0.02 (depending on scattering angle) or less. This result suggests that the large differences[2] in $A_y(\theta)$ for these two systems observed near 23 MeV are probably due to experimental difficulties.

[+] Supported in part by the Deutsche Forschungsgemeinschaft
[*] Supported by the U. S. Energy Research and Development Administration
[1] T. B. Clegg and W. Haeberli, Nucl. Phys. A95 (1967) 608
[2] See review article by R. L. Walter in Proceedings of the Fourth International Symposium on Polarization Phenomena In Nuclear Reactions, Ed. W. Grübler and V. König, Birkhaüser Verlag Basel 1976, p. 377

EFFECTS OF NON-MINIMAL INTERACTION CURRENTS ON NP BREMSSTRAHLUNG

W. Van Dijk
Dordt College, Sioux Center, Iowa 51250

M.A. Preston
McMaster University, Hamilton, Ontario

For nonlocal nucleon-nucleon interactions there is ambiguity in the coupling of the nucleons to the electromagnetic field. The usual prescription is to employ minimal coupling, but Heller[1] has pointed out that non-minimal currents can yield quite different results. In principle one ought to be able to obtain the correct bremsstrahlung matrix element by considering the elementary processes that result in the nuclear force, and then include photon emissions. In practice the realistic nuclear interaction has phenomenological components. In order to study differences in the bremsstrahlung cross sections for various couplings, we use a phenomenological interaction current formalism, in which one can obtain non-minimal and minimal couplings by varying a parameter. This parameter is interpreted as the range of the interaction current density. In this model calculation we consider only the coupling of the proton charge to the electromagnetic field. Two types of nonlocal nuclear interactions are employed, (1) Perey-Buck type[2] and (2) Serber exchange. For the Perey-Buck force we calculate the ratio of the bremsstrahlung matrix element when non-minimal coupling is used to the minimally coupled matrix element. The calculation assumes Harvard geometry. With this geometry and the Serber force, we calculate the bremsstrahlung cross section, including single and double scattering contributions, and interaction current contributions for various values of the non-minimality parameter. It is found that for low photon energies the kind of coupling is unimportant. We will present the effects of non-minimal coupling on the bremsstrahlung matrix element and cross-section at higher photon energies.

1. L. Heller, in The Two-Body Force in Nuclei, ed. by S.M. Austin and G.M. Crawley (Plenum, N.Y. 1972), p. 79.
2. F. Perey and B. Buck, Nucl. Phys. 32, 353 (1962).

ON A METHOD OF TREATING THE VELOCITY-DEPENDENT POTENTIAL

M. Wada and T. Obinata
The Physical Science Laboratories, College of Science and Technology
Nihon University, Funabashi 274, Japan

The nonstatic OBEP with retardation which contains the velocity-dependent central and tensor potentials was constructed by us.[1] Now, we propose the extended Green's method in order to deal with the velocity-dependent tensor potential. We assume the following potential, where $V^0(r)$ stands for the velocity-independent potential:

$$V(r) = V^0(r) - M^{-1}[\nabla^2 h_C(r) + h_C(r)\nabla^2] - M^{-1}[\nabla^2 h_T(r)S_{12} + h_T(r)S_{12}\nabla^2]. \quad (1)$$

The uncoupled Schrödinger equation with the potential (1) can be treated in the framework of the Green's method.[2] Because ∇^2 terms of the tensor potential make the coupled equation more complicated, we solve not for the ordinary radial wave functions, $u_J(r)$ and $w_J(r)$, but rather for

$$\begin{pmatrix} f_J(r) \\ g_J(r) \end{pmatrix} = \begin{pmatrix} \sqrt{\frac{J+1}{2J+1}} \cdot \sqrt{1+2h_{J-1}} & \sqrt{\frac{J}{2J+1}} \cdot \sqrt{1+2h_{J-1}} \\ -\sqrt{\frac{J}{2J+1}} \cdot \sqrt{1+2h_{J+1}} & \sqrt{\frac{J+1}{2J+1}} \cdot \sqrt{1+2h_{J+1}} \end{pmatrix} \begin{pmatrix} u_J(r) \\ w_J(r) \end{pmatrix} \quad (2)$$

where $h_{J-1} = h_C(r) + 2h_T(r)$ and $h_{J+1} = h_C(r) - 4h_T(r)$.

Modified wave functions, $f_J(r)$ and $g_J(r)$, satisfy an ordinary radial wave equation as follows:

$$\begin{pmatrix} D^{J-1} & D^{ND} \\ D^{ND} & D^{J+1} \end{pmatrix} \begin{pmatrix} f_J(r) \\ g_J(r) \end{pmatrix} = 0 \quad (3)$$

$$D^{J\pm1} = \frac{d^2}{dr^2} + k^2 - \frac{(J\pm1)(J+1\pm1)}{r^2} - M\,V_{J\pm1}, \quad D^{ND} = \frac{6\sqrt{J(J+1)}}{2J+1}\,M\,V_{ND}, \quad (4)$$

but with

$$V_{J\pm1} = \frac{1}{1+2h_{J\pm1}}\left[\frac{J+1}{2J+1}\,V_{J\pm1}^0 + \frac{J}{2J+1}\,V_{J\mp1}^0 \mp \frac{12J(J+1)}{(2J+1)^2}\,V_T^0\right]$$

$$-\left(\frac{dh_{J\pm1}/dr}{1+2h_{J\pm1}}\right)^2 \frac{1}{M} + \left(\frac{2h_{J\pm1}}{1+2h_{J\pm1}}\right)\frac{k^2}{M} \mp \frac{2J}{Mr^2}, \quad (5)$$

$$V_{ND} = \frac{1}{\sqrt{(1+2h_{J-1})(1+2h_{J+1})}}\left[\frac{1}{6}\left(V_{J+1}^0 - V_{J-1}^0\right) + \frac{1}{2J+1}\,V_T^0 + \left(1+h_{J-1}+h_{J+1}\right)\frac{2J+1}{3Mr^2}\right]. \quad (6)$$

ISSN: 0094-243X/78/569/$1.50

$V^0_{J\pm1}$ denote the effective potentials for the L=J±1 states, and V^0_T is the tensor part of $V^0(r)$. Then $u_J(r)$ and $w_J(r)$ are obtained from $f_J(r)$ and $g_J(r)$ through the inverse transformation of Eq. (2). The extended Green's method is applicable not merely to the two-nucleon system but to the nuclear matter and few-nucleon system.

1. T. Obinata and M. Wada, Prog. Theor. Phys. 53, 732 (1975).
2. A.M. Green, Nucl. Phys. 33, 218 (1963)

(αnp) 3-body model of ^{6}Li using hyperspherical harmonics*

K. K. Fang and E. L. Tomusiak
Saskatchewan Accelerator Laboratory,
University of Saskatchewan, Saskatoon, Canada   S7N 0W0

The method of hyperspherical harmonics[1,2] also called K-harmonics is applied to the calculations of bound states of ^{6}Li, the ground state ($J^\pi = 1^+$, -4.53 MeV) and the excited state ($J^\pi = 0^+$, -0.97 MeV). By assuming the interaction potentials (two-body) between the nucleon and the alpha particle[3], and the two nucleons[2], coupled equations for the wave functions, the hyper-radial part, are found according to the number of hyperspherical harmonics used for a given total angular momentum and the parity of the system. Binding energies and the wave functions are then numerically solved. The charge form factor of the ground state and the magnetic transition form factor (M-1) from $J^\pi = 1^+$ to $J^\pi = 0^+$ are discussed. Numerical results will be presented and comparison with the experiment is made. Due to the success of the calculation of 3-body to 3-body elastic scattering[2], we then present the by-product of this investigation, namely, α+N+P → α+N+P (elastic scattering).

References

1. Y. A. Simonov, Sov. J. Nucl. Phys. 3 (1966) 461.

2. K. K. Fang, Phys. Rev. C, March 1977.

3. S. Sack, L. C. Biedenharn and G. Breit, Phys. Rev. 93 (1954) 321.

* Work supported by the National Research Council of Canada.

THE CHARGE-ASYMMETRY OF TWO-NUCLEON INTERACTION DUE TO RADIATIVE CORRECTIONS

C. Yalçin, M. Halil and E. Aksahin

Physics Department, Middle East Technical University, Ankara, Turkey

The radiative corrections to the meson-nucleon vertices of I=0 and I=0 scalar, pseudoscalar and vector mesons are derived using the standard Feynman technique. For neutral mesons the radiative corrections give charge asymmetric contribution to the two-nucleon potential. Using the charge symmetry breaking contribution to the one-boson-exchange-potential models due to radiative corrections of scalar, pseudoscalar and vector mesons, we calculate the charge asymmetric energy Δn in ^{3}He-^{3}H nuclei. We find $\Delta n = 0.009$ and 0.007 MeV for different OBEP models. This is too small to explain the observed discrepancy.

The method of calculation we shall employ for finding radiative corrections to various neutral bosons is based on the original treatment of Feynman.[1] The electromagnetic corrections will be calculated to the lowest order of perturbation theory, and in order to avoid infrared and ultraviolet divergences we shall introduce the non-covariant parameters λ (photon mass) and Λ (cut-off), respectively. In a renormalizable theory Λ can be eliminated by introducing a counter-term into the Lagrangian and by redefining the physical quantities. On the other hand if we are dealing with a non-renormalizable theory, traditionally a sensible cut-off is provided by a comparison with experimental results. As for λ, it is customary among nuclear physicists to consider the soft photon mass equal to 10 MeV. After all the amplitude calculations, we must set $\lambda \to 0$ in order to recover gauge invariance, which unfortunately does not work for cases where charge mesons are exchanged. The primary assumption about the momentum transfer is that $|q|^2 \ll$ (a region which provides the major contribution to the vertex correction, where M is the nucleon mass. After the vertex function calculation, if one sets $|q| \to 0$ and then $\lambda \to 0$, this settles the problem in many cases. We find the following radiation correction for neutral, scalar, pseudoscalar and vector meson exchanges:

Neutral Scalar Bosons (σ^0, δ^0)

$$\Delta W = \frac{\alpha}{\pi} \frac{\mu}{4\pi^2} g_R^2 (1+\tau_{1z})(1+\tau_{2z}) \left[\frac{e^{-\mu r}}{\mu r} + \frac{1}{2}\left(\frac{\mu}{M_R}\right)^2 \left(1 + \frac{1}{\mu r}\right)\frac{e^{-\mu r}}{(\mu r)^2} 0(q^2/M_R^2)\right],$$

Pseudoscalar Bosons (π^0, η)

$$\Delta W = -\frac{2\alpha}{\pi} \cdot \frac{1}{4} (1+\tau_{1z})(1+\tau_{2z}) \left[f_{\pi^0}^2 \frac{e^{-\pi^0 r}}{r} + f_\eta^2 \frac{e^{-\mu r}}{r}\right],$$

Neutral Vector Bosons (ρ^0, ω, ϕ)

$$\Delta W(\omega) = \frac{7\alpha}{32\pi^2} \frac{(1+\tau_{1z})(1+\tau_{2z})}{4} \frac{g_\omega^2 e^{-\omega r}}{r} \left[1 - \frac{3}{2} \cdot \frac{\omega}{M} \cdot \frac{1}{Mr}\left(1 + \frac{1}{\omega r}\right)\right.$$
$$\left. + \frac{1}{4}\left(\frac{\omega}{M}\right)^2 + \frac{1}{6}\left(\frac{\omega}{M}\right)^2 \left(\sigma_1 \cdot \sigma_2\right) - \frac{1}{4}\left(\frac{\omega}{M}\right)^2 S_{12}\left(\frac{1}{3} + \frac{1}{\omega r} + \frac{1}{(\omega r)^2}\right)\right],$$

Same for ρ^0 and ϕ.

1. R.P. Feynman, Phys. Rev. 76, 769 (1949).

EXACT THREE DIMENSIONAL REDUCTION OF THE BETHE-SALPETER EQUATION

I. Zmora
Department of Physics
Ben Gurion University of the Negev, Beer Sheva 84120, Israel

A. Gersten*
Division de Physique Théorique,†
Institut de Physique Nucléaire, Paris‡

and

Laboratoire de Physique Théorique des Particules Elémentaires, Paris‡

The partial wave Bethe-Salpeter equation in the ladder approximation is converted into a set of two one-dimensional equations. The resulting reduced equations determine two off-energy, on-mass-shell amplitudes, one for the scattering of two positive energy and the other for two negative energy particles. The connection between the potentials of the reduced equations and the many particle intermediate states is displayed by their series expansion. Two versions of the equations which are convenient for numerical computations are also obtained and one of them is a one-channel equation which involves only the physical amplitude for the scattering of two positive energy particles. The potentials are determined by auxiliary equations which are two dimensional but nonsingular and exactly solvable by standard numerical methods. One of the results of the derivation of the reduced equations is a set of reduction relations which express the off-mass-shell amplitude as a functional of a restricted shell amplitude.

*On leave from the University of the Negev.
†Laboratoire associé au C.N.R.S.
‡Postal address: Université Pierre et Marie Curie, Tour 16-El
4, Place Jussieu, 75230 Paris Cedex 05, France

ISSN: 0094-243X/78/572/$1.50 Copyright 1978 American Institute of Physics

ELASTIC SCATTERING OF 1 GeV PROTONS FROM NUCLEI

Girish K. Varma and Larry Zamick
Serin Physics Laboratory, Rutgers - The State University
Freylinghuysen Road, Piscataway, New Jersey 08854

We obtain expressions for proton-nucleus elastic scattering amplitudes in the Glauber approximation which include the effects of spin dependence and Coulomb interferences. These effects are shown to be important for determining the neutron and matter radii for all target nuclei. The results for neutron radii are compared with theoretical predictions.

ISSN: 0094-243X/78/572/$1.50 Copyright 1978 American Institute of Physics

AUTHOR INDEX